Handbook of
Nonmedical Applications of
Liposomes

Theory and Basic Sciences

VOLUME I

Edited by

Danilo D. Lasic
MegaBios Corporation
Burlingame, California

Yechezkel Barenholz
Department of Biochemistry
Hebrew University–Hadassah Medical School
Jerusalem, Israel

CRC Press
Taylor & Francis Group
Boca Raton London New York

CRC Press is an imprint of the
Taylor & Francis Group, an **informa** business

Cover:

The cover depicts the artist's impression of colloidal amphiphilic systems. A droplet of disordered sponge phase (which has minimal surface and zero curvature) is spontaneously shedding off fragments of bilayers and normal and toroidal liposomes (at critical, entropy driven fluctuations). Some of the liposomes undergo fission process, budding off daughter vesicles. Created in mixed media by artist Alenka Dvorzak Lasic, Summer 1995.

CRC Press
Taylor & Francis Group
6000 Broken Sound Parkway NW, Suite 300
Boca Raton, FL 33487-2742

Reissued 2019 by CRC Press

A Library of Congress record exists under LC control number:

Publisher's Note
The publisher has gone to great lengths to ensure the quality of this reprint but points out that some imperfections in the original copies may be apparent.

Disclaimer
The publisher has made every effort to trace copyright holders and welcomes correspondence from those they have been unable to contact.

ISBN 13: 978-0-367-26096-5 (hbk)
ISBN 13: 978-0-367-26098-9 (pbk)
ISBN 13: 978-0-429-29144-9 (ebk)

Visit the Taylor & Francis Web site at http://www.taylorandfrancis.com and the
CRC Press Web site at http://www.crcpress.com

GENERAL PREFACE

To introduce liposomes, we shall look backwards to the origin of the name liposome. As shown in the last chapter of this series, in the fourth volume, liposomes were produced in scientific laboratories at least from 1847 on. The colloidal properties of lipid dispersions (emulsions, sols, micelles, etc.) were well known. However, it was only Bangham in the early 1960s who realized that they encapsulate part of the solvent and that their membrane presents a permeability barrier. So, they were at that time called "banghasomes", while the name "liposome" can occasionally be found in the literature between the World Wars as a description of colloidal fat (mostly triglyceride) particles. In the late 1960s, Weismann proposed the name liposomes for these structures. Due to the very general meaning of this word (i.e., fat body), Papahadjopoulos and others introduced the classification according to size and number of lamellae. In addition, and in contrast to this topological and structural classification, and following the development of liposomes as drug delivery vehicles, a new — functional — liposome classification could be proposed. With respect to functional characterization and the interaction characteristics, one can divide liposomes into those with nonspecific reactivity (conventional liposomes which are exposed to interactions of components from the suspending medium), inert liposomes (sterically stabilized or Stealth™ liposomes which have minimal reactivity with the medium), liposomes with specific reactivity (Stealth™ liposomes with attached specific ligands in which nonspecific repulsion is upgraded with specific antibodies or lectins), and liposomes that are very reactive under particular conditions. The latter are polymorphic liposomes, or "activosomes", which are topologically stable, but change their structure when triggered by a particular stimulus, such as polyions (with examples of cationic liposomes for DNA and antisense molecule delivery) or pH for pH-sensitive liposomes. Fusogenic and target-specific liposomes also fall into this category.

Liposomes are lipid bilayer-based colloidal particles which can encapsulate part of the aqueous medium into their interior, forming a "milieu interne" (internal microenvironment) which can be different from the extraliposomal medium. Nowadays, these assemblies are becoming a very important system in basic research, from theoretical concepts to serving as a tool, model, or reagent in basic studies of many processes. Parallel to numerous medical uses, a subject to which many books and reviews are dedicated, many other applications are being investigated. The aim of these four books is to present these lesser-known applications in depth. This series, *Handbook of Nonmedical Applications of Liposomes,* is divided into four rather different collections of related papers, so that many researchers will be able to find the most pertinent articles concentrated in a single volume or two.

The first volume presents theoretical studies on thermodynamics and molecular dynamics simulations, as well as some engineering applications of liposomes. The second volume presents liposomes as model systems for many biophysical, biochemical, and cellular processes, as well as protein studies using reconstitution into bilayer membranes. It concludes with a group of chapters related to the origin of life. The third volume commences with the "nuts and bolts" of "liposomology", from rational design, raw materials, large-scale preparation and active agent loading, stability, and stabilization, to sterility. It concludes with the use of liposomes as bioreactors and in catalysis. The fourth volume presents additional liposome applications in life sciences, with a group of chapters on gene therapy, cosmetics, diagnostics, and ecology.

Although these four volumes deal with nonmedical applications of liposomes, their content is also very relevant to their medical application, through many specific subjects such as all the design, preparation, loading, preservation, and characterization of liposomal formulations, gene delivery and diagnostics, as well as comparison with other recently developed colloidal carriers. Therefore, we believe that this series should appeal to everyone who has an interest in liposomes, whether from academia or from industry.

It was a pleasure for both of us to undertake the daunting project of putting this series together because obtaining such a broad spectrum of expertise required working with top academic and industrial researchers. We hope that this series will be useful to scientists from many disciplines and to students who are looking for new frontiers. From our personal experience as teachers, we know that such a series is missing, and we hope that this will fill the gap.

We would like to thank the many authors of this series who did such an excellent job in writing and submitting their papers to us. It was enjoyable to interact with them, and there is no question that the pressure we put on many of them was worthwhile.

Finally, we would like to thank Mrs. Beryl Levene for her excellent secretarial work and for sending so many letters to so many scientists and Mr. Sigmund Geller for his editorial assistance. CRC personnel should be acknowledged for their encouragement and patience, as well as their high level of proficiency.

Yechezkel Barenholz
Jerusalem, Israel

Danilo D. Lasic
Burlingame, California

INTRODUCTION TO VOLUME I

Liposomes and their precursors — myelin figures — have been studied extensively since the middle of the last century by optical microscopy. Currently, computer-aided optical microscopy is in the mainstream, paralleling theoretical studies of liposome shapes and their fluctuations. The liposome membrane is a two-dimensional surface floating in a three-dimensional space. Complex theoretical treatments, such as renormalization theory, can be used to describe their shapes. In the first approximation, the sole parameter characterizing vesicle properties is bending elasticity of the membrane, while adding structural or compositional heterogeneity into the model can explain membrane fusion and fission. After Canham described the shape of red blood cells by a similar treatment and Helfrich used his bending elasticity approach to describe vesicle shapes, the field revived in the 1980s, and today about ten groups are intensively studying vesicle shapes and their fluctuations theoretically and experimentally. The first three chapters by Svetina and Žekš, Seifert and Lipowsky, and Fourcade, Michalet, and Bensimon, give practically a complete picture of the state of the art in the field.

Similarly, the phase behavior of membranes was described by Marčelja 20 years ago, and at the same time, qualitative models for the effect of cholesterol on characteristics of membranes were introduced. Cholesterol was found to decrease order in crystalline bilayers and increase order in fluid bilayers. The concept of solid ordered, fluid disordered, and fluid ordered phases was introduced in the 1970s to describe this important phenomenon. Recent work by Mouritsen and colleagues represents a culmination of the understanding of membrane properties, and this chapter elegantly explains mechanical properties of bilayers as a function of bilayer composition. Computer simulation is used to present solutions of complex equations. Computer methods for calculations in this research are described in the chapter by Ben-Shaul and Fattal on molecular dynamics. Kinnunen, Huang, and Li, and Mason use similar concepts to discuss phase transitions and properties of bilayers with mixed hydrocarbon chains. Liposome aggregation, thermodynamics, interactions, spontaneous vesicle formation, and phase behavior are described in chapters by Nir, Marčelja, Corti and Cantù, and Tenchov and Koynova, who use the simple Clapeyron–Clausius equation to describe the behavior of bilayers in different solutions. Recent developments in synthetic organic chemistry resulted in the synthesis of various lipids conjugated with different polymers. Liposome properties can be critically altered and tailored by the attachment of various polymers on their surface. This can be used for selective, site-specific targeting, liposome destabilization, induction of fusion, or complexation of (macro)molecules, and Torchilin describes the attachment of polyethylene glycol to improve liposome stability. Such liposomes are very stable in the biological milieu and are extensively used in cancer chemotherapy. Another condition for their effectiveness is stable encapsulation of drugs into their interior, as will be discussed in Volume III. Furthermore, this research may have important consequences in the field of biomaterials where nonfouling, nonadsorbing surfaces are being sought. In addition, Winterhalter discusses the behavior of liposomes in electric fields, a recent and very powerful new development in electrobiology. For instance, optical tweezers and controlled membrane rupture are becoming very important tools in basic sciences. Recently, a powerful new method for visualization of colloidal structures, cryoelectron microscopy, was developed. Frederik describes the details of this technique, which is much less prone to sample preparation artifacts than are conventional freeze–fracture, thin-sectioning, and negative-stain electron microscopies. Scanning electron microscopy is not very useful, mostly due to lower sensitivity and cumbersome sample preparation, while the resolution of atomic force microscopy of lipid bilayers is still rather low.

Danilo D. Lasic
Burlingame, California

Yechezkel Barenholz
Jerusalem, Israel

THE EDITORS

Danilo D. Lasic, Ph.D., is Principal Scientist at MegaBios, Burlingame, California, where he is currently working with cationic liposomes in gene delivery. Previously, he was a senior scientist at Liposome Technology, Inc. (now SEQUUS), where he led studies for theoretical understanding of long-circulating liposomes as well as developed the first formulations for preclinical studies. In addition, he also actively participated in the scale-up of the preparation of Stealth™ liposomes laden with the anticancer agent, doxorubicin.

Dr. Lasic graduated from the University of Ljubljana, Slovenia, in 1975 with a degree in Physical Chemistry. He received his M.Sci. in 1977 from the University of Ljubljana. He obtained his Ph.D. at the Institute J.Stefan, Solid State Physics Department in Ljubljana in 1979. After postdoctoral work with Dr. Charles Tanford at Duke and Dr. Helmut Hausser at ETH Zurich, he was a research fellow at the Institute J.Stefan in Ljubljana, and a visiting lecturer in the Department of Chemistry and a visiting scientist in the Department of Physics at the Univerity of Waterloo in Canada. Then he joined Liposome Technology, Inc., in Menlo Park, California. Currently, he is studying cationic liposomes, DNA–liposome interactions, and DNA–lipid complexes for gene delivery and gene therapy.

He has published more than 120 research papers as well as a monograph on liposomes (*Liposomes: from Physics to Applications,* Elsevier, 1993) and co-edited a book, *Stealth Liposomes* (CRC Press, 1995). His best known papers are those dealing with thermodynamics and the mechanism of vesicle formation, stability of liposomes, the origin of liposome stability in biological environments, as well as the applications of drug-laden liposomes. He is a member of the editorial board of the *Journal of Liposome Research* and of *Current Opinion in Colloid and Interface Science.*

Yechezkel Barenholz, Ph.D., is Professor of Biochemistry and head of the Department of Biochemistry at the Hebrew University–Hadassah Medical School, Jerusalem, Israel. He received his undergraduate, M.Sc., and Ph.D. degrees in Biochemistry from the Hebrew University of Jerusalem in 1965, 1968, and 1971, respectively. While working on his Ph.D. thesis, he studied at the Animal Research Council Institute at Babraham, Cambridge, England with Dr. R. M. C. Dawson and Dr. A. D. Bangham. He has been on the permanent academic staff of the Hebrew University since 1974, and a professor there since 1982. He was also on the staff of the University of Virginia in Charlottesville from 1973 to 1976 and has remained a Visiting Professor at the University of Virginia since then. He was a Donders Chair Professor at the University of Utrecht in the Netherlands in 1992.

Professor Barenholz's interests are in basic and applied science. In basic research, he is involved in many fields related to the biochemistry and biophysics of lipids, especially sphingolipids, phospholipids, and sterols, and including synthesis, chemical and physical characterization, and the relationship between membrane lipid composition, structure, and function — with special focus on aging. In applied research, his main interests are in amphiphile-based drug carriers, especially liposomes — from the design of the drug carrier basic aspects through animal studies to clinical trials. Professor Barenholz was involved in clinical trials of Amphocil® an amphotericin B micelle-like assembly, and in the development of DOX-SL®, a doxorubicin remote-loaded sterically-stabilized liposome. At present he is also studying the applications of liposomes for vaccination and for gene therapy. Professor Barenholz is the author of over 200 publications and the editor of the special issue of *Chemistry and Physics of Lipids* on "Quality Control of Liposomes". He is on the editorial boards of *Chemistry and Physics of Lipids,* and the *Journal of Liposome Research.* Professor Barenholz is a recipient of the 1995 Kay Award for innovation.

THE CONTRIBUTORS

Yechezkel Barenholz
Department of Membrane Biochemistry
Hebrew University Medical School
Jerusalem, Israel

Avinoam Ben-Shaul
Department of Physical Chemisty
Fritz Haber Research Center
Hebrew University
Jerusalem, Israel

D. Bensimon
Ecole Normale Superieure
Laboratory of Physical Statistics
Paris, France

P. H. H. Bomans
EM Unit/Pathology
University of Limburg
Maastricht, The Netherlands

Laura Cantù
Department of Chemisty and Biochemisty
University of Milan Medical School
Milan, Italy

Mario Corti
Dipartimento di Elettronica
University of Pavia
Pavia, Italy

B. Dammann
Department of Physical Chemisty
Technical University of Denmark
Lyngby, Denmark

Deborah R. Fattal
Department of Physical Chemisty
Fritz Haber Research Center
Hebrew University
Jerusalem, Israel

H. C. Fogedby
Institute of Physics and Astronomy
Aarhus University
Aarhus, Denmark

B. Fourcade
Institute Laue Langevin
Grenoble, France

P. M. Frederik
EM Unit/Pathology
University of Limburg
Maastricht, The Netherlands

Ching-hsien Huang
Department of Biochemistry
Health Sciences Center
University of Virginia
Charlottesville, Virginia

J. H. Ipsen
Department of Physical Chemistry
Technical University of Denmark
Lyngby, Denmark

C. Jeppesen
Materials Research Laboratory
University of California
Santa Barbara, California

K. Jørgensen
Department of Physical Chemistry
Technical University of Denmark
Lyngby, Denmark

Paavo K. J. Kinnunen
Department of Medical Chemisty
Institute of Biomedicine
University of Helsinki
Helsinki, Finland

Rumiana Koynova
Institute of Biophysics
Bulgarian Academy of Sciences
Sofia, Bulgaria

Danilo D. Lasic
MegaBios Corporation
Burlingame, California

Shusen Li
Department of Biochemistry
Health Sciences Center
University of Virginia
Charlottesville, Virginia

Reinhard Lipowsky
Max-Planck-Institute of Colloids and Interfaces
Teltow-Seehof, Germany

S. Marčelja
Department of Applied Mathematics
Research School of Physical Sciences
Australian National University
Canberra, Australia

Jeffrey T. Mason
Department of Cellular Pathology
Division of Biophysics
Armed Forces Institute of Pathology
Washington, DC

X. Michalet
Ecole Normale Superieure
Laboratory of Physical Statistics
Paris, France

O. G. Mouritsen
Department of Physical Chemistry
Technical University of Denmark
Lyngby, Denmark

Shlomo Nir
Seagram Centre
Faculty of Agriculture
Hebrew University of Jerusalem
Rehovot, Israel

J. Risbo
Department of Physical Chemistry
Technical University of Denmark
Lyngby, Denmark

M. C. Sabra
Department of Physical Chemistry
Technical University of Denmark
Lyngby, Denmark

Udo Seifert
Max-Planck-Institute of Colloids and Interfaces
Teltow-Seehof, Germany

M. M. Sperotto
Department of Physical Chemistry
Technical University of Denmark
Lyngby, Denmark

M. C. A. Stuart
EM Unit/Pathology
University of Limburg
Maastricht, The Netherlands

Saša Svetina
Institute of Biophysics
Medical Faculty
University of Ljubljana
Ljubljana, Slovenia

Boris Tenchov
Institute of Biophysics
Bulgarian Academy of Sciences
Sofia, Bulgaria

Vladimir P. Torchilin
Center for Imaging and Pharmaceutical Research
Massachusetts General Hospital
Charlestown, Massachusetts

Mathias Winterhalter
Biophysical Chemisty
Biozentrum
Basel, Switzerland

Boštjan Žekš
Institute of Biophysics
Medical Faculty
University of Ljubljana
Ljubljana, Slovenia

M. J. Zuckermann
Department of Physics
McGill University
Montreal, Canada

TABLE OF CONTENTS

Handbook of
Nonmedical
Applications
of
Liposomes

Theory and Basic Sciences

VOLUME I

On the History of Liposomes

Danilo D. Lasic

CONTENTS

I. INTRODUCTION

It is very difficult to discuss such a delicate subject, especially in times when liposome literature is rather often plagued by plagiarism, eclectics, incompetent as well as politically tailored citations, and many biased and self-serving claims. To avoid some treacherous issues I shall mainly discuss events prior to the start of the liposome era as described in scientific papers. After the introduction I will only very briefly discuss the main path of evolution and then in slightly more detail will describe the development of sterically stabilized liposomes and the theoretical part, the two areas in which I was more involved.

Apart from the egocentric claims of some authors, there are really not many archives on which I could have based my review. The early years of the liposome revolution can be followed via Bangham's letters, some of which were published in his book *Liposome Letters*.[1] With regard to the developments which led to the discovery of liposomes and events thereafter, I strongly recommend two reviews by the discoverer himself[2,3] and two reviews by his postdoctoral fellow at that time and future collaborator, Papahadjopoulos.[4,5] Another attempt to deal with this subject is the co-editor's review.[6] This review is based mostly on my article in the newsletter *Liposome Technology* ("Liposomes before A.D. [Anno Domini] Bangham", 1993) and on numerous discussions with many of the liposome researchers. I was asked by several of these people to write an article on this delicate subject, and only the encouragement and positive reviews by undoubtedly the two most influential persons[8,9] of my book,[7] in which I also discussed historical aspects in the field, gave me the appropriate courage.

Before I start with the review of scientific literature I will speculate on the formation of liposomes in the natural environment on early Earth and then on the inadvertent preparation of liposomes by early man. The majority of the article will deal with the lipid- and liposome-related literature between 1847 and 1950 as well as between 1950 and 1963, and then I shall quickly review the main contributions afterward. In addition to biophysical and applied aspects, I shall briefly discuss the development of a theoretical understanding of the phenomena, concentrating mostly on the mechanism of vesicle formation and steric stabilization.

From physicochemical phenomena occurring on Earth as well as human activity on it, it is obvious that liposomes had to be created well before their official invention in A.D. 1964 by A.D. Bangham, in Babraham, near Cambridge, England. In addition, in the last 150 years numerous researchers were interested in lipid systems, and many earlier reports on the swelling behavior and other properties of lecithin films, lecithin suspensions/dispersions/colloidal solutions, and myelin figures were published.

0-8493-4731-9/96/$0.00+$.50

Despite all this work, there is no doubt that Alec Bangham was the first person to show that liposomes are spherical, self-closed lipid particles which encapsulate a fraction of the aqueous medium in which they are dispersed and that the lipid membrane represents a selective permeability barrier for charged and polar species.

II. LIPOSOMES IN NONLIVING NATURE

Apart from the fact that biological systems are constantly producing various lipid vesicles (cell metabolism — endocytosis, exocytosis, fusion, fission, secretory granules; signal transduction — synaptic vesicles) their formation also occurs in nonliving nature. Many water surfaces on Earth are covered with a monolayer of amphiphiles. Some are polar lipids which can form liposomes when dispersed in water under appropriate conditions. Water containing polar lipids is dried on rocks, sand, and other template surfaces. Upon rehydration, these dry lipid flakes swell and can self-close into spherical structures under the force generated by the waves or rain. Although lipids also could have been formed by chemical reactions in the primordial soup, Deamer showed that polar lipids also could have reached Earth on meteorites.[10] Their presence could have resulted in the formation of self-closed membranes, which evolved coherently with protein synthesis into the first cells. The beneficiary effect of artificial template surfaces was shown to improve formation of large single-lamellar liposomes.[11] Such processes constantly occur at water-land boundaries.

A. PREHISTORIC LIPOSOMES

We shall now speculate on the formation of liposomes by prehistoric people. It is very likely that the first liposomes were produced inadvertently when people began to use eggs in their cuisine and for personal care. Cro-Magnon man of the Paleolithic era was familiar with fire and simple ceramics. It is therefore likely that liposomes were first produced by cave people beating eggs in water, some 30,000 years ago, possibly in the preparation of a pre-omelette or simply, even earlier, by washing hair either dirtied by an egg or with a "prehistoric shampoo" containing perhaps eggs, ash, etc. The explanation is straightforward. Each egg yolk contains about a gram of lecithin, and agitating a suspension of lecithin in water can yield some multilamellar vesicles. Eggs also contain approximately 0.25 g of cholesterol, and it is possible that these liposomes contained cholesterol in their membranes.[12] One should add that many lipid and oil dispersions, emulsions, creams, and suspensions used in our everyday cuisine may also contain some liposomes.

Eggs and other natural products were also used for other purposes, such as in the preparation of dyes, inks, and similar colloidal systems. Gelatin and gum arabic were used 3000 years B.C. in Egypt in the preparation of colored inks as steric stabilizers for the hydrophobic pigment particles which otherwise tended to aggregate and flocculate. The possible generation of sterically stabilized liposomes during these activities cannot be ruled out.

III. SCIENTIFICALLY DOCUMENTED STUDIES OF AQUEOUS LIPID SYSTEMS

After this speculative introduction, we shall now look at the published literature on scientific investigations with lecithin, their suspensions, lipins (lipids), and myelin figures. Sometimes I shall use the original words of the authors to further contemplate the historical origin.

A. STUDIES FROM 1846 TO 1911

Lecithin was discovered in Paris in the middle of the last century by Gobley, who isolated it from brain and egg yolks and later named these fractions "lécithine" from the Greek for egg yolk, lekitos (λεκιθοσ). Its chemical formula was determined in 1868 by Strecker.

The swelling of lipids was first described by Virchow (1854), who observed the swelling when he transferred nerve core into an aqueous phase.[13] He later described the swelling of dried nerves in water. Since it was very likely that some liposomes detached from the hydrating mass, Virchow failed to recognize not only liposomes but also liquid crystals of myelin figures during his microscopic observation (Figure 1).

Figure 1, from Lehmann's book on liquid crystals (1911), showed the growth of myelin figures schematically.[14] Lehmann (Leipzig) and Reinitzer (Vienna), who were both studying anisotropic liquids,

Figure 1 A drawing of myelin figures as described by Virchow in 1854 and shown by Lehmann in 1911.[14]

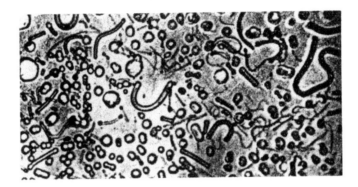

Figure 2 An optical micrograph of "Künstliche Zellen" by Lehmann (1911).[14] Today, from the description and appearance we may safely conclude that these were large multilamellar vesicles.

discovered liquid crystals around 1888. In 1911 Lehmann showed a micrograph of liposomes (Figure 2) and called them "Künstlische Zellen", artificial cells. Already in 1888, Thudicum in London recognized the importance of colloidal associates of phosphatides and wrote: "Amongst their (phosphatides) properties none are more deserving of further inquiry than those which may be described as their power of colloidation. Without this power brain as an organ would be impossible, as indeed the existence of all bioplasm is dependent on the colloidal state." The better we understand Nature and Life, the more we see how correct this statement is. The liquid crystalline state provides organization which allows fast exchange, and colloidal-sized particles have large carrying abilities and can be trafficked with the help of proteins.

B. STUDIES FROM 1912 TO 1949

In this period many interesting studies on the behavior and characteristics of lipids were undertaken. In one study,[15] lecithin suspensions were made by the ether injection technique to study the effects of electrolytes on the viscosity and osmotic behavior of these systems because such lipoids are so widely distributed in living cells. Changes in osmotic pressure were attributed to interactions of various electrolytes with particles. All the electrolytes were found to reduce osmotic pressure, which we can understand today by liposome aggregation and reduced activity of ions due to interactions with a large surface area colloid system. Similarly, electrolytes reduced the viscosity, while it was increased by lipoid solvents. Again, shielded electrostatic repulsion and aggregation/fusion can explain this phenomenon.

In another investigation, the size of particles in lecithin emulsions was measured by ultrafiltration. Different pressures were used for filtration through collodion membranes which had pores of 50 µm.[16] Although filtration is not extrusion and the authors reported mostly the percentage of retained material, we can see many parallels to the extrusion method of liposome preparation.

The isoelectric points of lecithin and sphingomyelin in an aqueous dispersion were measured in the mid-1930s. The results were in dispute until it was shown that freshly prepared lecithin rapidly becomes acidic upon standing, and it was therefore suggested that freshly prepared lecithin be used to obtain results which are likely not in error.[17]

Little work on the colloidal properties of lipids was done until the middle of this century. At that time colloids were still regarded as rather irreproducible systems and people struggled to explain their stability. We must remember that the Derjaguin-Landau-Verwey-Overbeek (DLVO) theory was first proposed in 1943 by Landau and Derjaguin. Since in these earlier times the gold standard for colloids was silver iodide, with a stability measured in decades, we can understand the low interest in colloidal properties of chemically unstable organic compounds. Research conducted between the wars (1918 to 1939) produced micrographs of myelin figures (Gicklhorn, 1932; Nageotte, 1936), a measurement of the isoelectric point of lecithin (6.7, in 1934), ideas about lecithin micelles, an estimate of the thickness of the bilayer of 5 nm (Trilliat, 1920s), and X-ray diffraction of bilayers (de Broglie, 1930s).

After microscopic and theoretical studies of liquid crystals by Friedel in Paris (1922), the first proposals for the bilayered structure of the cell membrane were laid down by Gortel and Wendel (1926). This was further improved by Danielli and Davson (1935), who proposed a bilayer sandwiched by two protein layers, and later by the Singer-Nicholson fluid mosaic model (1970).

During this period, people recognized that lecithin, even when taken orally, could reduce cholesterol levels in blood. Investigators were using crude soy lecithin dispersions to feed or inject cholesterol-fed rabbits in the dose range of 0.2 to 1.6 g lecithin per kilogram. Similar studies in the 1960s and 1970s were repeated in other animal models, such as quail and baboons, with better-defined preparations. To improve dispersal, Friedman also sonicated these dispersions in the late 1950s. These studies showed a significant reduction of atherosclerotic lesions. It is not exactly clear why no data on human studies exist, but it is quite possible that large doses of lipids ("lipid flash") knock out the body's defense mechanisms and influence blood chemistry. Furthermore, depleting cholesterol from blood cells can strongly affect their functioning. However, these effects can be minimized by appropriate lipid and dose selection. In any case, these studies had to produce liposomes, which were apparently thought of simply as lipin (lipid) colloid dispersions and hence not interesting enough for further studies.

One should mention that during this time period the name liposomes occasionally appeared in the scientific literature. It meant simply "fat bodies" (mostly triglycerides of colloidal size; a direct translation of the observed particle morphology into Greek) and was a more suitable application of the name than that used today. Since by "liposome" most of researchers mean "hollow spheres", the term "lipid vesicle" is more appropriate today. Studies during this period encompassed emulsion droplets, fat, and various low-density lipoprotein particles. Current liposome nomenclature was proposed in the first book edited on the subject (by Papahadjopoulos) in 1978, but since then many new, more (and less) ingenious names have penetrated into the literature, and there is a need to reduce this anarchy as well as to update the nomenclature with novel terms.

C. STUDIES FROM 1950 TO 1965

In the 1950s studies included research on lecithin dispersions and treatments by different physicochemical methods. In London Saunders et al., who sonicated phospholipid sols (sol means fluid gel, e.g., suspension/dispersion), studied the size and shape of the micelles and noted that the absence of oxygen during sonication is required to prevent lipid degradation. He also estimated the molecular weight of micelles to be 5 million Da, corresponding to a liposome with a diameter of 30 nm. Saunders et al. noted that solutions were clear and particles were symmetric, and they thought the sols were large disk-like micelles or coiled leaflets.[18,18a]

Today we know that these sols had to be multilammelar large vesicles (MLV) suspensions and that micelles were probably small unilamellar vesicles. It is interesting that Saunders et al. hydrated lipids by adding water to an ether solution of lipid and, after removing organic solvent, added more aqueous phase, which is very similar to the "REV" method for producing large multilamellar vesicles. In 1952, he also dissolved lecithin sols with PEGylated fatty acids, creating during the process probably the first sterically stabilized vesicles. Robinson (in London) measured the size of lecithin micelles by light scattering. He attributed the result (20 million Da) to large laminar (disk-like) micelles.[19]

A study of cholesterol solubilization in phospholipid micelles was also very interesting.[20] Cholesterol was dissolved with lecithin in chloroform and organic solvent was evaporated under a stream of nitrogen. Traces of organic solvents were removed in a vacuum desiccator and the residue was suspended in buffer during homogenization. The mixture was then exposed to ultrasonic irradiation and subsequently centrifuged at 35,000 rpm for 1 h. This is essentially a well-described preparation of liposomes by sonication!

In the late 1950s many researchers were studying blood clotting using lipid dispersions. This included the relationship between the colloidal state of phospholipids and clot formation. In these experiments, the lipid suspensions were probably large multilamellar liposomes and droplets of inverse hexagonal phase in excess water. Undoubtedly, in many of these studies during this time in which researchers were preparing hydrated lipid suspensions, liposomes were inadvertently produced. For instance, Bangham's paper in 1961 essentially describes a thin film hydration method for the preparation of "lipid emulsions".[21]

This short review of the work between the two wars shows that even before the formal liposome discovery, and in addition to numerous studies of myelin figures,[22,23] we had thin film hydration methods, sonication, extrusion, and homogenization methods, as well as the techniques for preparation of liposomes from reverse phases and by organic solvent injection already described in the literature! Of course, the products were not characterized and their morphology not anticipated.

Despite having written several very good papers, these scientists failed to realize that liposomes enclose an aqueous phase. Bangham and his co-workers challenged the concept of planar and/or "rolled up" (Swiss roll-type) micelles.

In addition to permeability studies, Alec Bangham (1965) and Demetrios Papahadjopoulos (1967) undertook a series of pioneering studies that established the physicochemical properties of liposomes. Before we move to the work which defined and characterized liposomes, we shall look into the studies which directly preceded it.

D. DISCOVERY OF LIPOSOMES

In the late 1950s several excellent scientists living in or near Cambridge, England studied the physical properties of blood, blood cells, and phospholipases. Bangham was primarily interested in the physical properties of different cellular constituents in blood. For these purposes he used electrophoretic measurements since he believed that surface properties alone determined their behavior. Rex Dawson was trying to understand the function of phospholipases. For this reason he needed pure phospholipids as substrates. As a consequence, the best-characterized lipid samples at that time were found in Dawson's laboratory. To his surprise, the purer the lipids the less enzymatic activity could be measured. In collaboration with Bangham, Dawson realized that the substrate had to be negatively charged to show enzymatic activity. Together they used electrophoresis of various lipid dispersions, and many measurements of mobility of these particles in various conditions were reported. Today we would call these multilamellar vesicles having a variety of lipid compositions.

Bangham wanted to characterize the lipid suspensions he had used to model the surfaces of cellular membranes. His main goal was to quantify dispersal through size and number of particles in order to estimate the surface area of various preparations. In collaboration with Horne he studied negatively stained preparations of briefly sonicated lecithin suspensions using electron microscopy, and they were the first to show the spherical, self-enclosed, multilamellar nature of dispersed particles. Bangham immediately realized the importance of the discovery and started to measure the permeability properties of the various membranes for various ions. The fact that he as well as Horne, an excellent electron microscopist, used to go for a sail in the North Sea coast of Anglia together probably sped up the discovery of liposomes.

Bangham's work showed that liposomes are (quite) impermeable to ions and large molecules, but permeable to water and small nonelectrolytes. Papers written by Bangham and his collaborators in 1964 and 1965 clearly established the existence of liposomes as multilamellar vesicles, characterized the properties of liposomes, and predicted the quick development of liposomes into a major system in biophysics, colloidal science, chemistry, and cell biology.[24,25] At that time, Papahadjopoulos had joined the group as a postdoctoral fellow, which eventually resulted in a prolific career in the field after establishing his own laboratory.

The subsequent development was very fast. One reason for this was that the state of the bilayers had been well characterized by Luzzati and Dervichian (in Paris), Chapman (in London), and several others who studied black lipid membranes. Luzzati and Husson studied phase diagrams of phospholipid water systems in the early 1960s,[26] while Stockenius investigated these liquid crystalline phases by electron microscopy.[27] In addition, phase diagrams of soap-water systems had been known since the 1920s because of the work of McBain and Lawrence in England. Here an interesting parallel can be drawn. While liposomes were extensively used before their real invention, the story of micelles is just the opposite.[28] Micelles were first proposed in 1913, but the concept was met with resistance; McBain reported: "So novel was this finding that when in 1925 some of the evidence for it was presented to the Colloid Committee for the Advancement of Science in London, it was dismissed by the chairman, a leading authority, with the words, 'Nonsense, McBain'."

During the 1970s a rapid development of various liposome preparation and characterization methods occurred. After describing a technique for liposome preparation by sonication in excellent, but still largely forgotten papers in 1967,[29,29a] Papahadjopoulos and collaborators also introduced the formation of liposomes from reverse phases[30] and by extrusion,[31] which is the best method for the preparation of homogeneous liposomes with the tightest size distribution. Barenholz and colleagues realized that one can down-size liposomes also by other high-energy treatments, and they introduced the French press extrusion.[32] The next step was the use of homogenizers.[33] In parallel, detergent depletion techniques, which were performed by Racker in 1974, were further developed by Hauser and others. In addition to the many biophysical and biochemical studies in the labs of Thompson, Tanford, and Nozaki, and many others, many novel lipids were introduced into liposome preparations by Barenholz and others, including nonionic lipids by Vanlerberghe in what eventually resulted in the first cosmetic formulations.

IV. LIPOSOMES IN DRUG DELIVERY

Before I conclude, I would like to mention the commencement of liposome applications in drug delivery. After Bangham's first paper on liposomes appeared, Schneider, a pharmacologist at The University of Birmingham, called Bangham and asked if he thought liposomes could be used to release encapsulated drugs slowly. After encouragement by Bangham, Schneider submitted a patent application on his research. However, Schneider's patent was denied due to the preexisting British patent issued to IG Farbenindustrie in 1934, which described aqueous dispersions of lecithin and cholesterol as effective drug carriers.[34] Other collaborators of Bangham, most notably Papahadjopoulos and Weissmann, continued to pursue these goals and, along with Ryman and Gregoriadis, helped establish the application of liposomes as a drug delivery system. Weissman also renamed the structures (at the time called by various names, including Banghasomes) liposomes. This area really experienced exponential growth as can be seen from the number of publications and patents (Figure 3). Despite numerous articles not many commercial products emerged in the next 25 years. Basically, only amphotericin B liposomes (AmBisome, Vestar) have been on the European market since the early 1990s (sales started in Sweden in December 1989), while Stealth™ doxorubicin from Liposome Technology may be available in the U.S. as well as in Europe before the end of 1995. In addition, Hansjoerg Eibl, one of the leading lipid synthetic chemists, who synthesized more than 5000 novel synthetic lipids, discovered a special phospholipid which by itself showed antitumor activity; it has been on the market in liposomal topical form since 1992. In addition to drug delivery studies, several researchers, including Kinsky and Alving, pursued the interaction of liposomes with the immune system for vaccination and (de)stimulation of the immune system. Another important discovery was that liposomes could be freeze-dried and reconstituted; this was shown first for large multilamellar liposomes by Swiss researchers and for unilamellar ones by Crowes. Here another parallel, concerning liposome patents, should be mentioned. All this work was patented, and several other patents on the lyophilization were issued in the early 1980s. On these grounds, The Liposome Company, which also licensed the Swiss (1979) patent, sued Vestar for infringement of patent rights. The judge, however, dismissed the case, again because of the preexisting literature by Racker (1974).

Another set of patents deals with another crucial issue in medical liposome applications — loading of the drug molecules into liposomes. It seems that the most important patents are the ones which deal with remote loading of some drugs into preformed liposomes by the use of special gradients across liposome membranes. Here, again, published information well precedes patents and can be traced back at least to the work of Crofts and Deamer (1972 and 1976), Forte (1978), and others. The pH gradient loading was further developed by Cullis, while Barenholz introduced the ammonium salt gradient method. Lasic showed that formation of intraliposomal precipitates can greatly enhance the loading capacity and stability of drug encapsulation.

The next substantial breakthrough was the realization that long-circulating liposomes are necessary for most applications. After pioneering work by Allen, Gabizon, and Papahadjopoulos and later by Martin, Storm, Woodle, and Lasic, and independently by Huang, Cevc, and Gregoriadis with their collaborators, the first formulations containing polymer-bearing lipids were tested for their enhancing antitumor efficacy in *in vivo* models by Mayhew and Lasic. Theoretical understanding of the processes was pursued by Lasic, Needham, Israelachvili, Sackmann, Safinya, and others. This behavior was found to be in excellent agreement with simple scaling concepts, as developed by de Gennes.[35]

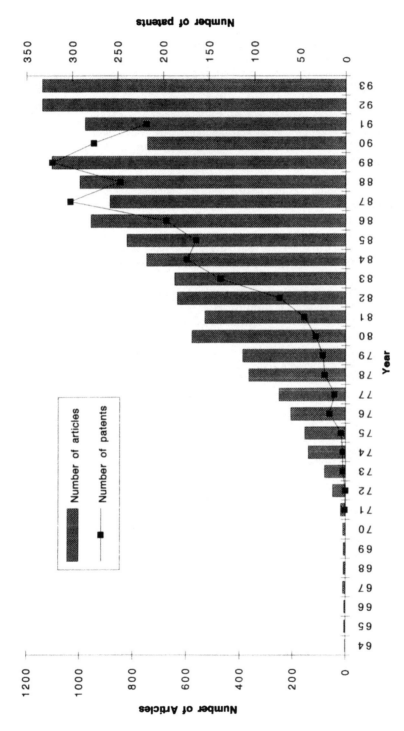

Figure 3 Number of liposome papers and patents per year after their discovery, as obtained from liposome data base "Lipodisk", from Avestin, Ottawa, Ontario.

Liposomes were first injected into humans in the early 1970s, mostly for safety, pharmacokinetics, and biodistribution studies. Drug-containing liposomes were injected into humans in 1977 by Gregoriadis and colleagues for enzyme replacement therapy, but the best-known early results in humans are from 1985 when Lopez-Berestein injected amphotericin B in large multilamellar liposomes into 12 patients, with very promising results. These early studies were reviewed by Zonneveld and Crommelin,[36] who with their collaborators performed basic studies of lipid and liposome stability. Barenholz and others, as well as all three liposome companies in the U.S. (Liposome Technology, Inc., The Liposome Company, and Vestar, plus Ciba Geigy and some others), concentrated on federal requirements for industrial production of injectable liposome formulations, including issues such as raw materials, chemical stability of lipids, large-scale production, and sterilization as well as regulatory issues.[37,38] Since then many phase I and II trials have been performed in anticancer therapy by Gabizon and many other clinical researchers. In 1989, several clinical trials in the therapy of cancer were described in the book *Liposomes*, edited by Fidler and Lopez-Berestein, while Reference 35 concentrates on the anticancer agent doxorubicin. Prerequisites for administration of liposomes into humans, in addition to a sterile and stable formulation, are safety and tolerability.[39]

Another important realization was that nucleic acids and oligonucleotides could be complexed with cationic liposomes and used in transfection. After the use of conventional liposomes for DNA encapsulation by Papahadjopoulos, Nicolau, and many others was proved to be rather cumbersome, Behr and Felgner discovered that DNA could be complexed with cationic liposomes and that such complexes could be effective in the introduction of genes into cells. Now several liposome reagents for transfection *in vitro* are commercially available and half a dozen small companies are working on liposomal delivery systems in gene therapy.

In conclusion, I would first like to strongly disagree with some claims that liposomes were discovered serendipitously. They were studied for many years as particulate suspensions in water prior to the discovery that they were closed, hollow vesicles, which was a result of an intelligent, self-consistent analysis. The same realization, as well as the importance of that unique topology of "fatty droplets", eluded many well-respected scientists. In addition to all the work mentioned above plus numerous other studies, I remember looking at myelin figures during biology class in high school in the late 1960s. Obviously, due to our curiosity, excitement, and/or inexperience, we had to have compressed cover slips on the slides, which undoubtedly detached some of the myelin figures and produced liposomes. However, neither we nor those of other generations who had been observing the same phenomenon for perhaps 50 years were interested in these "detached myelin figures".

V. THEORETICAL UNDERSTANDING OF LIPOSOME BEHAVIOR

The theoretical understanding of liposomes did not parallel the quick pace of improvements in their preparation procedures, characterization techniques, and applications. To my surprise, nothing was known about the mechanism of their formation, size distribution, thermodynamics, and kinetic effects in the preparation procedures of the early 1980s. While sophisticated theoretical analyses were employed to explain various spectroscopic and diffraction measurements, there were practically only two theoretical reports on liposome structure and size distribution. Saupe (1973) discussed lamellarity of liposomes by a Landau free energy expansion, while Helfrich (1978) discussed the size of sonicated vesicles by minimization of elastic bending energy and unfavorable edge interaction of open fragments. Theoretical beginnings of this approach stem from Zolcher's work on thermotropic liquid crystals and Franck's application of this model to smectic liquid crystals, where he defined three deformation modes of these lamellar phases: twist, splay, and bend. Assuming that the former two are zero in an isotropic two-dimensional membrane, Helfrich defined the well-known equation for bending energy of a bilayer.[40] These equations were used by Lasic to explain the dynamic processes involved in vesicle formation by detergent depletion techniques and different sizes of these liposomes.[41] Briefly, small mixed micelles of detergent-(phospho)lipid fuse upon removal of detergent because the system is minimizing its total edge energy, which is proportional to the perimeter of the disk-like micelles. This growth is opposed by entropy and a reduced amount of detergent molecules which can shield the edges. This forces large disk-like micelles to bend and eventually self-close into closed vesicles. This eliminates the unfavorable exposure of the edges but increases the bending energy of the system, and it was shown that liposomes are, in agreement with experimental observations, in a thermodynamically metastable stage.[42] Later this reasoning, which postulated an open bilayered fragment as an intermediate structure in the vesiculation process, was used to explain some other preparation techniques as well.[43] The formation of large multilamellar liposomes from swelling dried lipid bilayers was explained by the difference between the areas of polar heads on the outer and inner monolayers, which induces curvature, causes a mismatch between the opposing surfaces of a bilayer, and can result in budding off when the excess of outer monolayer surface area over the inner one exceeds some critical value.[7,44] This normally does not happen spontaneously if no crystal defects are present. It was shown, however, that because liposomes are thermodynamically at a higher energy level than the hydrated lamellar phase in excess water, some energy input is normally required to form liposomes, which therefore can be characterized as a kinetic trap. Agitating hydrating myelin figures causes rupture of long cylindrical lipids tubules, and open edges quickly reseal and form self-closed liposomes. The lipid packing parameter, as introduced by Israelachvili et al., is a very useful parameter to predict various geometries, although we must be aware that such a model is an approximation valid only for states of thermodynamic equilibrium — hence the inability of such models to explain the variable size of liposomes as a function of different experimental conditions, their overall instability, and (broad) size distribution of these structures.

Nowadays we believe that liposomes are produced either by self-closure of smaller bilayered fragments or by fission due to a surface area gradient between the two opposing monolayers, which can be induced by various gradients across the bilayer.[7] In addition, liposomes also can be formed spontaneously; this process resembles either a micelle formation process, a spontaneous reorganization of lipid molecules upon mixing appropriate substances, or fission from (disordered) cubic phases (sponge) in some surfactant systems near or at the critical point, where fluctuations are large. With the recent advent of this theory, these concepts were rigorously analyzed. We should mention that spontaneous formation of liposomes from oleic acid (ufasomes)[45] and other fatty acids[46] long preceded the development of catanionic liposomes.[47] The problem with spontaneous vesicles is that they are in equilibrium with their surroundings and therefore cannot exhibit stable encapsulation. Furthermore, upon application they simply change their phase, i.e., disintegrate, due to changes in the environment (travel in phase diagram). In contrast, kinetically trapped systems exhibit better stability. In most cases liposomes are characterized via their mean diameter; ideally, however, they should be characterized by their size distribution function,[48] by the distribution of lipids between them (if they are not single-component systems), and perhaps by their shape.

Recently, theoretical and experimental studies of various liposome shapes became very popular and vesicles of complex topology were observed. After fundamental papers by Canham (1968) on the shape of the red blood cell and by Helfrich on vesicle shapes, Svetina and Žekš (1984) developed the bilayer couple model which is today very popular among many theoreticians and experimentalists, who are employing more and more sophisticated theoretical (Lipowsky, Seifert, Boal, Webb, Vortis, de Gennes, etc.) and experimental (Sackmann, Evans, Vortis, Guedeau-Boudeville, Safinya, Waugh, Frederik, and many others) methods to study these systems. Further work will probably concentrate on measurements of Gaussian curvature, membrane fusion and fission, and understanding of equilibrium properties of lipids in water.[49,50]

In comparison with simple experimental systems, real samples are more complex. In general, liposomes are thermodynamically unstable, and therefore one cannot easily explain their size distribution, stability, and other characteristics. They are not in a state of equilibrium, and because kinetic parameters affect their formation and due to the fact that the free energy function is a very rough, pitted surface with many local free energy minima (Figure 4) it is very difficult to predict anything theoretically in real samples or have good reproducibility in experiments. Perhaps methods from nonlinear physics, chaotic systems, or spin glasses will be able to encompass some general behavior.

Apart from the very successful application of liposomes in cosmetics by L'Oreal in 1986 and by Christian Dior a year later (nowadays more than 500 cosmetic liposomal products worldwide account for more than $1 billion U.S. in sales) and some pharmaceutical-cosmetic formulations (the first example of this rapidly growing area was the antifungal topical cream Prevaryl®, launched in Switzerland in 1987), a quick analysis of the present situation shows a saturation in various liposome preparation methods, their characterization (with few exceptions such as cryo-electron microscopy), a few successful and promising medical applications, and their promising start in gene therapy.

One must be aware of the major differences between medical and cosmetic formulations, the standards with respect to reproducibility, sterility, stability, as well as efficacy activity being less strict in the latter.

FREE ENERGY

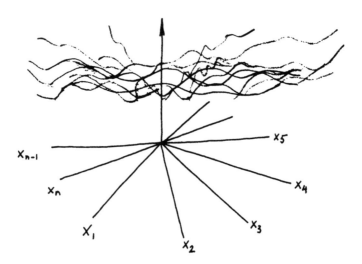

Figure 4 A schematic presentation of free energy of liposomes as an *n*-dimensional surface which depends on *n* variables, such as various concentrations (lipids, ions, inert molecules, impurities, specific agents, etc.), temperature, pressure, potentials, etc. Because there are many local minima and not a simple global minimum the final state(s) of the system depends on the kinetic conditions and is path dependent.

Only recently these problems, including reproducible large-scale production of stable and sterile formulations, were solved.[7,37,38] Medical applications of conventional liposomes have culminated in one liposomal formulation of the antifungal agent amphotericin and show promise in vaccines, with some promise in some diseases of liver, spleen, and yet unexploited parasitic disease of the immune system. Parasitic diseases are perhaps the easiest to treat, but Third World economies cannot support research or attract the interest of developed nations in this research. Sterically stabilized (Stealth™) liposomes will reach their culmination in anticancer chemotherapy, with a possibility for use in the treatment of infectious diseases and inflammation. However, targeting of these liposomes is, in my view, rather restricted due to geometrical limitations, generation of the immune response, and lack of consistent control of cellular uptake of liposomes, as well as demanding preparation. Cationic lipids and liposomes are currently the most rapidly developing field, with promising new applications in gene therapy.[50] In the future I foresee a far greater role for synthetic organic chemistry, which can induce various changes, (e.g., liposome disintegration, controlled release of its cargo, etc.) via controllably unstable bonds in lipids or polymers. Polymer-decorated liposomes will undoubtedly become very important, not only because of steric stabilization, but also because they can induce drug release, fusion with cells, etc. In addition to these, as this text will attest, many other nonmedical applications may become very important, such as diagnostics, biomaterials, ecology, and the food industry.

In conclusion, liposomes have become part of our everyday lives in less than 30 years, and we hope that with improved understanding new important applications will be achieved.

REFERENCES

1. **Bangham, A. D.,** Ed., *Liposome Letters,* Academic Press, New York, 1983.
2. **Bangham, A. D.,** Liposomes in Nuce, *Biol. Cell,* 17, 1, 1983.
3. **Bangham, A. D.,** Liposomes: the Babraham connection, *Chem. Phys. Lipids,* 64, 275, 1993.
4. **Papahadjopoulos, D.,** A personal view of liposomes: from molecular models to drug delivery, in *Medical Applications of Liposomes,* Yagi, K., Ed., Japan Scientific Society Press, Tokyo, 1986.
5. **Papahadjopoulos, D.,** Optimal liposomal drug action: from serendipity to targeting, in *Liposome Technology,* 2nd ed., Vol. 3, Gregoriadis, G., Ed., CRC Press, Boca Raton, FL, 1993, 1-14.
6. **Barenholz, Y.,** Liposome production: historic aspects, in *Liposome Dermatics,* Braun-Falco, O., Korting, H. C., and Maibach, H. I., Eds., Springer-Verlag, Berlin, 1992, 69-81.

7. **Lasic, D. D.,** *Liposomes: From Physics to Applications,* Elsevier, Amsterdam, 1993.

8. **Bangham, A. D.,** Book review, *Biophys. J.,* 67, 1358, 1994.

9. **Papahadjopoulos, D.,** Book review, *Langmuir,* 10, 4397, 1994.

10. **Deamer, D. W.,** Boundary structures are formed by organic components of the Murchinson carbonaceous chondrite, *Nature,* 317, 792, 1985.

11. **Lasic, D. D., Belic, A., and Valentincic, T.,** A simple method for the preparation of large unilamellar vesicles, *J. Am. Chem. Soc.,* 110, 970, 1988.

12. **Lasic, D. D.,** Liposomes, *Am. Sci.,* 80, 20, 1992.

13. **Virchow, R.,** Über das ausgebreitete Vorkommen einer dem Nervenmark analogen Substanz in den Thierische Geweben, *Virchows Arch. Pathol. Anat. Physiol.,* 6, 563, 1854.

14. **Lehmann, O.,** *Die flussige Kristalle,* Akademische Verlagsgesellschaft m.b.H., Leipzig, 1911.

15. **Thomas, A.,** A study of the effects of certain electrolytes and lipoid solvents upon the osmotic pressures and viscosities in lecithin suspensions, *J. Biol. Chem.,* 23, 359, 1915.

16. **Bechold, H. and Neuscloss, S. M.,** Ultrafiltrationstudien an Lecithinsol, *Kolloid. Z.,* 29, 81, 1921.

17. **Chain, E. and Kemp, I.,** The isoelectric points of lecithin and sphingomyelin, *Biochem. J.,* 28, 2052, 1934.

18. **Saunders, L., Perrin, J., and Gammack, D.,** Ultrasonic irradiation of some phospholipid sols, *J. Chem. Soc.,* 14, 567, 1962.

18a. **Attwood, D. and Saunders, L.,** A light scattering study of ultrasonically irradiated lecithin sols, *Biochim. Biophys. Acta,* 98, 344, 1965.

19. **Robinson, N.,** A light scattering study of lecithin, *J. Chem. Soc.,* 12, 1260, 1960.

20. **Fleischer, S. and Brierley, G.,** Solubilization of cholesterol in phospholipid micelles in water, *Biochem. Biophys. Res. Commun.,* 5, 367, 1961.

21. **Bangham, A. D.,** A correlation between surface charge and coagulant action of phospholipids, *Nature,* 192, 1197, 1961.

22. **Teitel, B. A.,** Note sur les figures myelinniques et leur interpretation physicochemique, *Arch. Roum. Pathol. Exp.,* 14, 53, 1947.

23. **Frey-Wyssling, A.,** *Submicroscopic Morphology of Protoplasm,* Elsevier, Amsterdam, 1953.

24. **Bangham, A. D. and Horne, W.,** Negative staining of phospholipids and their structural modification by surface active agents as observed in the electron microscope, *J. Mol. Biol.,* 8, 660, 1964.

25. **Bangham, A. D., Standish, M. M., and Watkins, J. C.,** Diffusion of univalent ions across the lamellae of swollen phospholipids, *J. Mol. Biol.,* 13, 238, 1965.

26. **Luzzati, V. and Husson, F.,** The structure of liquid crystalline phases of lipid water systems, *J. Cell Biol.,* 12, 207, 1962.

27. **Stockenius, W.,** Some EM observations on liquid crystalline phases in lipid water systems, *J. Cell Biol.,* 12, 221, 1962.

28. **Mittal, K. L. and Mukherjee, P.,** The wide world of micelles, in *Micellization, Solubilization, and Microemulsions,* Vol. 1, Mittal, K. L., Ed., Plenum Press, New York, 1977, 1-21.

29. **Papahadjopoulos, D. and Miller, N.,** Phospholipid model membranes. I. Structural characterization of hydrated liquid crystals, *Biochim. Biophys. Acta,* 135, 624, 1967.

29a. **Papahadjopoulos, D. and Miller, N.,** Phospholipid model membranes. II. Permeability properties of hydrated liquid crystals, *Biochim. Biophys. Acta,* 135, 639, 1967.

30. **Szoka, F. C. and Papahadjopoulos, D.,** Procedure for preparation of liposomes with large internal aqueous space and high capture by reverse phase evaporation, *Proc. Natl. Acad. Sci. U.S.A.,* 75, 4194, 1978.

31. **Olson, F., Hunt, C. A., Szoka, F., Vail, W. J., and Papahadjopoulos, D.,** Preparation of liposomes of defined size by extrusion through polycarbonate membranes, *Biochim. Biophys. Acta,* 557, 9, 1979.

32. **Barenholz, Y., Amselem, S., and Lichtenberg, D.,** A new method for preparation of phospholipid vesicles, *FEBS Lett.,* 99, 210, 1979.

33. **Mayhew, E., Lazo, R., Vail, W., King, J., and Green, A. M.,** Characterization of liposomes prepared using a microemulsifier, *Biochim. Biophys. Acta,* 557, 9, 1984.

34. **Bangham, A. D.,** Private communication, 1991. (See also Lasic, D. D., Les liposomes, la révolution ou la illusion, *Rêcherche,* 20, 904, 1989.)

35. **Lasic, D. D. and Martin, F.,** Eds., *Stealth Liposomes,* CRC Press, Boca Raton, FL, 1995.

36. **Zonneveld, G. M. and Crommelin, D. J. A.,** Liposomes: parenteral administration to man, in *Liposomes in Drug Delivery,* Gregoriadis, G., Ed., John Wiley & Sons, Chichester, 1988.

37. **Barenholz, Y.,** Ed., Quality control of liposomes, *Chem. Phys. Lipids,* 64, 1, 1993.

38. **Barenholz, Y. and Crommelin, D. J. A.,** Liposomes as pharmaceutical dosage forms, in *Encyclopedia of Pharmaceutical Technology,* Vol. 9, Swarbrick, J. and Boylan, C., Eds., Marcel Dekker, New York, 1993.

39. **Storm, G., Oussoren, C., Peeters, P. A. M., and Barenholz, Y.,** Tolerability of liposomes *in vivo,* in *Liposome Technology,* Gregoriadis, G., Ed., CRC Press, Boca Raton, FL, 1993.

40. **Helfrich, W.,** Elasticity of membranes, *Z. Naturforsch.,* 28c, 693, 1973.

41. **Lasic, D. D.,** A molecular model of vesicle formation, *Biochim. Biophys. Acta,* 692, 501, 1982.

42. **Lasic, D. D.,** On the thermodynamic (in)stability of liposomes, *J. Colloid Interface Sci.,* 140, 302, 1990.

43. **Lasic, D. D.,** A general model of vesicle formation, *J. Theor. Biol.,* 1987.
44. **Lasic, D. D.,** The mechanism of liposome formation, a review, *Biochem. J.,* 256, 1, 1988.
45. **Gebicki, J. M. and Hicks, M.,** Ufasomes are stable particles surrounded by unsaturated fatty acid membranes, *Nature,* 243, 232, 1974.
46. **Deamer, D. W.,** Liposomes from ionic single chain amphiphiles, *Biochemistry,* 17, 3759, 1978.
47. **Kaler, E. W., Murthy, A. K., Rodriguez, B. E., and Zaszadinski, J. A. N.,** Spontaneous vesicle formation in aqueous mixtures of single tailed surfactants, *Science,* 245, 1371, 1989.
48. **Lichtenberg, D., Friere, E., Schmidt, L. F., Barenholz, Y., Felgner, P. L., and Thompson, T. E.,** Effect of surface curvature on stability, thermodynamics, behavior, and osmotic activity of DPPC single lamellar liposomes, *Biochemistry,* 20, 3462, 1981.
49. **Helfrich, W.,** Pathways of vesiculation, *Prog. Surf. Sci.,* 95, 7, 1994.
50. **Lasic, D. D. and Papahadjopoulos, D.,** Liposomes revisited, *Science,* 267, 1275, 1995.

Chapter 1

Elastic Properties of Closed Bilayer Membranes and the Shapes of Giant Phospholipid Vesicles

S. Svetina and B. Žekš

CONTENTS

I. INTRODUCTION

Elastic properties of phospholipid membranes can be related to many aspects of the behavior of phospholipid vesicles such as their formation, stability, size, shape, fusion, budding processes, and so on. It is the purpose of this chapter to discuss the elastic properties of membranes forming closed surfaces from the point of view that phospholipid membranes are in general composed of two layers. Namely, these membranes are formed by two oppositely oriented monolayers which are in tight contact because of the hydrophobic effect, but are able to slide by each other in the lateral direction. This simple notion about the behavior of phospholipid membranes has in recent years led to the elucidation of rather specific elastic behavior of closed layered membranes with several interesting consequences which we shall try to reveal with respect to the shapes of giant phospholipid vesicles.

The chapter will be divided into three main sections. In Section II we shall show by which basic elastic modes it is possible to describe elastic deformations of a bilayer forming a closed surface. Section III will be devoted to the respective theoretical determination of shapes of unsupported phospholipid vesicles. We shall treat vesicles in a flaccid condition, i.e., vesicles with volume smaller than the maximum possible volume for the same membrane area. Because of the tendency to minimize the bending energy of their membranes, flaccid phospholipid vesicles in general incorporate water and tend to become spherical. However, water permeability of phospholipid membranes is small; therefore, because of their small area/volume ratio the giant vesicles (vesicles of several tens of micrometers) change their volume very slowly, and thus for typical observational times they remain flaccid. In Section IV we shall, by treating some selected examples, show how vesicle shapes are modified under the effect of externally applied forces. In the concluding remarks we shall indicate the importance of theoretical and experimental studies of phospholipid vesicle elastic properties and shapes for the understanding of the general properties of systems involving closed layered membranes.

0-8493-4731-9/96/$0.00+$.50
© 1996 by CRC Press, Inc.

The chapter is primarily based on the work of our laboratory. However, in the introductory part of each section we shall also give reference to other related work. We shall everywhere try to provide the relevant theoretical basis and at the same time to present the major consequences for the vesicle behavior.

II. ELASTIC PROPERTIES OF CLOSED BILAYER MEMBRANES

The general mechanical properties of phospholipid and cellular membranes have been extensively treated in several reviews.[1-4] Therefore, we shall concentrate here only on the fact that the phospholipid membrane is a bilayer. Namely, phospholipid membranes are formed by two oppositely oriented phospholipid monolayers which are in contact at the ends of their hydrocarbon chains and which have their polar heads oriented towards the two embedding aqueous solutions. Elastic behavior of such systems is governed by the connectedness of the two composing layers with respect to their relative lateral movement.[5] In phospholipid bilayers the two composing monolayers are in contact because of the hydrophobic effect but can slide by each other and are thus unconnected. The significance of this property for the elastic behavior of closed membranes was first recognized by Evans[5,6] and Helfrich,[7] who showed that there is a nonlocal contribution to the elastic energy of the membrane which is proportional to the square of the integral of the mean curvature over the whole membrane. More recently,[8] it has been shown that this nonlocal elastic energy corresponds to the relative stretching or relative expansion of the two membrane layers, meaning that at a given average membrane area one of the layers is expanded and the other compressed with respect to their equilibrium areas. The general expression for the elastic energy of the bilayer was derived from the elastic properties of the two monolayers constituting the bilayer where, by the use of the geometrical constraint that these two layers impose on each other, the contributions of the membrane area expansion and the relative expansion of layers were expressed as two independent elastic deformation modes.[8,9]

This section will emphasize two levels of the study of the elastic properties of a phospholipid bilayer. First, we shall demonstrate how elastic properties of a general closed bilayer derive from the elastic properties of the constituent monolayers. We shall ascribe to these monolayers elastic properties of a phospholipid monolayer in a liquid crystalline state. In the second part we shall treat the elastic properties of a monolayer and address the question of how membrane elastic properties depend on structural features of the constituent molecules. In the first example we shall consider the monolayer as an anisotropic continuum and in the second example we shall describe a possible molecular model.[10]

A. ELASTIC ENERGY OF CLOSED BILAYER MEMBRANES WITH UNCONNECTED MONOLAYERS

The elastic properties of a closed bilayer where the composing two layers are in contact but unconnected are governed by the elastic properties of these two layers and the geometrical constraint imposed by the contact. We assume that both phospholipid monolayers (denoted by the index m, where we shall use m = 2 for the outer monolayer and m = 1 for the inner monolayer) are in the liquid crystalline phase. Then the monolayer elastic energy is given as the sum of the area expansivity term, the bending term, and the Gaussian (saddle splay) bending term, respectively.

$$W_m = \frac{1}{2} \frac{K_m}{A_{0,m}} \left(A_m - A_{0,m} \right)^2 + \frac{1}{2} k_{c,m} \int \left(c_1 + c_2 - c_{0,m} \right)^2 dA_m + \bar{k}_{c,m} \int c_1 c_2 \, dA_m \qquad (1)$$

where in the area expansivity term A_m is the area of the neutral surface of the mth layer, i.e., the surface at which membrane area expansion and bending are decoupled. $A_{0,m}$ is the equilibrium value of this area, and K_m is the area expansivity modulus. In the bending terms c_1 and c_2 are the two principal curvatures (reciprocals of the principal radii of curvature) defined so that they are positive for a sphere, $c_{0,m}$ is the spontaneous curvature[11] of this layer, $k_{c,m}$ its bending modulus, and $\bar{k}_{c,m}$ is the Gaussian bending modulus. Integration is over the whole area of the neutral surface of the layer, A_m. Because we are limiting ourselves only to harmonic deformations and are thus taking into consideration only quadratic terms in expansion as well as in bending, we can integrate over A_m instead of over $A_{0,m}$, as should be done according to the definition of the bending energy.

The geometrical constraints that the two monolayers exert on each other arise as a consequence of the contact between them. This contact causes the distance between their neutral surfaces (h) to be fixed. Because this distance is much smaller than the dimensions of giant vesicles and the curvatures are varying slowly over the membrane, the area of the outer layer is related to the area of the inner layer up to the first order in h by the relation

$$A_2 = A_1 + h\,C \tag{2}$$

where C is the integral of the mean curvature over the whole membrane area,

$$C = \int (c_1 + c_2)\, dA \tag{3}$$

This integration is performed over the area of a reference surface within the membrane.

In view of constraint 2 it is appropriate to express[8,9] the elastic energy of the bilayer, which is the sum of the elastic energies of the two monolayers, in terms of independent elastic deformation modes, i.e., the area expansion of the reference surface of the membrane, the relative expansion of membrane layers, and the local and Gaussian membrane bending:

$$W = \frac{1}{2}\frac{K}{A_0}(A - A_0)^2 + \frac{1}{2}\frac{K_r}{A_0}(\Delta A - \Delta A_0)^2 + \frac{1}{2}k_c\int (c_1 + c_2 - c_0)^2\, dA + \bar{k}_c\int c_1 c_2\, dA \tag{4}$$

where in the area expansivity term A is the area of the reference surface of the bilayer. The position of this surface in the membrane is obtained in such a way that the area expansivity and the relative expansivity terms become independent elastic deformation modes. A_0 is the equilibrium value of the area of the reference surface, and K is the area expansivity modulus. In the relative expansivity term ΔA is the difference between the areas of the outer and the inner neutral surfaces $(A_2 - A_1)$, ΔA_0 is the corresponding equilibrium value $(A_{0,2} - A_{0,1})$, and K_r is the relative expansivity elastic modulus. In the bending term, k_c is the membrane bending modulus and c_0 is the spontaneous curvature. $\bar{k}_c$ is the Gaussian modulus of the bilayer.

The elastic behavior of closed membranes composed of unconnected layers (Equation 4) fundamentally differs from the elastic behavior of a single-layered membrane. In a bilayer there is an additional energy term which arises due to the possibility that the two layers can be expanded or compressed relative to each other at a constant value of the area of the reference surface of the membrane. It can be represented as the elastic energy which appears if the difference between the areas of the two layers varies from the corresponding equilibrium difference.

As becomes evident by inspecting Equation 4, the elastic behavior of the membrane composed of two unconnected layers is governed by four elastic moduli. These moduli and the equilibrium membrane parameters, i.e., A_0, ΔA_0, and the spontaneous curvature c_0, are determined by the elastic moduli and the equilibrium parameters of the composing layers. We have[8,9]

$$A = \frac{K_1/A_{0,1}}{K_1/A_{0,1} + K_2/A_{0,2}}A_1 + \frac{K_2/A_{0,2}}{K_1/A_{0,1} + K_2/A_{0,2}}A_2 \tag{5}$$

$$A_0 = \frac{K_1 + K_2}{K_1/A_{0,1} + K_2/A_{0,2}} \tag{6}$$

$$K = K_1 + K_2 \tag{7}$$

$$K_r = \frac{(K_1 + K_2)K_1 K_2}{A_{0,1}A_{0,2}(K_1/A_{0,1} + K_2/A_{0,2})^2} = \frac{A_0 K_1 K_2}{A_{0,1}A_{0,2}(K_1/A_{0,1} + K_2/A_{0,2})} \tag{8}$$

$$k_c = k_{c,1} + k_{c,2} \tag{9}$$

$$\bar{k}_c = \bar{k}_{c,1} + \bar{k}_{c,2} \tag{10}$$

and

$$c_0 = \frac{k_{c,1} c_{0,1} + k_{c,2} c_{0,2}}{k_{c,1} + k_{c,2}} \tag{11}$$

The membrane area expansivity modulus (Equation 7), membrane bending modulus (Equation 9), and membrane Gaussian modulus (Equation 10) are simply the sums of the corresponding moduli of the composing layers. The dependence of the elastic modulus K_r (Equation 8), which is a measure of the relative expansivity of the areas of the two layers, on monolayer parameters is more complex. For a symmetric bilayer (i.e., $K_1 = K_2$) it can be seen that if $A_{0,1} \approx A_{0,2}$ in Equation 8 then the value of K_r is one fourth of the value of the bilayer area expansivity modulus ($K_r = K/4$). The spontaneous curvature of the bilayer depends on the spontaneous curvatures of the two monolayers and their bending constants (Equation 11). In a symmetric bilayer the two bending constants are equal and the two spontaneous curvatures differ only in sign, so that the bilayer spontaneous curvature in this case is zero.

It is possible to reexpress the relative expansivity term of Equation 4 as

$$\frac{1}{2} \frac{k_r}{A_0} \left(C - C_0\right)^2 \tag{12}$$

where C, the integral of the mean curvature over the membrane reference surface (Equation 3), is given by Equation 2, i.e., $C = \Delta A/h$, and C_0 is the corresponding equilibrium value equal to $\Delta A_0/h$. The effect of relative expansion is in this representation expressed as a nonlocal bending and is measured by the nonlocal bending modulus k_r which is related to the relative expansivity modulus K_r by the equation

$$k_r = h^2 K_r \tag{13}$$

In contrast to k_r, the membrane bending modulus k_c (Equation 9) represents a local membrane property and therefore will be from here on called the local membrane bending modulus.

Several different experimental methods have been introduced for determinations of the area expansivity and bending moduli.[3,12,13] In Section IV we shall describe a method for measuring k_c and k_r of phospholipid membranes.[14]

B. MODELS FOR PHOSPHOLIPID MONOLAYERS

The values of elastic constants of a phospholipid monolayer, defined in Equation 1, and the position of the neutral surface of the monolayer ultimately depend on monolayer composition and properties of the constituent molecules. It is therefore of interest to obtain quantitative relationships between the microscopic properties of this system and its macroscopic elastic behavior. Models can be developed based on different levels of the microscopic picture of the system. Here we shall present two examples. In the first one a monolayer will be considered as a simple elastic shell. The second example will be a five-parameter molecular model of a monolayer (see the review of Petrov and Bivas[10]). Other similar models have been developed involving different specific details of phospholipid monolayers.[15,16] A more general thermodynamic treatment of elastic properties of interfaces can be found in Reference 17.

In order to illustrate a possible usage of monolayer models, we shall in two examples present as the result the ratio between the nonlocal and local bending constants, k_r/k_c (for the bilayer), which is a parameter significantly affecting (see Section III.D) the phospholipid vesicle shape. Although in general the two monolayers of a given bilayer can be different, we will give these results for a symmetric bilayer, i.e., a bilayer composed of monolayers of the same composition.

1. Elastic Shell Model

A phospholipid monolayer can be treated as an elastic shell which is in general inhomogeneous in the normal direction and when in equilibrium is, because of its asymmetry, spontaneously curved with mean curvature c_0. Let us consider an element of the shell with the area dA. If its mean curvature $c_1 + c_2$ is larger than c_0, the outer layers of the shell expand and the inner layers contract, while the area of the neutral

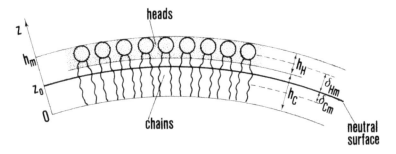

Figure 1 A schematic representation of the geometrical parameters of the model for a monolayer based on the division into the head part (shaded) and the chain part.

surface remains constant. The relative area increase of an infinitesimally thin layer with thickness dz at the distance $z - z_0$ from the neutral surface (Figure 1) equals $(c_1 + c_2 - c_0)(z - z_0)$, and therefore the bending energy of this element of the shell consists of the sum of the stretching energy contributions over the whole thickness of the monolayer (h_m):

$$dW_b = dA \, \frac{1}{2} \int_0^{h_m} \kappa(z)\left(c_1 + c_2 - c_0\right)^2 \left(z - z_0\right)^2 dz \qquad (14)$$

Here $\kappa(z)$ is the stretching elastic coefficient of the layer at coordinate z. From the definition of the bending energy (the second term in Equation 1) one obtains for the monolayer bending modulus

$$k_{c,m} = \int_0^{h_m} \kappa(z)\left(z - z_0\right)^2 dz \qquad (15)$$

The position of the neutral surface (z_0) is determined from the condition that the total lateral tension in a curved monolayer equals zero:

$$\int_0^{h_m} \kappa(z)\left(c_1 + c_2 - c_0\right)\left(z - z_0\right)dz = 0 \qquad (16)$$

from which it follows that

$$z_0 = \frac{\displaystyle\int_0^{h_m} \kappa(z)\, z\, dz}{\displaystyle\int_0^{h_m} \kappa(z)\, dz} \qquad (17)$$

In the above derivation we have implicitly taken into account that, because of the two-dimensional liquid character of monolayers, shear resistance does not exist and only the changes in areas of infinitesimally thin layers contribute to the bending elastic energy.

When we consider the lateral expansion of a monolayer, all the infinitesimal layers expand in the same manner, and the monolayer area expansivity coefficient K_m is simply the sum of the stretching elastic coefficients:

$$K_m = \int_0^{h_m} \kappa(z)\, dz \qquad (18)$$

Elastic properties of a monolayer, i.e., the bending modulus $k_{c,m}$ and the area expansivity coefficient K_m, therefore depend on the distribution of the stretching coefficient $\kappa(z)$. In the simplest case one can treat the phospholipid monolayer as a homogeneous material and there take $\kappa(z)$ to be constant. One obtains from Equations 15, 17, and 18

$$k_{c,m} = \frac{1}{12}\kappa h_m^3 \tag{19}$$

$$K_m = \kappa h_m \tag{20}$$

Using these results one obtains from Equations 8 and 9 for a symmetric bilayer the relative expansivity modulus $K_r = \kappa h_m/2$ and the bending modulus $k_c = \kappa h_m^3/6$, which leads to the relation between the nonlocal bending modulus k_r (Equation 13) and the bilayer bending modulus k_c: $k_r/k_c = 3$. Here we have taken into account the fact that the two homogeneous monolayers of thickness h_m are in contact, the neutral surfaces are in the middle of monolayers, and therefore the distance between the two neutral surfaces is equal to h_m.

In general we expect that the chain part of the monolayer is softer than the head part and therefore $\kappa(z)$ is an increasing function of z. The monolayer neutral surface is therefore not in the middle but rather is closer to the heads (Equation 17). For a given K_m the monolayer bending modulus is smaller than for a homogeneous monolayer and therefore also the bilayer local bending modulus k_c. The distance between the two neutral surfaces is larger and therefore the nonlocal bending modulus k_r is also larger than for homogeneous monolayers. Therefore,

$$\frac{k_r}{k_c} > 3 \tag{21}$$

A more realistic model for elastic properties of a monolayer can be obtained by assuming that the monolayer can be divided into the head part with $\kappa(z) = \kappa_H$ and the chain part with $\kappa(z) = \kappa_C$ (Figure 1). The corresponding thicknesses are h_H and h_C, respectively, where $h_H + h_C = h_m$. For such a stretching coefficient profile it is easy to obtain for the monolayer area expansivity (Equation 18)

$$K_m = \kappa_C h_C + \kappa_H h_H \tag{22}$$

and for the monolayer bending modulus

$$k_{c,m} = \frac{1}{12(\kappa_C h_C + \kappa_H h_H)}\left[\kappa_C^2 h_C^4 + \kappa_C \kappa_H \left(4h_C^3 h_H + 6h_C^2 h_H^2 + 4h_C h_H^3\right) + \kappa_H^2 h_H^4\right] \tag{23}$$

One can show that for κ_H larger that κ_C the ratio k_r/k_c is larger than 3, but it can be arbitrarily large if κ_H is much larger than κ_C and h_H is much smaller than h_C.

2. A Model of Two Connected Elastic Surfaces

An alternative approximative way to model the elastic properties of a phospholipid monolayer is to consider it[8,10] as a system of two elastic surfaces representing the headgroup and chain regions of a monolayer. This model on one hand is a simplification of the above shell model but on the other hand allows one to take explicitly into consideration some more details of the constituent molecules, e.g., their conicity.

In the model of two elastic surfaces the monolayer is again divided into two regions as described above (Figure 1). In the absence of shear elasticity, and assuming that an independent bending of the headgroup region or of the chain region of the monolayer costs no energy, the elastic properties of the surfaces representing these regions of the monolayer can be described in terms of stretching elasticity, whereby the elastic energy per molecule of the monolayer is[10]

$$\overline{W}_m = \frac{1}{2}K_H\overline{A}_{o,H}\left(\frac{\overline{A}_H}{\overline{A}_{o,H}} - 1\right)^2 + \frac{1}{2}K_C\overline{A}_{o,C}\left(\frac{\overline{A}_C}{\overline{A}_{o,C}} - 1\right)^2 \tag{24}$$

Here K_H and K_C are the elastic constants, $\overline{A}_H$ and $\overline{A}_C$ are the areas pertaining to the head and chain regions of a molecule, respectively, and $\overline{A}_{o,H}$ and $\overline{A}_{o,C}$ are the corresponding equilibrium areas.

It is convenient to reexpress the elastic energy of a monolayer in terms of the area expansivity in the plane of the membrane and in terms of bending. For this purpose the neutral surface is defined, i.e., the surface at which the expansivity and bending modes are uncoupled, meaning for instance that the neutral surface remains unstretched if a monolayer undergoes only bending. The corresponding equation for the elastic energy density, i.e., the elastic energy per unit area, in terms of the parameters of the presented model up to a constant reads[8]

$$\frac{\overline{W}_m}{\overline{A}_{o,m}} = \frac{1}{2}K_m\left(\frac{\overline{A}_m}{\overline{A}_{o,m}}-1\right)^2 + \frac{1}{2}k_{c,m}\left(c_1+c_2-c_{o,m}\right)^2 + \overline{k}_{c,m}c_1c_2 \tag{25}$$

where

$$\overline{A}_{o,m} = \frac{K_H + K_C}{K_H/\overline{A}_{o,H} + K_C/\overline{A}_{o,C}} \tag{26}$$

$$K_m = K_H + K_C \tag{27}$$

$$k_{c,m} = \left(\frac{h_m}{2}\right)^2 \frac{\overline{A}_{o,H}\overline{A}_{o,C}K_HK_C(K_H+K_C)}{\left(\overline{A}_{o,C}K_H + \overline{A}_{o,H}K_C\right)^2} = \delta_{H,m}\delta_{C,m}K_m \tag{28}$$

$$\overline{k}_{c,m} = \left(\frac{h_m}{2}\right)^2 \frac{\left(\overline{A}_{o,H}-\overline{A}_{o,C}\right)\left(\overline{A}_{o,C}K_H - \overline{A}_{o,H}K_C\right)K_HK_C}{\left(\overline{A}_{o,C}K_H + \overline{A}_{o,H}K_C\right)^2} \tag{29}$$

and

$$c_{o,m} = \frac{(-1)^m}{\left(h_m/2\right)^2} \frac{\left(\overline{A}_{o,H}-\overline{A}_{o,C}\right)\left(\overline{A}_{o,H}K_C + \overline{A}_{o,C}K_H\right)}{\overline{A}_{o,H}\overline{A}_{o,C}\left(K_H+K_C\right)} \tag{30}$$

The distances from the middle of headgroup and from the middle of chain regions of a lipid molecule to the neutral surface, respectively, are

$$\delta_{H,m} = \frac{h_m}{2}\frac{\overline{A}_{o,H}K_C}{\overline{A}_{o,C}K_H + \overline{A}_{o,H}K_C} \quad \text{and} \quad \delta_{C,m} = \frac{h_m}{2}\frac{\overline{A}_{o,C}K_H}{\overline{A}_{o,C}K_H + \overline{A}_{o,H}K_C} \tag{31}$$

$\overline{A}_m$ is the area of a lipid molecule at the neutral surface and $\overline{A}_{o,m}$ is the corresponding equilibrium area. c_1 and c_2 are the two principal curvatures of the neutral surface. $c_{o,m}$ is the spontaneous curvature. K_m, $k_{c,m}$, and $\overline{k}_{c,m}$ are the area expansivity and the bending and Gaussian bending moduli, respectively. The surfaces representing the head and the chain regions have been taken here to lie in the middle of these regions. The distance between them is thus $h_m/2$. The sign of a curvature is defined in such a way that the curvature of a sphere is positive if chains are closer to the center of the sphere than heads, about which it is taken care of by the factor $(-1)^m$ in Equation 30.

In order to obtain Equation 1, Equation 25 is integrated over the area of the neutral surface. The equilibrium area of the monolayer is defined as

$$A_{o,m} = N_m\overline{A}_{o,m} \tag{32}$$

where N_m is the number of molecules of the mth monolayer. In the derivation of Equation 1 it is also taken into consideration that the neutral surface is uniformly expanded, i.e.,

$$A_m / A_{o,m} = \overline{A}_m / \overline{A}_{o,m} \qquad (33)$$

The presented model represents the elastic properties of a phospholipid monolayer of a minimum complexity but still including its finite thickness and also packing features of phospholipid molecules. For instance, the value of spontaneous curvature basically arises due to the difference between the equilibrium areas of the headgroup and chain regions of the molecule, and the distances between these regions and the position of the neutral surface depend also on the elastic properties of these two regions (Equation 31).

The model of two elastic surfaces can give an estimate about which monolayer characteristics affect the ratio k_r/k_c. For a symmetric bilayer,

$$\frac{k_r}{k_c} = \frac{\left[h_C\left(2\overline{A}_{o,C}K_H + \overline{A}_{o,H}K_C\right) + h_H\overline{A}_{o,C}K_H\right]^2}{h_m^2 \overline{A}_{o,C}\overline{A}_{o,H}K_C K_H} \qquad (34)$$

For a given value of h_H/h_m the ratio k_r/k_c depends only on $A_{o,H}K_C/A_{o,C}K_H$. It can be shown that for each value of h_H/h_m there is a minimum value of the ratio k_r/k_c. For instance, if the thickness of the headgroup region is equal to the thickness of the chain region the minimum value of this ratio is 3. If it is, more realistically, about one third of the monolayer thickness, the minimum value of this ratio is about 4.4.

It has to be pointed out that the model of two elastic surfaces can only be used when it is possible to neglect the separate bending contributions of the head and chain regions. The model also presupposes that a monolayer has a constant thickness and thus a well defined distance from the neutral surface to the contact surface of the two monolayers. This is not necessarily true for phospholipid monolayers, which has to be considered with regard to the general validity of Equation 4.

III. SHAPES OF UNSUPPORTED FLACCID PHOSPHOLIPID VESICLES

A phospholipid vesicle with its volume (V) smaller than the maximum volume which the membrane of a given area (A) can enclose ($V_s = A^{3/2}/6\pi^{1/2}$) can be shaped in an infinite number of ways. However, membrane elastic energy, Equation 4, in general depends on shape and therefore a flaccid vesicle, if it is unsupported, attains an equilibrium shape which corresponds to the minimum value of this energy. Here it is possible to neglect the Gaussian bending energy because its integral over the vesicle membrane surface is a constant that depends only on vesicle topology and we shall treat only vesicles with spherical topology. Area expansivity constant (K) is large for phospholipid membranes and the problem can be consequently reduced to the minimization of the nonlocal and local bending energy terms at keeping membrane area constant. Originally, such an approach has been used by Canham,[18] who studied erythrocyte shapes by approximating them by the ovals of Cassini, and took into consideration only the local membrane bending energy term with zero spontaneous curvature. The inclusion of the spontaneous curvature enabled Deuling and Helfrich[19] to obtain a large variety of vesicle shapes for different values of relative vesicle volume v (v = V/V_s) and spontaneous curvature c_o. This, spontaneous curvature model has been largely applied in phospholipid vesicle determinations.[20-23] In an alternative theoretical approach known as the bilayer couple model[24,25] the layered structure of phospholipid membranes was taken into consideration in the sense that only those vesicle shapes are possible for which the areas of both membrane layers have given fixed values. Correspondingly, the equilibrium shape of a vesicle of a given volume can be obtained by minimizing the membrane local bending energy at a constant membrane area and a constant difference between the areas of the two layers (ΔA). The variety of vesicle shapes in this model is due to the different possible values of vesicle volume and of ΔA. The two described models differ in their consideration of the relative stretching (the nonlocal bending) term of Equation 4. The spontaneous curvature model ignores this term, which is equivalent to taking the value of k_r to be zero. In the bilayer couple model the minimization of membrane bending energy is performed at a constant difference between the layer areas, i.e., by assuming $\Delta A = \Delta A_o$, which is equivalent to taking the value of k_r to be infinite. In accordance with the recent experimental estimate[14] for the value of the nonlocal membrane bending constant k_r and the theoretical estimates presented in Section II, it is now commonly accepted that shapes of phospholipid vesicles correspond to a generalized bilayer couple model[26-31] in which it is assumed that this constant has a finite nonzero value, and the minimum is sought for the sum of the local and nonlocal bending energies at a constant membrane area and vesicle volume. The spontaneous curvature model and the bilayer couple model (which we shall from now on call the strict bilayer couple model) therefore represent two opposite limiting cases of the generalized bilayer couple model with $k_r \Rightarrow 0$ and $k_r \Rightarrow \infty$, respectively.

Theoretical studies of possible vesicle shapes based on the bilayer couple model revealed several interesting and significant properties of closed layered membranes. For instance, all the shapes corresponding to the extrema of the membrane elastic energy can be ascribed to different shape classes, where a class of shapes contains all shapes of the same symmetry which can be obtained in a continuous manner by continuously varying model parameters. The systematic analysis of shape classes and the corresponding phase diagrams has been performed for the strict bilayer couple model[23,25] and the spontaneous curvature model[23] and is now also evolving for the generalized bilayer couple model.[29-31] In the strict bilayer couple model the obtained phase diagram involves the class boundaries, which either originate from the symmetry breaking properties of the system or are caused by the geometrical constraints due to constant layer areas.[25,32,33] The corresponding limiting shapes can have a particularly important role in vesicle or cell vesiculation[34,35] and fusion processes.[36] A special problem in theoretical shape determinations is the problem of the stability of the shapes obtained on the basis of the variational principle. This mathematically complex problem has not yet been solved generally; however, recently it has been satisfactorily approached at least within some regions of the bilayer couple phase diagram.[30,37,38]

In this chapter we shall first present the mathematical basis for the variational determination of vesicle shapes. This will be done only for the case of the strict bilayer couple model, and a membrane with zero spontaneous curvature will be considered because, when treating other described models, the principles involved are the same. Namely, the results obtained by the strict bilayer couple model can be readily used when treating the general case of Equation 4.[25] A special section will be devoted to the determination and properties of limiting shapes. Then we shall show how the shapes of the strict bilayer couple model can be classified on the basis of their symmetry properties and will present some examples of shape classes. It will be demonstrated separately how the shapes obtained on the basis of the strict bilayer couple model can be used to predict the shapes within the generalized bilayer couple model. Finally, some unresolved problems will be indicated.

A. MINIMIZATION OF THE MEMBRANE BENDING ENERGY IN THE LIMIT OF THE STRICT BILAYER COUPLE MODEL

Here the problem is to find the minimum value of the bending energy at constant values of vesicle volume, membrane area, and the difference between the areas of the two membrane layers. In the most general manner the variational approach can be used, where the variational variable is the general shape of the vesicle. For technical reasons, up to now this approach has been applied only to axisymmetric shapes, where the problem can be reduced to the variation of the contour of the vesicle in the axial cross section. This method gives rise to a set of nonlinear differential equations which can be solved numerically.[23,25] The obtained solutions, which we shall call stationary solutions, are not necessarily stable.

In the search for possible nonaxisymmetric shapes the parametric variational principle can be applied. Here vesicle shape is described by a certain geometrical body defined by a finite number of parameters, and these parameters are varied to obtain the minimum membrane local bending energy at the required constraints for constant vesicle volume, membrane area, and difference between the areas of the two membrane layers. By using elliptically distorted Cassini ovals it was possible to demonstrate the existence of nonaxisymmetric vesicle shapes having in a part of the $v/\Delta a$ phase diagram lower bending energy than the lowest-energy axisymmetric shape.[39] In a more general fashion, a vesicle surface can be expressed as a series of spherical harmonics[38] and the shape obtained by the determination of the expansion coefficients by using the Ritz procedure.[30] The latter approach is general in the sense that it enables the calculation of both axisymmetric and nonaxisymmetric shapes, that its accuracy increases by increasing the number of included spherical harmonics, and that it can be used as well for the stability analysis of stationary shapes with respect to all possible deformations. Its only limitation is that the shapes must be representable by unique functions of the spherical angles.

In the following we shall restrict ourselves to the description of a variational method[25] for determining shapes of axisymmetric vesicles which is a modification of the one used first by Deuling and Helfrich[19] for the spontaneous curvature model. The problem is to find the extreme values of the membrane local bending energy

$$W_b = \frac{1}{2}k_c \int (c_1 + c_2)^2 \, dA^* \tag{35}$$

where k_c is the membrane bending elastic constant, c_1 and c_2 are the two principal curvatures, and integration is performed over the whole area of the neutral surface of the bilayer (A^*). The minimization procedure is to be carried out at fixed values of cell volume (V), area of the reference surface (A), and difference between the areas of the two membrane leaflets (ΔA).

The shape of a cell can therefore be obtained by minimizing the functional

$$G = W_b - \lambda(A^* - A) - \mu(V^* - V) - \nu(\Delta A^* - \Delta A) \tag{36}$$

Here A^*, V^*, and ΔA^* are the membrane area, cell volume, and difference in leaflet areas, respectively, and the three Lagrange multipliers (λ, μ, ν) are determined from the conditions

$$A^* = A, \qquad V^* = V, \qquad \Delta A^* = \Delta A \tag{37}$$

The expression for the membrane bending energy (Equation 35) is scale invariant; i.e., it has the property that for a given shape it does not depend on the vesicle size. It is therefore appropriate in the forthcoming analysis to choose the unit of length in such a way that the membrane area equals unity. If R_s is the radius of the sphere with the membrane area A,

$$R_s = (A/4\pi)^{1/2} \tag{38}$$

the two dimensionless curvatures are defined in the sense that

$$R_s c_1 \to c_1 \quad \text{and} \quad R_s c_2 \to c_2 \tag{39}$$

The relative volume $v = V/V_s$ has already been defined, and it is then also convenient to define analogously the relative difference between the areas of the two membrane leaflets,

$$\Delta a = \Delta A / \Delta A_s \qquad \Delta A_s = 8\pi h R_s \tag{40}$$

while relative area a = 1. In an analogous manner, the relative area element can be expressed as

$$da^* = dA^* / 4\pi R_s^2 \tag{41}$$

and $v^* = V^*/V_s$, $\Delta a^* = \Delta A^*/\Delta A_s$.

It is clear that the result of the minimization procedure does not depend on the value of the membrane bending constant, k_c. It is therefore appropriate to measure also the membrane bending energy and the energy functional G (Equation 36) relative to the bending energy of the sphere,

$$w_b = W_b / 8\pi k_c, \qquad g = G / 8\pi k_c \tag{42}$$

We then obtain for the dimensionless energy functional g the expression

$$g = \frac{1}{4}\int (c_1 + c_2)^2 \, da^* - \frac{1}{4}L(a^* - 1) - \frac{1}{6}M(v^* - v) - \frac{1}{2}N(\Delta a^* - \Delta a) \tag{43}$$

Here the new Lagrange multipliers L, M, and N are related to λ, μ, and ν as follows:

$$L = \lambda\frac{2R_s^2}{k_c}, \qquad M = \mu\frac{R_s^3}{k_c}, \qquad N = \nu\frac{2hR_s}{k_c} \tag{44}$$

It is easy to see from the structure of the energy function G (Equation 36) that the three Lagrange multipliers λ, μ, and ν represent the thermodynamically conjugated fields of A, V, and ΔA, respectively. Therefore in the equilibrium, when $W_b = W_b$ (A, V, ΔA), the Lagrange multipliers are given by

$$\lambda = \frac{\partial W_b}{\partial A}, \qquad \mu = \frac{\partial W_b}{\partial V}, \qquad \nu = \frac{\partial W_b}{\partial \Delta A} \tag{45}$$

On the other hand, one can conclude from Equation 43 that in equilibrium the bending energy $W_b = 8\pi k_c w_b$ ($\nu,\Delta a$) depends only on ν and Δa. As a consequence, the Lagrange multipliers can be shown to be interrelated as

$$A\lambda + \frac{3}{2}V\mu + \frac{1}{2}\Delta A\nu = 0 \tag{46}$$

or in the dimensionless form as

$$L + \nu M + \Delta aN = 0 \tag{47}$$

Following the procedure used by Deuling and Helfrich,[19] for a cell with an axisymmetric shape the corresponding contour $z = z(x)$, with z being the dimensionless coordinate along the rotational axis ($z/R_s \rightarrow z$) and x being the dimensionless distance from the axis ($x/R_s \rightarrow x$), can be obtained by minimizing the functional g (Equation 43) in which we take for the principal curvatures the principal curvature along the parallels, c_p, and the principal curvature along the meridians, c_m. The geometrical parameters of the cell are then given by the equations

$$a^* = \frac{1}{2}\int \frac{xdx}{\left(1-\left(xc_p\right)^2\right)^{1/2}} \tag{48}$$

$$v^* = \frac{3}{4}\int \frac{x^3 c_p dx}{\left(1-\left(xc_p\right)^2\right)^{1/2}} \tag{49}$$

$$\Delta a^* = \frac{1}{4}\int \frac{\left(c_p + c_m\right)xdx}{\left(1-\left(xc_p\right)^2\right)^{1/2}} \tag{50}$$

The integrals here go over all the branches of the contour function $z = z(x)$ from the minimal to the maximal value of x. The two principal curvatures of an axisymmetric object are interrelated:

$$c_m = c_p + x\frac{dc_p}{dx} \tag{51}$$

and the functional g can therefore be expressed as

$$g = \int \tilde{g}\left(x, c_p, \frac{dc_p}{dx}\right)dx \tag{52}$$

The variational procedure then leads to the Lagrange-Euler equation

$$\frac{\partial \tilde{g}}{\partial c_p} - \frac{d}{dx}\frac{\partial \tilde{g}}{\partial \left(dc_p / dx\right)} = 0 \tag{53}$$

which has the form of a second-order differential equation for $c_p(x)$:

$$\frac{x}{2\left(1-x^2c_p^2\right)}\left[xc_p\frac{dc_p}{dx}\left(2c_p+x\frac{dc_p}{dx}\right)+M+Lc_p+Nc_p^2\right]+\frac{d}{dx}\left(2c_p+x\frac{dc_p}{dx}\right)=0 \qquad (54)$$

This equation is to be solved numerically, where it is necessary to pay special attention to the behavior of the system close to the poles of the axially symmetric shapes ($x = 0$) and close to the extrema in the x-direction. At the poles the membrane surface must be a continuous and derivable function. The use of different parametrizations in the variational procedure leads in general to different differential equations for the determination of the axisymmetric vesicle shapes, which are of a higher order[40] than Equation 54. However, the application of the above requirement of smoothness gives rise to the identical second-order equation and consequently to the same vesicle shapes.[41,42]

After Equation 54 is solved for c_p, the contour $z(x)$ of the axisymmetric vesicle is obtained by performing the integration

$$z = z_0 \mp \int \frac{xc_p dx}{\left[1-\left(xc_p\right)^2\right]^{1/2}} \qquad (55)$$

where z_0 is the coordinate of a pole and the sign always changes when the extremum in x is reached.

B. POSSIBLE LIMITING SHAPES

One can also pose the question[25,32] of what are the shapes of a vesicle with given A and ΔA which have maximal volume, or analogously, what are the shapes of a vesicle with given A and V which have an extreme value of ΔA. In a dimensionless representation, this means that we are looking for a shape with extreme relative volume v* (Equation 49) with the conditions that the relative area a* (Equation 48) equals one and the relative area difference Δa^* (Equation 50) equals Δa. We therefore study the functional

$$g_v = v^* - \frac{1}{4}\tilde{L}(a^* - 1) - \frac{1}{2}\tilde{N}(\Delta a^* - \Delta a) \qquad (56)$$

where $\tilde{L}$ and $\tilde{N}$ are Lagrange multipliers for this problem. The Euler-Lagrange minimization procedure here leads to an algebraic equation for the curvature c_p:

$$\tilde{N}c_p^2 + \tilde{L}c_p - 6 = 0 \qquad (57)$$

which means that the curvature c_p is constant over the surface. From Equation 51 we see that in such a case c_m is also constant and equals c_p. We therefore conclude that the surface which corresponds to the solution of Equation 57 is a sphere or a section of a sphere. As Equation 57 has in general two different solutions, we conclude that the limiting shapes consist of spheres or of sections of spheres where only two different radii are allowed.

Let us analyze in more detail some simple limiting shapes. A simple possibility is that a limiting shape consists of two spheres with different radii. By requiring that the relative area of the cell equals one and that the relative area difference equals Δa, one can determine the two relative radii $r_1 \geq r_2$:

$$r_1 = \frac{1}{2}\left\{\Delta a + \left[2 - (\Delta a)^2\right]^{1/2}\right\} \qquad (58)$$

$$r_2 = \frac{1}{2}\left\{\Delta a - \left[2 - (\Delta a)^2\right]^{1/2}\right\} \qquad (59)$$

The maximum volume v_e is then given by

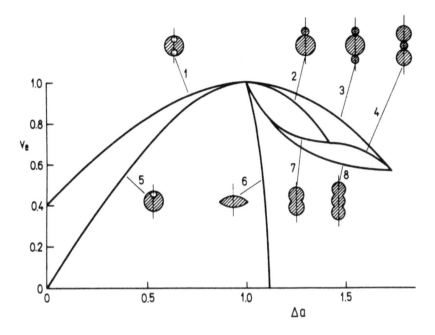

Figure 2 Some extreme relative volumes (denoted by v_e) in dependence on relative difference between the layer areas (Δa), together with corresponding limiting shapes. Lines 2 and 5 are obtained from Equation 60 and lines 6 and 7 from Equations 61 and 62. (From Svetina, S. and Žekš, B., *Eur. Biophys. J.*, 17, 101, 1989. With permission.)

$$v_e = \frac{1}{2}\Delta a\left[3 - (\Delta a)^2\right] \tag{60}$$

For $1 < \Delta a < 2^{1/2}$ the dependence $v_e(\Delta a)$ is line number 2 in Figure 2. For $\Delta a = 2^{1/2}$ one obtains a limiting shape composed of two equal spheres. With decreasing Δa, one radius increases and the other decreases until at $\Delta a = 1$ one radius equals zero and only one sphere remains.

For $0 < \Delta a < 1$ the dependence $v_e(\Delta a)$ is line number 5 in Figure 2. Here r_2 is negative, which means that the internal vesicle has oppositely oriented membrane. In the corresponding shape it is seen that by decreasing Δa the internal vesicle increases up to $\Delta a = 0$, where its radius becomes equal to the radius of the external vesicle.

Another simple possibility for the limiting shape is that it has a mirror plane symmetry and is composed of two equal sections of a sphere (Figure 2, lines 6 and 7). In this case one obtains a parametrical expression for $v_e(\Delta a)$ as

$$v_e = \frac{1}{2^{1/2}}\sin(\phi/2)\cos^2(\phi/2)\left[3 + \text{tg}^2(\phi/2)\right] \tag{61}$$

$$\Delta a = \frac{1}{2^{1/2}}\cos(\phi/2)\left[\pi/2 - \phi + 2\text{tg}(\phi/2)\right] \tag{62}$$

The parameter ϕ here gives the angular magnitude of the section of the sphere. For $0 < \phi < \pi/2$ the sections are smaller than half of the sphere, and these shapes correspond to line number 6 in Figure 2. For $\phi = \pi/2$ one obtains a single sphere ($\Delta a = 1$, $v_e = 1$), while for $\phi > \pi/2$ the sections are larger than half of a sphere (Figure 2, line number 7). Both these shapes involve an edge and therefore their bending energy is infinitely large.

Figure 2 also shows some other examples of limiting shapes and dependencies of their relative volumes on the relative leaflet area difference.

C. CLASSES OF SHAPES
The analysis has shown[25] that a large variety of shapes can be obtained by variationally searching for the extrema of Equation 36 and that these shapes can be assorted into different classes of shapes, where a class

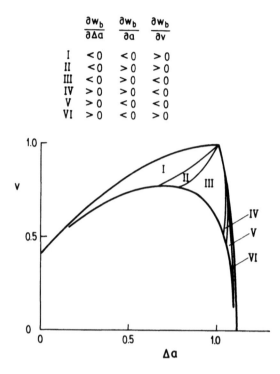

$$\frac{\partial w_b}{\partial \Delta a} \qquad \frac{\partial w_b}{\partial a} \qquad \frac{\partial w_b}{\partial v}$$

	$\frac{\partial w_b}{\partial \Delta a}$	$\frac{\partial w_b}{\partial a}$	$\frac{\partial w_b}{\partial v}$
I	< 0	< 0	> 0
II	< 0	> 0	> 0
III	< 0	> 0	< 0
IV	> 0	> 0	< 0
V	> 0	< 0	< 0
VI	> 0	< 0	> 0

Figure 3 Representation of the region in the v/Δa diagram belonging to the disk shape class. The region is bounded by three (bold) lines. The left line is line 1 of Figure 2. The right line is the line 6 of Figure 2. The bottom line represents shapes where the membrane makes a point contact at the two poles. The region is subdivided by thin lines into parts I to VI with different signs of the partial derivatives of the membrane bending energy with respect to membrane area, volume, and difference between the areas of the bilayer layers, as shown. Lines which divide these parts of the region are obtained by calculating shapes in which one of the Lagrange multipliers L, M, or N is taken to be zero. (From Svetina, S. and Žekš, B., *Eur. Biophys. J.*, 17, 101, 1989. With permission.)

of shapes contains all the shapes of the same symmetry which can be obtained in a continuous way if relative volume v and relative area difference Δa are continuously changing. We shall restrict ourselves to the discussion of vesicles with a spherical topology. An analogous discussion of toroidal vesicles was recently presented by Jülicher et al.[43] The classes of spheroidal vesicles identified to date have various symmetries. Shapes belonging to most of them are axially symmetric; however, a class of nonaxisymmetric shapes has been found as well.[30,38,39] Some of the axisymmetric classes also exhibit mirror equatorial symmetry. The values of v and Δa for different classes occupy different well-defined regions in the v/Δa diagram.[23,25] Classes are conveniently named after a certain characteristic shape of a given class. In Figures 3 to 5 regions are shown in the v/Δa diagram for the disk, cup, and pear class, respectively. The shapes of these three classes are axisymmetric. The disk class in addition exhibits equatorial mirror symmetry. Classes have well-defined boundaries. Some of them can be recognized as the lines representing v/Δa dependence of the limiting shapes. Other class boundaries mark a transition between shapes with different symmetries. Namely, the treated system exhibits symmetry breaking behavior: by continuously changing v and Δa in some classes it is possible to reach a point where the symmetry changes. The line connecting these points represents a boundary of a class of a lower symmetry. The shapes with higher symmetry become unstable at this line. The third boundary of the cup and disk classes appears because we included in these classes only shapes in which different parts of the membrane are not in contact. The corresponding class boundary is the line connecting the shapes in which the membrane makes a point

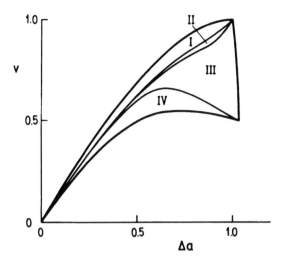

Figure 4 Representation of the region in the v/Δa diagram belonging to the cup shape class. The region is bounded by three (bold) lines. The left line is line 5 of Figure 2. The right line represents the values of v and Δa where the cup shapes attain an equatorial mirror symmetry (symmetry breaking line for symmetrical disk shapes). The bottom line represents shapes where the membrane makes a point contact at the two poles. The region is subdivided by thin lines into parts I to IV in the same manner as in Figure 3. (From Svetina, S. and Žekš, B., *Eur. Biophys. J.*, 17, 101, 1989. With permission.)

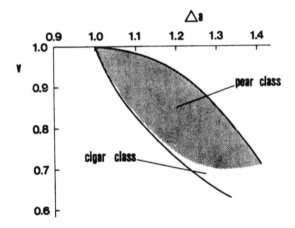

Figure 5 Representation of the region in the v/Δa diagram belonging to the pear shape class. The region is presented by the shaded area. The right boundary of the shaded area is line 2 of Figure 2. The left boundary of the shaded area represents the values of v and Δa where the pear shapes attain an equatorial mirror symmetry (symmetry breaking line for symmetrical cigar shapes). The narrow strip between the shaded area and the line on the left involves the stable shapes of the cigar class. The latter line connects the limiting shapes of this class of different relative volumes which are cylinders with two semispherical caps. (From Svetina, S. and Žekš, B., in *Proc. 11th School on Biophysics of Membrane Transport*, Vol. II, Kuczera, J. and Przestalski, S., Eds., Agricultural University of Wrocław, Poland, 1992, 115. With permission.)

contact at the two poles. In the disk and cup classes (Figures 3 and 4) it is also shown how they are subdivided into regions involving different signs of Lagrange multipliers, which shows that shapes in these subregions have different tendencies with respect to increase or decrease of the volume, membrane area, and layer area difference, respectively.

It is instructive to see how within the strict bilayer couple model a vesicle is expected to transform its shape between the two class limits. As the first example we show a series of shapes of the cup class.[33] It can be seen in Figure 6 how a vesicle with the relative volume v = 0.60 is changing its shape from a

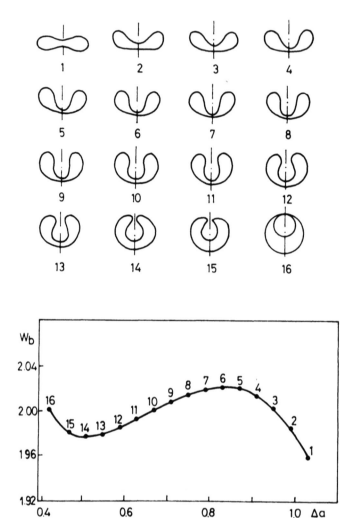

Figure 6 A series (1 to 16) of cup shapes at the relative volume v = 0.60 and at decreasing values of the relative difference between the areas of membrane layers (Δa), and the corresponding membrane local bending energies (denoted by w_b), obtained by solving Equation 54 for shapes 1 to 15 and Equations 58 to 60 for shape 16. (From Svetina, S. and Žekš, B., *Biomed. Biochim. Acta*, 44, 979, 1985. With permission.)

disk shape exhibiting mirror equatorial symmetry (the symmetry breaking point) to the limiting shape with one invaginated sphere. When Δa is decreasing, bending energy is first increasing, reaches a maximum and then a minimum, and finally attains in the limiting shape the expected value 2 (for two spheres). In another example (Figure 7), which involves the shapes belonging to the pear class,[44] it is shown for a vesicle with the relative volume v = 0.95 how by increasing Δa its shape changes from the cigar shape to the limiting shape composed of two linked spheres. The bending energy in this case increases with increasing Δa and again attains its maximum ($w_b = 2$) upon reaching a Δa value on the line of limiting shapes. It is important to realize that the limiting shapes are reached in a continuous manner, which means that their membranes are still a single closed surface and that the two spheres are connected by an infinitesimally small neck.

In many experimental procedures[28,45-47] it is possible by physical or chemical means to vary continuously the parameters of the system, i.e., v and Δa. The example shown in Figure 8 shows shape transformations within the cup class,[46] and the example shown in Figure 9 shows shape transformations within the pear class.[28] In both these cases shape transformations were obtained by continuously changing the temperature. The observed series of shapes correspond to the predicted series shown in Figures 6 and 7, respectively.

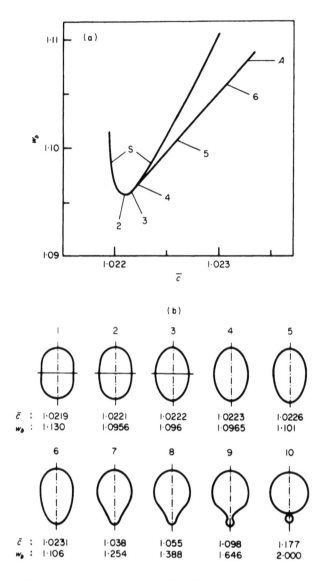

Figure 7 A series (1 to 3) of cigar shapes and a series (3 to 10) of pear shapes at the relative volume v = 0.95 and at increasing values of the relative difference between the areas of membrane layers Δa (denoted here by $\bar{c}$), and the corresponding local bending energies (denoted by w_b), obtained by solving Equation 54 for shapes 1 to 9 and Equations 58 to 60 for shape 10. The curve in (a) denoted by S is for cigar shapes and the curve denoted by A is for pear shapes. (From Svetina, S. and Žekš, B., *J. Theor. Biol.*, 146, 115, 1990. With permission.)

The regions in the v/Δa diagram of different classes partially overlap, as can be seen by comparing the class regions of the cup and disk classes in Figures 3 and 4. The shape of the class with the lowest bending energy at given v and Δa is in a stable equilibrium, whereas shapes of classes with higher bending energies are not stable, which has been confirmed by stability analysis.[30] It is now well established which stable shapes are predicted by the strict bilayer couple model in different parts of the v/Δa phase diagram.[23,25,30] In general, at lower Δa values prevail axisymmetric oblate shapes whereas at higher Δa values prevail axisymmetric prolate shapes. In the intermediate region the most stable shapes are nonaxisymmetric shapes of ellipsoidal symmetry.

D. GENERALIZED BILAYER COUPLE MODEL

The strict bilayer couple model is generalized by assuming that the relative expansivity modulus k_r has a finite value. Then the vesicle shapes are to be determined by minimizing at constant membrane area

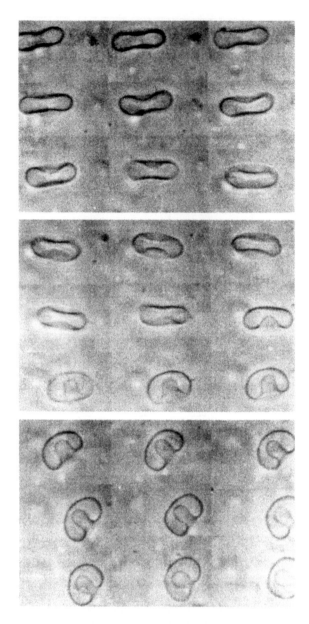

Figure 8 Disk-cup transition of a DMPC vesicle in pure water observed at increasing temperature. The temperature was varied from 41.9 to 42.4°C at a rate 0.02°C per image. The set of images starts at the left top and ends at the right bottom. (From Käs, J. and Sackmann, E., *Biophys. J.*, 60, 825, 1991. With permission.)

(a = 1) and vesicle volume (v) the sum of the local and nonlocal bending energies. The functional to be minimized is then

$$g' = \frac{1}{4}\int \left(c_1 + c_2 - c_0\right)^2 da * + \frac{k_r}{k_c}\left(\Delta a * - \Delta a_0\right)^2 - \frac{1}{4}L(a * - 1) - \frac{1}{6}M(v * - v) \tag{63}$$

Here the nonlocal bending energy (relative expansivity term) was also normalized with respect to the bending energy of a sphere. Δa_0 is the relative area difference for the unstretched membrane layers. We are also considering the general case of an asymmetrical membrane with a given spontaneous curvature c_0.

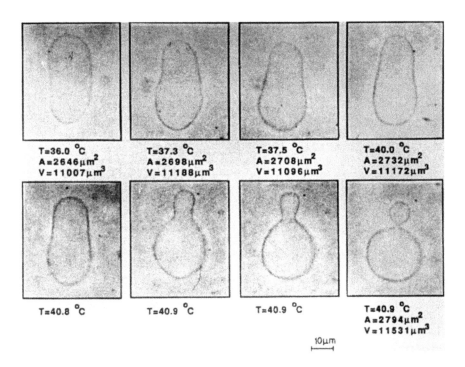

Figure 9 The shape transitions of a DMPC vesicle in pure water induced by heating in steps of 0.1°C. The temperatures (T) and the measured volumes (V) and areas (A) are indicated in the pictures. All the shapes for T ≤ 40.8°C are stable. Increase of temperature to 40.9°C causes a spontaneous transition to an outside budded shape. (From Käs, J., Sackmann, E., Podgornik, R., Svetina, S., and Žekš, B., *J. Phys. II (Paris)*, 3, 631, 1993. With permission.)

This functional (Equation 63) differs from the functional in Equation 43 by having a nonzero spontaneous curvature in the local bending term and by having the strict requirement for constant area difference replaced by the nonlocal bending term. In the analysis of the generalized bilayer couple model the results of the strict bilayer couple model can be used by assuming first that the vesicle has a given but not yet determined relative area difference Δa. The problem now reduces to the strict bilayer couple model with $\Delta a^* = \Delta a$, $a^* = 1$, and $v^* = v$; the shape is determined by Equation 54, and the corresponding local bending energy $w_b(v,\Delta a)$ can be (for $c_0 = 0$) evaluated as described in Section III.A. The bending energy, which for the generalized bilayer couple model is the sum of the local bending energy and relative expansivity terms, can then be expressed as

$$w(v,\Delta a) = w_b(v,\Delta a) + c_0 \Delta a + \frac{1}{4}c_0^2 + \frac{k_r}{k_c}\left(\Delta a - \Delta a_0\right)^2 \tag{64}$$

where Equations 2 and 3 and Equations 40 and 41 have been used to reexpress the spontaneous curvature part. To find the value of Δa, which corresponds to a stable vesicle shape, one must minimize the energy $w(v,\Delta a)$ with respect to Δa.

In order to illustrate the effect of the relative expansivity term we show in Figure 10 the Δa dependence of the bending energy w for $c_0 = 0$ for the cigar and pear class shapes for the relative volume $v = 0.85$. The ratio between the two bending constants is $k_r/k_c = 10$. This dependence is shown for different values of Δa_0. Whereas in the limit $k_r/k_c \Rightarrow \infty$, i.e., within the strict bilayer couple model, the bending energy of the pear class shapes (equal in this limit to the local bending energy w_b) is an increasing function of Δa (see Figure 7a for the low Δa part of this curve), at finite values of the ratio k_r/k_c the bending energy w may have extrema. The three noted extrema in the dependences of the bending energy on Δa shown in Figure 10 can be conveniently obtained by requiring that the first derivative of $w(v,\Delta a)$ (Equation 64, with $c_0 = 0$) with respect to Δa equals zero:

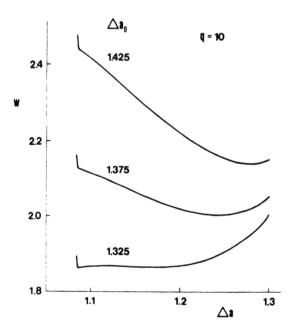

Figure 10   The sum of the membrane local (for $c_o = 0$) and nonlocal bending energies (denoted by w) as a function of the area difference Δa for the cigar class (steep part of the curves on the left) and pear class shapes is plotted for different values of Δa_o. The ratio between the nonlocal and local bending moduli (denoted by q) is chosen to be $k_r/k_c = 10$. (From Svetina, S. and Žekš, B., in *Proc. 11th School on Biophysics of Membrane Transport*, Vol. II, Kuczera, J. and Przestalski, S., Eds., Agricultural University of Wrocław, Poland, 1992, 115. With permission.)

$$dw_b / d\Delta a = -2\frac{k_r}{k_c}\left(\Delta a - \Delta a_o\right) \tag{65}$$

The solutions of Equation 65 can be obtained by graphically obtaining the cross sections of the left-hand and the right-hand sides of this equation (Figure 11). The left-hand side of Equation 65 is the dependence of the derivative $dw_b/d\Delta a$ on Δa (full line) which is obtained by solving the corresponding strict bilayer problem, and the right-hand side is a straight line (broken lines) with the slope $-2k_r/k_c$. The dependence of the derivative $dw_b/d\Delta a$ on Δa has a steep part which belongs to stable cigar class shapes. The U-shaped part of the $dw_b/d\Delta a$ dependence on Δa belongs to the pear class shapes. The number of stationary solutions of Equation 65 depends on the value of the second derivative of w_b on Δa ($d^2w_b/d\Delta a^2$) at the symmetry breaking point at its positive side ($\Delta a \Rightarrow + \Delta a_{s.b.}$), because this derivative is for the pear class the most negative at this point. If the absolute value of the slope $-2k_r/k_c$ of the line on the right-hand side of Equation 65 is larger than the absolute value of this derivative — i.e., $2k_r/k_c > -d^2w_b/d\Delta a^2(\Delta a \Rightarrow +\Delta a_{s.b.})$ — the left and the right sides of Equation 65 intersect only once for all values of Δa_o, and therefore we obtain at all Δa_o values a single stable solution. By increasing Δa_o the stable cigar class shapes transform at $\Delta a = \Delta a_{s.b.}$ continuously into the stable pear shapes. However, in cases where $2k_r/k_c < -d^2w_b/d\Delta a^2(\Delta a \Rightarrow +\Delta a_{s.b.})$ there is at increasing Δa_o a discontinuous transition from the cigar class shapes to the pear class shapes. It can be seen in Figure 11b that at Δa_o values higher than the one, at which the line $-2(k_r/k_c)$ $(\Delta a - \Delta a_o)$ becomes a tangent of the curve $dw_b/d\Delta a$, Equation 65 has three solutions, two of which are stable, i.e., the cigar one and, as evident from Figure 10, the one corresponding to the cross section of the line with the pear class branch of $dw_b/d\Delta a$ at the higher Δa value. By continuously increasing Δa_o the vesicle discontinuously transforms from the cigar to the pear shape at the Δa_o value at which the energies of the cigar and pear class stable shapes are equal.

It should be noted that the inclusion of the spontaneous curvature does not change the described behavior of the system with regard to the predicted discontinuous transitions. In the case of nonzero c_o we are looking for the solutions of the equation

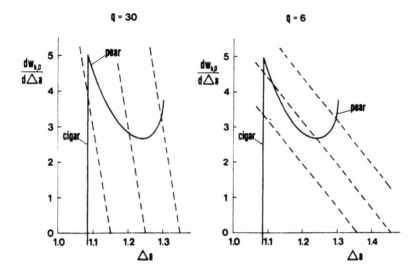

Figure 11 Graphical representation of the solutions of Equation 65 for the ratio between the nonlocal and local bending moduli (denoted by q) k_r/k_c = 30 (a) and k_r/k_c = 6 (b). The dependence of the derivative of the membrane bending energy w_b (denoted in the graph by $w_{b,o}$) for cigar class (the steep part) and pear class (the U-shaped part) is plotted as a function of Δa (full line). The solutions of Equation 65 are the cross sections of this dependence and the straight lines (broken lines) with the slope $-2k_r/k_c$ where in each case (a and b) three such lines are plotted, corresponding to different values of Δa_o. The respective Δa_o values can be obtained from the Δa values at which the lines intersect the abscissa. (From Svetina, S. and Žekš, B., in *Proc. 11th School on Biophysics of Membrane Transport*, Vol. II, Kuczera, J. and Przestalski, S., Eds., Agricultural University of Wrocław, Poland, 1992, 115. With permission.)

$$dw_b / d\Delta a = -2\frac{k_r}{k_c}\left(\Delta a - \Delta a_0\right) + c_0 \qquad (66)$$

which differs from Equation 65 in a constant term. All previous dependences on Δa_0 occur in this more general case at Δa_0 values shifted for $-c_0/(2k_r/k_c)$. It is of interest to note that the limit of the generalized bilayer couple model at which $k_r/k_c \Rightarrow 0$ is identical to the spontaneous curvature model in that it predicts the same shapes and the same differences between their energies.

The presented analysis indicates that the phase diagram of the generalized bilayer couple model involves in addition to v and Δa also the ratio k_r/k_c as a third dimension. In order to illustrate some features of this generalized phase diagram we present a two-dimensional cross section of the three-dimensional $v/\Delta a/(k_r/k_c)$ phase diagram for pear shapes (we take $c_0 = 0$) for the relative volume v = 0.85. Figure 12 shows which stable pear class shapes (locally stable shapes are included) are possible for different values of the ratio k_r/k_c. It can be seen that we obtain all pear shapes only for the values of k_r/k_c which are larger than the absolute value of one half of the second derivative of the local bending energy of pear shapes at the symmetry breaking point. If k_r/k_c is smaller than this derivative, not all shapes of the pear class can be realized. When k_r/k_c approaches zero only the shapes with Δa values larger than the Δa value at which the curve $dw_b/d\Delta a$ has a minimum can exist.

An analogous analysis was recently performed[30] for the phase diagram of shapes belonging to the nonaxisymmetric class and the neighboring shapes belonging to the disk and cigar classes at lower and higher Δa values, respectively. An interesting result of this analysis was that the nonaxisymmetric shapes were stable only above a certain critical value of the ratio k_r/k_c (the value is different for different relative volumes). If the ratio k_r/k_c is smaller than this critical value there is at continuously increasing Δa_0 a discontinuous transition from the disk shapes into cigar shapes.

E. SOME OPEN PROBLEMS
In most instances the observed phospholipid vesicle shapes and their changes correspond to the behavior expected from the described theoretical predictions. An opposite example is the discontinuous transition[28,46]

34

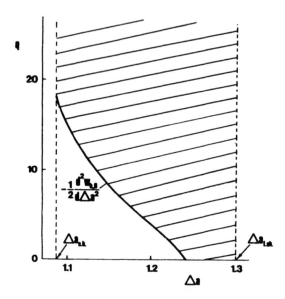

Figure 12 The shaded part of this diagram shows at which Δa values and which values of the ratio between the nonlocal and local bending moduli k_r/k_c (denoted by q) there are stable shapes of the pear class. $\Delta a_{s.b.}$ is the symmetry breaking area difference and $\Delta a_{l.sh.}$ is the area difference of the limiting shape. (From Svetina, S. and Žekš, B., in *Proc. 11th School on Biophysics of Membrane Transport,* Vol. II, Kuczera, J. and Przestalski, S., Eds., Agricultural University of Wrocław, Poland, 1992, 115. With permission.)

from the shape of the pear class with low Δa value to the shape of the pear class with a high Δa value, shown in Figure 9. Such transitions are not consistent with the predictions of the generalized bilayer couple model, where a discontinuous transition is expected to occur directly from the cigar to the high Δa pear shape. Attempts have been made[28] to describe the observed discontinuous shape transformation by additional energy terms, e.g., by invoking the dependence of the van der Waals attraction forces between different parts of the membrane on the vesicle shape, or by the inclusion of higher order energy terms in Δa. Neither of the proposed interpretations is fully satisfactory, and one should in the future look for other possibilities based, for instance, on a better understanding of monolayer microscopic elastic behavior, as was indicated at the end of Section II.

Bending energies of phospholipid vesicles are comparable to the thermal energy, and therefore their membranes exhibit intense thermal fluctuations. The presented treatment disregarded a possible contribution of membrane fluctuations to the vesicle free energy. Its role in vesicle shape formation has yet to be elucidated.

IV. SHAPES OF FLACCID PHOSPHOLIPID VESICLES UNDER THE INFLUENCE OF EXTERNAL FORCES

Under the influence of external forces shapes of vesicles and cells are affected by both these forces and the membrane internal force, i.e., membrane elasticity. Studies of the effect of external forces on vesicle shapes, especially if they are well defined, can therefore further increase our knowledge about the elastic properties of phospholipid membranes. A notable example is pipette aspiration of vesicles, which enabled the simultaneous determination of nonlocal and local membrane bending constants.[14] It has been well established that phospholipid vesicles change their shape under the influence of the external electric field, and this effect has also been used for the determination of membrane elastic constants.[48] Vesicles can change their shapes because of their interaction with the substrate[49] or in a shear flow.

Shape determination in the case of applied external forces is based on varying the corresponding thermodynamic potential of the system, which in general involves membrane elastic energy (Equation 4) and the potential energy of the external forces. In many instances the effect of external forces was studied when it could be treated as a small perturbation to the internal energy of the system. For instance, it was shown that a small external electric field deforms a spherical vesicle into a prolate ellipsoid.[50,51] A treatment of a linear response of a general vesicle shape to mechanical stresses was presented by Peterson.[52] However, in many instances the contribution due to the external forces surmounts the vesicle internal energy, and in such cases it is necessary to develop a generalization of the exact methods for shape determination of unsupported vesicles. We shall treat here as examples the effect of an external electric field on vesicle shape[53] and the determination of shape of a vesicle squeezed between two parallel plates.[54] In addition, we shall briefly describe the analysis of the method for the determination of the nonlocal bending constant by an experiment in which a vesicle is partly aspirated into a pipette and an external point force causes the formation of a thin microtube (tether).[55]

A. VESICLE IN AN EXTERNAL ELECTRIC FIELD

An external static electric field deforms phospholipid vesicles because of the electric Maxwell stresses acting on the membrane. The field tends to compress spherical vesicles laterally,[50,51] and the vesicles get elliptically deformed for small fields if the relative volume v is allowed to change. For larger fields the electric field cannot be treated as a perturbation, and more complicated shapes appear as a result of the minimization of the membrane bending energy in the presence of the electric field.[53,56] The problem can be divided into two parts. The first is the electric one. For a given vesicle shape one must solve the Laplace equation for the electric potential, with the condition that the electric field is equal to the external electric field (E_0) far away from the vesicle, while at the vesicle surface the normal component of the field equals zero, corresponding to a nonconductive membrane and to the condition that the current in the external ionic solution must be tangential at the surface. The mathematical procedure, which is based on the expansion in Legendre polynomials, gives as a result the tangential electric field on the surface of the vesicle.

The second part of the problem consists of finding the equilibrium shape. Let us consider the variation of the electrical free energy of the system, δW_{el}, defined as the work of the external forces which compensate the electrical forces for infinitesimally small displacements of the surface, $\delta \bar{u}$:

$$\delta W_{el} = -\int \bar{T} \, \delta \bar{u} \, dA \tag{67}$$

Here $\bar{T}$ is the electrical force per unit surface area. It is equal to the tensor product of the Maxwell's stress tensor[57] and the outside normal to the surface, $\bar{n}$:

$$\bar{T} = \varepsilon \varepsilon_0 \left[\bar{E}(\bar{E}\bar{n}) - \frac{1}{2} \bar{n} E^2 \right] \tag{68}$$

Here ε is the permittivity of the external solution. The first term in this expression equals zero because the electric field is tangential to the surface, and the variation of the electrical free energy can be expressed as

$$\delta W_{el} = \frac{1}{2} \varepsilon \varepsilon_0 \int E^2 \, \delta u_n \, dA \tag{69}$$

where $\delta u_n = \delta \bar{u} \cdot \bar{n}$ is the normal component of the displacement. This expression shows that the electrical free energy decreases for negative displacements and therefore the electric field tends to compress the vesicle. The pressure $\varepsilon \varepsilon_0 E^2 / 2$ is largest on the equator, where the electric field is the largest, and it equals zero on the poles.

For an axially symmetric shape with the symmetry axis z in the direction of the external electric field $\bar{E}_0$ one can express the normal component of the displacement, δu_n, by the variations of the principal curvatures along the parallels, δc_p, and obtain

$$\delta W_{el} = \pi \varepsilon \varepsilon_0 \int_0^{x_{max}} E^2 x \, dx \int_x^{x_{max}} \frac{x' dx'}{\left[1 - x'^2 c_p^2(x')\right]^{3/2}} \delta c_p(x') \tag{70}$$

By changing the order of integration and by introducing dimensionless quantities as in Section III we get

$$\delta\left(\frac{W_{el}}{8\pi\,k_c}\right) = \frac{\varepsilon\varepsilon_o E_o^2 R_s^3}{8\,k_c}\int_o^{x_{max}}\frac{x\,dx}{\left(1-x^2 c_p^2\right)^{3/2}}\,\delta\,c_p\int_o^x e^2(x')x'\,dx' \tag{71}$$

Here $e(x) = E/E_o$ is the tangential component of the electric field relative to E_o. Adding Equation 71 to the variation of the dimensionless elastic energy function (Equation 43) one obtains a differential equation with the same structure as Equation 54, except that the Lagrange multiplier M is modified and includes the electric field term:

$$M \rightarrow M - \frac{\varepsilon\varepsilon_o E_o^2 R^3}{k_c}\frac{1}{x^2}\int_o^x e^2(x')x'\,dx' \tag{72}$$

The external electric field influences the shape only in the combination

$$\tilde{E}_o^2 = \frac{\varepsilon\varepsilon_o E_o^2 R_s^3}{k_c} \tag{73}$$

which is proportional to the ratio of the electric and bending energies of the vesicle. It can be seen from expression 73 that the problem is no longer scale invariant and that the effect of the electric field is stronger for larger vesicles and for smaller bending constants.

Numerical analysis shows that not only spherical vesicles, which are allowed to change their volume or surface area, become elliptical in small external fields,[51] but also vesicles, which are not spherical in a zero field, have shapes similar to prolate ellipsoids in large external electric fields. In Figure 13 a vesicle is shown which has a prolate cigar shape in a zero field. In a large field the shape becomes more similar to a prolate ellipsoid. In Figure 14 the corresponding effect of the electric field on the length of the vesicle is presented. One observes that the effect of the electric field is negligibly small for the strict bilayer couple model where the area difference Δa cannot change ($k_r = \infty$), while the changes are larger when Δa is allowed to change ($k_r = 0$). The measurements of the electric field effects on the shapes of vesicles could therefore represent a method for determination of the nonlocal bending modulus k_r.

B. VESICLE SQUEEZED BETWEEN TWO PARALLEL PLATES

As a second example we determine the shape of a vesicle squeezed between two parallel plates (Figure 15, inset). This example differs from the first one in that the external force acts on the vesicle only over a part of its surface. Therefore special care has to be taken about the boundary conditions at the lines dividing different regions of the surface.

The equilibrium state of the vesicle squeezed between two parallel plates is obtained by varying the sum of the membrane bending energy and the potential energy of the force squeezing the vesicle. Here we shall assume the validity of the bilayer couple model; therefore the expression to be minimized is

$$W = W_b + FZ \tag{74}$$

where W_b is the bending energy given by Equation 35, F is the force, and Z is the distance between the two plates, i.e., vesicle height. It is assumed that the geometrical constraint which the plates exert on the vesicle causes there to be two circular sections where vesicle membrane is in contact with plates. We also assume that the vesicle has axial and equatorial mirror symmetries. Then we look for the minimum of the bending energy of the section of the vesicle which is not in contact with the plates, where in addition to the constraints for unsupported vesicles (Equation 37) we also have to require that the calculated vesicle height matches the required value. The functional to be minimized in this case is then

$$g' = g - \frac{1}{8}O(Z* - Z) \tag{75}$$

where the Lagrange multiplier O is proportional to the force at vesicle height Z. The Euler equation then reads, analogously to Equation 54, as

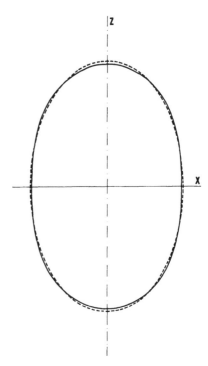

Figure 13 The effect of the electric field on a vesicle with a prolate cigar shape (shape 2 in Figure 7) is presented for v = 0.95. For $\tilde{E}_o$ = 12 and k_r = 0 the shape becomes similar to a prolate ellipsoid (broken line).

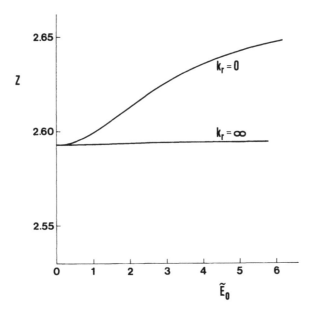

Figure 14 The effect of the electric field on the length (Z) of the vesicle from Figure 13 is shown for k_r = 0 and k_r = ∞, respectively.

$$\frac{x}{2(1-x^2c_p^2)}\left[xc_p-\frac{dc_p}{dx}\left(2c_p+x\frac{dc_p}{dx}\right)+M+Lc_p+Nc_p^2+\frac{O}{x^2}\right]+\frac{d}{dx}\left(2c_p+x\frac{dc_p}{dx}\right)=0 \qquad (76)$$

Because there is no interaction between the membrane and plates, the value of the curvature c_m at the two lines, where the vesicle detaches from the plates, is zero.

The resulting force/thickness dependence for a vesicle with the relative volume v = 0.95 and for several values of the relative area difference Δa is given in Figure 15. It can be noted that, for each Δa

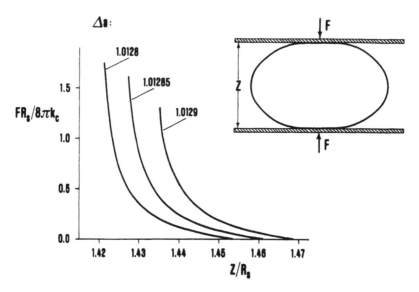

Figure 15 The dependence of the force squeezing a vesicle (F) on the distance between the plates (Z) at three values of the relative area difference Δa. The system is schematically represented in the inset. Force and distance are given in relative units. The relative volume of the vesicle is 0.95. The vesicle belongs to the class of disk shapes. Shapes are calculated by solving the Euler-Lagrange equation obtained by varying the free energy of the system at constant vesicle volume, constant membrane area, constant difference between the areas of the membrane layers, and constant distance between the plates. (From Svetina, S. and Žekš, B., in *Biomechanics of Active Movement and Division of Cells*, Akkas, N., Ed., Springer-Verlag, Berlin, 1994, 479. With permission.)

value, below a certain thickness the force increases very steeply. This can be related to the decreasing flaccidity of a vesicle.

C. DETERMINATION OF THE NONLOCAL BENDING CONSTANT BY THE TETHER PULLING EXPERIMENT

An experimental approach involving the formation of thin bilayer cylinders (tethers) from giant phospholipid vesicles has been developed[58] and analyzed[55] to enable the determination[14] of both local and nonlocal bending moduli. Here we shall briefly summarize the analysis of the tether experiment with emphasis on the stability of the system and indicate the results.

In the experimental procedure to be analyzed,[58] a vesicle is aspirated into a pipette, forming a spherical portion outside the pipette and a projection within the pipette. Then the vesicle is placed in contact with a glass bead which sticks to it; i.e., a vesicle-bead pair is formed. After reducing the aspiration pressure the bead falls due to the gravitational force away from the vesicle, forming a tether between the bead and the body of the vesicle. The pressure can be adjusted to stop the bead movement, which means that equilibrium can be established.

In this work the shape of the tethered vesicle is approximated by a simple geometrical model described in Figure 16. The analysis involves the determination of the elastic energy of the membrane for a given configuration of the tethered vesicle and the search for the configuration which corresponds to the minimum of the thermodynamic potential of the system (F_t), comprising besides the elastic energy (Equation 4) also the gravitational potential energy of the bead ($-fL_t$) and the contribution of the work done by the membrane due to hydrostatic pressure differences ($-\Delta p\pi R_p^2 L_p$). The significant contributions to the elastic energy of this system are[55] bending energy of the tether cylinder and the relative expansivity due to changes of the tether length; therefore the function to be minimized with respect to the system variables reads

$$F_t = \pi k_c \frac{L_t}{R_t} + \frac{2\pi^2 k_r}{A_o}\left(L_t - L_t^*\right)^2 - fL_t - \Delta p\pi R_p^2 L_p \tag{77}$$

Here $\Delta p = p_o - p_p$ (see Figure 16), and it is a positive quantity. In the relative expansivity term of Equation 77 it is taken into consideration (by introducing the parameter L_t^*) that the area difference of the

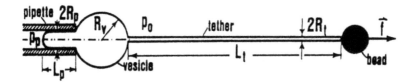

Figure 16 Schematic representation of the vesicle which is aspirated in a pipette and pulled with constant force in the opposite direction. The shape is described by five parameters: pipette radius (R_p), pipette projection length (L_p), vesicle radius (R_v), tether length (L_t), and tether radius (R_t). The pressure surrounding the vesicle is p_o, the pressure in the pipette is p_p, and the force pulling the tether is f. (From Svetina, S., Božič, B., Song, J., Waugh, R. E., and Žekš, B., in *The Structure and Conformation of Amphiphilic Membranes*, Lipowsky, R., Richter, D., and Kremer, K., Eds., Springer-Verlag, Berlin, 1992, 154. With permission.)

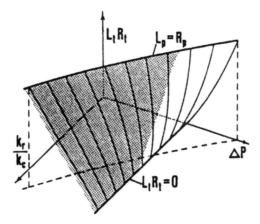

Figure 17 Points on the theoretically obtained surface represent possible tether configurations for a vesicle with relative volume v = 0.70 aspirated into the pipette with the relative radius R_p/R_s = 0.40 for different values of the ratio k_r/k_c and dimensionless parameter $\Delta P = 4\pi^2 R_s k_c \Delta p/(f + 4\pi^2 k_r L_t*/A)^2$. Only within the shaded part of the surface are the tether configurations stable. (From Svetina, S., Božič, B., Song, J., Waugh, R. E., and Žekš, B., in *The Structure and Conformation of Amphiphilic Membranes*, Lipowsky, R., Richter, D., and Kremer, K., Eds., Springer-Verlag, Berlin, 1992, 154. With permission.)

constitutive vesicle is not necessarily the same as the area difference of the aspirated vesicle with the spherical external part, i.e., that the possibility exists that the areas of the layers of the aspirated vesicle may be relatively expanded (or compressed) before the tether is made.

To obtain the equilibrium tether configuration the first derivatives of the free energy (Equation 77) are determined with respect to all the independent shape variables and set to zero while membrane area and vesicle volume are kept constant. A convenient way[59] to visualize the role of nonlocal bending in tether formation is represented in Figure 17. The stability analysis reveals[55] that the only stable tether configurations are those for which at a given value of k_r/k_c the derivative of the product R_tL_t with respect to the pressure difference is negative. It can therefore be depicted from Figure 17 that there is a certain critical value of the ratio k_r/k_c beyond which the tether configurations cannot exist.

Experiments were performed[14] with the lipid 1-stearoyl-2-oleoyl-phosphatidylcholine plus smaller amounts of the charged lipid 1-palmitoyl-2-oleoyl-phosphatidylserine (2% and 10%). Essentially, the length of the projection in the pipette and the aspiration pressure were measured as functions of the tether length. From such data for several vesicles the mean values obtained for the local and nonlocal bending moduli k_c and k_r are $1.2 \pm 0.2 \times 10^{-19}$ J and $4.1 \pm 2.4 \times 10^{-19}$ J, respectively, and they do not show any dependence on composition. The obtained ratio k_r/k_c = 3.3 is consistent with estimates presented in Section II.

V. CONCLUDING REMARKS

The behavior of phospholipid membranes is in many aspects dominated by their layered structure. It is now firmly established both theoretically and experimentally that, besides the area expansivity and local bending, the elasticity of closed membranes also involves the nonlocal contribution to the membrane bending energy, which appears because membrane layers respond to elastic stresses in an independent manner and are at the same time geometrically constrained by their contact. In bilayers the effect of nonlocal bending is directly related to the relative stretching of the two layers against each other. The nonlocal membrane bending contribution to the membrane elastic energy does not affect membrane behavior in the same way as the membrane spontaneous curvature. Spontaneous curvature is a local membrane property describing an inherent tendency of a piece of the membrane to be bent, and it essentially arises as a consequence of asymmetry in the physical properties of the membrane due to properties of its constituents and/or because of asymmetric external conditions. The origin of nonlocal bending is in a different occupancy of the two membrane layers.

Shapes of unsupported flaccid phospholipid vesicles are determined by the minimum of the sum of the local and nonlocal contributions to the membrane elastic energy, i.e., by the generalized bilayer couple model. In general, vesicle shapes depend on relative vesicle volume, relative difference between the layer areas, spontaneous curvature, and the ratio between the nonlocal and local bending moduli. In the limit when this ratio is infinitely large we obtain the strict bilayer couple model, and in the limit when this ratio is zero we obtain the spontaneous curvature model. The phase diagram of the generalized bilayer couple model can be described as a generalization of the phase diagram of the strict bilayer couple model.

Theoretical studies of shape transformations within the bilayer couple model[25] revealed several properties of closed layered membranes which may also be particularly significant at the level of cells; e.g., (1) the occurrence of continuous shape transformations between shapes belonging to different shape classes indicates possible symmetry breaking properties of such a system, and (2) the classes of shapes exhibit limiting shapes. The symmetry breaking properties can be the structural basis for the establishment of order in a developing biological system.[44] The limiting shapes can be the basic membrane property in the processes of membrane vesiculation and vesicle fusion.[36] It is notable that these notions about the behavior of closed layered membranes were confirmed by different studies on phospholipid vesicles.[45-47]

External forces may modify vesicle shapes. The treated examples show how the theory developed for freely suspended vesicles is to be extended for studies of the effect of the external forces. Special care has to be taken with regard to the boundary conditions and the corresponding application of the variational principle.

We have restricted our analyses to laterally homogeneous layered membranes. Membranes are not necessarily homogeneous if they are composed of more than one constituent. It is plausible to expect that the interaction of a given molecule with its neighbors depends on the local curvature. For different membrane constituents such a dependence could be different, and consequently the lateral distribution of membrane constituents should in general depend on membrane curvature. On the other hand, the lateral distribution of membrane constituents affects the shape.[60] Eventually, phase separation could occur as has already been proposed.[61,62]

The described studies of layered membranes can be considered as a specific part of more general problems of membrane conformations which recently became an interesting subject of physical studies.[63] The described specific properties of layered membranes may also have an important role in the functioning of biological systems. This had already been noted when the bilayer couple model was originally introduced to interpret red blood cell shape transformations.[64] The consequent theoretical studies showed that shapes depend only on a small number of parameters such as v, Δa, c_0, and k_r/k_c and that the interpretation of cell shape transformations must be conceptually divided into two separate parts, i.e., on one hand into the molecular processes which take into account the cell content and the composition of membrane layers and therefore determine the above parameters, and on the other hand into the shape attainment which is a simple consequence of membrane elasticity. Biological systems are compositionally and structurally complex, and it is usually difficult to prove such notions. The development of the techniques for obtaining well-characterized phospholipid vesicles makes it possible for the theoretical predictions described in this chapter to be verified in a strict manner. In this respect we can consider phospholipid vesicles also as an ideal testing system for the evaluation of different methodologies for studying properties of layered membranes in general. The knowledge about the behavior of simple phospholipid vesicles can help to elucidate which aspects of different membrane processes in cells are simple nonspecific consequences of properties of layered membranes and at which point it is necessary to look for the effects of specific biological structures.

REFERENCES

1. **Evans, E. A. and Hochmuth, R. M.,** Mechanochemical properties of membranes, in *Current Topics in Membranes and Transport,* Vol. 10, Bronner, F. and Kleinzeller, A., Eds., Academic Press, New York, 1978, 1.
2. **Evans, E. A. and Skalak, R.,** *Mechanics and Thermodynamics of Biomembranes,* CRC Press, Boca Raton, FL, 1980.
3. **Evans, E. and Needham, D.,** Physical properties of surfactant bilayer membranes: thermal transitions, elasticity, rigidity, cohesion, and colloidal interactions, *J. Phys. Chem.,* 91, 4219, 1987.
4. **Hochmuth, R. M. and Waugh, R. E.,** Erythrocyte membrane elasticity and viscosity, *Annu. Rev. Physiol.,* 49, 209, 1987.
5. **Evans, E. A.,** Bending resistance and chemically induced moments in membrane bilayers, *Biophys. J.,* 14, 923, 1974.
6. **Evans, E. A.,** Minimum energy analysis of membrane deformation applied to pipet aspiration and surface adhesion of red blood cells, *Biophys. J.,* 30, 265, 1980.
7. **Helfrich, W.,** Blocked lipid exchange in bilayers and its possible influence on the shape of vesicles, *Z. Naturforsch.,* 29c, 510, 1974.
8. **Svetina, S., Brumen, M., and Žekš, B.,** Lipid bilayer elasticity and the bilayer couple interpretation of red cell shape transformations and lysis, *Stud. Biophys.,* 110, 177, 1985.
9. **Svetina, S. and Žekš, B.,** The elastic deformability of closed multilayered membranes is the same as that of a bilayer membrane, *Eur. Biophys. J.,* 21, 251, 1992.
10. **Petrov, A. G. and Bivas, I.,** Elastic and flexoelectic aspects of out-of-plane fluctuations in biological and model membranes, *Prog. Surf. Sci.,* 16, 389, 1984.
11. **Helfrich, W.,** Elastic properties of lipid bilayers: theory and possible experiments, *Z. Naturforsch.,* 28c, 693, 1973.
12. **Evans, E. and Rawicz, W.,** Entropy-driven tension and bending elasticity in condensed-fluid membranes, *Phys. Rev. Lett.,* 64, 2094, 1990.
13. **Duwe, H. P., Kaes, J., and Sackmann, E.,** Bending elastic moduli of lipid bilayers: modulation by solutes, *J. Phys. (Paris),* 51, 945, 1990.
14. **Waugh, R. E., Song, J., Svetina, S., and Žekš, B.,** Local and non-local curvature elasticity in bilayer membranes by tether formation from lecithin vesicles, *Biophys. J.,* 61, 982, 1992.
15. **Kozlov, M. M. and Winterhalter, M.,** Elastic moduli for strongly curved monolayers. Position of the neutral surface, *J. Phys. II (Paris),* 1, 1077, 1991.
16. **Fischer, T. M.,** Bending stiffness of lipid bilayers. V. Comparison of two formulations, *J. Phys. II (Paris),* 3, 1795, 1993.
17. **Kozlov, M. M., Leikin, S. L., and Markin, V. S.,** Elastic properties of interfaces, *J. Chem. Soc. Faraday Trans. 2,* 85, 277, 1989.
18. **Canham, P. B.,** The minimum energy of bending as a possible explanation of the biconcave shape of the human red blood cell, *J. Theor. Biol.,* 26, 61, 1970.
19. **Deuling, H. J. and Helfrich, W.,** The curvature elasticity of fluid membranes: a catalogue of vesicle shapes, *J. Phys. (Paris),* 37, 1335, 1976.
20. **Luke, J. C.,** A method for the calculation of vesicle shapes, *SIAM J. Appl. Math.,* 42, 333, 1982.
21. **Peterson, M. A.,** An instability of the red blood cell shape, *J. Appl. Phys.,* 57, 1739, 1985.
22. **Miao, L., Fourcade, B., Rao, M., and Wortis, M.,** Equilibrium budding and vesiculation in the curvature model of fluid lipid vesicles, *Phys. Rev. A,* 43, 6843, 1991.
23. **Seifert, U., Berndl, K., and Lipowsky, R.,** Shape transformations of vesicles: phase diagram for spontaneous-curvature and bilayer-coupling models, *Phys. Rev. A,* 44, 1182, 1991.
24. **Svetina, S., Ottova-Leitmannova, A., and Glaser, R.,** Membrane bending energy in relation to bilayer couples concept of red blood cell shape transformations, *J. Theor. Biol.,* 94, 13, 1982.
25. **Svetina, S. and Žekš, B.,** Membrane bending energy and shape determination of phospholipid vesicles and red blood cells, *Eur. Biophys. J.,* 17, 101, 1989.
26. **Wiese, W., Harbich, W., and Helfrich, W.,** Budding of lipid bilayer vesicles and flat membranes, *J. Phys. Condens. Matter,* 4, 1647, 1992.
27. **Seifert, U., Miao, L., Döbereiner, H. G., and Wortis, M.,** Budding transition for bilayer fluid vesicles with area-difference elasticity, in *The Structure and Conformation of Amphiphilic Membranes, Springer Proceedings in Physics,* Vol. 66, Lipowsky, R., Richter, D., and Kremer, K., Eds., Springer-Verlag, Berlin, 1992, 93.
28. **Käs, J., Sackmann, E., Podgornik, R., Svetina, S., and Žekš, B.,** Thermally induced budding of phospholipid vesicles — a discontinuous process, *J. Phys. II (Paris),* 3, 631, 1993.
29. **Svetina, S. and Žekš, B.,** Elastic properties of closed layered membranes and equilibrium shapes of phospholipid vesicles, in *Proceedings of the Eleventh School on Biophysics of Membrane Transport,* Vol. II, Kuczera, J. and Przestalski, S., Eds., Agricultural University of Wrocław, Poland, 1992, 115.
30. **Heinrich, V., Svetina, S., and Žekš, B.,** Nonaxisymmetric shapes in a generalized bilayer-couple model and the transition between oblate and prolate axisymmetric shapes, *Phys. Rev. E,* 48, 3112, 1993.
31. **Miao, L., Seifert, U., Wortis, M., and Döbereiner, H. G.,** Budding transitions of fluid-bilayer vesicles: the effect of area-difference elasticity, *Phys. Rev. E,* 49, 5389, 1994.

42

32. Svetina, S. and Žekš, B., Bilayer couple hypothesis of red cell shape transformations and osmotic hemolysis, *Biomed. Biochim. Acta*, 42, 86, 1983.

33. Svetina, S. and Žekš, B., Bilayer couple as a possible mechanism of biological shape formation, *Biomed. Biochim. Acta*, 44, 979, 1985.

34. Svetina, S., Gros, M., Vrhovec, S., Brumen, M., and Žekš, B., Red blood cell membrane vesiculation at low pH and bilayer couple mechanism of red blood cell shape transformations, *Stud. Biophys.*, 127, 193, 1988.

35. Döbereiner, H. G., Käs, J., Noppl, D., Sprenger, I., and Sackmann, E., Budding and fission of vesicles, *Biophys. J.*, 65, 1396, 1993.

36. Svetina, S., Iglič, A., and Žekš, B., On the role of the elastic properties of closed lamellar membranes in membrane fusion, *Ann. N.Y. Acad. Sci.*, 710, 179, 1994.

37. Peterson, M. A., Deformation energy of vesicles at fixed volume and surface area in the spherical limit, *Phys. Rev. A*, 39, 2643, 1989.

38. Heinrich, V., Brumen, M., Heinrich, R., Svetina, S., and Žekš, B., Nearly spherical vesicle shapes calculated by use of spherical harmonics: axisymmetric and nonaxisymmetric shapes and their stability, *J. Phys. II (Paris)*, 2, 1081, 1992.

39. Kralj-Iglič, V., Svetina, S., and Žekš, B., The existence of non-axisymmetric bilayer vesicle shapes predicted by the bilayer couple model, *Eur. Biophys. J.*, 22, 97, 1993.

40. Hu, J. G. and Ou-Yang, Z. C., Shape equations of the axisymmetric vesicles, *Phys. Rev. E*, 47, 461, 1993.

41. Jülicher, F. and Seifert, U., Shape equations for axisymmetric vesicles: a clarification, *Phys. Rev. E*, 49, 4728, 1994.

42. Podgornik, R., Svetina, S., and Žekš, B., Parametrization invariance and shape equations of elastic axisymmetric vesicles, *Phys. Rev. E*, 51, 544, 1995.

43. Jülicher, F., Seifert, U., and Lipowsky, R., Phase diagrams and shape transformations of toroidal vesicles, *J. Phys. II (Paris)*, 3, 1681, 1993.

44. Svetina, S. and Žekš, B., The mechanical behaviour of cell membranes as a possible physical origin of cell polarity, *J. Theor. Biol.*, 146, 115, 1990.

45. Berndl, K., Käs, J., Lipowsky, R., Sackmann, E., and Seifert, U., Shape transformations of giant vesicles: extreme sensitivity to bilayer asymmetry, *Europhys. Lett.*, 13, 659, 1990.

46. Käs, J. and Sackmann, E., Shape transitions and shape stability of giant phospholipid vesicles in pure water induced by area-to-volume changes, *Biophys. J.*, 60, 825, 1991.

47. Farge, E. and Devaux, P., Shape changes of giant liposomes induced by an asymmetric transmembrane distribution of phospholipids, *Biophys. J.*, 61, 347, 1992.

48. Kummrow, M. and Helfrich, W., Deformation of giant lipid vesicles by electric fields, *Phys. Rev. A*, 44, 8356, 1991.

49. Seifert, U. and Lipowsky, R., Adhesion of vesicles, *Phys. Rev. A*, 42, 4768, 1990.

50. Helfrich, W., Deformation of lipid bilayer spheres by electric fields, *Z. Naturforsch.*, 29c, 182, 1974.

51. Winterhalter, W. and Helfrich, W., Deformation of spherical vesicles by electric fields, *J. Colloid Interface Sci.*, 122, 583, 1988.

52. Peterson, M. A., Linear response of the human erythrocyte to mechanical stress, *Phys. Rev. A*, 45, 4116, 1992.

53. Žekš, B., Svetina, S., and Pastushenko, V., The shapes of phospholipid vesicles in an external electric field — a theoretical analysis, *Stud. Biophys.*, 138, 137, 1990.

54. Svetina, S. and Žekš, B., Elastic properties of layered membranes and their role in transformations of cellular shapes, in NATO ASI Series, Vol. H 84, *Biomechanics of Active Movement and Division of Cells*, Akkas, N., Ed., Springer-Verlag, Berlin, 1994, 479.

55. Božič, B., Svetina, S., Žekš, B., and Waugh, R. E., The role of lamellar membrane structure in tether formation from bilayer vesicles, *Biophys. J.*, 61, 963, 1992.

56. Sokirko, A., Pastushenko, V., Svetina, S., and Žekš, B., Deformation of a lipid vesicle in an electric field: a theoretical study, *Bioelectrochem. Bioenerg.*, 34, 101, 1994.

57. Landau, L. D. and Lifshiz, E. M., *Electrodynamics of Continuous Media*, 2nd ed., Pergamon Press, Oxford, 1984.

58. Bo, L. and Waugh, R. E., Determination of bilayer membrane bending stiffness by tether formation from giant, thin-walled vesicles, *Biophys. J.*, 55, 509, 1989.

59. Svetina, S., Božič, B., Song, J., Waugh, R. E., and Žekš, B., Phospholipid membrane local and non-local bending moduli determined by tether formation from aspirated vesicles, in *The Structure and Conformation of Amphiphilic Membranes, Springer Proceedings in Physics*, Vol. 66, Lipowsky, R., Richter, D., and Kremer, K., Eds., Springer-Verlag, Berlin, 1992, 154.

60. Svetina, S., Kralj-Iglič, V., and Žekš, B., Cell shape and lateral distribution of mobile membrane constituents, in *Proceedings of the Tenth School on Biophysics of Membrane Transport*, Vol. II, Kuczera, J. and Przestalski, S., Eds., Agricultural University of Wrocław, Poland, 1990, 139.

61. Markin, V. S., Lateral organization of membranes and cell shapes, *Biophys. J.*, 36, 1, 1981.

62. Jülicher, F. and Lipowsky, R., Domain-induced budding of vesicles, *Phys. Rev. Lett.*, 70, 2964, 1993.

63. Lipowsky, R., The conformation of membranes, *Nature*, 349, 475, 1991.

64. Sheetz, M. P. and Singer, S. J., Biological membranes as bilayer couples. A molecular mechanism of drug-induced interactions, *Proc. Natl. Acad. Sci. U.S.A.*, 71, 4457, 1974.

Chapter 2

Shapes of Fluid Vesicles

Udo Seifert and Reinhard Lipowsky

CONTENTS

I. INTRODUCTION

Giant vesicles form spontaneously from lipid bilayers in aqueous solution. Even though they do not share the more complex architecture of a cell membrane, they exhibit an amazing variety of shapes. Shape transformations among these shapes can be induced by simple changes of external parameters such as temperature. As an example, the most prominent shape transformation, the budding transition, is shown in Figure 1 as it has been observed by video microscopy.[1] An increase in temperature transforms a

44

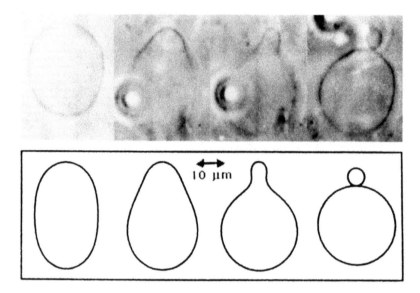

Figure 1 Budding transition induced by temperature which is raised from 31.4°C (left) to 35.8°C (right) in this sequence. The theoretical shapes have been obtained within the bilayer couple model assuming an asymmetric thermal expansion of the two monolayers. (Modified from K. Berndl et al., *Europhys. Lett.*, 13:659–664, 1990.)

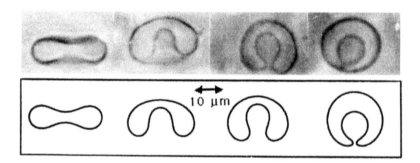

Figure 2 Discocyte-stomatocyte transition induced by temperature which is raised from 43.8°C (left) to 44.1°C (right) in this sequence. (Modified from K. Berndl et al., *Europhys. Lett.*, 13:659–664, 1990.)

quasi-spherical vesicle via thermal expansion of the bilayer to a prolate shape and then to a pear shape. Finally, a small bud is expelled from the vesicle. Upon further increase of temperature several additional buds can emerge. Usually the buds remain connected to the mother vesicle via narrow constrictions or necks. Another shape transformation, the discocyte-stomatocyte transition, is shown in Figure 2.

The theoretical understanding of these shape transformations is based on the important notion of bending elasticity introduced some 20 years ago by Canham,[2] Helfrich,[3,4] and Evans.[5] However, a systematic analysis of these so-called curvature models was performed only recently.[1,6-8] These studies indeed revealed a large variety of shapes which minimize the energy for certain physical parameters such as the enclosed volume and the area of the vesicle. These shapes are then organized in so-called "phase diagrams" in which trajectories predict how the shape transforms as, e.g., the temperature is varied.

Shape transformations are also predicted to arise in vesicles consisting of a bilayer with several components due to different mechanisms. If the two components form domains, the line energy associated with these domain boundaries can be reduced by budding of such a domain.[9] Even if the membrane is in the one-phase region, a temperature-induced budding process should lead to curvature-induced phase segregation since, in general, the two components couple differently to the local curvature.[10] So far these effects have not been verified experimentally, even though budding and fission have been observed in vesicles consisting of lipid mixtures.[11]

The interaction of vesicles with substrates (or other vesicles) also affects their shape. For weak adhesion, an adhesion transition driven by the competition between bending elasticity and adhesion energy is predicted.[12,13] For strong adhesion, the notion of an effective contact angle becomes applicable.[12] Experimentally, the adhesion of vesicles can be studied either with the micropipet aspiration technique[14] or by reflection interference microscopy.[15,16] The latter technique allows very precise length measurements and, thus, the deduction of the shape of adhering vesicles. In this way, a flat pancake can be distinguished from a quasi-spherical shape which is only slightly distorted by the adhesion to the substrate.

In this chapter, we describe the present understanding of these phenomena. In Section 2 we introduce various variants of the curvature model. Section 3 is devoted both to the presentation of phase diagrams for vesicles of spherical topology and to the comparison of theory and experiment for the budding transition. In Section 4, shape transformations and, in particular, budding are discussed for vesicles consisting of two-component membranes. Adhesion of vesicles to a substrate is considered in Section 5. Various aspects of the work presented here have also been treated in recent reviews (see References 17 to 19).

II. BENDING ELASTICITY AND CURVATURE MODELS

A. BASIC IDEA OF CURVATURE MODELS

In equilibrium, vesicles will acquire the shape at which their total energy is minimal since processes such as hydrodynamic flows, convection, or transport related to temperature gradients can be neglected. This energy arises from the bending elasticity of the bilayer, which can be expressed theoretically as a curvature energy for which various variants have been introduced during the last 25 years.

Common to all curvature models is the description of the vesicles as closed two-dimensional surface $R(s_1,s_2)$ embedded in three-dimensional space, which is justified by the length-scale separation between the thickness of the bilayer, which is in the nanometer range, and the typical size of the vesicles, which is in the micrometer range. Since the membrane is fluid, there is no fixed connectivity in the plane, and s_1 and s_2 are arbitrary internal coordinates. Such a surface can be locally characterized by its two radii of curvature, R_1 and R_2, which determine the mean curvature H and the Gaussian curvature K, defined as

$$H \equiv \left(1/R_1 + 1/R_2\right)/2 \tag{1}$$

and

$$K \equiv 1/\left(R_1 R_2\right) \tag{2}$$

Since the radii R_1 and R_2 carry a sign, saddle surfaces have vanishing mean curvature if the absolute value of the two radii of curvature are the same. Surfaces with $H = 0$ everywhere are called minimal surfaces.

B. SPONTANEOUS CURVATURE MODEL

The *spontaneous curvature model* introduced by Helfrich,[3] generalizing a somewhat simpler model by Canham,[2] is based on the assumption that the energy associated with bending the membrane can be expanded in H and K. For small deformations of a symmetric membrane, the most general expression symmetric in R_1 and R_2 up to a second order in $1/R_i$ is

$$F_{SC} \equiv (\kappa/2)\oint dA \left(2H - C_0\right)^2 + \kappa_G \oint dA \, K \tag{3}$$

This expression contains three material parameters: the bending rigidity κ, the Gaussian bending rigidity κ_G, and the spontaneous curvature C_0.

The spontaneous curvature C_0 reflects a possible asymmetry of the membrane, justified either by a different chemical environment on both sides of the membrane or by a different composition of the two monolayers. C_0 is usually assumed to be laterally homogeneous; this implies that it does not depend on the local shape of the membrane. So far, no measurements of C_0 are available for phospholipid vesicles.

C. BENDING RIGIDITY

The bending rigidity κ has been measured with a variety of techniques which can be classified into three categories. In the mechanical approach, using micropipet aspiration,[20-23] the response of the membrane to an applied force is measured, from which the bending rigidity is deduced.

In the flickering experiments, the bending rigidity is derived from mean-square amplitudes of thermally excited membrane fluctuations using phase contrast microscopy combined with fast image processing. This technique has been used with quasi-spherical vesicles,[24-30] shape fluctuations of tubular vesicles,[31] fluctuations of almost planar membrane segments,[32] and fluctuations of weakly bound vesicles.[15,33]

A third class of experiments combines the mechanical with the entropic approach. Since aspiration of the vesicle in the micropipet changes the area available for fluctuations,[34] the strength of the fluctuation can be controlled mechanically. From the relation between the area stored in the fluctuations and the suction pressure, which is related to the effective entropic tension, the bending rigidity can be deduced.[35] The same idea has also been used in a very different setup where quasi-spherical vesicles are elongated in an AC electric field.[36]

The values obtained by these techniques scatter strongly. Even for the same lipid, the bending rigidity differs by a factor of 2 using different methods. Sometimes the same technique applied by different research groups yields significantly different values. Whether these discrepancies have to be attributed to real physical effects such as impurities or to experimental problems needs further analysis. Likewise, the hypothesis put forward by Helfrich[37,38] that these discrepancies support evidence for a superstructure of the membrane on suboptical scales still awaits direct verification.

The bending rigidity for mixtures is particularly interesting given the fact that biological membranes always involve mixtures. Addition of cholesterol to fluid bilayers increases the bending rigidity (as well as the area compressibility modulus) significantly.[21,29,39] A significant decrease in the apparent bending rigidity of an order of magnitude follows the addition of a few (2 to 5) mol% of a short bipolar lipid (bola lipid) or small peptides (e.g., valinomycin).[29] Since the apparent bending rigidity then becomes comparable to the thermal energy T, these vesicles exhibit very strong shape fluctuations. Entropic terms then become relevant for the description of typical conformations. One must note, however, that bola lipid is much more soluble in water and thus will go in and out of the bilayer more rapidly.

The Gaussian bending rigidity is difficult to measure experimentally since it is a topological invariant given by $\oint dA K = 4\pi(1 - g)$, where g is the genus, i.e., the number of holes or handles of the vesicle. In principle, it can be inferred from measurements on ensembles of vesicles which are equilibrated with respect to topological changes. To our knowledge, there is no experimental record of a topology change in single-component phospholipid bilayer vesicles. Thus, one can conclude that the time scale for such changes is quite large. In this case, κ_G is neither accessible nor relevant for shape transformations. For vesicles containing several domains, on the other hand, the Gaussian bending rigidity should have observable effects even for a single vesicle.[40]

D. CONSTRAINTS ON AREA AND VOLUME

The minimum of F_{sc} for vesicles of spherical topology would be given by spheres of radius $R = 2/C_0$ if there were no further constraints present. In fact, the total area A as well as the enclosed volume V are constrained (or controlled) quantities on the time scale of typical shape transformations.

The area is constant for two reasons. First, there are essentially no free lipid molecules in solution, and thus, the number of lipid molecules in the membrane is constant. Second, elastic stretching or compression of the membrane would involve much more energy than pure bending. The enclosed volume is also basically constant even though the membrane is permeable to water. However, the inevitable presence of impurities or ions to which the membrane is impermeable fixes the volume by the condition that essentially no osmotic pressure builds up. Indeed, the osmotic pressure which could be sustained by the bending energy is tiny.[3]

E. BILAYER COUPLE MODEL

The spontaneous curvature model does not contain any explicit reference to the two leaflets which form the bilayer. It had been recognized early on that there is a further energetic constraint from the fact that the two monolayers do not exchange molecules on short time scales.[5,41,42]

An extreme version of this constraint was implemented in the so-called *bilayer couple model* introduced by Svetina and Žekš[43] based on much earlier work by Evans[5,42] and Sheetz and Singer.[44] Here it is postulated that not only the area of the vesicle but also the area difference ΔA between the two leaflets is constant and given by

$$\Delta A_0 = \left(N^+ - N^-\right)/\phi_0 \tag{4}$$

where N^+ and N^- are the numbers of lipid molecules within the two layers and ϕ_0 is their equilibrium density. The area difference ΔA can be expressed by the integrated mean curvature according to

$$\Delta A \approx 4d \oint dA \, H \tag{5}$$

for a separation of $2d$ between the (neutral surfaces of the) two monolayers up to corrections of the order of dH and $d^2 K$. Thus, this model has a constraint on the total mean curvature. In such a model any nonzero spontaneous curvature has become irrelevant and the energy is given by

$$G \equiv (\kappa/2) \oint dA (2H)^2 \tag{6}$$

together with the three constraints on A, V, and ΔA.

F. AREA-DIFFERENCE ELASTICITY MODEL

Starting from a somewhat more detailed description on the local scale which takes into account the relative compression and expansion of the two leaflets induced by bending the bilayer, it has been shown recently that the hard constraint on ΔA has to be relaxed. This leads to the *area-difference elasticity* (ADE) *model* with the energy[45-49]

$$W \equiv (\kappa/2) \oint dA (2H)^2 + \frac{\bar{\kappa}\pi}{8 A_0 d^2} \left(\Delta A - \Delta A_0\right)^2 \tag{7}$$

The nonlocal character of the second term in this energy arises from equilibration of the densities within each monolayer. Thus, the model implicitly assumes that the time scale of shape transformations is larger than the typical time scale for this equilibration.

The elastic constant $\bar{\kappa}$ in front of the second term is the so-called nonlocal bending rigidity, which turns out to be

$$\bar{\kappa} \equiv \alpha\kappa \equiv 2k^{(m)} d^2 / \pi \tag{8}$$

where $k^{(m)}$ is the area compression modulus of a monolayer. If one assumes that the monolayers are homogeneous sheets of thickness $2d$, the monolayer rigidity $\kappa^{(m)} = \kappa/2$ is related to $k^{(m)}$ through

$$\kappa^{(m)} = k^{(m)} d^2 / 3 \tag{9}$$

This leads to the important estimate of $\alpha = \bar{\kappa}/\kappa = 3/\pi$. However, from a theoretical point of view it is convenient to leave the value of α open and to treat it as an independent material parameter. Moreover, the ADE model may be applicable even if a simple relation like Equation 9 breaks down due to the internal structure of the monolayer. Likewise, the effective value of α for multilamellar vesicles can become larger since it is roughly quadratic in the number of bilayers[50] assuming a fixed intermembrane separation.

For $\alpha = 0$, one recovers the spontaneous curvature model from the area-difference elasticity model, provided one replaces $2H$ by $2H - C_0$ in Equation 7. This applies to the case where the two monolayers are asymmetric in the first place, which would leave an effective C_0 for the bilayer. The explicit derivation of the ADE model assuming monolayers with different C_0 requires some care in the definition of the neutral surfaces.[48,49,51] In fact, it turns out that the effect of a spontaneous curvature $C_0 \neq 0$ in the area-difference elasticity model is easy to determine theoretically since any $C_0 \neq 0$ can be mapped on a renormalization of the area difference ΔA_0.[48,49]

III. SHAPES OF MINIMAL ENERGY AND SHAPE TRANSFORMATIONS

A. THEORETICAL METHODS

The first step towards a systematic theory based on curvature models is to calculate the shapes of lowest energy. Since for phospholipid membranes the bending rigidity κ is large compared to the thermal energy T, thermal fluctuations around the shape of lowest energy can be ignored in such a first step. This holds for vesicles which are small compared to the persistence length $\xi_p \equiv a \exp(2\pi\kappa/T)$, where a is a molecular scale of the order of nanometers.[52] For phospholipid membranes, this length scale is astronomical. If the size of the vesicle becomes comparable to or larger than ξ_p, thermal fluctuations lead to irregular shapes which can theoretically be investigated by Monte Carlo simulation.[53]

In general, one obtains different types (or families or branches) of shapes of lowest energy in different regions of the parameter space. This division of the parameter space represents the so-called phase diagram of the model.

So far, the complete phase diagram has not been obtained for any curvature model since, in most cases, the minimization of the corresponding curvature energy has only been performed within a certain subspace of shapes. A significant simplification arises if one restricts the problem to axisymmetric shapes. Whether this restriction, which might look plausible, is indeed permissible has to be checked *a posteriori*. The canonical procedure for the calculation of axisymmetric shapes of lowest energy is as follows.[4,6-8,54-58] First, the constraints of fixed volume and area are added to the curvature energy via Lagrange multipliers. Then the axisymmetric shape is expressed in terms of its contour parameters. Stationarity of the shape leads to the Euler-Lagrange equations corresponding to the curvature energy (augmented with the constraints). These coupled nonlinear differential equations have to be solved numerically. The solutions (the so-called stationary shapes) can be organized in branches due to their symmetry properties. The most important branches are the prolate and oblate ellipsoids, which include the dumbbells and discocytes, respectively. Pear-shaped vesicles and stomatocytes bifurcate from the prolate and the oblate branch, respectively. Since the solutions to the shape equations contain local energy minima as well as unstable saddle points, one has to check the stability of each solution. Stability with respect to axisymmetric deformations follows from a close inspection of the bifurcation diagram.[8] Stability with respect to nonaxisymmetric perturbations requires additional work which has not yet been performed in a systematic way[56,59] except in the spherical limit.[27,60-63]

All the curvature models introduced above lead to the same set of stationary shapes since they only differ in global energy terms. However, the stability of these shapes and, thus, the phase diagram depends crucially on the specific model and the corresponding model parameters.

B. A SIMPLE MODEL: LOCAL CURVATURE ENERGY ONLY

To illustrate the general procedure described above, consider the spontaneous curvature model for $C_0 = 0$ together with the constraints on the area A and the enclosed volume V.[8] Since the curvature energy is scale invariant in this case, only the reduced volume

$$v \equiv V/\left(4\pi R^3/3\right) \tag{10}$$

enters, where

$$A \equiv 4\pi R^2 \tag{11}$$

defines the equivalent sphere radius R. Such a model is presumably the simplest model for a vesicle consisting of a symmetric bilayer. It should be applicable to membranes with fast flip-flop between the two monolayers.

For $v = 1$ the shape is necessarily a sphere, while for $v < 1$ the solutions to the shape equations are prolate ellipsoids and dumbbells, oblate ellipsoids and discocytes, and stomatocytes. Typical shapes as well as their energies are shown in Figures 3 and 4, taken from Reference 8. For $v \lesssim 1$ the prolates have the lowest energy. For $v < 0.65$ the energy of the oblates becomes smaller than the energy of the prolates. This amounts to a discontinuous transition from the prolates to the oblates with decreasing v. For any discontinuous transition, it is important to know for which value of the reduced volume a metastable shape loses its local stability. Approaching the transition from either the small or the large v values, the shapes

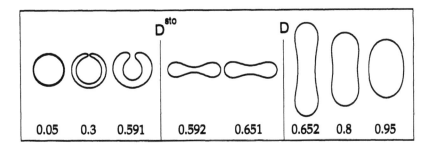

Figure 3 Shapes of lowest bending energy for spontaneous curvature $C_0 = 0$ and several values of the reduced volume v. D and D^{sto} denote the discontinuous prolate/oblate and oblate/stomatocyte transitions, respectively. All shapes have the same area. (Modified from U. Seifert, K. Berndl, and R. Lipowsky, *Phys. Rev. A*, 44:1182–1202, 1991.)

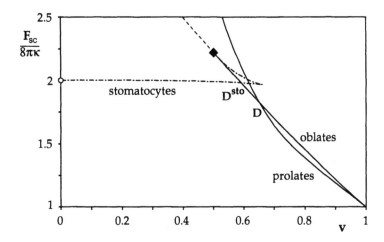

Figure 4 Bending energy F_{sc} as given by Equation 3 for $C_0 = 0$ (and $\kappa_G = 0$) as a function of the reduced volume v. Three branches of stationary shapes are displayed: the prolate, the oblate, and the stomatocyte branches. The latter bifurcates from the oblate branch. Its upper part, between the bifurcation and the cusp, corresponds to locally unstable shapes. Its lower part corresponds to locally stable shapes. The oblate branch beyond the diamond corresponds to self-intersecting (and thus nonphysical) states. D and D^{sto} are defined in caption of Figure 3. (Modified from U. Seifert, K. Berndl, and R. Lipowsky, *Phys. Rev. A*, 44:1182–1202, 1991.)

will be trapped in the local (metastable) minimum as long as the thermal energy is not sufficient to overcome the barrier for the transition towards the global minimum. An approximate stability analysis shows that the prolates are locally stable for all v whereas the oblates lose local stability with increasing v at some value v_* with $0.7 < v_* < 0.85$.[63] Thus, for large reduced volume, the oblates whose energy is shown in Figure 4 correspond to saddle points.

The energy diagram shown in Figure 4 directly reveals that for even smaller volume the oblates become unstable with respect to an up/down symmetry breaking for $v < 0.51$. This type of instability, which preserves the axisymmetry, can be read off from the bifurcation diagram. It is then also obvious that the upper part of the stomatocyte branch corresponds to the saddle point between a locally stable oblate and a locally stable stomatocyte. This energy diagram leads to a discontinuous transition between the oblate and the stomatocyte at $v \simeq 0.59$. The limits of metastability are at $v \simeq 0.66$ for the stomatocyte and $v \simeq 0.51$ for the oblate.

Two facts about this simple model should be emphasized. (1) The biconcave discocytes have the lowest energy in a narrow range of reduced volume v. Thus a negative C_0 is not required to obtain these red blood cell-like shapes, contrary to repeated claims in the older literature. (2) Budding does not occur in this model since pears do not show up as stationary shapes for vanishing spontaneous curvature.

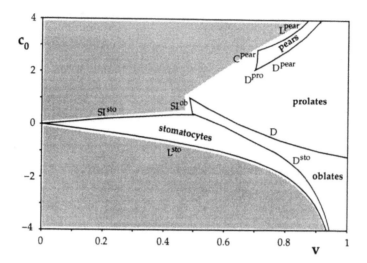

Figure 5 Phase diagram of the spontaneous curvature model. This phase diagram shows the shape of lowest bending energy as a function of the scaled spontaneous curvature c_0 and of the reduced volume v. The regions where the prolate/dumbbell, pear shapes, oblate/discocyte, and stomatocytes have lowest energy are separated by transition lines. The line C^{pear} denotes a continuous transition. All other transitions are discontinuous. The lines L^{sto} and L^{pear} correspond to limit shapes with infinitesimal neck. Beyond the lines SI^{ob} and SI^{sto} self-intersecting states occur. In the shaded area the shape of lowest energy has yet to be determined. (Modified from U. Seifert, K. Berndl, and R. Lipowsky, *Phys. Rev. A*, 44:1182–1202, 1991.)

C. PHASE DIAGRAM OF THE SPONTANEOUS CURVATURE MODEL

The phase diagram in the spontaneous curvature model is shown in Figure 5, taken from Reference 8. Apart from the reduced volume v, it depends on the scaled spontaneous curvature

$$c_0 \equiv C_0 R \tag{12}$$

It contains large regions of stomatocytes, oblates/discocytes, prolates/dumbbells, and pear-shaped vesicles. A negative C_0 favors stomatocytes, while a positive C_0 leads to pear-shaped vesicles which are involved in the budding transition.[7,8,57,64,65] In particular, the pear shapes end up at a curve L^{pear} along which the neck diameter of the pears tends towards zero. The shapes then consist of two spheres with radii R_1 and R_2 which are connected by a narrow neck. For such an "ideal" neck, one has the "kissing condition"[7,8,57,65]

$$1/R_A + 1/R_B = C_0 \tag{13}$$

Here, R_A and R_B denote the local radii of curvatures of the two adjacent segments extrapolated through the neck, which are necessarily equal at the poles in each individual segment. This kissing condition can be phenomenologically derived by the requirement that the energy density $(2H - C_0)^2$ is the same for the two adjacent segments. As a thorough mathematical analysis shows, it also holds for more complicated budding processes where spheres and prolates are connected by this type of neck.[66] Beyond the limiting line L^{pear}, the phase diagram becomes quite complicated since further branches of shapes appear which involve a prolate connected to a sphere or shapes of multiple segments.[7] In this region, additional energy terms such as attractive van der Waals interaction can become relevant.[67,68]

Two general features of the phase diagram of the spontaneous curvature model should be kept in mind. (1) Most transitions are discontinuous; i.e., one expects large hysteresis effects. (2) Even though one cannot exclude the possibility that somewhere in the phase diagram the ground state of the vesicle is nonaxisymmetric, so far there are no indications for such a region.

D. PHASE DIAGRAM OF THE BILAYER COUPLE MODEL

In the bilayer couple model, the phase diagram depends on the reduced area difference

$$\Delta a \equiv \Delta A / 16\pi R d \tag{14}$$

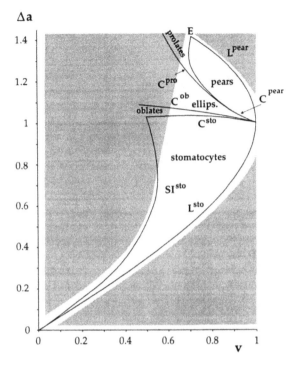

Figure 6 Phase diagram of the bilayer coupling model. This phase diagram shows the state of lowest energy as a function of the reduced area difference Δa and of the reduced volume v. C^{pear} denotes the line of continuous transitions between up-down symmetric prolate/dumbbell shapes and up-down asymmetric pear shapes. Likewise, C^{sto} denotes the locus of the continuous transitions between the oblate/discocyte shapes and the stomatocytes. L^{pear} and L^{sto} are limit curves which correspond to budding and the inclusion of a spherical cavity, respectively. In the region between the prolate/dumbbell and discocyte regime, nonaxisymmetric ellipsoids have the lowest energy. This region is separated by continuous transitions C^{pro} and C^{ob} from the corresponding axisymmetric shapes. E denotes the point where two spheres of equal radii are sitting on top of each other. Along the line SI^{sto} the two poles of the shape touch each other. In the shaded areas the shape of lowest energy has not been determined so far. (Modified from U. Seifert, K. Berndl, and R. Lipowsky, *Phys. Rev. A*, 44:1182–1202, 1991.)

and the reduced volume v. The phase diagram (Reference 8) is shown in Figure 6. Qualitatively speaking, ΔA takes over the role of the spontaneous curvature. A large ΔA promotes pear-shaped vesicles, while smaller values can lead to stomatocytes. However, two important differences to the phase diagram of the spontaneous curvature model arise. (1) All transitions found so far are continuous; i.e., there are no metastable states and there are no hysteresis effects. (2) A region of nonaxisymmetric ellipsoids emerges between the oblate and the prolate phases.[8,59,63]

E. PHASE DIAGRAM OF THE AREA-DIFFERENCE ELASTICITY MODEL

Since the area-difference elasticity model interpolates between the spontaneous curvature model and the bilayer couple model, its phase diagram also interpolates between the two phase diagrams shown above. In fact, the phase diagram of the area-difference elasticity model depends on three variables: the reduced volume v, the reduced equilibrium area difference

$$\Delta a_0 \equiv \Delta A_0 / 16\pi\,Rd \qquad (15)$$

and the parameter $\alpha = \bar{\kappa}/\kappa$ as discussed in Section 2.6. Since the phase diagrams of the two limiting cases are notably different, one can expect that the full three-dimensional phase diagram reveals a complex topology. So far, mainly two parts of this three-dimensional phase diagram have been investigated in some detail, namely the oblate/prolate transition[59] and the budding region,[45,48,49] for which we present in Figure 7 a section of the three-dimensional phase diagram in the region where budding occurs at a fixed value of $\alpha = 4$, chosen here for illustrative purposes.[49]

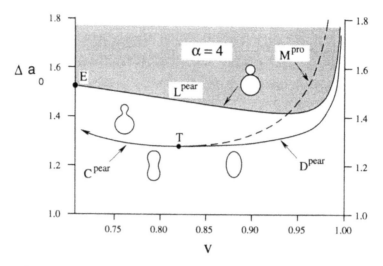

Figure 7 Phase diagram of the area-difference elasticity model in the budding region for rigidity ratio $\bar{\kappa}/\kappa = \alpha$ = 4 as a function of the equilibrium area difference Δa_0 and of the reduced volume v. The lines C^{pear} and D^{pear} denote a continuous and a discontinuous budding transition, respectively, separated by the tricritical point T. For a discontinuous transition, the prolates become infinitesimally unstable at the line M^{pro}. L^{pear} is the vesiculation curve where two spheres of different radii are connected by an ideal neck. At E the two spheres have equal size. (Modified from L. Miao et al., *Phys. Rev. E*, 49, 5389, 1994.)

Two different cases for the budding transition can be distinguished. (1) For a relatively small reduced volume v, the symmetry-breaking budding transition between the prolates and the pear-shaped vesicles is continuous. Weak pears are therefore stable, and an increase in either v or Δa_0 will progressively increase the neck diameter until at L^{pear} the limit line, or vesiculation line, is reached. Along this line the neck diameter has formally shrunk to zero. The shape then consists of two spheres sitting on top of each other. (2) For a large reduced volume the budding transition becomes discontinuous, with the line of instability extending well into the vesiculated region beyond the limiting line L^{pear}. Therefore, the theory for this value of α predicts a qualitative difference between budding at a small reduced volume and budding at a larger reduced volume.

At present the best estimate for α is $\alpha \simeq 1$, which is supported by derivations using standard elasticity theory as sketched in Section 2.6 as well as by experimental measurements.[23] For such a value, the area-difference elasticity model predicts a discontinuous budding transition irrespective of the reduced volume.

F. TEMPERATURE TRAJECTORIES

For a physical interpretation of the phase diagrams one has to relate the variables v, c_0, and m_0 to physical observables, which poses practical as well as fundamental problems. For an axisymmetric vesicle, the area and the volume can be inferred, in principle, from its contour if the orientation of the symmetry axis is known. However, thermal fluctuations as well as rotational diffusion of the vesicle limit the resolution. No direct measurements of the total mean curvature M which would yield ΔA (compare Equation 5) have been performed so far even though this quantity could also be obtained from the contour through higher derivatives.

As for the spontaneous curvature C_0 and the equilibrium area difference ΔA_0, the situation is different since these quantities cannot be measured directly. In a symmetric environment, however, one would expect that $C_0 = 0$. Likewise, according to Equation 4, the equilibrium area difference ΔA_0 depends on the number of molecules in the two monolayers, which is determined by the preparation process and thus not yet accessible to any direct measurement.

Given this somewhat unsatisfying situation, one has to be content with a knowledge of the relative variation of these parameters upon changing a physical control parameter. The most widely used parameter so far is the temperature T, which can be controlled within 0.1°C. An increase in temperature leads to a trajectory

$$\Delta a_0 = \left[v(T_0)/v \right] \Delta a_0(T_0) \tag{16}$$

parametrized by the initial point $(\Delta a_0, v) = (\Delta a_0(T_0), v(T_0))$ of the trajectory[1,8] corresponding to the initial value T_0. Such a trajectory holds for perfectly symmetric monolayers.

Even a tiny asymmetry of the order of $d/R \simeq 10^{-3}$ in the thermal expansion coefficients of the two monolayers leads to significant deviations from the form of Equation 16 for the temperature trajectory.[1,8] The physical basis for this surprising effect is the length-scale separation emphasized in Section 2. The relevant quantity for a change in shape is not only the absolute change in area, which mainly affects v, but also the variation of the area difference in the two monolayers, for which the relevant scale is of the order of dR. If the outer monolayer expands more than the inner one, the additional area accumulated in this outer layer will cause budding since the formation of buds increases the area difference. Likewise a greater increase in the area of the inner monolayer induces a transition to the discocytes and the stomatocytes.

This sensitive dependence of the thermal trajectory in the phase diagram indicates that it will be rather difficult in general to reproduce experiments on vesicle shape transformations. Presently, the available purity of the lipids does not exclude the presence of residual impurities which could result in an asymmetric expansion of the order of 10^{-3}. It would be highly beneficial to investigate this effect systematically by deliberately adding traces of a second (miscible) component to one of the monolayers. However, one then has to be aware of additional effects arising from the mixture (see Section 4).

G. COMPARISON WITH EXPERIMENTS

The theoretical phase diagrams should be compared with experiments. It turns out that the experiments on budding are still in partial conflict with the phase diagram of the ADE model for $\alpha \simeq 1$, which at present is the most plausible model from a theoretical perspective. However, the other variants of the curvature model do not yield any better agreement.

Experimentally, two apparently different scenarios for the budding transition can be distinguished. (1) For DMPC vesicles, a slow increase in temperature leads continuously to a pear with weak up-down asymmetry which upon further temperature change jumps to a vesiculated shape with a narrow neck.[1,69] (2) For SOPC vesicles in sucrose/glucose solutions, discontinuous budding is observed without the pear-shaped intermediates.[70] A similar sequence can also be obtained for DMPC vesicles provided they are kept under some tension before the heating starts.[69]

The second experimental scenario is compatible with the predictions of a discontinuous budding of the area-difference elasticity model for $\alpha \simeq 1$. To explain experimental scenario 1 within the area-difference elasticity model seems to be impossible. The apparent continuous transition from the symmetric shape to the pear with weak up-down asymmetry could be caused by a large α value, which is expected to apply for multilamellar vesicles if the bilayers are strongly coupled. However, the apparent transition from a wide neck to a narrow neck is not contained in any of the curvature models discussed so far. Therefore, even on this qualitative level, there remains a serious discrepancy between theory and experiment. Further energy terms beyond those contained in the area-difference elasticity model could, in principle, provide such a scenario.[68] However, one expects any corrections to Equation 7 arising from elastic interactions to be of the order of $d/R \simeq 10^{-3}$. This magnitude is not sufficient to cause a discontinuous transition between two different neck sizes. Whether or not the interplay of these correction terms and the large effect caused by impurities can lead to such a discontinuous transition remains to be investigated.

Another possible explanation for the discrepancy could be that the pears with weak up-down asymmetry correspond to long-lived dynamic fluctuations around a metastable shape.[49] Indeed, one expects very long relaxation times near an instability since the time scale $t_f \equiv \eta R^3/\kappa$ for long-wavelength bending fluctuations acquires an additional factor $1/(v_c - v)$ near an instability at $v = v_c$. Thus, the time scale for a shape fluctuation towards the unstable mode t_f then formally diverges due to "critical slowing down". In practice, this divergence will be cut off by the decay of the metastable shape as soon as the energy barrier becomes comparable to the thermal energy T. On the basis of this estimate, the "stable" pears would be long-lived fluctuations around a still metastable symmetric shape. Experiments to settle this issue are underway.

There are two effects which indicate that the area difference can be modified by factors other than temperature. First, Fargé and Devaux[71] showed that redistributing lipids from one monolayer to another by applying a transmembrane pH gradient induces shape transformations similar to those predicted theoretically as one increases ΔA_0. Second, the effect mentioned above that precooling leads to budding can, somewhat speculatively, also be related to changes in ΔA_0 using an intriguing suggestion by Helfrich and co-workers.[72] They proposed that the osmotically enforced flow of water through the membrane drags along lipid molecules. Since precooling also forces solvent to flow through the membrane, one could wonder whether such a treatment also causes an increase in ΔA_0. In the phase diagram, shown in Figure 7, this would shift the initial spherical vesicle upwards. A temperature trajectory starting at the sphere would then reach the budding line for a smaller temperature increase and the size of the buds should be smaller. So far, there are no systematic tests of this hypothesis. Obviously, quantitative experiments on the effect of the equilibrium area difference will be very helpful in assessing whether the area-difference elasticity model describes the main physics of these systems.

IV. LIPID MIXTURES

A. FLUID MEMBRANES CONSISTING OF SEVERAL COMPONENTS

In general, a lipid bilayer is composed of different types of molecules which may differ in their headgroups, in the length of their hydrocarbon chains, or in the number of unsaturated bonds within these chains. In such a multicomponent system, the composition can become laterally inhomogeneous within each monolayer and can be different across the two monolayers. As a result, the elastic parameters of the membrane such as the bending rigidity and the spontaneous curvature become inhomogeneous, which leads to a coupling between the composition and the shape. Moreover, there are energies associated with the composition variables such as entropy of mixing and the cost of an inhomogeneous composition profile. To categorize the various phenomena which arise from the competition between these energies and the curvature energy, it is helpful to distinguish two cases based on the phase diagram of a multicomponent bilayer.

Usually, the phase diagram of a multicomponent bilayer exhibits a homogeneous one-phase region at high temperature and a two-phase coexistence region at lower temperatures for a certain range of compositions.

In the two-phase region, domains are formed in which the lipid composition is different from the surrounding matrix.[73,74] The edges of these domains are characterized by an edge or line tension. Since the length of the domain boundary decreases if the domain buds, the competition between this edge tension and the curvature energy leads to *domain-induced budding*.[9,40,75] This budding process is discussed in Section 4.2 both for a flat membrane matrix and for vesicles.

Quite generally, these domains seem to be an essential feature of the spatial organization of biological membranes. For example, the plasma membranes of most cells contain specialized regions such as coated pits or various types of cell junctions. These domains represent an important theme of current research in structural biology.[76]

In the one-phase region, the ground state is a flat and laterally homogeneous membrane in the absence of a spontaneous curvature. However, any inhomogeneity in the composition, either laterally within a monolayer or between the two monolayers, induces a local spontaneous curvature if the two lipid species have a different molecular geometry. For an almost planar membrane, this leads to a coupling between bending fluctuations and composition fluctuations which decreases the bending rigidity.[77,78] For a nonspherical vesicle, this coupling between shape and composition causes *curvature-induced lateral phase segregation*,[10] as discussed in Section 4.3.

In the following section we focus on fluid lipid bilayers. Even though the mixture of two lipid components often leads to the coexistence of a fluid and a gel phase, there are several examples of the coexistence of two fluid phases. An especially important one is provided by mixtures of phospholipids and cholesterol, as has been established quite recently.[79-81] Fluid-fluid coexistence also occurs in the binary mixture of diethyl pyrocarbonate (DEPC) and DPPE[82] and in mixtures with partially unsaturated alkyl chains.[83]

Domain-induced budding and curvature-induced phase segregation as considered here must be distinguished from related but different phenomena such as striped phases,[84,85] spontaneous formation of small vesicles in mixtures of oppositely charged surfactants where phase separation occurs across the membrane,[86-89] and shape transformations induced by a coupling between a local in-plane order and the membrane normal for smectic-C vesicles.[90]

B. DOMAIN-INDUCED BUDDING
1. Phase Separation within Bilayers

The two coexisting phases of the lipid mixture will be denoted by α and β. Now consider a membrane which is initially prepared in a homogeneous state within the one-phase region and is then quenched into the ($\alpha\beta$) two-phase region. This leads to phase separation within the membrane, which can proceed in two different ways depending on the "depth" of the quench. (1) The membrane is quenched deep into the spinodal decomposition regime. In such a situation, many small domains will be formed initially. In general, these domains may grow via different aggregation mechanisms. In this way, one may finally attain a state in which the bilayer consists of relatively large domains. (2) The membrane is quenched into the nucleation regime. If the activation energy for the "critical" domain is sufficiently large, only one domain will be nucleated initially, and one may study the diffusion-limited growth of such a single domain.

2. Edge Energy versus Bending Energy

The edge of an intramembrane domain has an energy which is proportional to the edge length. Therefore, the domain has a tendency to attain a circular shape in order to minimize its edge energy.

The line tension, Σ_e, is equal to the edge energy per unit length. Its magnitude can be estimated as follows.[9] First, consider a domain in the lipid bilayer which extends across both monolayers. In this case, the edge of the domain represents a cut across the whole bilayer. The cross section of such a cut consists of three distinct regions: two hydrophilic headgroup regions of combined thickness $\simeq 1$ nm and an intermediate hydrophobic tail region of thickness $\simeq 4$ nm. These two regions can have distinct interfacial free energies per unit area. For three-dimensional fluid phases, a typical value for the interfacial free energy is $\simeq 10^{-2}$ J m^{-2}. If one assumes that this value is also applicable to the headgroup region and that the latter region gives the main contribution to the line tension, one obtains the crude estimate $\Sigma_e \simeq 10^{-17}$ J/μm. For a domain which extends only across one monolayer, the line tension is reduced by a factor of one half. For monolayers composed of a phospholipid-cholesterol mixture, experimental studies of the domain shape gave the estimate $\Sigma_e \simeq 10^{-18}$ J/μm.[91]

A *flat* domain will form a circular disk in order to attain a state with minimal edge length. For a circular domain with radius L, the edge energy, F_e, is given by

$$F_e \equiv 2\pi L \Sigma_e \qquad (17)$$

However, as far as the edge energy is concerned, a *flat* circular disk does *not* represent the state of lowest energy since the length of the edge can be further reduced if the domain forms a bud: the domain edge now forms the neck of the bud, and this neck narrows down during the budding process (see Figure 8).

Budding involves an increase in the curvature and thus in the bending energy of the domain. Therefore, the budding process of fluid membranes is governed by the competition between the bending rigidity κ of the domain and the line tension Σ_e of the domain edge. This competition leads to the characteristic *invagination length*, $\xi \equiv \kappa/\Sigma_e$.[9] Using the typical values $\kappa \simeq 10^{-19}$ J and $\Sigma_e \simeq 10^{-17}$ J/μm, one obtains $\xi \simeq 10$ nm for domains across the bilayer. For phospholipid-cholesterol mixtures, this length scale seems to be much larger. On the one hand, the bending rigidity κ was experimentally estimated to have the

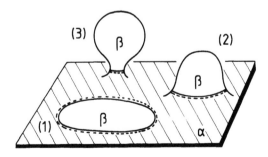

Figure 8 Budding of the membrane domain β embedded in the membrane matrix α. The domain edge is indicated by the full-broken line. The length of this edge decreases during the budding process from 1 to 3. (Modified from R. Lipowsky, *J. Phys. II (Paris)*, 2:1825–1840, 1992.)

relatively large value $\kappa \simeq 4 \times 10^{-19}$ J.[29] The line tension, on the other hand, seems to have the relatively small value $\Sigma_e \simeq 10^{-18}$ J/μm as mentioned above. This implies the invagination length $\xi \simeq 400$ nm. It will become clear in the following that the invagination length ξ provides the natural length scale for shape transformations of the domain.

3. A Simple Model

The competition between the edge and bending energies can be understood in the framework of a relatively simple model in which one assumes that the membrane matrix is flat and that the membrane domain forms a spherical cap with radius R. If the domain has surface area $A = \pi L^2$, mean curvature $C = 1/R$, and spontaneous curvature $C_{sp} = C_0/2$, its total energy $F = F_{SC} + F_\Sigma$ has the form

$$F/2\pi\kappa = \left(LC - LC_{sp}\right)^2 + \left(L/\xi\right)\sqrt{1-(LC/2)^2} \tag{18}$$

For $C_{sp} = 0$, such a model has also been studied in order to discuss the size of vesicles generated by sonification and to study the closure of open fluid membranes.[92-94] The energy F has several minima and maxima as a function of the reduced curvature LC.

Now consider a domain which is characterized by a fixed spontaneous curvature C_{sp} and fixed invagination length $\xi = \kappa/\Sigma_e$. The domain size L, on the other hand, changes with time by the diffusion-limited aggregation of molecules within the membrane and thus plays the role of a control parameter for the budding process.

For small L, the domain forms an incomplete bud corresponding to the minimum of F at small LC values. As L grows, the edge of the domain becomes longer and the energy of the incomplete bud is increased. At a certain critical size $L = L^*$, the incomplete bud has the same energy as the complete bud corresponding to a complete sphere, but both states are separated by an energy barrier. For the parameter values considered here, the energy barrier is typically large compared to the thermal energy $\simeq T$. In this case, the domain continues to grow in the incomplete bud state up to the limiting size $L = L^o$ at which this state becomes *unstable*. The simple model considered here leads to[9]

$$L^o = 8\xi / \left[1 + \left(4\xi|C_{sp}|\right)^{2/3}\right]^{3/2} \tag{19}$$

with $\xi = \kappa/\Sigma_e$ as before. Thus, as soon as the domain has grown to $L = L^o$, it *must* undergo a budding process.

4. Domain-Induced Budding of Vesicles

The same type of shape instability is found from a systematic minimization procedure for the energy of closed vesicles composed of two types of domains, α and β.[40] This energy consists of (1) the bending energies $F_{SC}^{(\alpha)}$ and $F_{SC}^{(\beta)}$ of the α and the β domains, respectively, and (2) the edge energy $F_e^{(\alpha\beta)}$ of the $(\alpha\beta)$ domain boundaries:

$$F = \oint dA^\alpha \frac{1}{2}\kappa^\alpha \left(2H - 2C_{sp}^\alpha\right)^2 + \oint dA^\beta \frac{1}{2}\kappa^\beta \left(2H - 2C_{sp}^\beta\right)^2 + \oint d\ell\, \Sigma_e \tag{20}$$

as appropriate for a surface consisting of α and β domains, where the last integral represents the line integrals along the domain boundaries.

The total energy F is minimized for given values of the domain areas A^α and A^β, of the pressure difference P or of the enclosed volume V, and of the line tension Σ_e. The area ratio

$$x \equiv A^\beta / \left(A^\alpha + A^\beta\right) \tag{21}$$

with $0 \le x \le 1$ is a measure for the size of the domain and again plays the role of a control parameter for budding.

Now assume that the vesicle contains two large domains, α and β, the sizes of which no longer change with time. This corresponds to the final state after the phase separation process has been completed. On

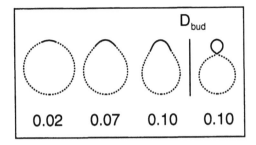

Figure 9 Vesicle shapes as a function of the relative domain area x for reduced line tension $\sigma_e = 7$. D_{bud} denotes a discontinuous transition between an incomplete and a complete bud. The shapes are axisymmetric; the α and the β domains correspond to the broken and the full contours, respectively. (Modified from F. Jülicher and R. Lipowsky, *Phys. Rev. Lett.*, 70, 2964–2967, 1993.)

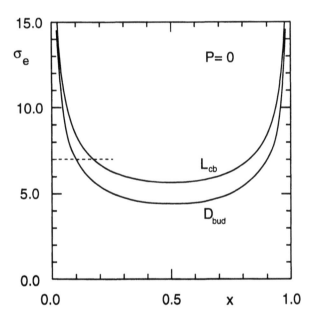

Figure 10 Phase diagram for domain-induced budding of a vesicle for pressure $P = 0$ across the membrane as a function of the reduced line tension σ_e and the relative domain area x. The vesicle undergoes a discontinuous budding transition along the line D_{bud} and attains a limit shape at L_{cb} with an infinitesimal neck connecting the bud to the vesicle. The dashed trajectory corresponds to Figure 9. (Modified from F. Jülicher and R. Lipowsky, *Phys. Rev. Lett.*, 70, 2964–2967, 1993.)

long time scales, water can permeate through the membrane, which leads to zero pressure difference, $P = 0$, provided the aqueous solution does not contain large molecules which lead to an osmotic pressure.

The equilibrium shapes of such a vesicle are shown in Figure 9. For simplicity, the α and the β domain are taken to have identical bending rigidities ($\kappa^\alpha = \kappa^\beta$) and to have no spontaneous curvature (i.e., $C_{sp}^\alpha = C_{sp}^\beta = 0$).

As shown in Figure 10, the corresponding phase diagram exhibits a line D_{bud} of discontinuous budding transitions and a line L_{cb} of limit shapes with an infinitesimal neck. The dashed line in Figure 10 corresponds to a vesicle with reduced line tension $\sigma_e = 7$ where

$$\sigma_e \equiv \Sigma_e R / \kappa^\beta = R / \xi \tag{22}$$

58

The corresponding energy and equilibrium shapes are shown in Figure 9 as a function of x. In practice, the infinitesimal neck should have a diameter which is of the order of the membrane thickness, $a \approx 5$ nm. The energy required to break such a neck is $2\pi a \Sigma_e$ and the time for thermally activated fission is $\sim \exp(2\pi a \Sigma_e/T)$.

If the vesicle membrane is composed of a phospholipid-cholesterol mixture, the invagination length seems to be $\xi \approx 400$ nm, as mentioned above, and the sequence of shape transformations shown in Figure 9 would then correspond to the vesicle size $R = 7\xi \approx 2.8$ μm, which is directly accessible to optical microscopy.

C. CURVATURE-INDUCED LATERAL PHASE SEGREGATION
1. Coupling between Composition and Shape
Even if there is no genuine phase separation in the flat membrane, the shape transformation of the closed vesicle can induce phase segregation within the membrane.[10] Suppose an initially spherical vesicle is subjected to a temperature increase. This change necessarily leads to deviations from the spherical shape and, thus, to a position-dependent curvature which induces a position-dependent composition. For a quantitative description, we introduce the composition (area fractions) of lipid A in the individual monolayers $x_A^i (=1 - x_B^i; i = in, out)$ and deviations $\delta x_A^i(s_1,s_2) \equiv x_A^i(s_1,s_2) - \bar{x}_A$ from the mean value $\bar{x}_A$. If this local deviation is different in the two monolayers, a local spontaneous curvature is induced according to

$$C_0(s) = \lambda\left[\delta x_A^{out}(s) - \delta x_A^{in}(s)\right] + \overline{C}_0 \equiv \lambda\phi(s) + \overline{C}_0 \tag{23}$$

where the phenomenological coupling constant λ has the dimensions of an inverse length.[77] For a rough estimate of its magnitude, assume that the lipid A has a cone-like shape with a radius of curvature R_A while lipid B has a cylindrical shape. The coupling constant λ is then of the order of $1/R_A$. Since R_A is determined by typical molecular dimensions, we will use $\lambda = 0.1$ nm for an estimate. We also allow for a systematic spontaneous curvature $\overline{C}_0$ which arises if the mean compositions $\bar{x}_A^i$ are different in the two monolayers.

2. Energy of a Two-Component Vesicle
The bending energy, F_1, of the two-component vesicle is then chosen as a generalization of the bending energy of a single-component vesicle as given by Equation 7. This leads to[10]

$$F_1 = (\kappa/2)\oint dA\left\{2H(s_1,s_2) - C_0\left[\phi(s_1,s_2)\right]\right\}^2 + \left[\pi\overline{\kappa}/(8Ad^2)\right](\Delta A - \Delta A_0)^2 \tag{24}$$

We have assumed for simplicity that neither the bending rigidities κ and $\overline{\kappa}$ nor the area of the vesicle, A, and area difference ΔA_0 depend on the composition.

Since the membrane does not show genuine phase separation, there is a free energy associated with the deviation of the composition from its mean value. For small deviations, this energy can be written in the form

$$F_2 = (\kappa/2)\epsilon \oint dA\left[\phi^2 + (\xi_c\nabla\phi)^2\right] \tag{25}$$

Here ξ_c is the correlation length for composition fluctuations, ∇ is the covariant gradient operator, and ϵ is a molecular energy, estimated below, divided by the bending rigidity.

Since the typical length scale for shape variations of large vesicles is in the micrometer range, while the typical correlation length ξ_c will be of the order of nanometers, the gradient term in F_2 will be, in general, much smaller than the ϕ^2 term and, thus, can be ignored. Under the constraint $\oint dA\phi = 0$, the total energy $F \equiv F_1 + F_2$, then becomes minimal for a composition profile $\phi(s_1,s_2)$ given by

$$\phi(s_1,s_2) = \frac{2\lambda}{\lambda^2 + \epsilon}\left[H(s_1,s_2) - \Delta A/(4dA)\right] \tag{26}$$

which shows that the local composition follows the deviation of the mean curvature $H(s_1,s_2)$ from its average value $\Delta A/(4dA)$. After inserting Equation 26 into F, the total energy becomes equivalent to the

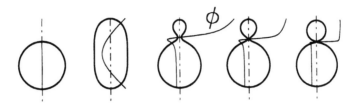

Figure 11 Curvature-induced lateral phase segregation. The spherical vesicle changes its shape as the reduced volume v decreases due to an increase in temperature. The thin curves show the composition ϕ. The reduced volume is given by $v = 1.0, 0.89, 0.86$, and 0.82 from left to right. At $v = 0.89$ the symmetric and the asymmetric shape have the same energy, indicating a discontinuous budding transition. The vesiculation line is reached with the last shape. (Modified from U. Seifert, *Phys. Rev. Lett.*, 70:1335–1338, 1993.)

energy W, as given by Equation 7, of a single-component vesicle with renormalized parameters κ, $\bar{\kappa}$, and ΔA_0.[10] The main physical effect is that the effective bending rigidity decreases for a two-component system, as has been previously derived.[77]

Once this mapping has been obtained, both the knowledge of the phase diagram of the area-difference elasticity model and shape calculations within this model can be used to obtain results for the two-component system. As an illustrative example, consider the thermal evolution of an initially spherical vesicle (with a homogeneous composition profile $\phi(s) = 0$) with increasing temperature (see Figure 11, taken from Reference 10). With increasing temperature, the reduced volume decreases and the shape becomes more prolate. The inhomogeneous curvature then induces a nontrivial composition profile along the contour. In the outer monolayer, the A molecules are enriched at the poles (if their enhancement in the outer layer leads to a positive spontaneous curvature, i.e., if $\lambda > 0$) while the B lipids are enriched along the equatorial region of the vesicle. For smaller v, the up/down asymmetric shapes have lower energy, leading to a discontinuous budding transition. These shapes finally end up at the vesiculation point. In the vesiculated state, the composition within each sphere becomes homogeneous again, with all the variation of the composition occurring in the neck. Thus, the shape change, i.e., in this case budding and vesiculation, leads to phase segregation.

3. Magnitude of Curvature-Induced Phase Segregation

A crude estimate for the magnitude of the curvature-induced phase segregation can be obtained as follows.[10] As explained above, a typical value for the coupling λ might be $\lambda \simeq 0.1/$nm. For the free energy density coefficient ϵ, one estimates $\epsilon \simeq \tau T/\kappa a^2$, where a is a molecular length of $\simeq 1$ nm. The reduced temperature $\tau \equiv (T - T_C)/T_C$ is the distance to the A-B critical point which separates the one-phase region from the two-phase coexistence region of the lipid mixture. For a mixture of DEPC and DPPE, this critical point is at $T_c \simeq 340$ K according to Reference 82. For a vesicle at a temperature of 10°C above T_c, τ becomes 0.03. This value leads to $\phi \simeq 20[H - \Delta A/(4dA)]$ nm using $T/\kappa \simeq 1/25$. With $[H - \Delta A/(4dA_0)] \simeq 1/R$, where R is the radius of the vesicle, the typical variation in the composition becomes of the order of 1% for vesicles with a radius $R \simeq 1$ µm but 10% for $R \simeq 100$ nm. The smaller the vesicles are the larger becomes the phase segregation.

Very close to the critical point, the argument given above that the correlation length is small compared to the size of the vesicle is no longer valid. In fact, using $\xi \simeq a/\tau^\nu$, where $a \simeq 1$ nm and $\nu = 1$ is the critical exponent of the two-dimensional Ising model, one obtains $\tau_* \simeq a/R$ as a crossover temperature. For $\tau \lesssim \tau_*$, the gradient term in F_2 can no longer be neglected.

V. ADHESION

A. CONTACT POTENTIAL AND CONTACT CURVATURE

A vesicle near a wall or substrate experiences various forces, such as electrostatic, van der Waals, and hydration forces. The typical range of these forces is several nanometers, which is small compared to the size of a large vesicle, which is of the order of several microns. Thus, in a first step, the microscopic interaction potential may be replaced by a contact potential with strength W if we are mainly interested in the gross features of such a vesicle bound to the wall.[12]

There are several experimental methods by which one can infer information about the effective interaction $V_{eff}(l)$ between two members or between a membrane and another surface which are separated by a thin liquid film of thickness l. These methods include X-ray diffraction on oriented multilayers,[95] the surface force apparatus,[96] micropipet aspiration,[97] and video microscopy of dilute systems.[98] The contact potential W is then identified with $|V_{eff}(l_0)|$, where l_0 is the mean separation of the membrane in the absence of an external pressure. The value of $W \simeq |V_{eff}(l_0)|$ depends on the strength of the shape fluctuations, which can be reduced by lateral tension or by immobilizing the membranes on solid surfaces. Therefore, the different experimental methods give, in general, different values for W which are in the range between 10^{-4} to 1 mJ/m^2. For rough estimates in this section, we will use the value of 1 mJ/m^2 for *strong* adhesion and the value of 10^{-4} mJ/m^2 for *weak* adhesion.

If the vesicle and the wall have the contact area A^*, the vesicle gains the adhesion energy

$$F_a = -WA^* \tag{27}$$

which must be added to the curvature energy. The balance between the gain in adhesion energy and the cost in curvature energy for a bound vesicle gives rise to several interesting phenomena.[12] A first manifestation of this balance is a condition for the contact curvature $1/R_1^*$ which is determined by[12]

$$1/R_1^* = (2W/\kappa)^{1/2} \tag{28}$$

The contact angle is necessarily π since any sharp bend would have an infinite curvature energy. This implies that the membrane is only curved in one direction and $1/R_2^* = 0$ along the line of contact. Therefore, the same boundary condition (Equation 28) also holds for a cylindrical geometry for which it also applies to polymerized membranes or solid-like sheets.[99]

The universal condition (Equation 28) for the contact curvature holds irrespective of the size of the vesicle and the nonlocal energy contribution. Thus, a measurement of $1/R_1^*$ yields a value of the contact potential once the bending rigidity κ is known. The range of contact curvatures to be expected from this relation spans from $1/R_1^* \simeq 1/10$ nm for strong adhesion (using $\kappa = 10^{-19}$ J) to $1/R_1^* \simeq 1$ μm^{-1} for weak adhesion. The latter value clearly is accessible by light microscopy. In fact, a measurement of the contact curvature has recently been performed using reflection interference contrast microscopy.[16]

B. ADHESION TRANSITION
1. Bound Shapes
Solving the shape equations for axisymmetric shapes with the boundary condition (Equation 28) leads to a variety of bound shapes which can be arranged in a phase diagram as in the case of free vesicles. The basic physics behind the competition between adhesion and curvature energy already becomes evident in the simplest nontrivial ensemble, which contains only the local bending energy with $C_0 = 0$, the adhesion energy, and a constraint on the total area. This corresponds to a situation where the volume can adjust freely. With decreasing strength of the contact potential W, the area of contact A^* also decreases and vanishes for $W = W_a$, with

$$W_a = 2\kappa / R^2 \tag{29}$$

At this value, the bound shape resembles the free shape corresponding to the same constraint, which is a sphere, except for the fact that the contact curvature $1/R_1^* = 2/R$ is *twice* the curvature of the sphere. However, the contact *mean* curvature $H^* \equiv (1/R_1^* + 1/R_2^*)/2 = 1/R$ is equal to the mean curvature of the sphere. For $W < W_a$, an attractive potential does not lead to a bound shape with finite area of contact. Thus, the vesicle undergoes a continuous adhesion transition at $W = W_a$.[12]

A somewhat more complex situation arises when, in addition, the enclosed volume is also kept constant. The phase diagram becomes two-dimensional and depends on v and the reduced potential strength

$$w \equiv WR^2/\kappa \tag{30}$$

This phase diagram is shown in Figure 12 together with some bound shapes.[13,100,101] Its main characteristic is the line of adhesion transitions $w_a(v)$ which separates bound from free states.

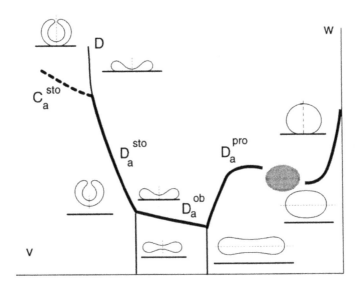

Figure 12 Schematic phase diagram with free and bound shapes at constant area and volume. The heavy lines show the adhesion transition at $W = W_a$, which can be discontinuous (D_a^{pro}, D_a^{ob}, and D_a^{sto}) or continuous (C_a^{sto}). In the dashed region, nonaxisymmetric bound shapes are relevant. The dashed straight lines across the shapes denote the axis of symmetry. (Modified from U. Seifert and R. Lipowsky, In D. Beysens, N. Boccara, and G. Forgacz, Eds., *Dynamical Phenomenon at Interfaces, Surfaces, and Membranes*, Nova Science, New York, 1991, p. 295–304.)

The four-dimensional phase diagram for adhesion, including the full area-difference elasticity energy, has not yet been studied. In analogy to the free case, one expects that the nonlocal energy favors continuous adhesion transitions and the occurrence of nonaxisymmetric shapes.

2. Possible Experimental Verification
Two very different approaches are conceivable in order to observe the adhesion transition experimentally. (1) Changing the temperature will affect both the reduced volume and the scaled adhesion potential $w = W R^2/\kappa$ via the area expansion. A temperature decrease also decreases w and increases v. Therefore, a bound vesicle may become free upon cooling provided its initial state at the higher temperature is already sufficiently close to the adhesion transition. Likewise osmotic deflation or inflation which does not affect w can induce a crossing of the adhesion transition in the phase diagram. (2) A more indirect but quite elegant confirmation of the theory described above could make use of the characteristic size-dependence of the adhesion transition as expressed in Equation 29. This relation implies that, for fixed W, in an ensemble of vesicles only those vesicles with $R > R_a \equiv (2\kappa/W_a)^{1/2}$ are bound to the substrate.

3. Beyond the Contact Potential
Further insight into the character of the adhesion transition arises from two ramifications of the simple picture presented so far. First, consider the influence of the small but finite range of the potential.[102,103] We assume that the potential can be characterized by a depth W and a range Z_0. It turns out that the free state discussed above for $W < W_a$ corresponds for such a potential to a pinned state where the vesicle adheres to the wall but keeps its free shape.[13,103] The energy gain associated with such a pinned state is

$$\Delta F \sim -W R Z_0 \tag{31}$$

for small Z_0/R. Second, we have to compare this energy with the thermal energy $\sim T$ to decide whether it is sufficient to pin the vesicle to the substrate. This comparison leads to a transition between the pinned state and the free state at

$$W_p \sim T/(R Z_0) \quad \text{for} \quad Z_0 \ll R \lesssim R_c \equiv (\kappa/T)Z_0 \tag{32}$$

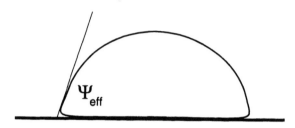

Figure 13 Bound shape for strong adhesion with effective contact angle Ψ_{eff}. The shape resembles a spherical cap except close to the line of contact. (Modified from U. Seifert and R. Lipowsky, *Phys. Rev. A*, 42:4768–4771, 1990.)

The length scale R_c arises from the consistency requirement that $W < W_u$, which was assumed when using the estimate (Equation 31) for ΔF. The breakdown of this relation for $R > R_c$ indicates that large vesicles will not enter the pinned regime because the energy gain ΔF of such a pinned state would be smaller than the thermal energy T. These large vesicles unbind at values of the potential depth W for which the analysis at $T = 0$ predicts bound vesicles with a finite contact area. The critical strength for the unbinding of these large vesicles is bounded above by the critical strength

$$W_u \sim T^2 / \left(\kappa Z_0^2 \right) \tag{33}$$

for the unbinding of open, almost planar membranes.[104,105] If the enclosed volume does not change during the adhesion process, an effective tension builds up which will decrease the value of the critical strength for unbinding. A refined theory of the unbinding of large vesicles which takes into account the fluctuations in the presence of constraints has yet to be worked out.

The crossover length R_c which separates the two qualitatively different regimes of unbinding can be estimated as follows. With the typical values $Z_0 = 4$ nm and $\kappa/T = 25$, we find $R_c = 0.1$ μm, which is just below optical resolution. It might, however, be shifted to larger values using multilamellar vesicles because κ is proportional to the number of bilayers.

C. EFFECTIVE CONTACT ANGLE FOR STRONG ADHESION

The adhesion transition takes place for $R = R_a = (w_a \kappa/W)^{1/2}$, where w_a is a numerical coefficient of $O(1)$ which depends on the constraints. For large vesicles with $R >> R_a$, i.e., for strong adhesion $W >> W_a$, the shape of the bound vesicle approaches a simple limit shape. If only the area is constrained, this limit shape is a pancake with an energy[100]

$$F \approx -2\pi W R^2 + 2\pi g (2\kappa W)^{1/2} R \tag{34}$$

with the dimensionless coefficient $g \simeq 2.8$. If in addition the volume is constrained, the vesicle becomes a spherical cap for strong adhesion, as shown in Figure 13. In both cases, an *effective contact angle* Ψ_{eff}

$$W = \Sigma \left(1 + \cos \Psi_{eff} \right) \tag{35}$$

where Σ is the (numerical) value of the Lagrange multiplier for the area constraint. For the pancake, one has $\Psi_{eff} = 0$.

For strong adhesion, the energetic competition which determines the conformation does not involve the balance between curvature energy and the adhesion energy, but rather the balance between the adhesion energy and an elastic stretching of the membrane. This elastic energy reads

$$F_k \equiv (k/2)(A - A_0)^2 / A_0 \tag{36}$$

where k is the area compressibility modulus of the order of 10^2 mJ/m^2 and A_0 is the equilibrium area. Such an extended model leads to the same shape equations as the model with a hard area constraint. Even the boundary condition (Equation 28) remains unchanged. The phase diagram, however, changes due to the additional energy. Balancing the adhesion energy with the stretching term leads to the adhesion-induced stretching of the order of

$$\left(A - A_0\right)/A_0 \simeq W/k \tag{37}$$

The relative area expansion $(A - A_0)/A_0$ thus becomes 0.01 for strong adhesion. Identifying $k(A - A_0)/A_0$ with an elastic tension Σ_{el}, we find the relation $\Sigma_{el} \sim W$ for strong adhesion. As soon as W exceeds the lysis tension, which is of the order of 1 to 5 mJ/m^2, the bound vesicle ruptures.

D. ADHESION-INDUCED FUSION AND RUPTURE

So far, an isolated vesicle at a wall has been considered. If more and more bound vesicles cover the wall, they will come into contact and may fuse. For free vesicles, fusion of two vesicles with equal area $A \equiv 4\pi R^2$ (but no constraint on the volume) leads to a gain in energy $\Delta F_{fv} = 8\pi\kappa + 4\pi\kappa_G$. If two bound vesicles fuse, the gain in energy is always larger and satisfies $\Delta F_{bv} \geq \bar{g}\pi\kappa + 4\pi\kappa_G$ with $\bar{g} \simeq 8.3$.[100] For large R, this energy gain behaves as $\Delta F_{bv} \approx 4\pi g(\sqrt{2}-1)(\kappa W)^{1/2}R$, where Equation 34 has been used. Thus, adhesion favors fusion.

As the size of the fused vesicle increases, its shape becomes more like the shape of a pancake. If the elastic tension exceeds the threshold for lysis, the pancake ruptures and becomes an open bound disk. The threshold for lysis typically occurs for $(A - A_0)/A_0 \simeq 0.03$. From Equation 37 one then derives that the adhesion potential has to be stronger than $0.03 k$ for adhesion-induced rupture to happen. Note, however, that this argument provides only an upper limit on the adhesion energy required for rupture since it neglects the fact that in the pancake conformation rupture may happen more easily along the strongly curved rim.

After the bound vesicle has ruptured, its conformation becomes an open disk. Such a bound disk has an energy[100]

$$F_{bd} = -4\pi W R^2 + 4\pi\Sigma_e R \tag{38}$$

where Σ_e is the edge tension along the circumference of the bound disk. A comparison of the bound disk energy (Equation 38) with the energy of a pancake (Equation 34) shows that for $R \gg R_{bd} \equiv \Sigma_e/W$ the bound disk always has lower energy (irrespective of the value of the other parameters). For phospholipid bilayers, we find with the typical value $\Sigma_e = 5 \times 10^{-20}$ J/nm the length scale $R_{bd} = 50$ nm for strong adhesion and $R_{bd} = 500$ μm for weak adhesion.

A recent experiment has shown that a lamellar structure can form at the air-water interface of a vesicle suspension.[106] The energetic considerations discussed above immediately lead to a scenario where vesicles adhere to the wall, fuse at the wall, and rupture. Finally, the open disks will also fuse, thus forming a bilayer parallel to the wall. The same experiment has also revealed that the activation barriers involved in these processes depend sensitively on temperature.

VI. A BRIEF PERSPECTIVE ON RELATED SYSTEMS

In this chapter we have focused on lipid vesicles in their fluid state. Their various shapes and shape transformations arise from the energetic competition between the curvature energy and (1) the geometrical constraints (for free single-component systems), (2) the line tension of domains (for domain-induced budding), or (3) the coupling between curvature and local composition (for curvature-induced phase segregation), or (4) the adhesion energy (for bound vesicles).

Thus, the concept of bending elasticity introduced some 20 years ago leads both to a qualitative description of the wide range of vesicle phenomena observed experimentally and to quantitative theoretical predictions which could inspire future experiments. The emphasis in future work should be on detailed comparisons between experiment and theory.

Similar concepts with suitable modifications can be and have been applied to related systems. We close with a brief and subjective selection of some of these topics.

As a consequence of the conformal invariance of the local curvature energy, vesicles of nonspherical topology exhibit qualitative new behavior such as a large region of nonaxisymmetric ground states for vesicles with one hole[107,108] and even a degeneracy of the ground state which leads to a "conformal diffusion" for shapes with at least two holes,[109] as discussed in detail by Fourcade et al. in Chapter 3 of this book.

We focused on the calculation of the shape of lowest energy. Since membranes are very soft, most of these shapes undergo thermal flickering. For quasi-spherical shapes, the mean square amplitudes of these fluctuations can be calculated and used to extract the bending rigidity from measurements, as mentioned in Section II.C. For a nonspherical mean shape, a determination of the fluctuations is highly nontrivial, and work in this direction is scarce.[110]

Dynamic shape fluctuations of these membranes are overdamped due to the surrounding viscous fluid.[111] For quasi-spherical vesicles, the experimentally observed relaxation times for the long-wavelength modes are in the range of seconds, which fits with a simple theoretical picture where the dissipation is in the bulk fluid only.[27] A more refined theory, which takes into account the coupling between the local density of lipids and the local curvature of the membrane, shows that friction between the chains[112] is an important contribution to the dissipation and indeed dominates the relaxation times at a wavelength below roughly 1 μm.[113]

Dynamic fluctuations of bound membranes can be measured with reflection interference microscopy.[15,33] Theoretical work on the dynamics of bound membranes[114,115] reveals several crossover length scales in the dispersion relation which contains important information about the interaction parameters between membrane and substrate.

A different class of membranes includes polymerized membranes, membranes in the gel state, and membranes with an attached polymer network, such as the red blood cell membrane.[116] Due to the fixed connectivity within these membranes, they can sustain shear deformations. Moreover, since bending typically involves stretching and shearing of the membrane, the in-plane displacements, i.e., phonons, are coupled to the bending modes. This coupling leads to an increase of the effective bending rigidity for long-wavelength bending modes.[117,118]

The quest to understand the shape of the red blood cell and its transformations has indeed been one of the motivations for the early work on vesicles. However, it is not yet even clear which role the network plays in determining the resting shape of the red blood cell. For a theoretical calculation of the shape of minimal energy, one would need to specify the stress-free conformation of the network. On a more local scale, one expects that the flickering of such a compound membrane should exhibit a crossover scale L_*.[119,120] For wavelengths $L < L_*$ the shape fluctuations are fluidlike, whereas the finite shear modulus of the network only affects the fluctuations of wavelengths $L > L_*$. The crossover length has been estimated both from the nonlinear elastic terms for almost planar membranes[119] and from the linear theory of closed vesicles[120] and yields a value which is somewhat larger than the mesh size of the network. This estimate is consistent with the naive expectation that the shape fluctuations of the red blood cell are governed by bending rigidity alone below this mesh size.

A comprehensive theoretical model for the red blood cell membrane has not yet emerged. Steps in such a direction are provided by a continuum theory in which the network is modeled as an ionic gel[121,122] as well as by recent computer simulations.[123,124] Whether computer simulations of more sophisticated membrane structures, more advanced elasticity theories, or a combination of both will finally be most useful in order to understand this membrane remains to be seen.

REFERENCES

1. **K. Berndl, J. Käs, R. Lipowsky, E. Sackmann, and U. Seifert.** Shape transformations of giant vesicles: extreme sensitivity to bilayer asymmetry. *Europhys. Lett.*, 13:659–664, 1990.
2. **P. B. Canham.** The minimum energy of bending as a possible explanation of the biconcave shape of the human red blood cell. *J. Theoret. Biol.*, 26:61–81, 1970.
3. **W. Helfrich.** Elastic properties of lipid bilayers: theory and possible experiments. *Z. Naturforsch.*, 28c:693–703, 1973.
4. **H. J. Deuling and W. Helfrich.** The curvature elasticity of fluid membranes: a catalogue of vesicle shapes. *J. Phys. (Paris)*, 37:1335–1345, 1976.
5. **E. A. Evans.** Bending resistance and chemically induced moments in membrane bilayers. *Biophys. J.*, 14:923–931, 1974.
6. **S. Svetina and B. Žekš.** Membrane bending energy and shape determination of phospholipid vesicles and red blood cells. *Eur. Biophys. J.*, 17:101–111, 1989.
7. **L. Miao, B. Fourcade, M. Rao, M. Wortis, and R. K. P. Zia.** Equilibrium budding and vesiculation in the curvature model of fluid lipid vesicles. *Phys. Rev. A*, 43:6843–6856, 1991.

8. U. Seifert, K. Berndl, and R. Lipowsky. Shape transformations of vesicles: phase diagrams for spontaneous-curvature and bilayer-coupling models. *Phys. Rev. A,* 44:1182–1202, 1991.

9. R. Lipowsky. Budding of membranes induced by intramembrane domains. *J.Phys. II (Paris),* 2:1825–1840, 1992.

10. U. Seifert. Curvature-induced lateral phase segregation in two-component vesicles. *Phys. Rev. Lett.,* 70:1335–1338, 1992.

11. H.-G. Döbereiner, J. Käs, D. Noppl, I. Sprenger, and E. Sackmann. Budding and fission of vesicles. *Biophys. J.,* 65:1396–1403, 1993.

12. U. Seifert and R. Lipowsky. Adhesion of vesicles. *Phys. Rev. A,* 42:4768–4771, 1990.

13. U. Seifert and R. Lipowsky. Adhesion and unbinding of vesicles. In D. Beysens, N. Boccara, and G. Forgacs, Eds., *Dynamical Phenomena at Interfaces, Surfaces, and Membranes,* Nova Science, New York, 1991, p. 295–304.

14. E. A. Evans. Analysis of adhesion of large vesicles to surfaces. *Biophys. J.,* 31:425–432, 1980.

15. A. Zilker, H. Engelhardt, and E. Sackmann. Dynamic reflection interference contrast (RIC-) microscopy: a new method to study surface excitations of cells and to measure membrane bending elastic moduli. *J. Phys. (Paris),* 48:2139–2151, 1987.

16. J. Rädler and E. Sackmann. Imaging optical thicknesses and separation distances of phospholipid vesicles at solid surfaces. *J. Phys. II (Paris),* 3:727–748, 1993.

17. E. Sackmann. Molecular and global structure and dynamics of membranes and lipid bilayers. *Can. J. Phys.,* 68:999–1012, 1990.

18. R. Lipowsky. The conformation of membranes. *Nature,* 349:475–481, 1991.

19. M. Wortis, U. Seifert, K. Berndl, B. Fourcade, L. Miao, M. Rao, and R. K. P. Zia. Curvature-controlled shapes of lipid-bilayer vesicles: budding, vesiculation and other phase transitions. In D. Beysens, N. Boccara, and G. Forgacs, Eds., *Dynamical Phenomena at Interfaces, Surfaces and Membranes,* Nova Science, New York, 1991, pages 221–236.

20. E. A. Evans. Bending elastic modulus of red blood cell membrane derived from buckling instability in micropipet aspiration tests. *Biophys. J.,* 43:27–30, 1983.

21. E. Evans and D. Needham. Physical properties of surfactant bilayer membranes: thermal transitions, elasticity, rigidity, cohesion, and colloidal interactions. *J. Phys. Chem.,* 91:4219–4228, 1987.

22. L. Bo and R. E. Waugh. Determination of bilayer membrane bending stiffness by tether formation from giant, thin-walled vesicles, *Biophys. J.,* 55:509–517, 1989.

23. R. E. Waugh, J. Song, S. Svetina, and B. Žekš. Local and nonlocal curvature elasticity in bilayer membranes by tether formation from lecithin vesicles. *Biophys. J.,* 61:974–982, 1992.

24. M. B. Schneider, J. T. Jenkins, and W. W. Webb. Thermal fluctuations of large quasi-spherical bimolecular phospholipid vesicles. *J. Phys.,* 45:1457–1472, 1984.

25. H. Engelhardt, H. P. Duwe, and E. Sackmann. Bilayer bending elasticity measured by Fourier analysis of thermally excited surface undulations of flaccid vesicles. *J. Phys. Lett.,* 46:L395–L400, 1985.

26. I. Bivas, P. Hanusse, P. Bothorel, J. Lalanne, and O. Aguerre-Chariol. An application of the optical microscopy to the determination of the curvature elastic modulus of biological and model membranes. *J. Phys.,* 48:855–867, 1987.

27. S. T. Milner and S. A. Safran. Dynamical fluctuations of droplet microemulsions and vesicles. *Phys. Rev. A,* 36:4371–4379, 1987.

28. J. F. Faucon, M. D. Mitov, P. Meleard, I. Bivas, and P. Bothorel. Bending elasticity and thermal fluctuations of lipid membranes. Theoretical and experimental requirements. *J. Phys. (Paris),* 50:2389–2414, 1989.

29. H. P. Duwe, J. Käs, and E. Sackmann. Bending elastic moduli of lipid bilayers: modulation by solutes. *J. Phys. (Paris),* 51:945–962, 1990.

30. P. Meleard, J. F. Faucon, M. D. Mitov, and P. Bothorel. Pulsed-light microscopy applied to the measurement of the bending elasticity of giant liposomes. *Europhys. Lett.,* 19:267–271, 1992.

31. M. B. Schneider, J. T. Jenkins, and W. W. Webb. Thermal fluctuations of large cylindrical phospholipid vesicles. *Biophys. J.,* 45:891–899, 1984.

32. M. Mutz and W. Helfrich. Bending rigidities of some biological model membranes as obtained from the Fourier analysis of contour sections. *J. Phys. (Paris),* 51:991–1002, 1990.

33. J. Rädler. Über die Wechselwirkung fluider Phospholipid-Membranen mit Festkörperoberflächen. Ph.D. thesis, TU-München, Munich, 1993.

34. W. Helfrich and R.-M. Servuss. Undulations, steric interaction and cohesion of fluid membranes. *Nuovo Cimento,* 3D:137–151, 1984.

35. E. Evans and W. Rawicz. Entropy-driven tension and bending elasticity in condensed-fluid membranes. *Phys. Rev. Lett.,* 64:2094–2097, 1990.

36. M. Kummrow and W. Helfrich. Deformation of giant lipid vesicles by electric fields. *Phys. Rev. A,* 44:8356–8360, 1991.

37. W. Helfrich. Hats and saddles in liquid membranes. *Liq. Cryst.,* 5:1647–1658, 1989.

38. W. Helfrich and B. Klösgen. Some complexities of simple liquid membranes. In D. Beysens, N. Boccara, and G. Forgacs, Eds., *Dynamical Phenomena at Interfaces, Surfaces and Membranes,* Nova Science, New York, 1992, pages 211–220.

39. J. Song and R. E. Waugh. Bending rigidity of SOPC membranes containing cholesterol. *Biophys. J.,* 64:1967–1970, 1993.

40. F. Jülicher and R. Lipowsky. Domain-induced budding of vesicles. *Phys. Rev. Lett.*, 70:2964–2967, 1993.

41. W. Helfrich. Blocked lipid exchanges in bilayers and its possible influence on the shape of vesicles. *Z. Naturforsch.*, 29c:510–515, 1974.

42. E. A. Evans. Minimum energy analysis of membrane deformation applied to pipet aspiration and surface adhesion of red blood cells. *Biophys. J.*, 30:265–284, 1980.

43. S. Svetina and B. Žekš. Bilayer couple hypothesis of red cell shape transformations and osmotic hemolysis. *Biochim. Biophys. Acta*, 42:86–90, 1983.

44. M. P. Sheetz and S. J. Singer. Biological membranes as bilayer couples. A molecular mechanism of drug-erythrocyte interactions. *Proc. Natl. Acad. Sci. U.S.A.*, 71:4457–4461, 1974.

45. U. Seifert, L. Miao, H.-G. Döbereiner, and M. Wortis. Budding transition for bilayer fluid vesicles with area-difference elasticity. In R. Lipowsky, D. Richter, and K. Kremer, Eds., *The Structure and Conformation of Amphiphilic Membranes*, Vol. 66 of *Springer Proceedings in Physics*, Springer-Verlag, Berlin, 1991, pages 93–96.

46. W. Wiese, W. Harbich, and W. Helfrich. Budding of lipid bilayer vesicles and flat membranes. *J. Phys. Cond. Matter*, 4:1647–1657, 1992.

47. B. Bozic, S. Svetina, B. Žekš, and R. E. Waugh. Role of lamellar membrane structure in tether formation from bilayer vesicles. *Biophys. J.*, 61:963–973, 1992.

48. L. Miao. Equilibrium Shapes and Shape Transitions of Fluid Lipid-Bilayer Vesicles. Ph.D. thesis, Simon Fraser University, Burnaby, 1992.

49. L. Miao, U. Seifert, M. Wortis, and H.-G. Döbereiner. Budding transitions of fluid-bilayer vesicles: the effect of area-difference elasticity. *Phys. Rev. E*, 49:5389–5407, 1994.

50. S. Svetina and B. Žekš. The elastic deformability of closed multilayered membranes is the same as that of a bilayer membrane. *Eur. Biophys. J.*, 21:251–255, 1992.

51. S. Svetina, M. Brumen, and B. Žekš. Lipid bilayer elasticity and the bilayer couple interpretation of red cell shape transformations and lysis. *Stud. Biophys.*, 110:177–184, 1985.

52. P. G. de Gennes and C. Taupin. Microemulsions and the flexibility of oil/water interfaces. *J. Phys. Chem.*, 86:2294–2304, 1982.

53. G. Gompper and D. M. Kroll. Inflated vesicles: a new phase of fluid membranes. *Europhys. Lett.*, 19:581–586, 1992.

54. J. T. Jenkins. Static equilibrium configurations of a model red blood cell. *J. Math. Biol.*, 4:149–169, 1977.

55. J. C. Luke. A method for the calculation of vesicle shapes. *SIAM J. Appl. Math.*, 42:333–345, 1982.

56. M. A. Peterson. An instability of the red blood cell shape. *J. Appl. Phys.*, 57:1739–1742, 1985.

57. K. Berndl. Formen von Vesikeln. Diplomarbeit, Ludwig-Maximilians-Universität, München, 1990.

58. F. Jülicher and U. Seifert. Shape equations for axisymmetric vesicles: a clarification. *Phys. Rev. E*, 49:4728–4731, 1994.

59. V. Heinrich, S. Svetina, and B. Žekš. Nonaxisymmetric vesicle shapes in a generalized bilayer-couple model and the transition between oblate and prolate axisymmetric shapes. *Phys. Rev. E*, 48:3112–3123, 1993.

60. Z.-C. Ou-Yang and W. Helfrich. Instability and deformation of a spherical vesicle by pressure. *Phys. Rev. Lett.*, 59:2486–2488, 1987.

61. Z.-C. Ou-Yang and W. Helfrich. Bending energy of vesicle membranes: general expressions for the first, second and third variation of the shape energy and applications to spheres and cylinders. *Phys. Rev. A*, 39:5280–5288, 1989.

62. M. A. Peterson. Deformation energy of vesicles at fixed volume and surface area in the spherical limit. *Phys. Rev. A*, 39:2643–2645, 1989.

63. V. Heinrich, M. Brumen, R. Heinrich, S. Svetina, and B. Žekš. Nearly spherical vesicle shapes calculated by use of spherical harmonics: axisymmetric and nonaxisymmetric shapes and their stability. *J. Phys. II (Paris)*, 2:1081–1108, 1992.

64. W. Wiese and W. Helfrich. Theory of vesicle budding. *J. Phys. Cond. Matter*, 2:SA329–SA332, 1990.

65. U. Seifert. Shape transformations of free, toroidal and bound vesicles, *J. Phys. Colloq.*, 51C7:339–344, 1990.

66. B. Fourcade, L. Miao, M. Rao, M. Wortis, and R. K. P. Zia. Scaling analysis of narrow necks in curvature models of fluid lipid-bilayer vesicles, *Phys. Rev. E*, 49:5276–5286, 1994.

67. R. Bruinsma. Growth instabilities of vesicles. *J. Phys. II (Paris)*, 1:995–1012, 1991.

68. J. Käs, E. Sackmann, R. Podgornik, S. Svetina, and B. Žekš. Thermally induced budding of phospholipid vesicles — a discontinuous process. *J. Phys. II (Paris)*, 3:631–645, 1993.

69. J. Käs and E. Sackmann. Shape transitions and shape stability of giant phospholipid vesicles in pure water induced by area-to-volume changes. *Biophys. J.*, 60:825–844, 1991.

70. H.-G. Döbereiner. Private communication.

71. E. Fargé and P. F. Devaux. Shape changes of giant liposomes induced by an asymmetric transmembrane distribution of phospholipids. *Biophys. J.*, 61:347–357, 1992.

72. E. Boroske, M. Elwenspoek, and W. Helfrich. Osmotic shrinkage of giant egg-lecithin vesicles. *Biophys. J.*, 34:95–109, 1981.

73. H. Gruler. Chemoelastic effects of membranes. *Z. Naturforsch.*, 30c:608–614, 1975.

74. C. Gebhardt, H. Gruler, and E. Sackmann. On domain structure and local curvature in lipid bilayers and biological membranes. *Z. Naturforsch.*, 32c:581–596, 1977.

75. R. Lipowsky. Domain-induced budding of fluid membranes. *Biophys. J.*, 64:1133–1138, 1993.

76. **M. Glaser.** Lipid domains in biological membranes. In *Current Opinion in Structural Biology. Curr. Biol.*, 3:475–481, 1993.

77. **S. Leibler.** Curvature instability in membranes. *J. Phys. (Paris)*, 47:507–516, 1986.

78. **S. Leibler and D. Andelman.** Ordered and curved meso-structures in membranes and amphiphilic films. *J. Phys. (Paris)*, 48:2013–2018, 1987.

79. **D. Needham, T. J. McIntosh, and E. Evans.** Thermomechanical and transition properties of dimyristoylphosphatidy-choline/cholesterol bilayers. *Biochemistry*, 27:4668–4673, 1988.

80. **M. R. Vist and J. H. Davis.** Phase equilibria of cholesterol/dipalmitoylphosphatidylcholine mixtures: ^2H nuclear magnetic resonance and differential scanning calorimetry. *Biochemistry*, 29:451–464, 1990.

81. **M. Bloom, E. Evans, and O. G. Mouritsen.** Physical properties of the fluid lipid-bilayer component of cell membranes: a perspective. *Q. Rev. Biophys.*, 24:293–397, 1991.

82. **S. H.-W. Wu and H. M. McConnell.** Phase separations in phospholipid membranes. *Biochemistry*, 14:847–854, 1975.

83. **M. Bloom.** The physics of soft, natural materials. *Phys. Can.*, 48(1):7–16, 1992.

84. **D. Andelman, T. Kawakatsu, and K. Kawasaki.** Equilibrium shape of two-component unilamellar membranes and vesicles. *Europhys. Lett.*, 19:57–62, 1992.

85. **T. Kawakatsu, D. Andelman, K. Kawasaki, and K. Taniguchi.** Phase transitions and shape of two component membranes and vesicles. I. Strong segregation limit. *J. Phys. II (Paris)*, 3:971–997, 1993.

86. **E. W. Kaler, A. K. Murthy, B. E. Rodriguez, and J. A. N. Zasadzinski.** Spontaneous vesicle formation in aqueous mixtures of single-tailed surfactants. *Science*, 245:1371–1374, 1989.

87. **S. A. Safran, P. Pincus, and D. Andelman.** Theory of spontaneous vesicle formation in surfactant mixtures. *Science*, 248:354–356, 1990.

88. **S. A. Safran, P. A. Pincus, D. Andelman, and F. C. MacKintosh.** Stability and phase behavior of mixed surfactant vesicles. *Phys. Rev. A*, 43:1071–1078, 1991.

89. **E. W. Kaler, K. L. Herrington, A. K. Murthy, and J. A. N. Zasadzinski.** Phase behavior and structures of mixtures of anionic and cationic surfactants. *J. Phys. Chem.*, 96:6698–6707, 1992.

90. **F. C. MacKintosh and T. C. Lubensky.** Orientational order, topology, and vesicle shapes. *Phys. Rev. Lett.*, 67:1169–1172, 1991.

91. **D. J. Benvegnu and H. M. McConnell.** Line tension between liquid domains in lipid monolayers. *J. Phys. Chem.*, 96:6820–6824, 1992.

92. **W. Helfrich.** The size of bilayer vesicles generated by sonication. *Phys. Lett.*, 50A:115–116, 1974.

93. **P. Fromherz.** Lipid-vesicle structure: size control by edge-active agents. *Chem. Phys. Lett.*, 94:259–266, 1983.

94. **D. H. Boal and M. Rao.** Topology changes in fluid membranes. *Phys. Rev. A*, 46:3037–3045, 1992.

95. **R. P. Rand and V. A. Parsegian.** Hydration forces between phospholipid bilayers. *Biochim. Biophys. Acta*, 988:351–376, 1989.

96. **J. Marra and J. N. Israelachvili.** Direct measurements of forces between phosphatidylcholine and phosphatidyle-thanolamine bilayers in aqueous electrolyte solutions. *Biochemistry*, 24:4608–4618, 1985.

97. **E. Evans.** Adhesion of surfactant-membrane covered droplets: special features and curvature elasticity effects. *Colloids Surfaces*, 43:327–347, 1990.

98. **R.-M. Servuss and W. Helfrich.** Mutual adhesion of lecithin membranes at ultralow tensions. *J. Phys. (Paris)*, 50:809–827, 1989.

99. **L. D. Landau and E. M. Lifshitz.** *Elastizitätstheorie.* Akademie-Verlag, Berlin, 1989.

100. **R. Lipowsky and U. Seifert.** Adhesion of vesicles and membranes. *Mol. Cryst. Liq. Cryst.*, 202:17–25, 1991.

101. **R. Lipowsky and U. Seifert.** Adhesion of membranes: a theoretical perspective. *Langmuir*, 7:1867–1873, 1991.

102. **E. A. Evans.** Detailed mechanics of membrane-membrane adhesion and separation. I. Continuum of molecular cross-bridges. *Biophys. J.*, 48:175–183, 1985.

103. **U. Seifert.** Adhesion of vesicles in two dimensions. *Phys. Rev. A*, 43:6803–6814, 1991.

104. **R. Lipowsky and S. Leibler.** Unbinding transitions of interacting membranes. *Phys. Rev. Lett.*, 56:2541–2544, 1986.

105. **R. Lipowsky and B. Zielinska.** Binding and unbinding of lipid membranes: a Monte Carlo study. *Phys. Rev. Lett.*, 62:1572–1575, 1989.

106. **G. Cevc, W. Fenzl, and L. Sigl.** Surface-induced X-ray reflection visualization of membrane orientation and fusion into multibilayers. *Science*, 249:1161–1163, 1990.

107. **U. Seifert.** Vesicles of toroidal topology. *Phys. Rev. Lett.*, 66:2404–2407, 1991.

108. **F. Jülicher, U. Seifert, and R. Lipowsky.** Phase diagrams and shape transformations of toroidal vesicles. *J. Phys. II (Paris)*, 3:1681–1705, 1993.

109. **F. Jülicher, U. Seifert, and R. Lipowsky.** Conformal degeneracy and conformal diffusion of vesicles. *Phys. Rev. Lett.*, 71:452–455, 1993.

110. **M. A. Peterson, H. Strey, and E. Sackmann.** Theoretical and phase contrast microscopic eigenmode analysis of erythrocyte flicker. I. amplitudes. *J. Phys. II (Paris)*, 2:1273–1285, 1991.

111. **F. Brochard and J. F. Lennon.** Frequency spectrum of the flicker phenomenon in erythrocytes. *J. Phys. (Paris)*, 36:1035–1047, 1975.

112. **E. Evans, A. Yeung, R. Waugh, and J. Song.** Dynamic coupling and nonlocal curvature elasticity in bilayer membranes. In R. Lipowsky, D. Richter, and K. Kremer, Eds., *The Structure and Conformation of Amphiphilic Membranes,* Vol. 66 of *Springer Proceedings in Physics,* Springer-Verlag, Berlin, 1991, pages 148–153.

113. **U. Seifert and S. A. Langer.** Viscous modes of fluid bilayer membranes. *Europhys. Lett.,* 23:71–76, 1993.

114. **U. Seifert.** Dynamics of a bound membrane. *Phys. Rev. E,* 49:3124–3127, 1994.

115. **M. Kraus and U. Seifert.** Relaxation modes of an adhering bilayer membrane. *J. Phys. (Paris),* 4:1117–1134, 1994.

116. **T. L. Steck.** Red cell shape. In W. Stein and F. Bronner, Ed., *Cell Shape Determinants, Regulation, and Regulatory Role,* Academic Press, New York, 1989, pp. 205–246.

117. **D. R. Nelson and L. Peliti.** Fluctuations in membranes with crystalline and hexatic order. *J. Phys. I (Paris),* 48:1085–1092, 1987.

118. **D. Nelson.** The statistical mechanics of crumpled membranes. In H. E. Stanley and N. Ostrowsky, Eds., *Random Fluctuations and Pattern Growth,* Kluwer Academic, Norwell, MA, 1988, pages 193–215.

119. **R. Lipowsky and M. Girardet.** Shape fluctuations of polymerized or solidlike membranes. *Phys. Rev. Lett.,* 65:2893–2896, 1990.

120. **S. Komura and R. Lipowsky.** Fluctuations and stability of polymerized vesicles. *J. Phys. II (Paris),* 2:1563–1575, 1992.

121. **B. T. Stokke, A. Mikkelsen, and A. Elgsaeter.** The human erythrocyte membrane skeleton may be an ionic gel. I. Membrane mechanochemical properties. *Eur. Biophys. J.,* 13:203–218, 1986.

122. **B. T. Stokke, A. Mikkelsen, and A. Elgsaeter.** The human erythrocyte membrane skeleton may be an ionic gel. II. Numerical analyses of cell shapes and shape transformations. *Eur. Biophys. J.,* 13:219–233, 1986.

123. **D. H. Boal, U. Seifert, and A. Zilker.** Dual network model for red blood cell membranes. *Phys. Rev. Lett.,* 69:3405–3408, 1992.

124. **D. H. Boal.** Computer simulation of a model network for the erythrocyte cytoskeleton. *Biophys. J.,* 67:521–529, 1994.

Chapter 3

Vesicles of Complex Topology

B. Fourcade, X. Michalet, and D. Bensimon

CONTENTS

I. INTRODUCTION

Surfaces of complex topology are observable in a variety of multicomponent mixtures such as surfactant-water-oil systems (bicontinuous or cubic phase),[1] amphiphilic molecules (such as egg lecithin) swelled in water,[2] or even natural systems such as plastids in plant Golgi apparatus.[3] All these systems are characterized by a fluid membrane dividing the space into symmetric or asymmetric compartments, possibly filled with the same kind of solvent. Membranes can also be randomly pierced by holes or interconnected by passages of a scale ranging from nanometers to micrometers. Holes and passages appear spontaneously during the swelling stage; they can also be created by artificial means, such as via strong electric fields[4,5] or osmotic pressure.[6] These passages can be described as topological defects whose number and position might fluctuate in time.[7,8] In all cases, the surface cannot be continuously transformed into a sphere without tearing and pasting the membrane so that it has a more complex topology than the sphere. The surface of vesicles can be easily studied by optical microscopy, and this allows a precise geometrical description on the micrometer scale.

Because there already exist reviews on vesicles of spherical topology,[9,10] this chapter deals with the equilibrium shape problem of vesicles of complex topology.[11-13] It is intended to parallel the more theoretically oriented chapter by Seifert and Lipowsky (Chapter 2),[14] where recent developments of the vesicle shape problem are summarized. As reported by other authors in this book, vesicles are closed bilayers made up of phospholipids. In the fluid state, the vesicles are subject to a set of constraints (volume, total area, and mean curvature) and their shapes result from the minimization of a curvature or bending elastic energy. We will hereafter report on vesicles of nonspherical topology equivalent to spheres with handles,* these surfaces being observable by phase contrast microscopy. This includes not only one-holed tori, symmetric or asymmetric, but also surfaces of more complex topology, such as two-holed tori or lattices of passages. For $g = 1,2$ surfaces, our experimental observations are connected to recent results in differential geometry,[15] since these vesicles are examples of closed surfaces minimizing absolutely the curvature energy. For $g = 1$, this is the case of the Clifford torus, and for $g = 2$, they correspond to the so-called Lawson surfaces.[16]

---

* The number of which gives the topological genus g.

0-8493-4731-9/96/$0.00+$.50
© 1996 by CRC Press, Inc.

This chapter starts with a short theoretical summary. It recalls the different approaches and it sets up the notations. After a short description of the experimental methods, we deal first with the case of partially polymerized toroidal surfaces, since these vesicles have distinctive properties. For unpolymerized vesicles, we address the case of the toroidal genus, genus 2, and even higher genus surfaces, where comparisons with theoretical results can be made. Temperature is used as a control parameter to scan through the phase diagram and, in particular, through the bifurcation lines where the axisymmetry is broken. These observations can also be directly supported by numerical simulations. The experimental shapes are optically digitized to build a triangular mesh, and these digitized shapes can be used as input to Surface Evolver, a computer program written by K. A. Brakke,[17] to check whether the observed shapes are stable solutions in one of the different curvature models summarized in SL. For toroidal vesicles, one finds good agreement with the theoretical predictions. One observes the different families of equilibrium shapes predicted by the models which are characterized by the shape of their meridian cross section. However, for some of the nonaxisymmetric ones, we show that they must be interpreted as being in metastable states. In temperature-controlled experiments, the axisymmetry for circular tori is broken at the so-called Clifford torus, and large fluctuations can be observed in some cases. These fluctuations are reminiscent of the conformal symmetry which leaves the bending Hamiltonian invariant, and for higher topology the effect can be more spectacular. Indeed, we provide experimental examples of the so-called conformal diffusion effect,[18,19] where a vesicle undergoes slow shape changes at constant energy and with constant constraints. In the last part, we concentrate on the problem of necks or passages connecting lamellae of membranes. Experimentally, the effect of fluctuations is much more pronounced for these systems. Using an electrostatic analogy, we derive the interaction between necks, which we show to behave as free particles with hard-core repulsion.[20]

II. THEORETICAL SUMMARY

In this section we recall the theoretical results on the equilibrium shapes of free vesicles summarized in Chapters 1 and 2. L. Duplantier was the first to recall the Willmore conjecture in the context of the physics of vesicles.[21] This mathematical conjecture states that the *minimum minimorum* of the curvature energy[22]

$$F_0 = \frac{1}{2} \oint dS (C_1 + C_2)^2 \tag{1}$$

for genus $g = 1$ is obtained for the so-called Clifford torus. This torus is characterized by a ratio of the radii of its generating circles:* $r/R = 1/\sqrt{2}$.[23] Although experiments provide examples of this conjecture (see Section IV), the shape of a vesicle results from a trade-off between a curvature energy and a set of constraints so that a toroidal vesicle is not necessarily the Clifford torus, but is a member of one of the different families of equilibrium shapes (see Figure 1 and Figure 2 for some examples). Up to now, three models have been proposed. We summarize their properties to compare with experimental observations.

1. In the spirit of a Landau functional approach, and following Canham[24] and Helfrich,[25] the spontaneous curvature (SC) model states that the shapes are minimal energy solutions of a curvature Hamiltonian with the constraints of a given volume V and surface A lumped together into a dimensionless constant corresponding to the reduced volume constraint $v = 6\sqrt{\pi} \, V/A^{3/2}$. A spontaneous curvature term C_0 breaking the inside-outside symmetry is generally taken into account such that the bending energy reads

$$F_{SC} = \frac{1}{2} \kappa \oint dS (C_1 + C_2 - C_0)^2 \tag{2}$$

where κ defines the bending modulus, which is typically of the order of 10 $k_B T$. Shapes are thus determined by the two parameters v and $c_0 = C_0 R_0$, where $R_0 = (A/4\pi)^{1/2}$ sets the scale of the equivalent surface sphere.

---

* One of the two circles is the meridian cross section, whose center, rotating around the symmetry axis, generates the other one.

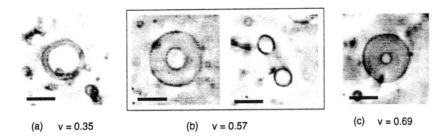

(a) v = 0.35   (b) v = 0.57   (c) v = 0.69

Figure 1 Three examples of circular tori with reduced volume $0.3 < v < 0.71$. In the intermediate case, $v = 0.57$; two characteristic views of such a torus are shown. Bar indicates 10 µm.

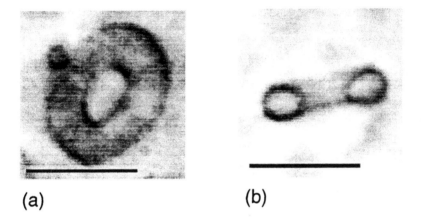

(a)   (b)

Figure 2 Two views of an axisymmetric vesicle with pear-shaped meridian cross section. Bar indicates 10 µm.

2. Originating from the work of Sheetz and Singer,[26] Evans,[27] and others,[28-30] the bilayer coupling (BC) model introduces another constraint on the difference of area ΔA between the two monolayers making up the bilayer. Because each individual monolayer keeps a constant number of molecules over many hours, and possibly days, it is relevant to consider that ΔA is constant on short time scales. To leading order in the bilayer thickness d, ΔA can be made dimensionless and is related to the mean curvature:

$$m = \frac{\Delta A}{dR_0} = \frac{1}{R_0}\oint dS\left(C_1 + C_2\right) \tag{3}$$

In this model, the equilibrium shapes are determined by the two parameters v and m.

3. This latter constraint can be relaxed by introducing a second elastic energy term for the area difference between the two leaflets.[31-33] An optimal leaflet area difference m_0 is introduced and the total energy reads

$$F_{ADE} = \frac{1}{2}\kappa\oint dS\left(C_1 + C_2 - C_0\right)^2 + \frac{1}{2}\alpha\kappa\left(m - m_0\right)^2 \tag{4}$$

This is the area difference elasticity model (ADE). According to Reference 33, α is a phenomenological parameter which depends on physicochemical parameters and is of the order of unity for bilayer membranes. When α approaches infinity, one recovers the hard constraint of the BC model, but in the other limit, $\alpha \rightarrow 0$, one recovers the spontaneous curvature model case. The ADE model is less restrictive than the previous two in that the shapes are now a function of three parameters (v, c_0, m_0).

When $c_0 = 0$ the bending term is invariant under a *continuous* set of transformations corresponding to the conformal group in three dimensions. This includes not only the rotations, the translations, and a global rescaling, but also inversions with respect to a sphere. The special conformal transformations can be labeled by a vector $\mathbf{a} = (a_x, a_y, a_z)$ and they transform any point $\mathbf{r}$ of the surface as[34]

$$\mathbf{r} \rightarrow \frac{\dfrac{\mathbf{r}}{r^2} + \mathbf{a}}{\left(\dfrac{\mathbf{r}}{r^2} + \mathbf{a} \right)^2} \tag{5}$$

Minimization of the curvature energy thus leads to a degenerate ground state of different shapes related by an inversion. For low topology surfaces, the constraints of constant reduced volume, of constant reduced mean curvature, or a spontaneous curvature term break this degeneracy. However, for surfaces of higher genus ($g \geq 2$), this degeneracy might still persist in some regions of the phase diagram (v,c_0) or (v,m).[35]

Because of Equation 3 all three models have the same Euler equations so that the *linear* stability problem is the same for all cases. In the SC model, c_0 must be taken into account in the bending energy, but in the latter two cases, c_0 is a Lagrange multiplier which must be adjusted self-consistently in the ADE model. However, for the same reduced volume v, the total energy depends on the model and the respective phase diagrams are different. This has been extensively studied in the case of the budding phenomenon for $g = 0$ surfaces[36-39] and, more recently, for higher genus vesicles.

As shown by Seifert[40] and Jülicher et al.,[35] the general feature of all phase diagrams for toroidal shapes is that there exists a second-order bifurcation line separating axisymmetric toroidal solutions from stable nonaxisymmetric ones.[35,40] In a schematic way, this line divides the low reduced volume section, where one expects various families of axisymmetric tori with different cross sections, from the high reduced volume section, where the conformally transformed surfaces of axisymmetric solutions are the lowest energy solutions. In the first two of the preceding models, the Clifford torus sits on this line, but in the ADE model, there is a gap between the Clifford torus and nonaxisymmetric tori, such that almost circular tori of reduced volume larger than $v_{Clifford} = 0.71$ are in principle observable. Nonaxisymmetric tori, conformal transforms of circular ones, are a particular case of the well-known Dupin cyclides.[41,42]

Other stable shapes are model dependent. They are characterized by the shape of their meridian cross section. For example, in the SC model, with $c_0 = 0$ and no constraint on m, tori with a sickle-shaped meridian cross section (hereafter called spheroidal tori) are minimum energy solutions for $0 < v < 0.57$, but the branch of tori with an almost circular cross section (hereafter called circular tori) is stable for $0.57 < v < 0.71$. For the BC model (i.e., with zero spontaneous curvature and a constraint on m) there are discoidal, stomatoidal, and spheroidal axisymmetric families of solutions in the (v,m) planes solutions from stable nonaxisymmetric ones.[35,40] For all of these branches, the hole diameter can shrink to zero. However, in this model, no discoidal or stomatoidal nonaxisymmetric shapes have been calculated up to now. The ADE model predicts also that these three types of solutions are minimal energy shapes in different regions of the (v,m_0) phase space (at fixed c_0).

III. EXPERIMENTAL METHODS

A. PREPARATION OF VESICLES

As described in other chapters, vesicles are made of one or more closed bilayers. Mostly prepared with multicomponent mixtures of natural phospholipids swelled in water using various procedures, they can also be prepared with very pure assays of synthetic surfactant. We used pure synthetic "standard" phospholipids purchased from Avanti Polar Lipids, Inc. (DMPC, DPPC, DOPC) as well as a polymerizable one, $DC_{8,9}PC$.

For the first ones (DMPC, DPPC, DOPC), we used crystallized phospholipids (several milligrams) deposited on a petri dish. A droplet of deionized water (at a temperature $T > T_m$) swells the lipids, which are then spread over the bottom of the dish. A further excess amount of heated deionized water (several milliliters) permits the separation of vesicles from the bulk lamellar phase (complete swelling occurs after several hours).

The latter one ($DC_{8,9}PC$) can be swelled in a similar way. Vesicles undergo a morphological transition toward tubules (diameter: 1 μm; length: some 100 μm) when cooled below the melting temperature of the chains ($T_m = 42°C$). This transition is reversible, which means that one can form vesicles by reheating tubules above T_m. As cooling cannot be avoided between the preparation and the observation stages, tubules can be considered as an intermediate step toward the formation of vesicles. Another, simpler method to prepare tubules has been proposed.[43-45]

B. OBSERVATION AND ANALYSIS

Observation is made at constant temperature using a phase contrast microscope coupled to an image analysis system. Vesicles are best seen when the tangent plane to their contour in the focal plane of the microscope is parallel to the optical axis. Since vesicles are free to rotate under Brownian motion, pictures can be taken from different viewpoints.

It is also possible to digitize the experimental pictures and build a geometrical model of the observed surface. Using the Surface Evolver computer program of K. A. Brakke,[17] it is possible to get the geometrical parameters of the vesicle and to determine by a direct minimization of the bending energy whether the observed shape is a stable, or metastable, solution of one of the curvature models described in Chapter 2.

Finally, in order to test the theoretical predictions concerning the various branches of solutions, we performed temperature change experiments. This permits one to vary the reduced volume of the vesicles, their volume being almost constant when their area changes as

$$\frac{dA}{dT} = \gamma A \qquad (6)$$

with $\gamma \approx 5.0 10^{-3}$ for most of the phospholipids.

IV. PARTIALLY POLYMERIZED VESICLES

$DC_{8,9}PC$ is a polymerizable phospholipid. Partial polymerization can be induced in the gel phase by shining UV light during a time period ranging from several seconds to 10 min. Below $T_m = 43°C$, the vesicles break up into shards of winding sheets. The helical structure of these ribbons results from an intrinsic bending force in the tilted chiral bilayer.[46,47]

The effect of polymerization depends on the time during which the vesicles are exposed to UV radiation. If the irradiation lasts more than 12 min, the rigid structures keep their form when reheated back above T_m. This can be interpreted as a percolation of the polymerized net on the whole surface of the vesicle. For shorter exposure time, however, the needles reverse to vesicles fluctuating with amplitudes which are only slightly reduced with increasing exposure time. Among shapes of spherical topology, one also observes tori with the aspect ratio of the Clifford torus.[48] They are axisymmetric with slight fluctuations of the hole around its equilibrium position. These experiments might be interpreted as a partial relaxation of the volume constraint due to leakage of the polymerized vesicle. The vesicle is then free to sit at the *absolute minimum* of the bending energy. Experimentally, nonaxisymmetric vesicles (Dupin cyclides) are also observed, but less often: as they also correspond to the *absolute minimum* of the bending energy, they are compatible with this theoretical picture[49] (see also Reference 50).

V. UNPOLYMERIZED MEMBRANES

A. EQUILIBRIUM SHAPES IN THE AXISYMMETRIC CASE

Following the experimental procedures introduced earlier, one observes at constant temperature a rich variety of toroidal shapes. These experiments are done with freshly prepared vesicles. The generic toroidal surface has a circular meridian cross section with a reduced volume ranging from 0.3 to 0.71 ± 0.03 (see Figure 1). These circular tori are solutions in all of the above models. For an exact circular meridian cross section, their reduced volume is related to the mean curvature by

$$v_{ci} m_{ci} = 3\pi \qquad (7)$$

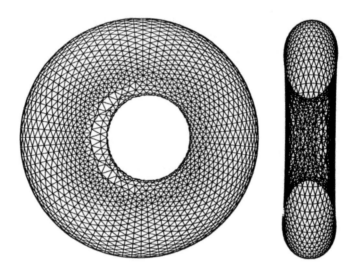

Figure 3 Numerically equilibrated surface, starting from a digitized surface obtained from Figure 2. The slight axisymmetry breaking reproduces the experimental one, due to thermal fluctuations. $F_0/8\pi\kappa = 2.35$, as calculated by the Surface Evolver program.

In spontaneous curvature model with $c_0 = 0$, circular tori are metastable states for $v \leq 0.57$. Discoidal toroidal surfaces are also observable. They are always metastable in case 1, but they can be minimal energy shapes in the BC model (case 2). An example is shown in Figure 2. This vesicle has a reduced volume of $v = 0.51$ and a mean curvature of $M/4\pi = 1.38$. According to Reference 35, such a discoidal vesicle is indeed a minimal energy shape for this set of parameters. However, spheroidal tori predicted in cases 2 and 3 have not been observed, but similar surfaces of a higher genus made of two spheres connected by three passages have been observed (see later Figure 16).

From the numerical point of view, one can test the stability of experimental discoidal tori. The shape is first digitized and then relaxed to a minimum energy surface. If model 1 is used with a zero spontaneous curvature, the shape relaxes to the quasi-circular family. In contrast, if a constraint on m is taken into account (or with $c_0 \neq 0$), it has been determined that the contour of the meridian cross section remains of the discocyte type (see Figure 3).

B. BIFURCATION TO THE NONAXISYMMETRIC CASE

All the theoretical phase diagrams predict a region where the equilibrium shape is nonaxisymmetric. In particular, changing the reduced volume of a vesicle sitting near the transition line allows one to observe this transition. A natural candidate is the Clifford torus with a reduced volume $v = 0.71$. This torus is exactly on the transition line in the SC model as well as for the BC case. In case 3, the ADE model, one can in principle observe almost circular tori for $v \leq 0.77$ (case $\alpha = 1$).[35]

Figure 4 shows three views of this torus at different temperatures. As explained in the experimental section, a decrease in temperature amounts to an increase in the reduced volume. Vesicles are observed at constant temperature after the temperature has been decreased by 10°C. This allows one to determine all the geometrical parameters, such as the reduced volume, i.e., the eccentricity e of the hole. For nonaxisymmetric shapes, we have used the Dupin cyclides parametrization to get v from e. For $T = 50$°C, the torus is axisymmetric. When the temperature decreases, the hole is shifted off the center due to the increase of the reduced volume. Since the minimal temperature accessible for this phospholipid is of the order of 25°C, we have not been able to move the hole off the center such that the shape looks like a sphere with a small handle. (This is because of the small thermal expansion of $DC_{8,9}PC$, the thermal expansion of DMPC is at least two times larger.) This transition is reversible as predicted in models 1 and 2. In all experiments on circular tori, the bifurcation took place at a reduced volume of $v = 0.71 \pm 0.03$. For $\alpha = 1$, the ADE model predicts that the transition could also take place for $v < v_{Clifford}$ (with $m_0 > 1$) or for $v > v_{Clifford}$ (with $m_0 < 1$). None of these cases has been observed.

Varying the temperature can also be done for vesicles which are spontaneously nonaxisymmetric, such as the Dupin cyclides. These tori are characterized by circular lines of curvature but, as seen in Figure 5, the two meridian cross sections have a different radius. In all cases, the effect of decreasing the

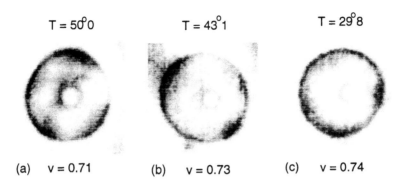

T = 50°0   T = 43°1   T = 29°8

(a)   v = 0.71   (b)   v = 0.73   (c)   v = 0.74

Figure 4 Axisymmetry breaking of the Clifford torus due to a temperature variation. This transformation is reversible. Bar indicates 10 μm.

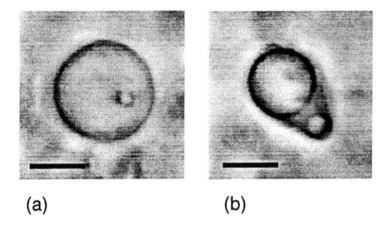

(a)   (b)

Figure 5 Two views of a Dupin cyclide with reduced volume $v = 0.80$. The two meridian cross sections are circular. Bar indicates 10 μm.

temperature is to increase the eccentricity. The maximum reduced volume we have been able to obtain in this case is of the order 0.92.

C. NONAXISYMMETRIC CASE

As in the axisymmetric case, the meridian cross section of a nonaxisymmetric torus is not necessarily circular. Nonaxisymmetric stomatoidal or discoidal tori, obtained by conformal transformation of stable or metastable axisymmetric shapes, are expected. This transformation has been used by Seifert[51] and by Jülicher et al.[35] to test the stability of axisymmetric solutions with respect to axisymmetry-breaking perturbations.

Figure 6 shows a nonaxisymmetric stomatoidal torus. This solution is predicted by none of the models. However, using the Surface Evolver program, it is possible to check that it is a metastable solution, with bending energy $F_0/8\pi\kappa = 1.61$, mean curvature $m/4\pi = 1.05$, and reduced volume $v = 0.70$. For this set of parameters, one expects in the BC model an axisymmetric stomatoidal torus sitting near the bifurcation line from nonaxisymmetric to axisymmetric shapes. A small spontaneous curvature may explain the deviation of this line towards the low reduced volume tori. In the framework of the ADE model, and for $\alpha = 1$, the situation is more complex. Stomatoidal tori are not minimal solutions near the bifurcation line. For this set of parameters, this torus can only be interpreted as being in a metastable state.

Figure 7 shows another example of a nonaxisymmetric torus. This discoidal shape has a reduced volume of $v = 0.52$ with a mean curvature $m/4\pi = 1.11$ as calculated by the Surface Evolver program. As in the preceding example, the phase diagram of Reference 35 for the BC model predicts for this set of parameters that the stable shape should be an axisymmetric discoidal torus.

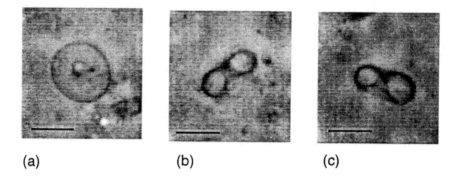

Figure 6 Nonaxisymmetric stomatoidal torus. (a) Top view. (b) Front view. (c) Side view, showing the nonaxisymmetry. Bar indicates 10 μm.

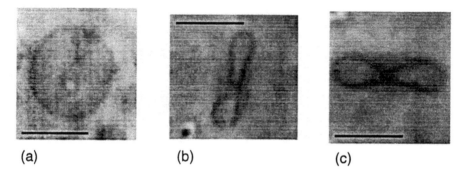

Figure 7 Nonaxisymmetric discoidal torus. (a) Top view. (b) Front view. (c) Side view, showing the nonaxisymmetry. Bar indicates 10 μm.

To clarify this point, the contour shape of Figure 7 can be numerically digitized and then numerically equilibrated. For the BC model, there is a nonaxisymmetric torus with a discoidal cross section and with the same mean curvature and reduced volume as the experimental one (see Figure 8). The resulting shape has, however, an energy of $F_0/8\pi\kappa = 2.25$, which is larger than the axisymmetric discoidal torus of the same reduced volume. This observation shows that a knowledge of metastable states is crucial to interpretation of experiments on vesicles.

VI. FLUCTUATIONS AROUND STABLE SHAPES

Surfaces of complex topology are more sensitive to fluctuations than those of spherical topology. This can already be seen in the case of genus 1, where the Clifford torus plays a special role. Because of its symmetry, this torus is almost conformally invariant with respect to a shift of the hole. At *constant* reduced volume, a second-order stability analysis shows that the energy increases as

$$\delta\left(\frac{F_b}{2\pi^2\kappa}\right) = \frac{1}{4}a_n^2\sum_{n\geq0}\left(2 - 3n^2 + n^4\right) \tag{8}$$

for $c_0 = 0.$[52] In Equation 8, $(a_n)_{n\geq0}$ labels the hierarchy of axisymmetry-breaking perturbations, where n corresponds to the broken symmetry: $n = 1$ moves the hole off the center, $n = 2$ is an elliptical deformation, and $n = 3$ produces a star.[52,53]* The conformal mode corresponds to $n = 1$, and this shows that the energy increases in proportion to e^4, instead of the usual e^2 potential, when the hole moves off the center. Therefore,

---

* The complete hierarchy of modes can be calculated from a stereographic projection from the three-dimensional sphere on the Euclidean space. In Reference 52 the set of modes was incomplete; this has been corrected in Reference 53.

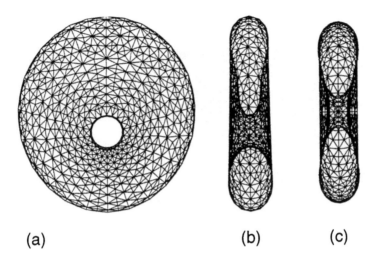

(a)   (b)   (c)

Figure 8 Numerically equilibrated surface, starting from a digitized surface obtained from Figure 7. $F_0/8\pi\kappa =$ 2.25, as calculated by the Surface Evolver program.

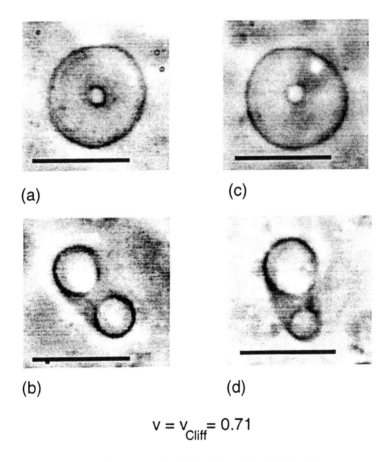

(a)   (c)

(b)   (d)

$$v = v_{Cliff} = 0.71$$

Figure 9 Fluctuations of a Clifford torus. Bar indicates 10 μm.

one expects large fluctuations. For some of the Clifford tori, this is actually the case. Figure 9 shows cut views of the same torus at different times (the membrane is in the fluid state and there is no partial polymerization). The typical amplitude for these fluctuations changes the aspect ratio by 20%, and this effect is much more pronounced than for common vesicles, where thermal fluctuations affect the aspect ratio by

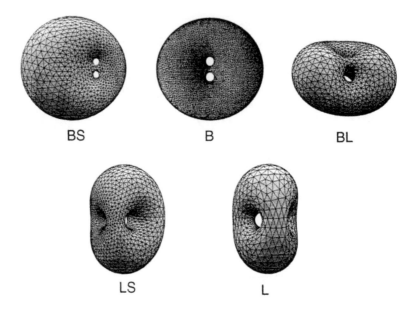

Figure 10 Some examples of genus 2 surfaces which minimize F_0 absolutely. They are conformal transformed one from the others.

a factor of the order of 5 to 10%. When the same stability analysis is performed on nonaxisymmetric Dupin cyclides, one can show that the energy increases in proportion to e^2. This shows that the Clifford torus is the only $g = 1$ surface for which the conformal symmetry gives an observable effect.

VII. VESICLES OF TOPOLOGICAL GENUS $G = 2$

Because a circular torus is highly symmetric,[52,54] there is only one branch of nontrivial special conformal transformations. This branch is characterized by the distance of the inversion center with respect to the symmetry axis. Other positions of the inversion center amount to rescaling, rotations, or translations. Surfaces of higher topology cannot be axisymmetric, so placing the inversion center along the x, y, or z axis yields three specific branches of shapes with the same bending energy. Mathematicians have already provided examples of surfaces which extremize the bare bending energy of Equation 1. Figure 10 shows examples of genus 2 Lawson surfaces which are all conformal images of one another. According to Kusner et al.,[55,56] these surfaces play a role analogous to that of the Clifford torus for $g = 1$ tori. They are absolute minima of the bending energy for genus 2 surfaces.

As for $g = 1$ surfaces, the set of genus 2 surfaces can be characterized by a two-parameter phase diagram, (v, c_0) or (v, m). All shapes of Figure 10 are actually observable. Figure 11 shows a type B Lawson surface. Figure 12 provides an example of type LS with a threefold symmetry axis (a genus $g > 1$ surface can have at most a $g + 1$ symmetry axis). Figure 13 is an example of a type BL surface.

The specificity of high genus surfaces is that the ground state can be degenerate even with the constraints of constant reduced volume and constant mean curvature (the same remark holds in all models).[18] Because the conformal group in three dimensions is indexed by three parameters, two of them are enough to satisfy the constraints and the last one generates a whole family. The conformal diffusion of Jülicher et al.[35] takes place in a low symmetry region of the phase diagram, and it is characterized by entropic diffusion of the shape along a branch of degenerate solutions with a time scale set by the viscosity of the solution.

Figure 14 shows this phenomenon for a LS vesicle. Different views are taken at intervals of 30 s such that the focal plane corresponds to the equatorial symmetry plane. The three piles parallel to the principal axis undergo a scale change on a slow time scale characteristic of the conformal diffusion but remain circular. Figure 15 provides an example similar to the one published in Reference 35. It is a BS-type vesicle for which, in the first view, the focal plane does not correspond to a symmetry plane. However, after 20 s, the shape becomes symmetric with respect to a plane passing through the principal axis.

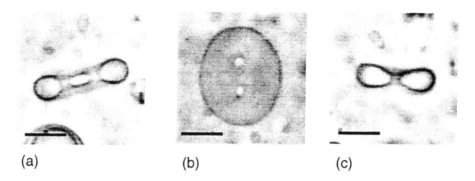

(a)   (b)   (c)

Figure 11 Button vesicle. Its numerically calculated geometrical parameters are $v = 0.52$, $m/4\pi = 1.17$. Bar indicates 10 µm.

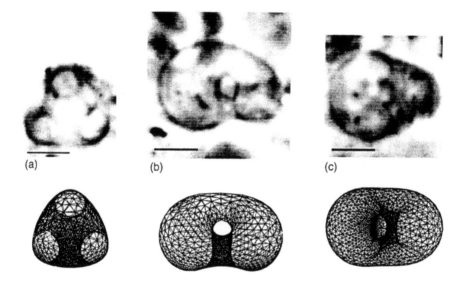

(a)   (b)   (c)

Figure 12 LS-type vesicle. Three views of the same vesicle can be compared to a numerical approximation of the surface computed by the Surface Evolver program. Bar indicates 10 µm.

Other genus $g = 2$ surfaces are observable. In agreement with the theoretical predictions, they do not all exhibit the conformal diffusion phenomenon. A paradigmatic surface of this type is presented in Figure 16. It can be viewed as two concentric spheres connected by three passages. By enlarging one of these necks, this shape is transformed into a torus with two holes. Its genus is therefore 2. Fluctuations around equilibrium shapes will be studied in the next section. It suffices to notice that these passages fluctuate strongly, their distance varying by a factor 2.

VIII. HIGHER GENUS SURFACES AND GIANT FLUCTUATIONS

Higher genus surfaces similar to the one of Figure 16 are also observable. They are of the same type but with a larger number of passages connecting the two concentric spheres. It is an experimental fact that all these low reduced volume surfaces (the volume corresponds to the region included between the two spheres) fluctuate strongly on a 1 s time scale. This time scale contrasts with the longer one necessary for the conformal diffusion to take place, thereby indicating that there is a restoring force. A model surface to study this effect has been described in Reference 20. A simple geometry corresponds to two lamellae connected by a periodic network of passages. Thus the system is divided into elementary cells with one or more holes per cell. We impose periodic boundary conditions on each cell and, for simplicity, we take the geometry of a square.

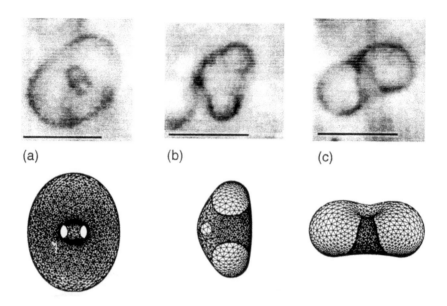

Figure 13 BL-type vesicle. Three views of the same vesicle can be compared to a numerical approximation of the surface computed by the Surface Evolver program. Bar indicates 10 μm.

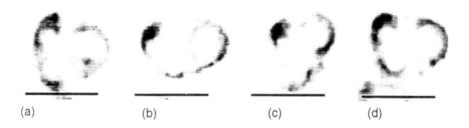

Figure 14 LS-type vesicle exhibiting conformal diffusion. Snapshots are some 20 seconds apart and correspond to different geometrical shapes. No Brownian motion takes place during this observation. Bar indicates 10 μm.

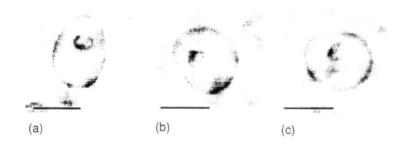

Figure 15 BS-type vesicle exhibiting conformal diffusion. Snapshots are some 20 seconds apart and correspond to different geometrical shapes, as can be easily seen: view (a) shows a symmetry plane, but none exists in view (b) or (c), which correspond to different geometrical shapes. Bar indicates 10 μm.

If the radius a of each hole is much smaller than the lattice spacing $2L$, a one-holed elementary cell can be decomposed into a core domain and a slightly bent surface $z(x,y)$, where the mean curvature can be approximate by $H = Dz$. This approximation fails in the core region, $r = \sqrt{x^2 + y^2} \approx a$, where the two radii of curvature are large but are of opposite signs. This surface divides the space into two

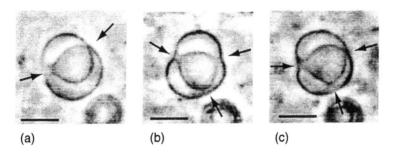

Figure 16 Fluctuating vesicles of genus 2. Snapshots are separated by some seconds. Arrows indicate the positions of necks connecting the two quasi-spherical parts of the vesicle. Bar indicates 10 μm.

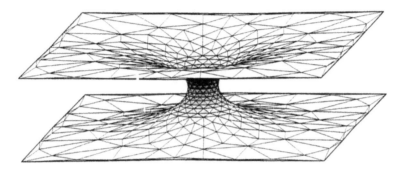

Figure 17 Elementary cell of a periodic network of holes.

compartments and, because there is no minimal surface ($H = 0$) of this type, the hole diameter must be kept to a finite value by imposing a pressure difference between the two sides. A more convenient way to consider this problem is to impose the value of a as a boundary condition. To leading order in a, both ensemble give the same result.[20]

In the latter ensemble, there are analogies between an electrostatic problem and the curvature problem, where the electrostatic potential corresponds to the height variable $z(x,y)$. Since the total electrical charge must be zero the charge $- a/2$ borne by each hole must be compensated for by a positive background. For a square lattice, the equilibrium shape problem is cast into the Poisson equation

$$\Delta z = -2\pi \frac{a}{L^2} \tag{9}$$

and, as a result, the curvature energy of each cell scales as a/L^2.

This analogy can be pushed further for more complex surfaces where the holes do not have the same "electrical charge" or for multilammelar membranes dividing the space into different compartments. Here it suffices to note that the holes have an inner core which matches the external constant curvature region. The radius of this matching domain sets the scale where two holes interact with a hard-core repulsion. A matching analysis gives an effective core radius at which this effect takes place which scales as a lg (a/L), L being the periodicity of the lattice.[59]

IX. CONCLUSION

Although the shape problem for phospholipidic vesicles had been tackled 20 years ago, recent developments, both theoretical and experimental, have shown new facets. Among them, vesicles of complex topology have been observed recently, and these studies have exemplified new phenomena related to curvature energy and to their symmetries.

Fluid but partially polymerized vesicles provide first an illustration of the well-known Willmore conjecture on minimal surfaces. When relaxing the constraint of a constant reduced volume, the only torus which can be observed corresponds to the absolute minimum of the bending energy. In principle, because of the conformal symmetry, this minimum is degenerate, but the constraints break this degeneracy for genus $g = 1$ surfaces. Large experimentally observed fluctuations show, however, the importance of this symmetry for dynamic phenomena.

As predicted in all models, experimental observations of vesicles of complex topology show a variety of different families. Even in the simplest case of toroidal shapes, we have provided examples not only of almost circular tori, but also, although more rarely, of tori with a discoidal or stomatoidal cross section. These observations have also shown that nonaxisymmetric families, where the hole is shifted off the center, are also observable for values of the reduced volume where axisymmetric solutions have been calculated. Not all of these surfaces are conformal transforms of axisymmetric ones. However, when the experimental observation is analyzed by a numerical optimization of an approximate triangular mesh, these shapes converge to nearby solutions whose energy is slightly larger than the axisymmetric ones. At the very least, this shows that metastable states are important and that not all families of equilibrium shapes are conformal transforms of axisymmetric ones. It also raises the interesting and open question of denumbering the number of metastable states as a function of the constraints (v, c_0) or (v, m). A bifurcation from an axisymmetric to a nonaxisymmetric shape is evidenced when temperature is varied. By varying the reduced volume, the bifurcation takes place in the neighborhood of the Clifford torus. This is consistent with all the models, though the ADE model predicts that this transition might take place at larger reduced volumes. For a reduced volume larger than 0.71, one scans reversibly through the Dupin cyclides shapes, conformal transforms of the Clifford torus.

The experimental study of the phase diagram requires the knowledge of geometric parameters such as area, volume, and average mean curvature. These are in principle accessible by imaging techniques, and one usually assumes that they are characteristic of a vesicle and that they are time independent. Although it is known that flip-flop of phospholipid molecules is a very slow process compared with the lateral diffusion time scale, it can play a role in the equilibrium shape problem. This discussion is consistent with the generic observation of circular tori for newly, but not freshly, prepared solutions of vesicles. All these observations are in agreement with the three models for the shape problem if metastable states are taken into account. However, one should distinguish between the short time scale observations, where the bilayer couple model may be relevant, and the longer time scale situations, where the spontaneous curvature or the ADE model might apply.

Mathematicians have provided examples of surfaces of higher topology which are experimentally accessible. One of the most spectacular effects for genus $g \geq 2$ surfaces is the possible degeneracy of whole minimal families even in the presence of constraints. In this review we have provided examples of vesicles exhibiting the so-called conformal diffusion process, confirming thereby one more time the importance of the curvature energy. Finally, low reduced volume vesicles are very sensitive to thermal fluctuations. Holes or passages fluctuate violently. This effect is best illustrated by an electrostatic analogy where the necks bear an electrical charge. Total neutrality is related to curvature so that one can compute the equilibrium shapes. Dynamic fluctuations can also be understood in this approach, since the interaction between holes is controlled by the overlap domain between their inner core and an outer domain with an almost constant curvature. This point of view may be seen as an intermediate step towards surfaces of more complex topology such as the sponge phases.

REFERENCES

1. **J. Charvolin and J. F. Sadoc.** Periodic systems of frustrated fluid films and "bicontinuous" cubic structures in liquid crystals. *J. Phys. (Paris),* 48:1559–1569, 1987.
2. **W. Helfrich.** Amphiphilic mesophases made of defects, in *Physics of Defects,* North-Holland Publishing, Amsterdam, 1981, pages 715–755.
3. **M. E. S. Gunning and M. W. Steer.** *Ultrastructure and the Biology of Plant Cells,* Edward Arnold, Ltd., London, 1975.
4. **W. Harbich and W. Helfrich.** Alignment and opening of giant lecithin vesicles by electric fields. *Z. Naturforsch. A,* 34:1063–1065, 1979.
5. **M. Winterhalter and W. Helfrich.** Effect of voltage on pores in membranes. *Phys. Rev. A,* 36(12):5874–5876, 1987.
6. **C. Taupin, M. Dvolaitzky, and C. Sauterey.** Osmotic pressure induced pores in phospholipids vesicles. *Biochemistry,* 14(21):4771, 1975.
7. **D. Roux, C. Coulon, and M. E. Cates.** Sponge phases in surfactant solutions. *J. Phys. Chem.,* 96(11):4174–4187, 1992.

8. **D. A. Huse and S. Leibler.** Phase behavior of an ensemble of noninteracting random films. *J. Phys. (Paris)*, 49(4):605, 1988.
9. **R. Lipowsky.** The conformation of membranes. *Nature*, 349:475–481, 1991.
10. **D. Nelson, T. Piran, and S. Weinberg,** Eds. *Statistical Mechanics of Membranes and Surfaces.* World Scientific, Singapore, 1989.
11. **X. Michalet.** Etude Expérimentale de Vésicules Phospholipidiques de Genre Topologique Non-Sphérique. Ph.D. thesis, Paris VII, 1994.
12. **X. Michalet and D. Bensimon.** Observation of stable shapes and conformal diffusion in genus 2 vesicles, *Science*, 269:666–668, 1995; Vesicles of toroidal topology: observed morphology and shape transformations, *Journal de Physique*, 5:263–287, 1995.
13. **B. Klösgen.** Private communication.
14. **U. Seifert and R. Lipowsky.** Shapes of fluid vesicles, *Nonmedical Applications of Liposomes: Theory and Basic Sciences*, Vol. 1, Lasic, D. and Barenholz, Y., Eds., CRC Press, Boca Raton, FL, 1986, chap. 2.
15. **U. Pinkall and I. Stirling.** Willmore surfaces. *Math. Intell.*, 9(2):38–43, 1987.
16. **H. B. Lawson, Jr.** Complete minimal surfaces in S^3. *Ann. Math.*, 92:335–374, 1970.
17. **K. A. Brakke.** The surface evolver. *Exp. Math.*, 1:141–165, 1992.
18. **F. Jülicher, U. Seifert, and R. Lipowsky.** Conformal degeneracy and conformal diffusion of vesicles. *Phys. Rev. Lett.*, 71(3):452–455, 1993.
19. **F. Jülicher.** Die Morphologie von Vesikeln. Ph.D. thesis, Universität zu Köln, Köln, Germany, 1993.
20. **X. Michalet, D. Bensimon, and B. Fourcade.** Fluctuating vesicles of nonspherical topology. *Phys. Rev. Lett.*, 72:168–171, January 1994.
21. **B. Duplantier.** Exact curvature energies of charged membranes of arbitrary shapes. *Phys. A*, 168:179–197, 1990.
22. **T. J. Willmore.** *Total Curvature in Riemannian Geometry.* Ellis Horwood, New York, 1982.
23. **Y. C. Ou-Yang.** Anchor ring-vesicle membranes. *Phys. Rev. A*, 41(8):4517–4520, April 1990.
24. **P. B. Canham.** The minimum energy of bending as a possible explanation of the biconcave shape of the human red blood cell. *J. Theor. Biol.*, 26:61, 1970.
25. **W. Helfrich.** Elastic properties of lipid bilayers: theory and possible experiments. *Z. Naturforsch.*, 28c:693–703, 1973.
26. **M. P. Sheetz and S. J. Singer.** Biological membranes as bilayer couples. A molecular mechanism of drug-erythrocyte interaction. *Proc. Natl. Acad. Sci. U.S.A.*, 71(11):4457–4461, November 1974.
27. **E. A. Evans.** Bending resistance and chemically induced moments in membrane bilayers. *Biophys. J.*, 14:923–931, 1974.
28. **S. Svetina and B. Žekš.** The elastic deformability of closed multilayerer membranes is the same as that of a bilayer membrane. *Eur. Biophys. J.*, 21:251–255, 1992.
29. **V. Kralj-Iglic, S. Svetina, and B. Žekš.** The existence of non-axisymmetric bilayer vesicle shapes predicted by the bilayer couple model. *Eur. Biophys. J.*, 22:97–103, 1993.
30. **R. E. Waugh, J. Song, S. Svetina, and B. Žekš.** Local and non local curvature elasticity in bilayer membranes by tether formation from lecithin vesicles. *Biophys. J.*, 61:974–982, April 1992.
31. **W. Wiese, W. Harbich, and W. Helfrich.** Budding of lipid bilayer vesicles and flat membranes. *J. Phys. Condensed Matter*, 4:1647–1657, 1992.
32. **L. Miao.** Equilibrium Shapes and Shape Transitions of Fluid Lipid-Bilayer Vesicles. Ph.D. thesis, Simon Fraser University, Burnaby, British Columbia, Canada, 1992.
33. **L. Miao, U. Seifert, M. Wortis, and H. G. Döbereiner.** Budding transitions of fluid-bilayers vesicles: the effect of area difference elasticity. *Phys. Rev. E*, 49(6):5389, 1994.
34. **C. Itzykson and J. M. Drouffe.** *Statistical Field Theory,* Vol. 2. Cambridge University Press, Cambridge, 1989. See pages 505 and forward.
35. **F. Jülicher, U. Seifert, and R. Lipowsky.** Phase diagrams and shape transformations of toroidal vesicles. *J. Phys. II (Paris)*, 3:1681, September 1993.
36. **L. Miao, B. Fourcade, M. Rao, M. Wortis, and R. K. P. Zia.** Equilibrium budding and vesiculation in the curvature model of fluid lipid vesicles. *Phys. Rev. A*, 43(12):6843–6856, 1991.
37. **U. Seifert, K. Berndl, and R. Lipowsky.** Shape transformations of vesicles: phase diagram for spontaneous curvature and bilayer coupling model. *Phys. Rev. A*, 44:1182, July 1991.
38. **K. Berndl, J. Käs, R. Lipowsky, E. Sackmann, and U. Seifert.** Shape transformations of giant vesicles: extreme sensitivity to bilayer asymmetry. *Europhys. Lett.*, 13(7):659–664, 1990.
39. **B. Fourcade, L. Miao, M. Rao, M. Wortis, and R. K. P. Zia.** Scaling analysis of narrow necks in curvature models of fluid lipid-bilayer vesicles. *Phys. Rev. E*, 49(6):5276–5286, 1994.
40. **U. Seifert.** Vesicles of toroidal topology. *Phys. Rev. Lett.*, 66(18):2404–2407, 1991.
41. **D. Hilbert and S. Cohn-Vossen.** *Geometry and Imagination.* Chelsea, New York, 1983.
42. **M. Kléman.** Energetics of the focal conics of smectic phases. *J. Phys. (Paris)*, 38:1511–1518, 1977.
43. **P. Yager and P. E. Schoen.** Formation of tubules by a polymerizable surfactant. *Mol. Cryst. Liq. Cryst.*, 106:371–381, 1984.
44. **J. M. Schnur, R. R. Price, P. E. Schoen, P. Yager, J. M. Calvert, J. Georger, and A. Singh.** Lipid-based tubule microstructures. *Thin Solid Films*, 152:181–206, 1987.

45. **P. Yager, R. R. Price, J. M. Schnur, P. E. Schoen, A. Singh, and D. G. Rhodes.** The mechanism of formation of lipid tubules from liposomes. *Chem. Phys. Lipids,* 46:171–179, 1988.

46. **J. V. Selinger and J. M. Schnur.** Theory of chiral lipid tubules. *Phys. Rev. Lett.,* 71(24):4091–4094, 1993.

47. **W. Helfrich and J. Prost.** Intrinsic bending force in anisotropic membrane made of chiral molecules. *Phys. Rev. A,* 38(6):3065, September 1988.

48. **M. Mutz and D. Bensimon.** Observation of toroidal vesicles. *Phys. Rev. A,* 43(8):4525–4527, April 1991.

49. **B. Fourcade, M. Mutz, and D. Bensimon.** Experimental and theoretical study of toroidal vesicles. *Phys. Rev. Lett.,* 68(16):2551–2554, April 1992.

50. **Y. C. Ou-Yang.** Selection of toroidal shape of partially polymerized membranes. *Phys. Rev. E,* 47(1):747–749, January 1993.

51. **U. Seifert.** Conformal transformations of vesicle shapes. *J. Phys. A. Math. Gen.,* 24:L573, 1991.

52. **B. Fourcade.** Theoretical results on toroidal vesicles. *J. Phys. II (Paris),* 2:1705–1724, 1992.

53. **J. M. Drouffe and A. C. Maggs.** Personal communication.

54. **R. Mosseri, J. F. Sadoc, and J. Charvolin.** Some remarks on the shape of toroidal vesicles. In R. Lipowsky, D. Richter, and K. Kremer, Eds. *The Structure and Conformation of Amphiphilic Membranes,* Springer-Verlag, Berlin, 1992, pp. 97–100.

55. **R. Kusner.** Comparison surfaces for the Willmore problem. *Pac. J. Math.,* 138:317–345, 1989.

56. **L. Hsu, R. Kusner, and J. Sullivan.** Minimizing the squared mean curvature integral for surfaces in space forms. *Exp. Math.,* 1(3):191–207, 1992.

57. **A. G. Petrov and M. M. Koslov.** Curvature elasticity and passages formation in lipid bilayer lattice of passages. *C. R. Acad. Bulg. Sci.,* 37(9):1192–1194, 1984.

58. **J. Goos and G. Gompper.** Topological defects in lamellar phases: passages, and their fluctuations. *J. Phys. I (Paris),* 3:1551–1567, 1993.

59. **T. Charitat and B. Fourcade.** Lattices of passages connecting membranes, *Phys. Rev. Lett.* (submitted).

Chapter 4

Computer Simulation of the Thermodynamic and Conformational Properties of Liposomes

B. Dammann, H. C. Fogedby, J. H. Ipsen, C. Jeppesen, K. Jørgensen,
O. G. Mouritsen, J. Risbo, M. C. Sabra, M. M. Sperotto, and M. J. Zuckermann

CONTENTS

I. INTRODUCTION

A. MOTIVATION FOR THE PRESENT REVIEW

Liposomes are mesoscopic molecular aggregates that typically consist of 10^4 to 10^6 molecules for liposomal diameters in the range of 300 to 3000 Å. Liposomes are therefore finite systems which are far from the thermodynamic limit and hence may display distinct finite-size effects in their thermodynamic behavior.[1,2] Still, they are many-particle systems which exhibit correlated and cooperative phenomena in a sufficiently effective manner to make it meaningful to talk about their thermodynamic state and to describe their physical properties in terms of phase equilibria and by means of thermodynamics functions.

0-8493-4731-9/96/$0.00+$.50

These properties include the specific heat and the lateral compressibility, as well as continuum-mechanical moduli such as the bending rigidity.

The many-particle character of liposomes is the central issue of the present review, which seeks to clarify the manner in which their conformational and thermodynamic properties are controlled by the cooperative and fluctuating modes in the lipid bilayer that makes up the liposome. These modes, which are highly nontrivial consequences of the many-body nature of the system, are of importance for the spatial configuration and topology as well as the molecular compositional organization of the lipid bilayer, and they are hence crucial for controlling liposome stability and permeability.

Liposomes are examples of soft condensed matter[3] which is itself characterized by a high degree of deformability and by thermally renormalized material properties. Indeed, liposomes are complex structured fluids with a liquid crystalline character that have aspects of both order and disorder. Their properties are strongly influenced by entropic effects.

In this chapter we shall make an attempt to bring together two separate viewpoints of liposomes, one of which deals with the molecular structure and composition of essentially planar bilayers and the other which considers the bilayer and liposome as a flexible fluid surface possibly characterized by some intrinsic continuum-mechanical properties. These two viewpoints translate into two different length scales, one corresponding to the microscopic and nanoscopic length scale and the other which operates in the continuum limit on the scale of the entire liposome and which neglects the small-scale structure of the system. By bringing the two viewpoints together one should ultimately be able to provide a full description of the generic properties of liposomes.

In this chapter we will not be material-specific in our description of liposome properties, but we shall instead concentrate on the general properties of the system. We will, however, also present some results on the physicochemical properties of specific classes of lipid material. Our description is of a theoretical nature but with reference to pertinent experimental results. The emphasis will be on the use of computer-simulation techniques to derive the properties of liposomes using a set of simple statistical mechanical models formulated in terms of different sets of mechanical variables, depending on which properties are under consideration. A computer-simulation approach is particularly suited for the calculation of the conformational and the thermodynamic properties of liposomes as these are controlled by collective phenomena. The simulations are capable of fully accounting for the fluctuations as well as the total entropy of the liposome systems. As will be seen below, the soft nature of liposomes implies that their properties to a substantial extent are determined by entropic factors.

B. LIPID BILAYERS, LIPOSOMES, AND MODEL MEMBRANES

Lipid bilayer aggregates form spontaneously from lipid molecules in aqueous solutions.[4] The stability of the bilayers is controlled by several different and often competing intra- and interbilayer interactions, which all represent weak physical forces.[5] Under those conditions that are physiologically relevant, the bilayer material is soft and is characterized by a considerable degree of elastic deformability and lateral area compressibility.[6] Under suitable conditions, the bilayers are formed as unilamellar structures that normally close onto themselves with the topological shape of a sphere, although other topological forms have been observed. Small closed bilayers are usually called vesicles whereas larger systems are referred to as liposomes (see other chapters of the present volume).

Lipid vesicles and liposomes are model systems of the lipid-bilayer component of biological membranes, and their physics and physical chemistry have been studied extensively[4] in order to investigate the structure and dynamics of cell membranes. However, liposomes are not only of interest as models of biological membranes; they are also of substantial technological importance for medical applications.[7] Furthermore, they are worthy of study in their own right. The present review deals with nonmedical applications of liposomes and presents liposomes as a unique realization of the soft condensed state of matter.[3] A motivation for undertaking a pure science study of soft matter is that there are interesting new and fundamental physics to be learned from the study of soft materials, particularly materials such as liposomes, which embody the unique material design principles that Nature herself has developed by evolutionary processes over the last 3 to 4×10^9 years.[8]

C. PHASE TRANSITIONS IN LIPOSOMES

Lipid bilayers undergo a series of phase transitions[9] driven by temperature and composition as well as a variety of external fields, such as hydrostatic and osmotic pressure, pH, and electric fields. Some of these transitions transform the bilayer into a nonbilayer phase[10] such as a cubic or hexagonal phase, and they may in this way destabilize the liposomes. Other transitions involve structural changes within the bilayer state — for example, chain melting, which occurs during the main phase transition (see below). Others types of transitions can result in changes in the surface morphology and even the overall topology of liposomes.

In this chapter we shall first be concerned with the main bilayer transition. This transition is temperature driven and takes the bilayer from a solid-ordered or crystalline (gel) phase with translational order and a high degree of lipid acyl-chain order to a liquid-disordered or fluid (liquid-crystalline) phase with translational disorder and a certain degree of lipid acyl-chain disorder. The liposome remains in its bilayer state during this transition. Second, we shall describe transitions that involve morphological changes of closed lipid bilayers induced by osmotic gradients (the inflation-deflation transition) and transitions between different topological shapes of closed lipid bilayers.

Phase transitions have dramatic consequences for global liposome properties such as the specific heat, the compressibility, and the permeability. On the microscopic and nanoscopic level, phase transitions imply lipid-domain formation and dynamic heterogeneity due to thermal and compositional fluctuations.[11,12] For many-component liposomes, phase transitions imply phase equilibria and phase-separation phenomena, i.e., global static phase separation, which may strongly alter the lateral molecular organization of the liposome and its conformational properties. Hence, phase transitions and phase equilibria are a major concern when the stability of liposomes is at issue. Moreover, phase transitions and, in particular, critical phenomena are of substantial theoretical interest since they can be described by general scaling behavior[13] and are possibly subject to classification in terms of universality classes.

II. FUNDAMENTAL MECHANICAL VARIABLES IN THE DESCRIPTION OF LIPID MEMBRANES

A. PLANAR LIPID BILAYERS IN TWO DIMENSIONS: TRANSLATIONAL AND CONFORMATIONAL VARIABLES

The formulation of a useful theoretical model for lipid bilayers is a balance between physical realism and computational feasibility. This balance becomes quite delicate when phase transitions are at issue because it requires that large assemblies of molecules are taken into account. With our present knowledge about interatomic potentials in hydrocarbon systems it is possible to write down a rather realistic model of a lipid bilayer membrane. However, it would be virtually impossible to calculate any properties near the phase transition of such a model due to its complexity. In contrast, it is possible to design highly simplified models which are analytically tractable, though this is sometimes at the expense of physical realism. The virtue of the analytically solvable models is that they have provided us with valuable information about those features of the interaction potentials which are necessary in order to produce the characteristics of the transition and those which are only marginally relevant. Such information is indispensable when the next level of more realistic but still computationally tractable models is approached. The single most important lesson to be learned from the analytical studies[14] is that although the lipid acyl-chain conformational statistics actually drive the transition, it is the excluded-volume interactions which dominate the phase behavior.

A theoretical treatment of any complex molecular system requires an identification of those mechanical variables which are relevant for the phenomenon under consideration. Obviously it is usually impossible as well as undesirable to treat any given phenomenon using the full set of variables. Most of the variables therefore have to be "frozen out" in the calculation. In models of the main transition, a planar lamellar membrane geometry is assumed and the lipid molecules are confined to a plane by anchoring the level of the glycerol backbones. Second, all but the acyl-chain degrees of freedom are usually "frozen out"; i.e., the flexibility of the polar headgroup is neglected. Third, the two acyl chains of a diacyl lipid molecule are often assumed to be independent. This leaves us with a set of fundamental variables, $\{m_i, \vec{r}_i\}$, where m denotes the conformational state of the acyl chain and $\vec{r}$ denotes the translational variables of the chain.

Even this dramatic reduction in the degrees of freedom of the membrane is not sufficient to permit the construction of models which are computationally tractable in the transition region. This may sound surprising, but it can be appreciated by considering the many unsolved problems regarding the transition in as simple a system as a two-dimensional particle system with purely translational degrees of freedom.

Hence, most approaches introduce dramatic assumptions about at least one of the two variables, m or $\bar{r}$. Different strategies which have been followed put emphasis on different variables: (1) conformational variables and frozen translational variables, (2) conformational variables and approximate translational variables, and (3) approximate conformational variables and translational variables. Only strategies (1) and (2) at the moment have led to reliable theoretical predictions for the main transition of a type which can be quantitatively compared with experiments.

In this section on planar lipid bilayers we shall only be concerned with microscopic interaction models that focus on the acyl-chain conformational variables and therefore freeze in the translational variables (strategy 1 above). Other strategies involving positional variables are discussed in References 15 through 20. The simplest possible way of freezing out the translational variables is to adopt a lattice approximation in which each acyl chain is positioned on a regular (e.g., triangular) two-dimensional lattice. The multistate Pink model[21-23] used extensively in our work builds on such a lattice approximation. Within the description of the Pink model it is furthermore assumed that the conformational properties of a single acyl chain can be described by a small number of selected conformational states corresponding to the mapping of the three-dimensional acyl-chain conformations onto a finite, discrete set of projected two-dimensional coarse-grained variables. The number of states included in a multistate model depends on the level of detail required. In Section III.A we describe how the mechanical variables can be included in the formalism.

B. RANDOM SURFACES, FLUCTUATING LIPID BILAYERS, AND THE THIRD DIMENSION: MEMBRANES AS GENERALIZED POLYMERS

Even though a tractable analysis of the phase transitions taking place in a lipid bilayer at the moment calls for a two-dimensional lattice description, as discussed above, the bilayers are in fact highly flexible and fluctuate strongly in the "third dimension".[24] Lipid bilayer membranes belong to the very interesting class of random surfaces which play an important role both in soft condensed matter and in quantum gravity and string theory.[25,26]

Owing to the small surface tension a lipid bilayer exhibits large thermal fluctuations at ambient temperatures. In order to understand the physical properties of a thermally fluctuating bilayer and in order to identify the relevant mechanical variables of the bilayer it is useful to consider the strong resemblance between the properties of membranes and those of polymers.[27]

The basic conformation of a fluctuating polymer is given by the statistics of the erratic path of a particle undergoing Brownian or random motion. Consequently, the linear size, R (e.g., the radius of gyration, R_G), of a polymer is proportional to the square root of the number of units or monomers, N, in the polymer. Generally, this property is characterized by a scaling law, $R \sim N^\nu$, where ν is known as the critical exponent for a random walk, $\nu = 1/2$. Since the polymer cannot intersect itself, i.e., including self-avoidance or excluded volume effects, the polymer coil swells, and the exponent ν is slightly larger than 1/2. Mean-field-theory arguments and renormalization group studies of polymers show that ν depends on the dimension of the environment. For polymers in two dimensions, ν is known exactly and has the value 3/4. The critical dimension for polymers is four; above four dimensions self-avoidance is irrelevant, and ν takes the random-walk value 1/2. It is important to note that polymers are always at the critical point exhibiting scale invariance. Apart from a local short-range stiffness due to steric hindrance, the scale-invariant fluctuations are entirely entropic.

A fluctuating membrane is in many respect a two-dimensional generalization of a polymer, but the description of a membrane is more subtle. In order to extract the thermodynamics and critical properties of a fluctuating membrane in solution, we would have to evaluate the partition function Z. This requires a summation over all membrane conformations, including different topologies weighted with the corresponding Boltzmann factor, $\exp(-\mathcal{H}/k_B T)$. In other words, one needs to calculate $Z = \sum \exp(-\mathcal{H}/k_B T)$, where $\mathcal{H}$ is the Hamiltonian or energy of the configuration, k_B the Boltzmann constant, and T the absolute temperature. The thermodynamics and critical properties follow from the partition function since the free energy F is given by $F = -k_B T \log Z$.

Unlike the polymeric case, the membranes typically fall into two classes:[26] (1) solid membranes or polymerized membranes, which can sustain lateral shear and are the direct analogies of polymers; and (2) fluid membranes, which have in-surface fluid degrees of freedom and cannot sustain a shear. Lipid bilayers belong to the class of fluid membranes. In Section III.B we shall describe how the Hamiltonian for a fluctuating membrane is constructed.

III. THEORY AND MODELS OF LIPID BILAYER MEMBRANES

A. LATTICE MODELS OF THE MAIN PHASE TRANSITION AND PHASE EQUILIBRIA IN PLANAR LIPID BILAYERS

Having identified the fundamental variables of the problem, the next step is to couple the variables by means of a "model" in which the physical interactions are included. There are two conceptually different ways of doing this. One operates on the level of a phenomenological free energy. This free energy is basically an expansion (Landau type) of the energy in terms of certain average values of the variables which are termed the order parameters. The interactions of the system are reflected in the mathematical structure of this expansion and the expansion parameters. The other way of constructing a model proceeds via a microscopic interaction model which implies a Hamiltonian function defined in terms of the chosen relevant microscopic mechanical variables. The Hamiltonian includes the correct interaction potentials, possibly parametrized in terms of interaction constants which may be unknown. Microscopic models require a full statistical mechanical treatment of the related many-body problem whose solution usually must be found using computer simulation.

In the Pink lattice model[21-23,28] referred to in Section II.A the conformational chain variables are coupled by anisotropic van der Waals interactions in the spirit of anisotropic liquid crystals. The lattice approximation automatically takes the excluded-volume interactions into account. The interactions between the hydrophilic moities are modelled by a Coulomb-like force or more simply by an effective intrinsic lateral pressure. The Pink model allows for a series of ten conformational states of the acyl chains of the lipid molecules. The ten-state model provides a reasonably accurate description of the phase-behavior of pure lipid bilayers and the associated density fluctuations since it accounts for the most important conformational states of the lipid chains as well as their mutual interactions and statistics. In the Pink model the bilayer is considered to be composed of two monolayer sheets which are independent of each other. Each monolayer is represented by a triangular lattice. The model is therefore a pseudo-two-dimensional lattice model which neglects the translational modes of the lipid molecules and focuses on the conformational degrees of freedom of the acyl chains. The model is illustrated in Figure 1. Since we will be extending the formalism to include several molecular species, we label the lipid variables corresponding to a particular lipid species "A". Each acyl chain can take on one of ten conformational states m, each of which is characterized by an internal energy E_m^A, a hydrocarbon chain length d_m^A (corresponding to half a bilayer, $2d_m^A = d_L$), and a degeneracy D_m^A, which accounts for the number of conformations that have the same area A_m^A and the same energy E_m^A, where $m = 1, 2, ..., 10$. The ten states can be derived from the all-*trans* state in terms of *trans-gauche* isomerism. The state $m = 1$ is the nondegenerate gel-like ground state, representing the all-*trans* conformation, while the state $m = 10$ is a highly degenerate excited state characteristic of the melted or fluid phase. The eight intermediate states are gel-like states containing kink and jog excitations satisfying the requirement of low conformational energy and optimal packing. The conformation energies, E_m^A, are obtained from the energy required for a *gauche* rotation (0.45×10^{-13} erg) relative to the all-*trans* conformation. The values of D_m^A are determined by combinatorial considerations.[22] The chain cross-sectional areas, A_m^A, are reciprocally

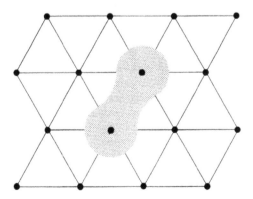

Figure 1 Schematic illustration of the two-dimensional lattice model used to describe planar lipid bilayers. The position on the triangular lattice of a diacyl phospholipid is indicated by its cross-sectional area.

related to the values of d_m^A since the volume of an acyl chain varies only slightly with temperature changes. The saturated hydrocarbon chains are coupled by nearest-neighbor anisotropic forces which represent both van der Waals and steric interactions. These interactions are formulated in terms of products of shape-dependent nematic factors. The lattice approximation automatically accounts for the excluded volume effects and to some extent for that part of the interaction with the aqueous medium which assures bilayer integrity. An effective lateral pressure, II, is included in the model.

The Hamiltonian (which represents the total energy of a microconfiguration) for the pure lipid bilayer can then be written

$$\mathcal{H}^A = \sum_i \sum_{m=1}^{10} \left(E_m^A + \Pi A_m^A \right) \mathcal{L}_{im}^A - \frac{J_A}{2} \sum_{\langle i,j \rangle} \sum_{m,n=1}^{10} I_m^A I_n^A \mathcal{L}_{im}^A \mathcal{L}_{jn}^A \tag{1}$$

where J_A is the strength of the van der Waals interaction between neighboring chains and $I_m^A I_n^A$ is an interaction matrix which involves both distance and shape dependence.[22] $\langle i,j \rangle$ denotes nearest-neighbor indices on the triangular lattice. The Hamiltonian is expressed in terms of site occupation variables $\mathcal{L}_{im}^A$: $\mathcal{L}_{im}^A = 1$ if the chain on site i is in state m; otherwise $\mathcal{L}_{im}^A = 0$. The model parameters J_A and II are chosen in such a way as to reproduce the transition temperature and transition enthalpy for a pure DPPC bilayer.[23] The values for other phospholipids are then determined by direct scaling.[29,30]

Extensions of the ten-state Pink model in Equation 1 have recently been made by Zhang et al.[31,32] in order to study effects of acyl-chain mismatch among lipid species of the same type and effects due to interactions between the two monolayers of the bilayer. The lipid mismatch model[31] of the main phase transition in lipid bilayer membranes was introduced to clarify the nature of this transition, which is still controversial, both experimentally[2] and theoretically.[28]

The ten-state Pink model for a pure lipid bilayer was also extended to the description of binary lipid mixtures[29,30,33] by explicitly incorporating a mismatch term which accounts for the incompatibility of acyl chains of different hydrophobic lengths of the two species. The Hamiltonian for a binary mixture of the two lipid species A and B is written

$$\mathcal{H} = \mathcal{H}^A + \mathcal{H}^B + \mathcal{H}^{AB} \tag{2}$$

where the two first terms describe the interaction between like species and the last term denotes the interaction between different species. The composition of the mixture is given by $x_B = \sum_m \langle \mathcal{L}_{im}^B \rangle = 1 - x_A$. The interaction between different lipid species is described by the Hamiltonian

$$\mathcal{H}^{AB} = \frac{-J_{AB}}{2} \sum_{\langle i,j \rangle} \sum_{m,n=1}^{10} \left(I_m^A I_n^B \mathcal{L}_{im}^A \mathcal{L}_{jn}^B + I_m^B I_n^A \mathcal{L}_{im}^B \mathcal{L}_{jn}^A \right)$$

$$+ \frac{\Gamma_{AB}}{2} \sum_{\langle i,j \rangle} \sum_{m,n=1}^{10} \left(\left| d_{im}^A - d_{jn}^B \right| \mathcal{L}_{im}^A \mathcal{L}_{jn}^B + \left| d_{im}^B - d_{jn}^A \right| \mathcal{L}_{im}^B \mathcal{L}_{jn}^A \right) \tag{3}$$

The first term in $\mathcal{H}^{AB}$ describes the direct van der Waals hydrophobic contact interaction between different acyl chains. The corresponding interaction constant is taken to be the geometric average $J_{AB} = \sqrt{J_A J_B}$. Γ_{AB} in the second term of Equation 3 represents the mismatch interaction and has been shown to be "universal" in the sense that its value does not depend on the lipid species in question for non-ideal mixtures.[30] The value of the mismatch parameter used in the simulation approach to the statistical mechanics of the model in Equation 3 was found to be $\Gamma_{AB} = 0.038$ erg/Å.[29]

The model presented above for binary lipid mixtures can be readily modified to account for a mixture involving cholesterol in the case where only the interaction between the sterol and the conformational degrees of freedom of the lipid acyl chains is taken into account.[34,35] The Hamiltonian for the mixture of lipid of species A and cholesterol (C) is written

$$\mathcal{H} = \mathcal{H}^A + \mathcal{H}^C \tag{4}$$

where

$$\mathcal{H}^C = \Pi A^C \sum_i \mathcal{L}_i^C - \frac{J_A}{2} \sum_{\langle i,j \rangle} \sum_{m,n=1}^{10} I_m^A I_C \left(\mathcal{L}_{im}^A \mathcal{L}_j^C + \mathcal{L}_i^C \mathcal{L}_{jn}^A \right) - \frac{J_A}{2} \sum_{\langle i,j \rangle} I_C^2 \mathcal{L}_i^C \mathcal{L}_j^C \tag{5}$$

Here $\mathcal{L}_i^C = 0,1$ denotes the occupation variables for the cholesterol molecules, A^C is the molecular cross-sectional area of cholesterol, and $I_C = 0.45^{34}$ is a phenomenological interaction parameter.

A modification of the Hamiltonian for the ten-state Pink model has been proposed to describe the effects of certain drugs and insecticides (denoted foreign molecules) on lipid bilayer properties.[36-39] The model is intended to describe the effects of those types of foreign molecules that do not change the transition enthalpy of the bilayer. The adsorbed foreign molecules in that case can be considered as interstitial impurities whose available locations are the centers of the triangles formed by three neighboring lipid acyl chains. The sites available to the foreign molecules hence constitute a honeycomb lattice embedded in the triangular lattice formed by the acyl chains. The requirement that the lipid chains and the foreign molecules occupy separate lattices and hence cannot mix results automatically in the absence of an entropy of mixing. It is assumed that the interstitial impurities interact with the three neighboring lipid acyl chains as well as with other impurities occupying neighboring interstitial sites on the honeycomb lattice. The Hamiltonian consists of terms describing the pure lipid-bilayer interactions, the lipid-foreign molecule interactions, and the interactions between neighboring foreign molecules, respectively; i.e., for the lipid species A interacting with the foreign molecule D

$$\mathcal{H} = \mathcal{H}^A + \mathcal{H}^D \tag{6}$$

where

$$\mathcal{H}^D = \sum_i \sum_m \sum_\ell J_{AD}^m I_m^A \mathcal{L}_{im} \mathcal{L}_\ell^D - \frac{J_{DD}}{2} \sum_{\langle \ell,k \rangle} \mathcal{L}_\ell^D \mathcal{L}_k^D \tag{7}$$

$\mathcal{L}_\ell^D = 0,1$ is an interstitial site occupation variable. J_{AD}^m and J_{DD} are interaction constants for the lipid-foreign molecule and the foreign molecule-foreign molecule interactions, respectively, as described in References 36 through 39. The concentration, x, of foreign molecules in the membrane is not conserved since partitioning between the membrane and the aqueous phase is assumed to occur. The use of a grand canonical ensemble is therefore appropriate and x is controlled by a chemical potential, μ. The effective Hamiltonian then becomes

$$\mathcal{H}_{grand} = \mathcal{H} - \mu \sum_\ell \mathcal{L}_\ell^D \tag{8}$$

Several extensions of the ten-state Pink model have been proposed in order to account for lipid-protein interactions in a specific fashion which depends on the conformational states of the lipid chains. Here we consider a particular model in which the lipid-protein interactions have been incorporated into the microscopic Pink model by identifying part of the interaction parameters in terms of hydrophobic matching between the hydrophobic length of the lipid chain and hydrophobic protein length. This is implemented in the spirit of the phenomenological mattress model of lipid-protein interactions.[40-42] The lipid-protein interactions were included in the model by assuming that the hydrophobic membrane-spanning part of the protein or polypeptide molecule is a stiff, rod-like, and hydrophobically smooth object with no appreciable internal flexibility. In this way the protein is characterized geometrically only by a cross-sectional area, A_P (or circumference ρ_P), and a hydrophobic length d_P. The protein can occupy one or more sites of the lipid lattice depending on its actual hydrophobic circumference.

The Hamiltonian of the microscopic version of the mattress model for a lipid bilayer of species A interacting with a protein P is now written[43-45]

$$\mathcal{H} = \mathcal{H}^A + \mathcal{H}^{AP} \tag{9}$$

where

$$\mathcal{H}^{AP} = \Pi A_P \sum_i L_{Pi} + \frac{\Gamma_{AP}}{4}\left(\frac{\rho_P}{z}\right)\sum_{\langle i,j\rangle}\sum_{m=1}^{10}\left(\left|2d_{im}^A - d_P\right|\mathcal{L}_{im}^A L_j^P + \left|2d_{jm}^A - d_P\right|\mathcal{L}_{jm}^A L_i^P\right)$$

$$-\frac{J_{AP}}{4}\cdot\left(\frac{\rho_P}{z}\right)\sum_{\langle i,j\rangle}\sum_{m=1}^{10}\left(\min\left(2d_{im}^A, d_P\right)\mathcal{L}_{im}^A L_j^P + \min\left(2d_{jm}^A - d_P\right)\mathcal{L}_{jm}^A L_i^{\ P}\right) \tag{10}$$

The parameter J_{AP} is related to the direct lipid protein van der Waals-like interaction which is associated with the interfacial hydrophobic contact of the two molecules, while the parameter Γ_{AP} is related to the hydrophobic effect. $L_i^P = 0,1$ is the protein occupation variable. $\mathcal{L}_{im}^A$ and L_i^P satisfy a completeness relation at each lattice site, $\sum_m \mathcal{L}_{im}^A + L_i^P = 1$. The model in Equation 10 can be readily extended to binary lipid mixtures using Equations 2 and 3. The appropriate values of the lipid-protein interaction parameters, Γ_{AP} and J_{AP}, are discussed in Reference 46.

B. MODELS INVOLVING MEMBRANE CURVATURE, BENDING ELASTICITY, AND TOPOLOGY

We shall now describe how the large-scale properties of a fluctuating, fluid membrane are described in terms of an appropriate Hamiltonian that couples the mechanical variables of the surface. Clearly, the free energy of a fluctuating fluid membrane cannot depend on a specific parametrization of the membrane surface, and the Hamiltonian $\mathcal{H}$ must therefore be constructed in terms of surface invariants. The first term in $\mathcal{H}$ is proportional to the surface area and the surface tension, γ. The second term in $\mathcal{H}$ characterizes the bending fluctuations or undulations of the membrane and is given by the surface integral of the square of the mean curvature, G. Finally, there is a term in $\mathcal{H}$ given by the surface integral of the Gaussian curvature, K. The Hamiltonian thus takes the following general form in the case of a fluid membrane

$$\mathcal{H} = \gamma \int dA + \frac{\kappa}{2}\int dA\, G^2 + \frac{\overline{\kappa}}{2}\int dA\, K \tag{11}$$

Here dA is the surface element, γ the surface tension, κ the bending rigidity, and $\overline{\kappa}$ the Gaussian bending rigidity. In terms of the radii of curvature, R_1 and R_2, we have $G = 1/R_1 + 1/R_2$ and $K = 2/(R_1R_2)$. In the case of an interface or meniscus separating two coexisting phases, the surface tension dominates and tends to stretch the membrane, i.e., to minimize the area. However, it is a characteristic of biological membranes that the surface tension is vanishingly small, $\gamma \simeq 0$, and the Hamiltonian can therefore be written[47]

$$\mathcal{H} = \frac{\kappa}{2}\int dA\left(\frac{1}{R_1} + \frac{1}{R_2}\right)^2 + \frac{\overline{\kappa}}{2}\int dA\,\frac{2}{R_1R_2} - p\int dV \tag{12}$$

where we have also included a $p - V$ term for the case of a closed surface.

It is important to note that for biological membranes κ and $\overline{\kappa}$ are of the order k_BT. Consequently, the membrane exhibits large thermodynamic bending fluctuations or undulations.

The task of evaluating the partition function Z now amounts to summing over (1) all spatial configurations or conformations of the membrane, (2) all possible topologies of the membrane, and (3) all internal coordinate systems or metrics, since in the evaluation of $\mathcal{H}$ we clearly have to choose a specific parametrization of the surface,

$$Z = \sum_{\text{configurations}} \sum_{\text{metric}} \sum_{\text{topology}} \exp\left(-\mathcal{H}/k_BT\right) \tag{13}$$

The Gauss-Bonnet theorem implies that the Gaussian term in $\mathcal{H}$ is a topological invariant, i.e., $\int dA\, K = 4\pi\chi$, where χ is the Euler characteristic, $\chi = 2(n - g)$, with n being the number of components (distinct surfaces) and g the number of handles. For example, for a spherical topology $\int dA\, K = 8\pi$ and for a toroidal topology $\int dA\, K = 0$. Despite these simple relations it is still a rather formidable task to compute Z, and we generally have to resort to numerical Monte Carlo simulations in order to study the scaling properties of fluctuating membranes. In the numerical simulation of a fluid membrane, the number of variables treated has to be reduced to a finite number. This is usually done by triangulation of the surface as described in Section IV.D.

IV. COMPUTATIONAL METHODS

A. MONTE CARLO TECHNIQUES FOR LATTICE MODELS OF PLANAR BILAYERS

We do not intend to give a detailed account of standard Monte Carlo simulation techniques applied to lattice models of lipid bilayers. The reader is referred to References 23 and 48 for further details. We shall instead provide a description of some recent techniques of data analysis that are particularly useful for the characterization of phase equilibria in lipid bilayers and lipid-protein systems when combined with Monte Carlo simulations. These techniques are also useful for nonlattice simulations of membrane models.

The fundamental problem in numerical investigations of cooperative phenomena is associated with an unambiguous assessment of the nature of phase transitions as well as a determination of the precise location of the phase boundaries in a phase diagram. Numerical simulations share this problem with laboratory experiments. The root of the problem is that most approaches do not function on the level of the free energy but rather in terms of derivatives of the free energy, such as densities and order parameters (first derivatives) or response functions (second derivatives). In the case of one-component systems these derivatives are seldom known with sufficient accuracy to discern between, for example, a fluctuation-dominated first-order phase transition and a continuous transition (critical point). In the case of phase coexistence in, e.g., a two-component system, there is no direct way of relating features in the response functions to the precise location of equilibrium phase boundaries.[30] However, by use of novel techniques[49] it is possible numerically to circumvent this problem and obtain access to that part of the free energy which is necessary to locate the phase equilibria. These techniques involve calculation of distribution functions (histograms) of thermodynamic functions, e.g., order parameters (composition) or internal energy, thermodynamic reweighting of the distribution functions in order to locate the phase transition or the phase equilibria, and then a subsequent analysis of the size dependence of the reweighted distribution functions by means of finite-size scaling theory.

B. HISTOGRAM AND REWEIGHTING TECHNIQUES

In order to use the finite-size scaling method effectively to study phase equilibria in mixtures it is a prerequisite that a multidimensional distribution function, $\mathcal{P}(\{y_i\}_i, T, L)$, can be calculated very close to coexistence for a set of thermodynamic variables, $\{y_i\}_i$. L is the linear extension of the two-dimensional bilayer lattice which therefore consists of $N = L^2$ particles. For one-component cases, e.g., pure lipid bilayers, this set of thermodynamic variables, $\{y_i\}_i$, contains the internal energy and the order parameter, which usually is taken to be the bilayer area. For a two-component system, the set includes in addition the relative chemical potential, μ, for the two species. The distribution function for any given variable y_j can be obtained from the complete distribution function as follows:

$$\mathcal{P}(y_j, T, L) = \sum_{i(\neq j)} \mathcal{P}(\{y_i\}_i, T, L) \tag{14}$$

Without loss of generality we now focus on the joint distribution function (or histogram), $\mathcal{P}(E, x, T, L)$, for the internal energy, E, and the composition, x. This distribution function is now, for a given system size L, calculated very accurately close to coexistence (see below for detection of coexistence). For a given set of thermodynamic parameters, (T, μ), the distribution function is defined by

$$\mathcal{P}_\mu(E, x, T, L) = \frac{n(E, x, L)\exp\left[-(E + \mu x)/k_B T\right]}{\sum_{E,x} n(E, x, L)\exp\left[-(E + \mu x)/k_B T\right]} \tag{15}$$

where $n(E, x, L)$ is the density of states which is independent of temperature.

If $\mathcal{P}(E, x, T, L)$ is calculated with sufficient accuracy for a specific set, (T, μ), it is possible to determine the distribution function for a nearby set of parameters, $T', \mu')$, by employing the reweighting method of Ferrenberg and Swendsen.[50] The method is based on standard thermodynamics and it leads to the reweighted distribution function

$$\mathcal{P}_{\mu'}(E, x, T', L) = \frac{\mathcal{P}_{\mu}(E, x, T, L) \exp\left[-\left(\frac{1}{k_B T'} - \frac{1}{k_B T}\right)E - \left(\frac{\mu'}{k_B T'} - \frac{\mu}{k_B T}\right)x\right]}{\Sigma_{E,x}\mathcal{P}_{\mu}(E, x, T, L) \exp\left[-\left(\frac{1}{k_B T'} - \frac{1}{k_B T}\right)E - \left(\frac{\mu'}{k_B T'} - \frac{\mu}{k_B T}\right)x\right]} \quad (16)$$

The two one-dimensional distribution functions can be readily derived from the combined two-dimensional distribution function, $\mathcal{P}(E, x, T, L)$, as follows:

$$\mathcal{P}(E, T, L) = \sum_x \mathcal{P}(E, x, T, L) \quad (17)$$

$$\mathcal{P}(x, T, L) = \sum_E \mathcal{P}(E, x, T, L) \quad (18)$$

These distribution functions allow us to calculate average quantities, such as internal energy, order parameter, and response functions, for any set of thermodynamic parameters, provided that the originally sampled distribution function is accurate enough to contain sufficient statistical information on the relevant part of the phase space. Data sets that are very dense in, e.g., temperature, can therefore be obtained, which is a prerequisite for locating the transition point and for performing a detailed finite-size scaling analysis at the transition point.

C. DETECTION OF PHASE EQUILIBRIA BY FINITE-SIZE SCALING

Based on the distribution functions described above, phase equilibria can be examined by a powerful method proposed by Lee and Kosterlitz.[51] This method, which is based on finite-size scaling analysis of distributions derived from computer simulations, constitutes an unambiguous technique for numerically detecting first-order transitions and phase coexistence.[49] The method involves calculation of a certain part of the free energy using the distribution functions in Equations 17 and 18. Free-energy-like functions, e.g., $\mathcal{F}(x, T, L)$ can be defined as follows from these distribution functions as

$$\mathcal{F}(x, T, L) \sim -\ln \mathcal{P}(x, T, L) \quad (19)$$

The quantity $\mathcal{F}$ differs from the bulk free energy by a temperature- and L-dependent additive quantity. However, the shape of $\mathcal{F}(x, T, L)$ at fixed T and L is identical to that of the bulk free energy, and furthermore $\Delta\mathcal{F}(\mu, T, L) = \mathcal{F}(x, T, L) - \mathcal{F}(x', T, L)$ is a correct measure of free-energy differences. At a first-order transition, $\mathcal{F}(x, T, L)$ has pronounced double minima corresponding to two coexisting phases at $x = x_1$ and $x = x_2$ separated by a barrier, $\Delta\mathcal{F}(\mu, T, L)$, with a maximum at x_{max} corresponding to an interface between the two phases. The height of the barrier measures the interfacial free energy between the two coexisting phases and is given by

$$\Delta\mathcal{F}(\mu, T, L) = \mathcal{F}(x_{max}, T, L) - \mathcal{F}(x_1, T, L) = \gamma(\mu, T)L^{d-1} + O(L^{d-2}) \quad (20)$$

where d is the spatial dimension of the system ($d = 2$ in the present case). $\gamma(\mu, T)$ is the interfacial free-energy density or interfacial tension. Therefore, $\Delta\mathcal{F}(\mu, T, L)$ increases monotonically with L at a first-order transition (or at phase coexistence) corresponding to a finite interfacial tension. The detection of such an increase is an unambiguous sign of a first-order transition and two-phase coexistence. In contrast, $\Delta\mathcal{F}(\mu, T, L)$ approaches a constant at a critical point, corresponding to vanishing interfacial tension in the thermodynamic limit. In the absence of a transition, $\Delta\mathcal{F}(\mu, T, L)$ approaches zero.

The iterative computational method for detecting coexistence proceeds as follows: a trial value of the chemical potential at coexistence, μ_m, for a given temperature is guessed or estimated from a short simulation on a very small lattice. The accurate value of μ_m for that system size is then determined by a long simulation (typically 10^6 Monte Carlo steps per lattice site) at this trial value of μ and an accurate value of μ_m is determined by the reweighting technique (see Equation 16). If the trial value turns out to be too far from coexistence to allow a sufficiently accurate sampling of the two phases, a new trial value is estimated from the long simulation. The system size is then increased and the value of μ_m for the previous system size is used as a trial value for the long simulation on the larger system. This rather lengthy iterative procedure assures a very accurate determination of the chemical potential at coexistence and hence leads to an accurate determination of the compositional phase diagram via the finite-size scaling analysis.

The Ferrenberg-Swendsen reweighting technique becomes more troublesome as the size of the system studied increases because the distribution functions are narrower for larger systems. The method works best for small systems which are subject to large fluctuations, allowing more information about a larger part of the phase space to be sampled. A particular problem is associated with the present application of the reweighting to mixtures, which involves an extra thermodynamic variable, the chemical potential. Reweighting of data obtained for one value of the chemical potential to another value gives a larger uncertainty for the larger system simply because it is the extensive composition which is involved in the Hamiltonian and hence in the reweighting procedure (see Equation 16). This implies that a knowledge of the value of the chemical potential is required at coexistence with an uncertainty which decreases approximately as L^{-2}. Another bottleneck of a more physical nature is the time, τ, associated with the crossing of the free-energy barrier between the two phases. This time increases exponentially with the linear dimension of the two-dimensional model system, $\tau \sim \exp(\gamma L)$. In order to facilitate the reweighting, the actual simulation time has to be much larger than τ. For a given model, this places a sharp upper bound for the system sizes which can be studied by these simulation techniques.

D. TRIANGULATION OF FLUCTUATING FLUID MEMBRANES

In order to treat fluid interfaces and membranes described by the continuum model in Equation 12 by computer simulation it is necessary to reformulate the model in terms of a finite number of variables, i.e., to discretize the model. Such a discretization is implemented via a triangulation of the surface. In order to determine thermodynamic properties of the discretized model it is necessary to sample over the totality of possible triangulations. During the sampling, the triangulation is updated dynamically in a manner that keeps the bilayer in the fluid phase. This is done by allowing the number of nearest-neighbor contacts between the nodes of the triangulation to fluctuate. The self-avoidance in the discrete model is enforced by choosing appropriate constraints involving a hard-core potential.

The triangulation may be seen as a patchwork of triangles sewn together in order to interpolate the surface. An example of a triangulation of a closed surface of spherical topology is shown in Figure 2a. To use this description we need the real-space position of the vertices where the corners of the triangles meet. We will also, for a given vertex, need a list of neighboring vertices, with which it is forming sides in triangles. In this way we can interpolate any reasonably smooth mathematical surface to any given accuracy since it is only a question of employing a sufficient numbers of vertices. For a given surface and for a chosen accuracy the choice of triangulation is not unique. Nevertheless, for a chosen accuracy a given triangulation is an appropriate interpolation for a whole class of surfaces. We shall assume that the measure of each of these classes is the same for each triangulation, although this is a point subject to considerable discussion.

We need a prescription for calculating the various terms in the energy function Equation 12 as well as other properties of a discretized surface. We shall first consider the mean-curvature elastic energy, $\mathcal{H}_{bend}$, and use the fact that it can be reformulated as

$$\mathcal{H}_{bend} = \frac{\kappa}{2} \int_A dA \left(\frac{1}{r_1} + \frac{1}{r_2} \right)^2$$

$$= \frac{\kappa}{2} \int_A dA \, D^a \vec{n} \, D_a \vec{n} + \kappa \int_A dA \frac{1}{r_1 r_2}$$

$$\approx \kappa \sum_{\langle \Delta_a \Delta_b \rangle} \left(1 - \vec{n}_{\Delta_a} \vec{n}_{\Delta_b} \right) + \kappa 2\pi \chi \tag{21}$$

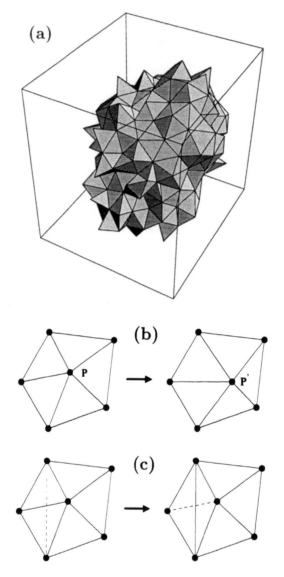

Figure 2 (a) Triangulation of a closed membrane surface. (b) Update of triangulated network by moving a vertex without changing the connectivity. (c) Update of triangulated network by the flip operation, which changes the connectivity and assures fluidity.

where $\vec{n}$ denotes surface normals which in the triangulated surface are associated with the elementary triangles. D_a is the gradient in curvelinear coordinates. The last term can be easily evaluated from the triangular network using the Euler relation, $\chi = N - N_L/3$, where N is the number of vertices and N_L is the number of tethers connecting the vertices in the triangulation of the surface.

Since we have triangulated surfaces with a fixed number of vertices and with a closed topology in mind (e.g., a sphere or a toroid), we can also assign a volume, V, to the region enclosed by the surface expressed as a sum over all triangles on the surface. Using Gauss's theorem in three dimensions we arrive at

$$V = \frac{1}{3} \int_V \mathrm{d}^3 \vec{r} \, \vec{\nabla} \vec{r} = \frac{1}{3} \oint_A \mathrm{d}A \, \vec{n} \cdot \vec{r} = \frac{1}{3} \sum_\Delta A_\Delta \vec{n}_\Delta \cdot \vec{r}_\Delta \tag{22}$$

where $\vec{n}_\Delta$ is the outward normal for triangle Δ, $\vec{r}_\Delta$ is a point on the triangle (e.g., one of the three bounding vertices), and A_Δ is the area of the triangle.

Within this framework there is no microscopic length scale by which one can associate the surface with an area since the bending energy is invariant under a dilation of the surface. There are obvious problems with this description. First, the individual molecules in a membrane should have a finite positive area given by the minimization of the full free energy for the surface. This is equivalent to equating the effective surface tension to zero, thus having the spatial conformations of the surface controlled by the bending energy alone. Another problem is that nothing prevents the triangulated surface from penetrating itself. In order to represent a physical membrane, the surface should, of course, be forced to avoid itself, and other surfaces as well as surface conformations with "overhangs" should be allowed. The self-avoidance of the membrane surface is a source of stability, for example, in a multilamellar stack of membranes.[52]

Both of the problems mentioned above can be solved by associating with each vertex of the triangulation a sphere of diameter d and then forcing all configurations to meet the criteria that no two spheres can overlap and that the distance between two vertices is limited to $\sqrt{2}d$ if the vertices form a side on the surface. Then each triangle will have an area between $\sqrt{3}/4 \cdot d^2$ and $\sqrt{3}/2 \cdot d^2$ and will acquire a thermal average value in that range.

Until now the consideration that the membrane is fluid has not entered the discussion. One way to view the in-plane fluidity of the membrane is that there is no preferred coordinate system in the membrane, i.e., there are no elastic restoring forces acting upon any displacement of a molecule in the plane of the surface. Fluidity would also entail that the molecules only have short-range correlations and that the identity of neighboring molecules is not static as in a solid. To allow this degree of freedom for the triangulated surface the connectivity of the vertices has to be dynamic. Therefore the triangulation of the surface is updated dynamically during the computer simulation as described below in Section IV.E.

E. UPDATING METHODS FOR TRIANGULATED NETWORKS

As described above, we have two sets of variables to update during the simulation of a triangulated fluid surface, namely the vertex position in three-dimensional space and the connectivity of the vertices. When creating data structures to encompass this information the efficiency and ease of updating them have to be considered. The vertices can simply be numbered $V_i, i = 1,\ldots,N$. The sides of the triangles (also called links) can likewise be number $L_i, i = 1,\ldots,2(N - \chi)$, where χ depends on the topology of the surface, i.e., $\chi = 2$ for a sphere and $\chi = 0$ for a toroid. The real-space positions can be organized as a matrix, $X(i,j)$, for $i = 1,\ldots,N$ and $j = 1,2,3$. The connectivity can be organized as a single vector, $T(i), i = -2(N - \chi),\ldots,2(N - \chi)$, by viewing the links L_i as oriented quantities such that a link L_i has the two numbers $\pm i$ associated with it. T is now assigned values such that, given a link number i (either positive or negative), we can find all the links in the two triangles sharing L_i: $i_1 = T(i); i_2 = T(i_1)$ and $i_3 = T(-i); i_4 = T(i_3)$. Since this is true for all i, T must satisfy $T(T(T(i))) = i \forall i$. It is possible to have the same sense of orientation for each triangle; i.e., by using T we recover i, i_1, i_2 and i, i_3, i_4 in a cyclic fashion. By using this oriented structure we have a consistent way of calculating outward normal vectors, although we need a structure to recover vertex numbers from link numbers to get access to real-space coordinates. Another useful structure is the reverse map, which for a given vertex number recovers a link connected to it.

With this setup we can use standard Metropolis Monte Carlo updating procedures:

1. Choose an initial configuration for the surface consistent with the self-avoidance.
2. Propose a change in the configuration that is consistent with the self-avoiding constraints mentioned previously.
3. Calculate the change in energy, ΔE, and accept the new configuration with probability min $[1, \exp(-\Delta E/k_B T)]$.
4. Measure quantities like area, volume, and energy.

Steps 2 to 4 are repeated as many times as desired. Usually there will be a thermalization period where points 2 and 3 are iterated a number of times before any measurements are performed in 4. In Step 2 we alternate between two updating algorithms, one for moving the real-space positions, as in Figure 2b, and one for changing the connectivity, as in Figure 2c. In the case of updating the real-space positions, a vertex is randomly selected and a trial position is calculated by adding to the current position a random vector in the cube $[-s,s] \times [-s,s] \times [-s,s]$. We then have to make sure that moving the vertex to this new position will not cause overlapping of spheres. To speed up this part of the algorithm a cell-list algorithm is used (details can be found in, e.g., Reference 53).

Updating of the connectivity proceeds by randomly choosing a link and then performing a "flip" as illustrated in Figure 2c. During the flip procedure, two vertices are each losing a side whereas two others are each gaining one side. To keep the surface regular we must require that no vertex has fewer than three sides connecting it to other vertices and that any two vertices cannot be connected more than once to each other. In the flip update we must also make sure that the length of the new side is less than or equal to the hard-core distance, $\sqrt{2}d$. To update T in accordance with the new connectivity we need only change $T(\pm i), T(i_1), T(i_2), T(i_3), T(i_4)$, which makes T a very compact data structure.

V. THERMODYNAMIC AND CONFORMATIONAL PROPERTIES OF LIPOSOMES

A. MAIN PHASE TRANSITION AND DYNAMIC BILAYER HETEROGENEITY

The nature of the main transition in one-component lipid bilayer membranes is a topic of continuing controversy.[2,28,30,54] It is usually assumed that it is of first order.[2,4,28] There is, however, no clear-cut experimental proof of this, and the usual assumption of a first-order transition is very much due to the tradition which developed after the original finding by Chapman et al.[55] of a very intense and narrow specific-heat peak with a large heat content for phospholipid bilayers at their main transition temperature, T_m. This finding was later accurately quantified in the classical high-sensitivity differential-scanning calorimetry study by Albon and Sturtevant.[56] Nevertheless, there is no single piece of experimental observation which reveals the kind of discontinuities that are usually associated with first-order behavior. In fact, the most recent and accurate calorimetric study[2] leads to the conclusion that the transition is at best weakly first order and that possibly there is no transition at all in a strict thermodynamic sense; i.e., the bilayer transition may be close to a critical point. Whether it is on the side of the critical point corresponding to a (weakly) first-order transition or on the side where there is no phase transition in a strict thermodynamic sense may critically depend on details of the particular system. The current picture of a lipid bilayer at its main transition is hence one of a strongly fluctuating system.[28] The strong fluctuations are signalled on the macroscopic level by dramatic peaks in the response functions and strong variations in the order parameters. Microscopically, the fluctuating state is manifested in terms of the formation of clusters and domains leading to a heterogeneous lateral structure which is dynamically maintained. It is interesting to note a well-established fact, experimentally as well as theoretically, that the appearance of the main bilayer transition is strongly dependent on small perturbations, such as imperfections, impurities, sample morphology, and transbilayer interactions, as well as environmental conditions. It has recently been pointed out[30] that if a phase transition is absent in a one-component lipid bilayer it may well be restored when a second lipid component is introduced even in a very small amount.

The variation of the acyl-chain conformational order parameter, $\langle S \rangle = \frac{1}{2}\langle 3\cos^2\theta - 1 \rangle$, as determined from a Monte Carlo simulation on the ten-state Pink model of a $DC_{16}PC$ lipid bilayer is shown in Figure 3. It is observed that the order decreases very rapidly, although in a continuous fashion, in a narrow temperature region around a temperature T_m. These numerical data are in rather close accordance with experimental measurements based on deuterium nuclear magnetic resonance (NMR) spectroscopy.[54] Hence it appears that if the transition in $DC_{16}PC$ is of first order it is strongly smeared by thermal fluctuations. The nature of the main transition has recently been investigated numerically by using reweighting and finite-size scaling techniques in conjunction with the ten-state Pink model and certain modifications of this model.[30-32,57] In Figure 4 we present the results for the free-energy function $\mathcal{F}(A,T,L)$ as a function of linear system size, L, for a series of different phospholipids with different acyl-chain lengths. Here A is the lipid-bilayer area per molecule. The data are obtained from simulations on the Pink model in Equation 1. The results for each system size refer to a temperature at which the two minima in the free energy are equally deep. It appears from this figure that for chain lengths shorter than $n_C = 18$ there is no phase transition in the thermodynamic limit (the free-energy barrier vanishes with increasing system size), whereas for longer chains the free-energy barrier increases, signalling the occurrence of a first-order phase transition. $DC_{18}PC$ is close to being at the border line. In all cases the transitional behavior is strongly influenced by thermal density fluctuations. The fluctuations increase as the chain length is decreased.[58]

A macroscopic consequence of these strong fluctuations is illustrated in Figure 5 in terms of the specific heat. The specific heat has a strong anomaly at the transition. As the chain length is decreased the intensity of the specific heat is increased in the wings, in good agreement with experimental findings.[2,58] Similarly, other response functions, e.g., the lateral compressibility, display anomalies at the transition.

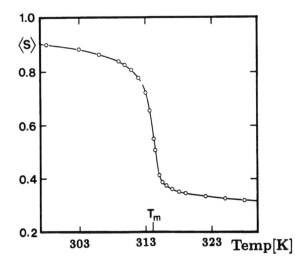

Figure 3 Average segmental acyl-chain order parameter, $\langle S \rangle = \frac{1}{2}\langle 3 \cos^3\theta - 1 \rangle$, as a function of temperature for $DC_{16}PC$ bilayers. $T_m = 314$ K is the transition temperature.

During the transitional process in the lipid bilayer (whether it be a phase transition or not) certain elements of order develop, as illustrated in Figure 6 in the case of $DC_{16}PC$. This figure shows that the bilayer in the transition region develops a substantial degree of dynamic heterogeneity.[12,58-60] Specifically, for the bilayer in the conformationally ordered gel phase, domains or clusters of chains are formed which are conformationally disordered. Conversely, in the conformationally disordered fluid phase, gel domains are formed in the fluid matrix. The domain formation is a dynamic phenomenon and could be characterized as a kind of nano-scale dynamic phase separation. Still, the overall symmetry of the mother phase is maintained; e.g., in the fluid phase the fluidity is maintained on length scales larger than the average gel domain size. While the individual lipid domains appear and disappear as they have a finite lifetime depending on their size, the average domain size is an equilibrium property. The average domain size varies strongly with temperature, and it exhibits a distinct maximum at the transition as shown in Figure 7. This maximum matches the maxima found in response functions, such as the specific heat and the lateral compressibility.[58] The occurrence of the domains implies that the thermal density fluctuations are correlated in space with a coherence length corresponding to the linear domain size.

The domain formation illustrated in Figure 6 may be seen as a dynamic state of order in disorder.[12] Any membrane-associated process that takes place on time scales less than the lifetime of a domain (typically of the order of 10^{-4} sec) and on length scales less than or comparable with the domain size will be influenced by this order-in-disorder phenomenon.[61] The domain sizes, and hence the dynamic bilayer heterogeneity, can be modified by choosing lipid species with different chain lengths, as illustrated quantitatively in Figure 7 and pictorially in Figure 8 for the series $DCn_C PC$, with $n_C = 12,14,16,18,20$, at the same relative temperature, T/T_m, in the fluid phase. The figures show that the longer the acyl chain, the smaller are the lipid domains (except right at the phase transition), reflecting the fact that the short-chain lipid bilayer is closer to a critical point and sustains the strongest thermal density fluctuations.

B. PHASE EQUILIBRIA AND LOCAL STRUCTURE IN BINARY LIPID BILAYERS

The possible absence of a phase transition in certain one-component lipid bilayers poses some intriguing problems with regard to the existence of phase equilibria in lipid bilayers incorporated with other species, such as other lipids, cholesterol, or proteins. It has recently been shown[30] by a computational approach of the type described in the present review that a first-order transition, which is absent in a one-component lipid bilayer, may be induced by mixing in small amounts of another species in such a way as to approach the critical points and the related closed coexistence regions. Figure 9 shows some results from simulations on the model in Equations 2 and 3 of binary lipid mixtures of saturated phospholipid acyl chains of different hydrophobic length in the specific case of $DC_{14}PC$-$DC_{18}PC$ mixtures.[30] The size dependence of the free-energy function in Equation 19 as a function of composition $x_{DC_{18}PC}$ is displayed in Figure 9a. The family of distribution functions is shown at the appropriate size-dependent chemical potential values

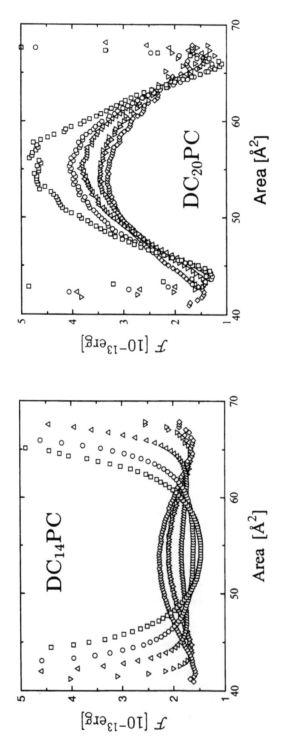

Figure 4 Free-energy function, $\mathcal{F}(A)$, in Equation 19 determined at the phase transition (or at the temperature where $\mathcal{F}$ is symmetric) for $DC_{14}PC$ and $DC_{20}PC$ bilayers for different system sizes, $N = L \times L$. $L = 4(\Diamond)$, $6(\nabla)$, $10(\triangle)$, $15(\circ)$, $20(\square)$.

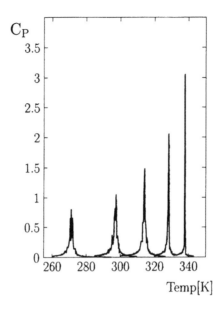

Figure 5 Specific heat per lipid molecule (in units of 10^{-13} erg) for $DC_{12}PC$, $DC_{14}PC$, $DC_{16}PC$, $DC_{18}PC$, and $DC_{20}PC$ (from left to right) lipid bilayers in the transition region.

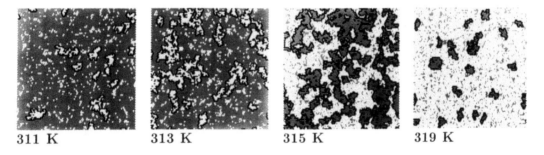

311 K   313 K   315 K   319 K

Figure 6 Lipid-domain formation and dynamic heterogeneity in the transition region of $DC_{16}PC$ bilayers. The transition temperature T_m = 314 K. The pictures illustrate configurations obtained from computer-simulation calculations on a system with 5000 lipid molecules. Gel and fluid regions are denoted by gray and light regions, respectively, and the interfaces of the lipid domains are highlighted in black.

corresponding to phase coexistence. It is seen that, as the system size is increased, there is a dramatic increase in the free-energy barrier between the two phases and that the composition of the mixture at the given thermodynamic conditions is given by $x^g_{DC_{18}PC} = 0.95$ and $x^f_{DC_{18}PC} = 0.42$. This is unambiguous evidence of phase coexistence in the thermodynamic limit and also shows that the interfacial tension tends towards a nonzero value. From data of this type the entire phase diagram can be determined very accurately, as shown in Figure 9b. This phase diagram shows that, even though bilayers of the two pure components do not have a phase transition in a strict thermodynamic sense, a transition is induced by mixing the two components. This leads to two critical mixing points in the phase diagram. Between these two critical points there is a pronounced two-phase coexistence region. The phase diagram found by these techniques is rather close to experimental results. It should be remarked that for all practical experimental purposes the displacement of the critical points from the axes of the phase diagram is probably of marginal relevance, in particular for the more nonideal mixtures.

The dynamic membrane heterogeneity and the lipid-domain formation phenomena described for one-component lipid bilayers involve gel domains. It has often been stated that such phenomena are not relevant under physiological conditions since gel domains are rarely, if ever, observed in functional biological membranes. However, dynamic lipid-domain formation in lipid-bilayer membranes need not involve gel-like domains. In fact, the liquid correlations in many component lipid mixtures naturally lead

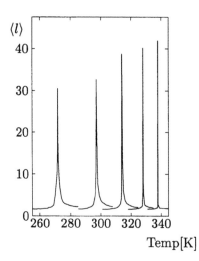

Figure 7 Average lipid domain size $\langle \ell \rangle$ (in units of number of lipid molecules) for $DC_{12}PC$, $DC_{14}PC$, $DC_{16}PC$, $DC_{18}PC$, and $DC_{20}PC$ (from left to right) lipid bilayers in the transition region.

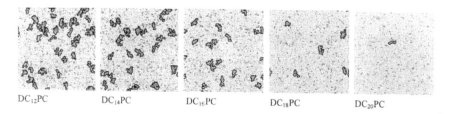

$DC_{12}PC \qquad DC_{14}PC \qquad DC_{16}PC \qquad DC_{18}PC \qquad DC_{20}PC$

Figure 8 Change in lipid-domain formation and dynamic heterogeneity by changing the lipid material. Typical membrane configurations are shown for $DC_{12}PC$, $DC_{14}PC$, $DC_{16}PC$, $DC_{18}PC$, and $DC_{20}PC$ bilayers at the same relative temperature, $T/T_m = 1.016$, in the fluid phase. The configurations are obtained from computer-simulation calculations on a system with 5000 lipid molecules. Gel and fluid regions are denoted by gray and light regions, respectively, and the interfaces of the lipid domains are highlighted in black.

to a local dynamic domain-like structure that is controlled by the compositional fluctuations. The compositional fluctuations are characterized by a lateral coherence length which intimately depends on the underlying phase diagram of the mixture.

For convenience we consider a series of mixtures of homologous saturated diacyl phosphatidylcholines that differ only with respect to their acyl-chain length. This will make it possible, in a rather transparent fashion, to vary the degree of nonideality of the mixture by changing the difference in acyl-chain length of the two species.[29] The generic nature of the binary lipid phase diagrams relevant for the mixtures considered here is illustrated in Figure 10 in the case of the $DC_{14}PC$-$DC_{18}PC$ and the $DC_{12}PC$-$DC_{18}PC$ mixtures. Both of these mixtures are highly nonideal, and we shall not be concerned with the possible absence of a phase transition for the pure lipids discussed above. Whereas the $DC_{14}PC$-$DC_{18}PC$ mixture within the temperature range of interest exhibits full miscibility within the fluid and gel phases and only has a broad gel-fluid coexistence region, the considerably more nonideal $DC_{12}PC$-$DC_{18}PC$ mixture develops peritectic behavior with a pronounced gel-gel coexistence region.[29] The $DC_{16}PC$-$DC_{18}PC$ system has a more ideal monotectic mixing behavior than $DC_{14}PC$-$DC_{18}PC$ and $DC_{16}PC$-$DC_{20}PC$, whereas, e.g., $DC_{12}PC$-$DC_{18}PC$ and $DC_{12}PC$-$DC_{20}PC$ (considered below) also display peritectic behavior.

The description provided by the phase diagrams in Figure 10 is purely thermodynamic and contains no information on the local structure of the mixture. Local structure and domain-formation phenomena can be revealed from computer-simulation calculations on microscopic models that not only can reproduce

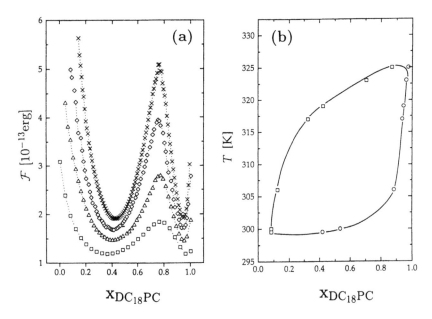

Figure 9 Monte Carlo computer-simulation data for a $DC_{14}PC$-$DC_{18}PC$ mixture. (a) Size dependence of the free-energy function, $\mathcal{F}(x,L)$, evaluated at coexistence at temperature $T = 319$ K. The linear sizes of the systems correspond to $L = 5$ ($\square$), 7 ($\triangle$), 9 ($\diamond$), and 11 ($\times$). (b) Phase diagram.

phase diagrams like those in Figure 10, but which can simultaneously provide information on fluctuations within the thermodynamic phases.[29] In Figure 11 are shown typical configurations of three binary mixtures, $DC_{14}PC$-$DC_{18}PC$, $DC_{12}PC$-$DC_{18}PC$, and $DC_{12}PC$-$DC_{20}PC$, arranged according to increasing degree of nonideality. They are all in the thermodynamic fluid phase at the same (equimolar) composition and the same temperature, $T = 338$ K. These configurations show that the fluid phase sustains a considerable degree of dynamic local structure. The lateral structure corresponds to a local compositional demixing by which chains of the same species (length) tend to cluster together. Strong lateral density fluctuations have been observed experimentally in the $DC_{14}PC$-$DC_{18}PC$ mixture by the use of inelastic neutron scattering.[62] The configurations in Figure 11 show that the larger the difference between the acyl-chain lengths, the more pronounced is the local structure.

A quantitative description of the local structure in the binary lipid mixtures is provided by the static pair correlation function[29] $g(R)$, which at a given concentration measures the probability that an acyl chain, in a given state of a given type of lipid, will be found at a distance R from the acyl chain in another given state of another given type of lipid. In the case of correlations between disordered (fluid) acyl-chain states of $DC_{n_c}PC$ this function is written $g_{ff}(DC_{n_c}PC)(R)$ and is shown in Figure 12 as a function of acyl-chain separation, R, for equimolar composition of the three different mixtures in Figure 11. The data in Figure 12 clearly show that the larger the chain-length difference between the different lipid species, i.e., the more nonideal the mixture, the more long ranged is the correlation, in accordance with the visual appearance of the configurations in Figure 11. This effect refers to isothermal conditions, and it should be kept in mind that the transition temperature of the pure lipid bilayer increases with acyl-chain length. A similar type of local ordering in binary lipid mixtures in the fluid phase has recently been found by Huang et al.[63,64] using computer simulation on a simple nonideal mixture model accounting for the electrostatic interactions in a mixture of phosphatidylserine and phosphatidylcholine. The coherence length of the compositional fluctuations and the local structure has a strong temperature dependence, as illustrated in Figure 13 in the case of an equimolar $DC_{14}PC$-$DC_{18}PC$ mixture at two different temperatures in the fluid phase. The figure shows that as the gel-fluid coexistence region is approached (see Figure 10) the correlation becomes more long-ranged.

Similarly, lipid phases involving gel domains sustain a substantial amount of local order and dynamic heterogeneity within these domains.[12,29] Furthermore, interfaces between fluid and gel regions display a considerable degree of structure.[29]

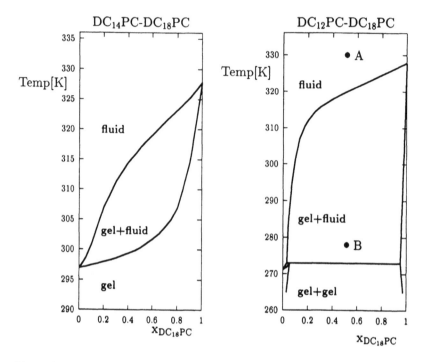

Figure 10 Phase diagrams for $DC_{14}PC$-$DC_{18}PC$ and $DC_{12}PC$-$DC_{18}PC$ lipid mixtures.

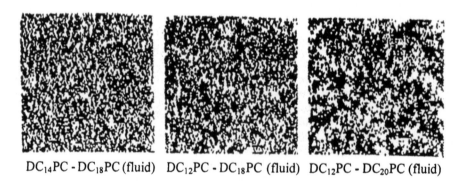

$DC_{14}PC$ - $DC_{18}PC$ (fluid) $DC_{12}PC$ - $DC_{18}PC$ (fluid) $DC_{12}PC$ - $DC_{20}PC$ (fluid)

Figure 11 Local structure in three different fluid lipid mixtures, $DC_{14}PC$-$DC_{18}PC$, $DC_{12}PC$-$DC_{18}PC$, and $DC_{12}PC$-$DC_{20}PC$, at the same temperature, T = 338 K, and the same composition, x = 0.5. The two lipid species in each snapshot are indicated by white and black areas. The configurations are derived from computer-simulation calculations on systems with 5000 lipid molecules.

C. BILAYERS INCORPORATED WITH CHOLESTEROL

Addition of foreign molecules to lipid bilayers may in some cases have a dramatic influence on both the phase equilibria and the bilayer dynamic heterogeneity. Cholesterol, drugs, insecticides, as well as polypeptides and proteins couple strongly to the heterogeneity, which may be either suppressed or strongly enhanced. Cholesterol as well as some drugs and insecticides act as emulsifiers and interfacially active agents by altering the lateral structure of the lipid bilayer. The interaction of proteins and polypeptides with lipids takes advantage of the lateral density and compositional fluctuations, and the coherence length of the perturbation on the lipid matrix due to the presence of proteins is directly related to the correlation length characterizing the dynamic membrane heterogeneity.

The phase diagram of the $DC_{16}PC$-cholesterol system is shown in Figure 14. The topology of this phase diagram seems to be generic for mixtures of different phospholipids with cholesterol.[65-67] Cholesterol can in fact decouple the positional variables and the conformational variables of the lipid acyl chain

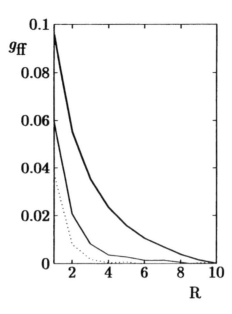

Figure 12 Pair correlation function g_{ff}, for fluid lipid states of the low-melting lipid in three different mixtures, $DC_{14}PC$-$DC_{18}PC$ (dotted line), $DC_{12}PC$-$DC_{18}PC$ (thin line), and $DC_{12}PC$-$DC_{20}PC$ (heavy line) (see Figure 11). Data are given at a fixed temperature, $T = 338$ K, and a fixed composition, $x = 0.5$, for varying acyl-chain separation, R (in units of lattice spacings).

and at high concentrations stabilize a phase that lacks translational order, i.e., a liquid, with high lateral diffusion and at the same time characterized by a high degree of conformational order. This unique phase, which only occurs at cholesterol contents above 20 mol%, has been termed the liquid-ordered phase.[68] Hence, at high concentrations cholesterol induces both order and disorder in a phospholipid bilayer. The effect of cholesterol incorporated into pure lipid bilayers in large amounts, ≳20%, is to suppress the thermal fluctuations. However, it has been shown both experimentally and theoretically that cholesterol at low concentrations increases the fluctuations in the neighborhood of the phase transition.[34,69]

Computer simulations on the model of cholesterol-lipid interactions in Equations 4 and 5 have shown that the cholesterol molecules effectively become interfacially active and tend to promote the lipid-domain formation phenomena and, consequently, to enhance the dynamic membrane heterogeneity.[34,69] This effect is illustrated in Figure 15, which shows the isothermal effects on a $DC_{16}PC$ lipid bilayer as cholesterol is introduced in increasing amounts up to 7.5 mol%. In this concentration range, the transitional behavior in the acyl-chain variables follows that of the narrow solid-ordered/liquid-disordered coexistence region (see Figure 14) and the three-phase line. The membrane configurations in Figure 15 demonstrate that the domain formation is enhanced by the presence of cholesterol. At the same time cholesterol tends to accumulate at the domain interfaces.[34] The strong enhancement of lipid fluctuations at low cholesterol contents has been confirmed by a number of experiments.[69,70]

D. BILAYERS INCORPORATED WITH POLYPEPTIDES AND PROTEINS

The major effect of the incorporation of integral proteins and transmembrane polypeptides into membranes is a dramatic change in the phase equilibria. Usually, proteins are predominantly soluble in the fluid lipid-bilayer phase, and a protein-induced phase separation occurs in the gel phase.[71] The presence of the proteins leads to structural changes in the adjacent lipid molecules as well as a modification of the local composition. One of the mechanisms proposed to relate protein-induced lipid-bilayer phase equilibria to the fundamental physical properties of the lipid-protein interfacial contact is based on a hydrophobic matching[40,41,71] between the lipid bilayer and the hydrophobic region of the protein. This hydrophobic matching concept has enjoyed considerable success in predicting phase diagrams for lipid bilayers reconstituted with specific proteins.[71] We shall describe here some results obtained from computer-simulation calculations on the mattress model[46] in Equations 9 and 10, which is based on the hydrophobic matching principle. According to the concept of hydrophobic matching, a major contribution to protein-lipid interactions is controlled by a hydrophobic matching condition which requires the

$$DC_{14}PC–DC_{18}PC \text{ (fluid)}$$

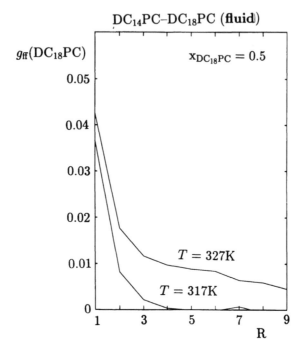

Figure 13 Pair correlation function, g_{ff}, for fluid lipid states of $DC_{18}PF$ in a $DC_{14}PC\text{-}DC_{18}PC$ mixture at equimolar concentration shown at two different temperatures in the fluid phase, $T = 317$ K and 327 K. Results are given for varying acyl-chain separation, R (in units of lattice spacings).

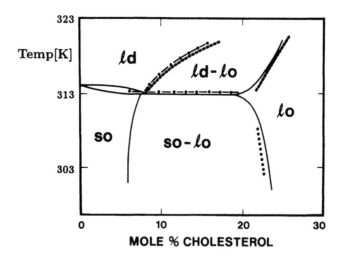

Figure 14 Phase diagram of $DC_{16}PC$-cholesterol mixtures as obtained from a variety of experimental techniques.[68] The different lipid phases are denoted solid-ordered (**so**), liquid-disordered (**ℓd**), and liquid-ordered (**ℓo**).

hydrophobic lipid-bilayer thickness to match the length of the hydrophobic domain of the integral membrane protein. From the results described above on the occurrence of membrane heterogeneity in phospholipid bilayers of different kinds (see Figures 6, 8, and 11) it may be anticipated that integral membrane proteins which, via a hydrophobic matching condition, couple to the membrane lipid acyl-chain order (area density) and/or the membrane composition are going to influence the degree of heterogeneity. Conversely, a certain degree of membrane heterogeneity will couple to the conformational state of the individual proteins as well as to the aggregational state of an ensemble of proteins.

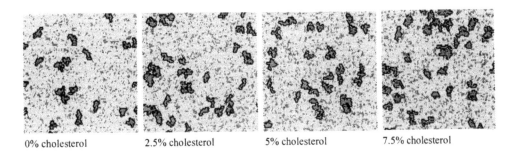

| 0% cholesterol | 2.5% cholesterol | 5% cholesterol | 7.5% cholesterol |

Figure 15 Typical lipid-domain configurations in the fluid phase of $DC_{16}PC$-cholesterol mixtures at a temperature $T = 319$ K in the fluid phase. Gel and fluid regions are denoted by gray and light regions, respectively, and the interfaces of the lipid domains are highlighted in black. The cholesterol molecules are denoted by o.

First we demonstrate how numerical simulations on the model in Equations 9 and 10 can be used to model the gel-fluid coexistence loop and the lower critical mixing point[45] found in lipid-polypeptide mixtures. The simulation results are in good agreement with the experimental data of Morrow et al.[72,73] The polypeptide considered is a transmembrane amphiphilic polypeptide with a hydrophobic length which closely matches the fluid hydrophobic thickness of $DC_{16}PC$ bilayers. The phase diagram, as obtained by the use of histogram techniques in combination with a finite-size scaling analysis, is shown in Figure 16 along with the calculated transition enthalpy, $\Delta H(x)$. The experimental data are also reproduced on this figure. The lipid-polypeptide interactions of the present microscopic model are seen to reproduce the characteristic closed coexistence loop of a binary mixture with a lower critical mixing point.

The microscopic version of the mattress model, Equation 10, has been used to systematically study the lateral distribution of proteins and polypeptides in model membranes as controlled by temperature (i.e., mixing entropy) and direct lipid-protein interactions. By a computer-simulation calculation[44] the relative importance for protein aggregation of different contributions to the lipid-protein interactions has been studied, specifically the attractive lipid-protein hydrophobic contact interaction and the repulsive interaction due to the possible mismatch between lipid bilayer and protein hydrophobic thicknesses. The results suggest that the formation of protein aggregates in the membrane plane is predominantly controlled by the strength of the direct van der Waals-like lipid-protein interaction. It is found that, whereas the mismatch is of prime importance for determining the phase equilibria, a mismatch may not be the only reason for protein aggregation within each of the individual phases: depending on the strength of the van der Waals-like interaction associated with the direct lipid and protein hydrophobic contact, the proteins may remain dispersed in the fluid phospholipid bilayer, even if the mismatch between the lipid and protein bilayer thicknesses is as high as 12 Å. The type of data on which the above conclusions are based is exemplified in Figure 17 in the case of a $DC_{16}PC$ bilayer. The figure shows the phase diagram and a typical bilayer configuration in the gel-fluid phase separation region in the case of a small, rather short protein, $d_P = 24$ Å, at a low concentration, $x_P = 0.095$. The phase diagram indicates that massive phase separation occurs below T_m. In this figure the solidus line is so close to the temperature axis that it cannot be discerned. In the phase-separated region, the proteins are dissolved almost exclusively in the fluid-like regions of the bilayer. This is due to the fact that the protein length is closely matched to the fluid bilayer thickness and dissolution of proteins in the gel phase would therefore be very costly. However, since the attractive interaction between the lipids and the proteins in this case is assumed to be very low, the solubility of the protein is also low in the fluid phase, and therefore one might expect a tendency for protein aggregation to occur within the coexistence region at low temperatures, where the entropy is low. The simulation results indicate that, when the value of the lipid protein interaction parameter, J_{AP}, is sufficiently small, protein aggregates form in the fluid region of the phase diagram just above the phase boundary due to dynamic aggregation induced by the lipid density fluctuations. This is a highly nontrivial effect caused by the dramatic density fluctuations accompanying the main transition (see Figure 6).

We now consider a more complex situation where larger proteins of the size of bacteriorhodopsin are dissolved in a binary lipid mixture. In the binary mixture, compositional fluctuations arise in the fluid phase (see Figure 11), with a substantial coherence length which may be picked up by a protein which has different degrees of hydrophobic matching with the two different lipid species.[74] The protein induces

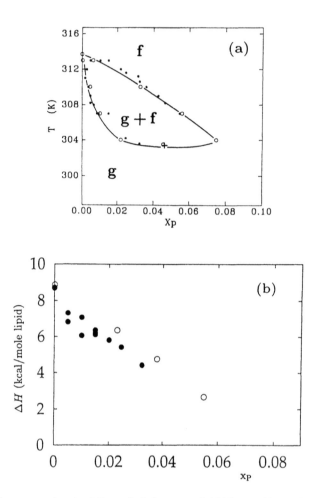

Figure 16 Results from computer-simulation calculations on a lipid bilayer with amphiphilic polypeptides. (a) Theoretical T-x_P phase diagram (o). The experimental data (•) for mixtures of $DC_{16}PC$ with Lys_2-Gly-Leu_{16}-Lys_2-Ala-amide[72] are shown for comparison. The critical mixing point is indicated by +. The solid line connecting the theoretical points is drawn as a guide to the eye. (b) Corresponding transition enthalpy, ΔH, per lipid molecule as a function of peptide concentration, x_P, as obtained from simulations (o) and from experiments (•).

a certain lipid compositional profile around itself, and an overlap in compositional profiles from adjacent proteins may result in an effective lipid-mediated protein-protein interaction which can give rise to an attractive force and hence protein aggregation. The mattress model for lipid-protein interactions, Equations 9 and 10, has recently been used[74] in conjunction with the model for binary mixtures of phospholipids with different acyl chain lengths, Equations 2 and 3, to study the thermodynamic properties as well as the local compositional structure in the case of $DC_{12}PC$-$DC_{18}PC$ mixtures incorporated with a protein of a size corresponding to that of bacteriorhodopsin. The model simulations have been inspired by recent experimental work on the effect of bacteriorhodopsin on lipid bilayers of varying thickness.[75] Results from Monte Carlo simulations of this model of bacteriorhodopsin in equimolar $DC_{12}PC$-$DC_{18}PC$ mixtures have been obtained for a protein/lipid (P/L) molar ratio of 0.025. Figure 18a shows the results for a specific-heat scan with and without bacteriorhodopsin incorporation. Due to the extreme degree of nonideality of the $DC_{12}PC$-$DC_{18}PC$ mixture, the liquidus and solidus (three-phase line) of the phase diagram clearly manifest themselves in two well-separated thermal anomalies in the specific heat. In the presence of bacteriorhodopsin the gel-fluid coexistence region is getting narrower and the specific-heat anomalies decrease in intensity. The lateral distribution of the proteins in two typical cases is shown in Figure 18b and c. The two cases correspond to the gel-gel coexistence region (b) and the gel-fluid coexistence region (c). The higher solubility, caused by a better hydrophobic matching of bacteriorhodopsin in the gel phase to the lipid species with the shorter chain length, $DC_{12}PC$, is clearly seen in Figure 18b.

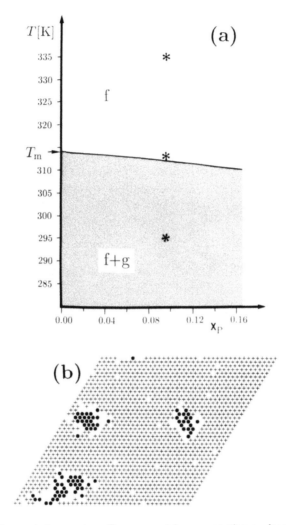

Figure 17 (a) Phase diagram in temperature, T, versus protein concentration, x_P, for a mixture of $DC_{16}PC$ lipids and small proteins with a hydrophobic length of 24 Å and a very weak hydrophobic protein-lipid interaction. T_m = 314 K is the transition temperature of the pure lipid bilayer. The labels **f** and **g** refer to the fluid and gel lipid phases, respectively, and the shaded region **f + g** indicates the fluid-gel coexistence region. (b) Snapshot of a typical microconfiguration of the lattice at T = 295 K. The proteins are indicated by dots, and gel and fluid lipid regions are denoted by gray (+) and white areas, respectively.

In the gel-fluid coexistence region, where bacteriorhodopsin is expelled from the predominantly $DC_{18}PC$ gel phase due to poor matching, the protein in the fluid region prefers to be wetted by the well-matched fluid $DC_{18}PC$ lipids. The fluid $DC_{12}PC$ lipids, which are too short, are repelled by the protein. The local lipid compositional structure found around bacteriorhodopsin in the fluid phase shows that there is a statistical dynamic lipid annulus around the protein. The well-matched fluid $DC_{18}PC$ lipids are enriched in concentration near the protein, whereas there is a concomitant depletion of the poorly matched fluid $DC_{12}PC$ lipids.

E. INTERACTION OF LIPID BILAYERS WITH DRUGS AND INSECTICIDES

A number of chemical compounds that are used to modify membrane-associated physiological functions have a strong and nonspecific influence on the physical properties of lipid bilayers. Examples include drugs like anesthetics and various insecticides.[36-38] It is likely that the potency of these agents is related to their capacity to alter the structure and the dynamics of the lipid bilayer, in particular the degree of order and heterogeneity.[38,60] Figure 19 shows the effects of a water-soluble drug on the dynamic lipid domains in a $DC_{16}PC$ lipid bilayer in the transition region. The results are obtained from computer-simulation

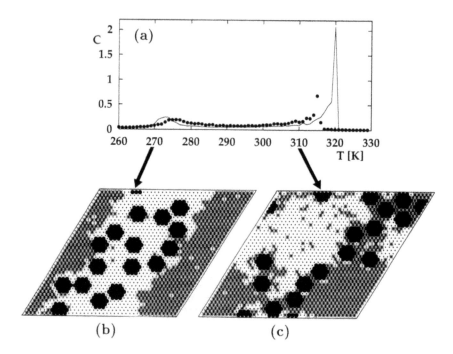

Figure 18 (a) Computer-simulation data for the specific heat of a $DC_{12}PC$-$DC_{18}PC$ mixture without (——) and with (●) a dispersion of bacteriorhodopsin in a molar protein-lipid ratio of 0.025. (b) Typical microconfiguration of the mixture at $T = 270$ K. The proteins are shown as solid hexagons. • denotes gel $DC_{12}PC$ lipids and * denotes gel $DC_{18}PC$ lipids. Empty spaces denote lipid molecules in the fluid state. (c) Typical microconfiguration at $T = 310$ K. The proteins are shown as solid hexagons. • denotes fluid $DC_{12}PC$ lipids, * denotes fluid $DC_{18}PC$ lipids, and empty spaces denote lipid molecules in gel states.

calculations on the model in Equations 6 to 8 in which the drug acts as an interstitial impurity which couples selectively to kink-like acyl-chain conformations. Since these conformations are generated predominantly at lipid-domain interfaces, the drug has a strong tendency to adsorb at the interfaces and to lower the interfacial tension. This in turn leads to a dramatic increase in the amount of interface and to a much more ramified interface structure, as seen in Figure 19. Since the drug in this case leads to a freezing-point depression, the effect of the lipid order being destroyed by the drug is most pronounced in the gel phase. Concomitantly with the alterations in the interface morphology there is a marked tendency for the drug molecules to accumulate at the interface. For the specific model calculation leading to Figure 19 the local concentration of the drug in the interfaces can be larger than the global concentration.[37,38] It may be interesting to consider this effect in relation to the discussion of clinical concentrations of, e.g., drugs used for anesthesia.[76] The full power of the ability of drugs to modify the global order in lipid bilayers[38] is illustrated in Figure 20 in the case of an equimolar binary mixture of $DC_{14}PC$ and $DC_{18}PC$ in the gel-fluid phase-separation region (see Figure 10). The figure shows that the drug under isothermal conditions completely abolishes the phase separation and induces a fluid phase with a considerable degree of local order. The former macroscopic gel phase is emulsified and appears as gel domains in a very loosely connected network.

The alteration of lipid order and domain formation due to the presence of water-soluble drugs in the bilayer has a number of dramatic consequences for the macroscopic properties of the bilayer. In particular, the proliferation of the interfaces in the transition region implies that the partition coefficient develops a maximum at the transition, rather than the classical step anomaly found for impurities which is described by a simple solubility difference in the two lipid phases.[36] Examples of molecules which may affect membrane order, as discussed above, are local and general anesthetics, such as dibucaine[77,78] and halothane,[79] and insecticides like lindane.[80] A variety of experiments on such systems indicate certain effects of the presence of the drugs, such as broadening of the specific heat and occurrence of anomalies in the passive ion permeability and the partition coefficient, effects that from the computer-simulation work reported above are known to be consequences of changes in the local domain structure.[37,38] The

A

B

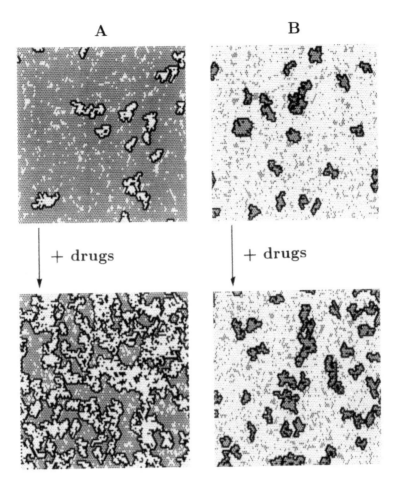

Figure 19 Effects on bilayer domain organization and order due to a water-soluble drug like halothane. Results are shown for a DC$_{16}$PC bilayer at temperatures (a) slightly below and (b) slightly above the transition temperature of the pure lipid bilayer, $T_m = 314$ K. The amount of drug dissolved in the bilayer varies significantly in the transition region. The configurations are obtained from computer-simulation calculations on a system with 5000 lipid molecules. Gel and fluid regions are denoted by gray and light regions, respectively, and the interfaces of the lipid domains are highlighted in black.

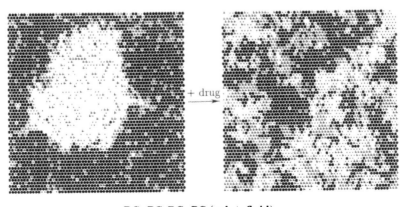

DC$_{14}$PC–DC$_{18}$PC (gel + fluid)

Figure 20 Effect on lateral organization in the gel-fluid phase-coexistence region of a DC$_{14}$PC-DC$_{18}$PC mixture due to the presence of a water-soluble drug like halothane. The configurations are obtained from computer-simulation calculations on a system with 1800 lipid molecules. Gel (predominantly DC$_{18}$PC chains) and fluid (predominantly DC$_{14}$PC chains) regions are denoted by dark and light regions, respectively.

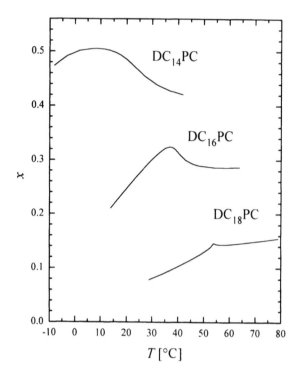

Figure 21 Partition coefficient, $x(T)$, of a water-soluble insecticide like lindane dissolved in DC$_{14}$PC, DC$_{16}$PC, and DC$_{18}$PC bilayers.

computer-simulation results hence suggest that the experimental findings can be interpreted in terms of changes in the dynamic lipid order.

As a specific example of the ability of the computer-simulation approach to describe the remarkable fluctuation-induced enhanced adsorption of certain substances to lipid bilayers, we show in Figure 21 the calculated partition coefficient for a model of lindane interacting with lipid bilayers of different thicknesses.[39] A clear peak in the partition coefficient is observed near the phase transition of the pure lipid bilayer. The shorter the lipid acyl-chain length, the more pronounced is the enhanced adsorption. This result is in accordance with experimental measurements.[80]

F. PASSIVE PERMEABILITY OF LIPID BILAYERS

One of the more dramatic consequences of strong density and compositional fluctuations in lipid bilayer membranes is the occurrence of enhanced passive transbilayer permeability of various species, such as ions, water, and larger amphiphilic species.[81] This phenomenon is likely to be of substantial importance for the understanding of liposome stability and for a clarification of the circumstances under which liposomes may be used as tight containers for drug-delivery purposes.[7]

A general model for passive permeation of ions through lipid bilayers has been suggested which builds on the assumption that the passage predominantly takes place at interfaces between lipid domains formed in the bilayer due to dynamic lateral heterogeneity.[81,82] A computer simulation approach of the type described in the present review to the cooperative effects in liposomes makes it possible to perform a quantitative estimate of the measure of heterogeneity and lipid-domain interface formation under different circumstances, e.g., near a phase transition and in the presence of other membrane components such as cholesterol and drugs, as well as polypeptides and proteins.

The bilayer configurations as presented in, e.g., Figures 6 and 8 suggest that the bilayer plane at any given time can be seen as being partitioned into three regions: the bulk, the domains (clusters), and the interfaces bounding the two, corresponding to three fractional contributions to the total bilayer area — the fractional bulk area, $a_b(T)$, the fractional cluster area, $a_c(T)$, and the fractional interfacial area, $a_i(T)$;[82] i.e., $a_b + a_c + a_i = 1$. The computer simulations using the Pink model for pure lipid bilayers (Equation 1) show that $a_b(T)$ displays a minimum at the transition whereas both $a_c(T)$ and $a_i(T)$ have pronounced maxima. The interfacial region is found to be strongly dominated by chains in intermediate conformational states.[82] This observation suggests that the interfaces are soft and that they act as a sink for excited acyl chain conformations. The softness of the interfaces is of importance for prolonging the lifetime of the clusters, and it is moreover crucial for the way in which other membrane components, such as cholesterol, protein, and drugs, distribute themselves in the heterogeneous bilayer membrane structure.

For simplicity, we shall only describe the permeability model in the case of a pure lipid bilayer, where only the temperature variation of the fractional areas is relevant. The model can be easily extended to account for more complex situations.[81] The basic assumption of the model is[82] that the bulk and cluster regions of the bilayer each have their characteristic rates of transbilayer diffusion which are both considerably less than the diffusion rate in the interface. When a permeating particle such as an ion collides with the bilayer it will therefore have very different chances of crossing the bilayer depending on the structure of that part of the bilayer on which it impinges. In the case of a closed liposome of volume $V(T)$ with a certain internal concentration of permeating particles, $c(t) = N(t)/V(T)$, the number of particles leaving the liposome at time t, $dN(t)$, can be expressed as

$$dN(t) \propto c(t)v(T)A(T)P(T)\,dt \qquad (23)$$

where $v(T)$ is the velocity of the impinging particle, $A(T)$ is the internal area of the liposome, and $P(T)$ is the probability that a particle crosses the bilayer once it hits it. From kinetic theory we know that the velocity of the particles has a Maxwellian distribution and hence $v(T) = (k_B T/2\pi m)^{1/2}$, where m is the particle mass. Substitution of this expression for $v(T)$ into Equation 23 gives

$$dN(t) \propto CA(T)^{-1/2} N(t)T^{1/2}P(t)\,dt \qquad (24)$$

with

$$C = 3\left(\frac{2k_B}{m}\right)^{1/2} \qquad (25)$$

It is now assumed that the probability, $P(T)$, of a particle crossing the bilayer can be written as a sum of three terms:

$$P(T) = a_b(T)p_b + a_c(T)p_c + a_i(T)p_i \qquad (26)$$

where a_b, a_c, and a_i are the fractions of the bilayer area occupied by the bulk, the clusters, and the interfaces, respectively. p_b, p_c, and p_i denote the corresponding regional probabilities of transfer. As indicated by Equation 26, only the area fractions, a, are considered to be temperature dependent. Integration of Equation 24 then leads to the simple expression for the fraction, $f(t)$, of particles retained in the liposome:

$$f(t,T) = \frac{N(t)}{N(t=0)} = \exp\left[-CA(T)^{-1/2}T^{1/2}P(T)t\right] \qquad (27)$$

In Figure 22 we show some results for Na[+] permeation obtained from this permeability model using as input the computer-simulation data on temperature dependence for the different liposome fractional areas. A comparison is given between the relative permeability for three different acyl-chain lengths. In order to compare results for the different systems we have used

$$R(T) = A(T)^{-1/2}T^{1/2}P(T) \qquad (28)$$

as a measure of the permeability. It is straightforward to obtain similar predictions in the case of other ionic species. The data clearly demonstrate that there is a significant enhancement of permeability when the acyl-chain length is decreased. The theoretically predicted passive ion permeability is in good agreement with experimental data.[82]

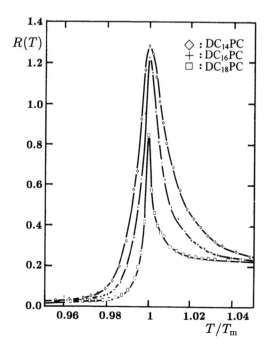

Figure 22 Relative passive ion permeability, $R(T)$, of Na$^+$ ions across DC$_{14}$PC, DC$_{16}$PC, and DC$_{18}$PC bilayers.

Cholesterol in large concentrations is known to have a strong effect on bilayer permeability. In lower concentrations, the effect of cholesterol on the permeability of a simple lipid bilayer is, however, much more subtle and somewhat surprising.[34,35,69] The effect of low levels of cholesterol on lipid bilayer permeability can be calculated using the computer-simulation results described in Section V.C above. The permeability is found to be enhanced by the presence of small amounts of cholesterol[34] in accordance with recent experimental measurements.[69]

We will briefly describe the effects on bilayer passive permeability due to the presence of foreign molecules which intercalate interstitially in the lipid-bilayer matrix (see Section V.E). Examples of classes of compounds covered by this description are manifold, ranging from general anesthetics like halotane, to local anesthetics like cocaine derivatives, to organochloric pesticides like lindane and DDT. By using the model of bilayer permeability as described above, the results shown in Figure 23 have been obtained for permeability, $R(T)$, in the case of a drug like halothane.[38] For comparison the permeability of the pure lipid bilayer is also shown. It is observed that the drugs make the bilayer much more leaky over a broader temperature range. This is a consequence of the ability of the drugs to enhance the fluctuations and create more leaky interfaces in the heterogeneous bilayer state. The inclusion of cholesterol in large concentrations in the model[38] allows us to partially reverse the effect of drug-induced permeability enhancement as cholesterol is a rigid molecule which has a tendency to squeeze the drug out of the lipid bilayer.

Pink and Hamboyan[83] studied a version of the model of the passive permeability of bilayers which applies to bilayers incorporated with transmembrane proteins, such as bacteriorhodopsin. The results of the calculations are that, whereas there is only a small effect due to the protein on the permeability at temperatures above the transition, there is a depression of the permeability at the transition and a considerable enhancement of the permeability at a wide range of temperatures below the transition. It should be remarked that the interpretation of these results may be complicated by the fact that the phase diagram of the lipid-protein system exhibits dramatic phase separation phenomena below the transition temperature of the pure lipid bilayer.

G. EFFECTS ON MEMBRANE CONFORMATIONS DUE TO BENDING RIGIDITY

It has been known for some time that an unsupported and unconstrained fluid, flexible surface is unstable with respect to collapse into a ramified object consisting of connected thin cylindrical tubules.[84] The driving force for this instability is the configurational entropy of the surface. The collapsed surface

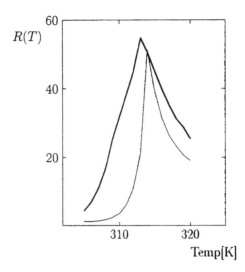

Figure 23 Temperature dependence of relative passive transmembrane permeability, $R(T)$, of ions for a pure DC$_{16}$PC lipid bilayer (thin solid line) and a DC$_{16}$PC lipid bilayer (heavy solid line) interacting with a water-soluble drug like halothane.

displays physical properties characteristic of a branched polymer. Most surfactant surfaces in biological systems are not subject to this branched polymer instability, at least not on the length scales relevant for biological cells, ~1 μm. A major reason for this is that biological membranes are rigid with respect to bending since their bending moduli fall into the range $\kappa \sim 1$ to $20k_B T$.[85]

The properties of self-intersecting "fluid" surfaces with mean-curvature elasticity have been the subject of a substantial number of studies. Analytical considerations suggest that rigid surface conformations with long-range correlations between surface normals are unstable for $T \neq 0$, and the large-scale properties beyond a certain persistence length

$$\xi_p = a \exp\left(\frac{4\pi\kappa}{sk_B T}\right) \tag{29}$$

leave the surface completely flexible; i.e., it will crumple and obtain the characteristics of a self-intersecting branched polymer.[86-91] In Equation 29 a is a microscopic cut-off length, κ is the bending modulus, and s is a numerical constant of order unity. This suggests that even if the surface has a significant bending rigidity this will not prevent the branched polymer instability from manifesting itself when the surface is large enough, simply because the bending modulus κ is being thermally renormalized corresponding to a softening of the surface. So far it has not been possible to confirm this picture by computer simulations of discretized models of self-intersecting surfaces with finite bending rigidity. Instead a second-order phase transition is observed for a finite value of κ.[92-95] However, except for minor modifications, the simulations give a good description of self-avoiding surfaces with bending rigidity in the sense that the mean-curvature elasticity leaves the membrane semiflexible at length scales smaller than a certain persistence length, while the large-scale conformations have the characteristics of self-avoiding branched polymers.[84]

A few studies have been made for triangulated random surfaces with mean-curvature elasticity and partial self-avoidance.[93,96,97] In both Reference 93 and Reference 97 peaks in the specific heat are observed for finite κ. However, opposite conclusions about the presence of a phase transition were reached. Flexible triangulated surfaces with full self-avoidance and $\kappa = 0$ were analyzed in Reference 98 and full consistency with branched-polymer behavior was found. In the following we discuss simulations of flexible triangulated surfaces with full self-avoidance and $\kappa \neq 0$.[99] The simulation data are consistent with the presence of an intermediate length scale which displays properties in overall agreement with the persistence length obtained from approximate analytical treatments.

Our model is that in Equation 12 with $p = 0$ and $\bar{\kappa} = 0$. The basic quantities for characterization of a self-avoiding surface are its spatial extension, shape, area, and volume. The area, A, for a surface is obtained by summation of the areas of the elementary triangles, while the volume is calculated as described in Section IV.D.

Information about the surface shape and extension can, similarly to polymer systems, be obtained from Euclidean invariants of the gyration tensor:

$$\mathcal{T}^{\alpha\beta} = \frac{1}{2N^2} \sum_{i,j=1}^{N} \left(X_i^\alpha - X_j^\alpha \right) \left(X_i^\beta - X_j^\beta \right) \tag{30}$$

Of particular interest is the radius of gyration, $R_g^2 = \mathrm{Tr}\, \mathcal{T}$, that describes the overall size of the surface.

Typical surface configurations for varying bending rigidity, as obtained from the simulations, are displayed in Figure 24. The results from the simulations are expressed in a compact way in Figure 25, where V/ξ_p^3 is plotted against R_g/ξ_p in a double-logarithmic representation. The scaling factor, ξ_p, is adjusted for each value of κ. The fact that the different curves can be mapped onto a single curve suggests a universal relationship

$$V = \xi_p^3 \, f\left(\frac{R_g}{\xi_p} \right) \tag{31}$$

where f is a scaling function that expresses the asymptotic properties of the membrane. Figure 25 indicates that f displays a crossover between two scaling regimes:

$$f(x) \sim x^3 \text{ for } x < 1$$
$$\sim x^2 \text{ for } x > 1 \tag{32}$$

The crossover is consistent with the picture outlined above: for $R_g \gg \xi_p$ the surface volume scales as a branched polymer, while for $R_g \ll \xi_p$ the surface tends to behave as a compact object. The scaling parameter, ξ_p, can thus be identified with a persistence length. In Figure 26 is shown the dependence of ξ_p on κ in a semilogarithmic plot. For $\kappa/k_B T > 0.5$, an approximate linear relation holds, $\log \xi_p \approx \bar{\beta}_2 \kappa / k_B T + \text{constant}$, where $\bar{\beta}_2 = 4.0 \pm 0.2$. For $\kappa/k_B T < 0.5$, ξ_p has a much slower variation with κ. The dependence of κ on the persistence length, ξ_p, is in good accordance with the prediction from analytical calculations on rigid membranes.[87-89] Here we have found that bending rigidity stabilizes the membrane shape on length scales up to a persistence length, and we find no support in the data for the presence of a phase transition for self-avoiding surfaces with finite bending rigidity.

H. INFLATION-DEFLATION TRANSITION OF LIPOSOMES IN OSMOTIC PRESSURE GRADIENTS

Closed fluid membranes, represented by triangulated random surfaces, undergo as a function of pressure, p (osmotic pressure difference between inside and outside), a conformational transition from a low-pressure (deflated, small-volume) phase to a high-pressure (inflated, large-volume) phase. The low-pressure phase is characterized by a branched polymer behavior, and a typical configuration is shown in Figure 27a. However, above a certain critical pressure, p^*, the closed surface looks a crumpled balloon, as illustrated in Figure 27b.

The inflation-deflation transition has been studied by means of Monte Carlo simulations.[100-102] Most recently[102] a simulation study has been performed on the discretized version of the Hamiltonian in Equation 12 with $\kappa = \bar{\kappa} = 0$. With the help of Ferrenberg-Swendsen reweighting[50] of the probability distributions (see Section IV.B) the pressure-volume phase diagram can be constructed from simulations in the vicinity of the critical pressure p^*. The resulting p-V phase diagram for different system sizes is shown in Figure 28. The dramatic change in the vesicle volume over a narrow pressure interval seen in

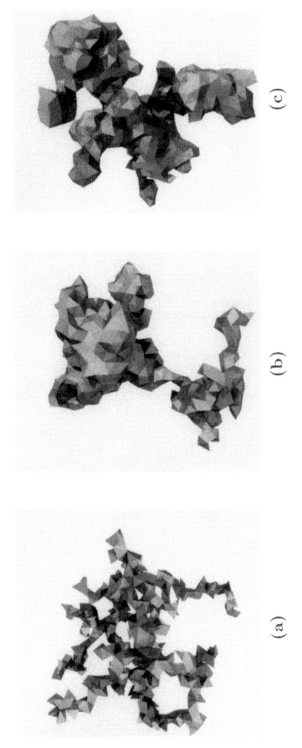

Figure 24 Conformations of triangulated random surfaces with the same area resembling lipid bilayer membranes in the fluid state. (a) Branched-polymer configuration of a random closed surface of spherical topology with zero bending rigidity, $\kappa = 0$. (b) Short-scale compact and large-scale branched-polymer configuration of a random closed surface of spherical topology with a small bending rigidity, $\kappa/k_B T = 0.75$. (c) Short-scale compact and indication of large-scale branched-polymer configuration of a random closed surface of spherical topology with a larger bending rigidity, $\kappa/k_B T = 1$.

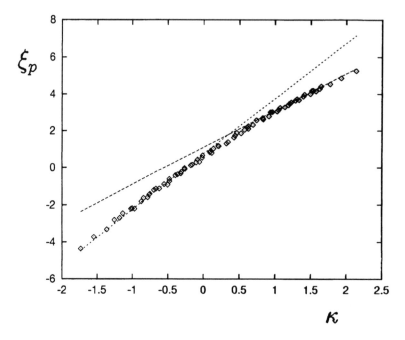

Figure 25 Double-logarithmic plot of scaled volume, V/ξ_p^3, vs. scaled gyration radius, R_G/ξ_p, for a random spherical surface with finite bending rigidity (see Equation 31). The two dashed lines have slopes of 3 and 2, respectively.

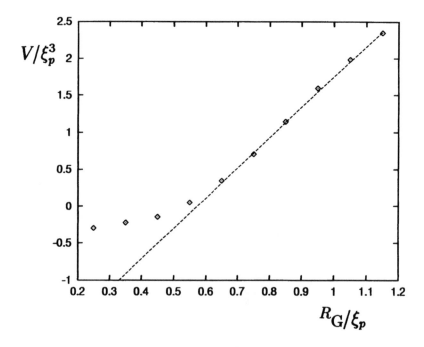

Figure 26 Semilogarithmic plot of persistence length, ξ_p, vs. bending rigidity, κ, for spherical random surfaces with finite bending rigidity. ξ_p is only determined up to a multiplicative factor. The dashed regression line corresponds to $\log \xi_p = \tilde{\beta}_1 \kappa/k_B T$, where $\tilde{\beta}_1 = 4.0 \pm 0.2$.

(a)

(b)

Figure 27 Inflated, crumpled state of a random closed surface of spherical topology and zero bending rigidity. The inflated state is induced by an osmotic pressure, p. (a) $p/k_BT = 0.13$. (b) $p/k_BT = 0.14$.

Figure 28 indicates the presence of a phase transition in the thermodynamic limit $N \to \infty$ from a low-pressure deflated phase to a high-pressure inflated phase. The plots of volume versus pressure allow an approximate determination of the critical pressure p^*.

A more precise determination of the critical pressure can be obtained by evaluating the total volume fluctuations and estimating the critical pressure from the position of the maximum of the fluctuations. Figure 29 shows the resulting p^* for different system sizes. p^* is well described by a power law in N:

$$p^* \sim N^{-\alpha} \tag{33}$$

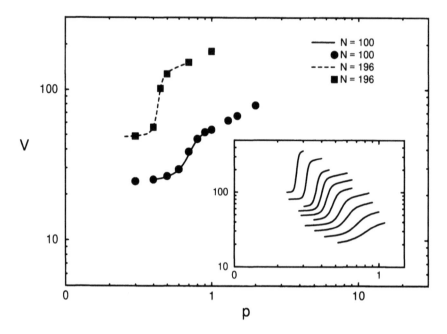

Figure 28 Volume averages, $\langle V \rangle$, for closed random surfaces with zero bending rigidity obtained from simulations at different pressures plotted versus osmotic pressure, p, for two different system sizes ($N = 100$ and $N = 196$). The lines are the results from the Ferrenberg-Swendsen reweighting for a single simulation at $p = 0.7$ for $N = 100$ and $p = 0.445$ for $N = 196$. The insert shows the volume values obtained using the Ferrenberg-Swendsen reweighting technique on simulation data obtained near the critical pressure for $N = 81, 100, 121, 144, 169, 196, 225, 256, 324,$ and 400.

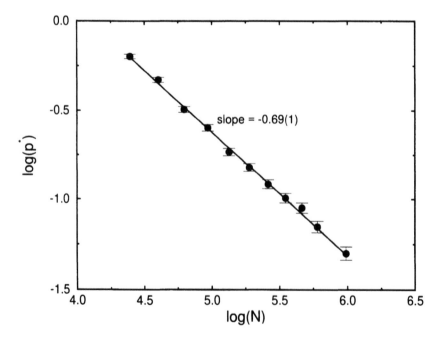

Figure 29 Plot of $\log p^*$ vs. $\log N$ for closed random surfaces with zero bending rigidity. The values for p^* are obtained from the peak in the compressibility $\langle (\Delta V)^2 \rangle$ which was calculated using the Ferrenberg-Swendsen reweighting technique. The exponent is obtained from a linear fit to be -0.69 ± 0.01.

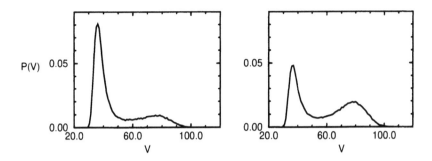

Figure 30 The volume distribution function, $\mathcal{P}(V)$, for closed random surfaces with zero bending rigidity plotted for two different pressure values for a system of size $N = 144$. The sharp peak in the $\mathcal{P}(V)$ distribution belongs to the inflated phase whereas the broad peak belongs to the branched polymer phase.

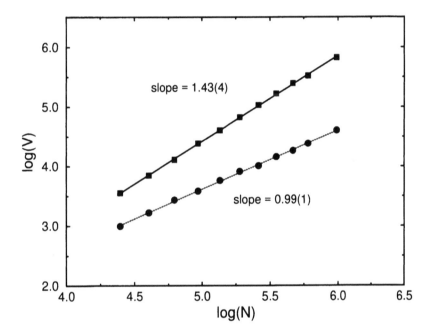

Figure 31 The volumes V_- (●) and V_+ (□) corresponding to the two peaks in $\mathcal{P}(V)$ at p^* shown in Figure 30 plotted versus N in a double-logarithmic representation. The regression lines correspond to $V_\pm \propto N^{\beta_\pm}$, with $\beta_- = 0.99 \pm 0.01$ and $\beta_+ = 1.43 \pm 0.04$.

with $\alpha = 0.69 \pm 0.01$. This exponent value is in disagreement with the previous findings, $\alpha = 0.5$[100] and $\alpha = 1$.[101] The Ferrenberg-Swendsen reweighting technique[50] can also be applied to construct the probability distribution, $\mathcal{P}(V)$, at specific pressures near the anticipated critical pressure, p^*. $\mathcal{P}(V)$ shows a double-peak structure and as a first approximation can be written as the superposition of contributions from the inflated and the deflated phase,

$$\mathcal{P}(V) \simeq a_- \mathcal{P}_-(V) + a_+ \mathcal{P}_+(V) \tag{34}$$

as suggested by Figure 30. The maximum probability volumes for the two peaks V_- and V_+ can be easily estimated, and the results are shown in Figure 31. We find the following relation to hold:

$$V_\pm \sim N^{\beta_\pm} \tag{35}$$

where the associated scaling exponents are $\beta_- = 0.99 \pm 0.01$ and $\beta_+ = 1.43 \pm 0.04$. A similar analysis can be carried out on the data for the radius of gyration.[102]

The above analysis hence confirms that fluid, flexible vesicles undergo an abrupt conformational change between a deflated and an inflated regime at some pressure value, $p*$, which decreases with the system size according to a power law (Equation 33). In the deflated regime, $p < p*$, the simulation data are consistent with branched polymer configurations of the surface and thus support the general expectation that unpressurized, fluid membrane conformations are controlled by the branched polymer fixed point, i.e., the stable $\kappa = 0$ crumpling fixed point.[96,100]

I. MEMBRANE TOPOLOGY

In the previous sections we have studied the large-scale conformational properties of a single flexible fluid membrane and how it can be stabilized by bending rigidity and external pressure. The overwhelming structural richness of fluid membranes is further determined by their ability to undergo topological changes, e.g., budding and fusion of membranes. Fluid membranes will in general form an ensemble of vesicles. However, a single membrane is also subject to topological instabilities which may sometimes lead to the development of convoluted structures and handles. It has been subject to debate whether a large, fluid, flexible membrane with varying topology will undergo spontaneous compaction.[103] This is an important issue for our understanding of the structure of fluid surfaces and membranes, particularly in relation to their function.

So far, only a few theoretical studies have been performed of self-avoiding surfaces with free topology, i.e., with fluctuating Euler characteristic, χ. In References 104 to 106, studies of ensembles of self-avoiding random surfaces with free topology indicate that these systems in general exhibit a very rich phase behavior. Plaquette surface models are naturally produced as the high-temperature limit of lattice gauge theories. This analogy has recently been used in a real-space renormalization-group study of a lattice gauge theory to show that $V \sim R_G^3$ for self-avoiding surfaces with free topology, indicating that they belong to a different universality class than self-avoiding surfaces with the topology of a sphere.[103] This result is very exciting because it has far-reaching consequences for our understanding of fluid interface systems, but it is puzzling because it has been demonstrated that lattice animals fall into the universality class of branched polymers irrespective of the possibility of loop formation.[107] The analytical result[103] has recently been tested by computer-simulation calculations on self-avoiding randomly triangulated surfaces with free topology.[108]

The topology of a single closed surface is fully determined by its number of handles, g, or equivalently by its Euler characteristic, $\chi = 2 - 2g$. Besides an estimate for the volume exponent, d_v, the computer simulations provide estimates for the exponents for the area, d_A, and the Euler characteristics, $d\chi$, where the exponents are defined by

$$\langle A \rangle \sim R_G^{d_A} \tag{36}$$

$$\langle V \rangle \sim R_G^{d_v} \tag{37}$$

$$\langle g \rangle \sim R_G^{d_\chi} \tag{38}$$

Besides the surface updating described in Section IV.E, the simulation procedure must also allow for topological changes of the surface. The move classes for the topological excitations are restricted to involve pairs of non-neighboring triangles with the formation of a narrow neck. The "cut" operation will identify such necks by random inspection of triangles on the surface, cut the six relevant links, and "sew" the two triangular holes in the surface to keep the surface closed. The "cut" operation is allowed if the surface can be kept as a single component, thus enhancing χ by 2. The "fuse" operation, which reduces χ by 2, is performed by identifying random elementary triangles on the surface, which is eligible for fusion of surface components, e.g., the orientation, and the conditions for self-avoidance must be inspected.

An example of a spatial configuration of an ensemble of fluid surfaces with free topology is shown in Figure 32. A number of closed surfaces of different topology can be identified, including some small vesicles. The genus number of the large object in Figure 32 is rather large: $g = 22$. The results from an analysis of a single closed surface with a fluctuating number of handles are shown in Figure 33, where $\langle A \rangle$, $\langle V \rangle$, and $\langle g \rangle$ are plotted against R_G^2 in a double-logarithmic representation. The figure shows that self-avoiding random surfaces with free topology display an asymptotic scaling behavior for large system

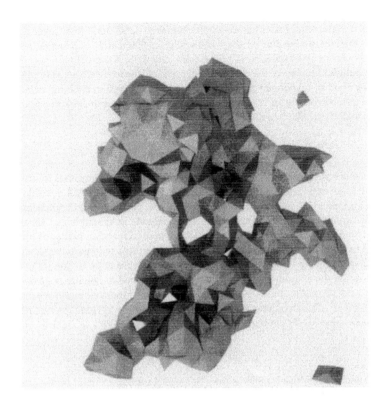

Figure 32 An ensemble of closed random surfaces with $\kappa/k_B T = 1$ and free topology. The large object has $g = 22$ handles.

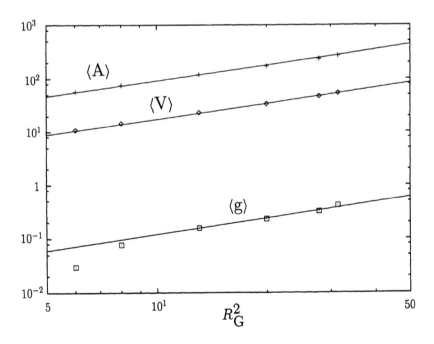

Figure 33 Double-logarithmic plot of $\langle A \rangle$, $\langle V \rangle$, and $\langle g \rangle$ versus R_G^2 for a self-avoiding random surface with free topology. The slopes of the solid lines lead to the exponents $d_A = 1.94$, $d_V = 1.96$, and $d\chi = 2.0$ (see Equations 36 to 38). The data are obtained for system sizes corresponding to $N = 49, 64, 100, 144, 196$, and 225.

sizes. This result is stable with respect to different choices of "cut" and "fuse" rates as well as the value of the hard-core diameter used to enforce the self-avoidance. The scaling of exponents of interest can now be estimated to be $d_A = 1.94 \pm 0.08$, $d_V = 1.96 \pm 0.06$, and $d\chi = 2.0 \pm 0.1$.

These results support the theory that self-avoiding random surfaces with free topology indeed belong to the universality class of branched polymers. The entropy involved in creating, deleting, and redistributing handles on a self-avoiding surface is thus asymptotically insignificant compared with the configurational entropy underlying the branched-polymer instability.[84]

VI. FUTURE PERSPECTIVES

The quantitative study of the physical chemistry of lipid bilayers and liposomes by computer-simulation techniques is still in an early stage. Current computer technology has made it possible to analyze simple models, but there is still a long way to go before it will be practicable to treat realistic models that include a molecular description of the in-plane molecular membrane organization at the same time as the conformational complexity of the soft bilayer sheet. It is, however, only a matter of time before it should become possible to simulate models of flexible membrane surfaces with particles that carry some degree of molecular structure, such as small polymers grafted to a flexible surface. Studies of such systems will lead to information on the coupling between bilayer curvature and molecular conformations and should ultimately lead to a description of the thermal renormalization of the continuum-mechanic parameters of flexible elastic bilayers. Results along these lines are likely to be helpful for the molecular design of liposomes of importance for technological and medical applications.[7]

ACKNOWLEDGMENTS

This work was supported by the Danish Natural Science Research Council under grant 11–0065–1, by the Danish Technical Research Council under grant 16–5039–2, by Jenny Vissings Fond, by the Carlsberg Foundation, by le FCAR du Quebec under a center and team grant, and by NSERC of Canada. A sabbatical stay in Denmark for MJZ was supported by the Danish Research Academy under grant S940049.

REFERENCES

1. **Biltonen, R. L. and Lichtenberg, D.** The use of differential scanning calorimetry as a tool to characterize liposome preparations. *Chem. Phys. Lipids* 64, 129, 1993.
2. **Biltonen, R. L.** A statistical-thermodynamic view of cooperative structural changes in phospholipid bilayer membranes: their potential role in biological function. *J. Chem. Thermodyn.* 22, 1, 1990.
3. **de Gennes, P. G.** Soft matter. *Rev. Mod. Phys.* 64, 645, 1992.
4. **Cevc, G. and March, D.** *Phospholipid Bilayers. Physical Principles and Models.* Wiley-Interscience, New York, 1987.
5. **Israelachvili, J.** *Intermolecular and Surface Forces.* Academic Press, New York, 2nd ed., 1992.
6. **Bloom, M., Evans, E., and Mouritsen, O. G.** Physical properties of the fluid-bilayer component of cell membranes: a perspective. *Q. Rev. Biophys.* 24, 293, 1991.
7. **Lasic, D. D.** *Liposomes: From Physics to Applications.* Elsevier, Amsterdam, 1993.
8. **Bloom, M. and Mouritsen, O. G.** In *The Evolution of Membranes.* Biophysics Handbook on Membranes, Vol. 1 (Lipowsky, R., Ed.), Elsevier, Amsterdam, 1984, p. 89.
9. **Kinnunen, P. and Laggner, P., Eds.** Phospholipid phase transitions. Topical issue of *Chem. Phys. Lipids,* Vol. 57, 1991.
10. **Cullis, P. R. and Hope, M. J.** Physical properties and functional roles of lipids in membranes. In *New Comprehensive Biochemistry. Biochemistry of Lipids, Lipoproteins and Membranes,* Vol. 20 (Vance, D. E. and Vance, J. E., Eds.), Elsevier, Amsterdam, 1991, 1.
11. **Mouritsen, O. G. and Jørgensen, K.** Micro-, nano-, and meso-scale heterogeneity of lipid bilayers and its influence on macroscopic membrane properties. *Mol. Membr. Biol.* 12, 15, 1995.
12. **Mouritsen, O. G. and Jørgensen, K.** Dynamical order and disorder in lipid bilayers. *Chem. Phys. Lipids* 73, 3, 1994.
13. **Goldenfeld, N.** *Lectures on Phase Transitions and the Renormalization Group.* Addison-Wesley, New York, 1992.
14. **Nagle, J. F.** Theory of the main lipid bilayer phase transition. *Annu. Rev. Phys. Chem.* 31, 157, 1980.
15. **Mouritsen, O. G., Ipsen, J. H., Jørgensen, K., Sperotto, M. M., Zhang, Z., Corvera, E., Fraser, D. P., and Zuckermann, M. J.** Computer simulation of phase transitions in Nature's preferred liquid crystal: the lipid bilayer membrane. In *Computer Simulation of Liquid Crystals* (Luckhurst, G. F., Ed.), Kluwer Academic, Amsterdam (in press).

16. **Fraser, D. P., Zuckermann, M. J., and Mouritsen, O. G.** Simulation technique for hard-disk models in two dimensions. *Phys. Rev. A* 42, 3186, 1990.
17. **Fraser, D. P., Zuckermann, M. J., and Mouritsen, O. G.** Theory and simulations for hard-disk models of binary mixtures of molecules with internal degrees of freedom. *Phys. Rev. A* 43, 6642, 1991.
18. **van der Ploeg, P. and Berendsen, H. J. C.** Molecular dynamics of a bilayer membrane. *Mol. Phys.* 49, 233, 1983.
19. **Heller, H., Schaefer, M., and Schulten, K.** Molecular dynamics simulation of a bilayer of 200 lipids in the gel and the liquid-crystal phases. *J. Phys. Chem.* 97, 8343, 1993.
20. **Scott, H. L.** Computer aided methods for the study of lipid chain packing in model membranes and micelles. In *Molecular Description of Biological Membrane Components by Computer Aided Conformational Analysis,* Vol. 1 (Brasseur, R., Ed.), CRC Press, Boca Raton, FL, 1990, 123.
21. **Doniach, S.** Thermodynamic fluctuations in phospholipid bilayers. *J. Chem. Phys.* 68, 4912, 1978.
22. **Pink, D. A., Green, T. J., and Chapman, D.** Raman scattering in bilayers of saturated phosphatidylcholines. Experiment and theory. *Biochemistry* 19, 349, 1980.
23. **Mouritsen, O. G.** Computer simulation of cooperative phenomena in lipid membranes. In *Molecular Description of Biological Membrane Components by Computer Aided Conformational Analysis,* Vol. 1 (Brasseur, R., Ed.), CRC Press, Boca Raton, FL, 1990, 3.
24. **Meunier, J., Langevin, D., and Boccara, H., Eds.** *Physics of Amphiphilic Layers.* Springer-Verlag, Berlin, 1987.
25. **Nelson, D. R., Piran, T., and Weinberg, S., Eds.** *Statistical Mechanics of Membranes and Surfaces.* World Scientific, Singapore, 1989.
26. **Lipowsky, R.** The conformation of membranes. *Nature* 349, 475, 1991.
27. **de Gennes, P.-G.** *Scaling Concepts in Polymer Physics.* Cornell University Press, London, 1991.
28. **Mouritsen, O. G.** Theoretical models of phospholipid phase transitions. *Chem. Phys. Lipids* 57, 179, 1991.
29. **Jørgensen, K., Sperotto, M. M., Mouritsen, O. G., Ipsen, J. H., and Zuckermann, M. J.** Phase equilibria and local structure in binary lipid bilayers. *Biochim. Biophys. Acta* 1152, 135, 1993.
30. **Risbo, J., Sperotto, M. M., and Mouritsen, O. G.** Theory of phase equilibria and critical mixing points in binary lipid bilayers: free energy, enthalpy, specific heat, and interfacial tension. *J. Chem. Phys.,* 103 (in press).
31. **Zhang, Z., Laradji, M., Guo, H., Mouritsen, O. G., and Zuckermann, M. J.** Phase behavior of pure lipid bilayers with mismatch interaction. *Phys. Rev. A* 46, 7560, 1992.
32. **Zhang, Z., Zuckermann, M. J., and Mouritsen, O. G.** Effect of intermonolayer coupling on the phase behavior of lipid bilayers. *Phys. Rev. A* 46, 6707, 1992.
33. **Sperotto, M. M. and Mouritsen, O. G.** Lipid enrichment and selectivity of integral membrane proteins in two-component lipid bilayers. *Eur. Biophys. J.* 22, 323, 1993.
34. **Cruzeiro-Hansson, L., Ipsen, J. H., and Mouritsen, O. G.** Intrinsic molecules in lipid membranes change the lipid-domain interfacial area: cholesterol at domain boundaries. *Biochim. Biophys. Acta* 979, 166, 1989.
35. **Zuckermann, M. J., Ipsen, J. H., and Mouritsen, O. G.** Theoretical studies of the phase behavior of lipid bilayers containing cholesterol. In *Cholesterol and Membrane Models* (Finegold, L. X., Ed.), CRC Press, Boca Raton, FL 1993, 223.
36. **Jørgensen, K., Ipsen, J. H., Mouritsen, O. G., Bennett, D., and Zuckermann, M. J.** A general model for the interaction of foreign molecules with lipid membranes: drugs and anaesthetics. *Biochim. Biophys. Acta* 1062, 227, 1991.
37. **Jørgensen, K., Ipsen, J. H., Mouritsen, O. G., Bennett, D., and Zuckermann, M. J.** The effects of density fluctuations on the partitioning of foreign molecules into lipid bilayers: application to anaesthetics and insecticides. *Biochim. Biophys. Acta* 1067, 241, 1991.
38. **Jørgensen, K., Ipsen, J. H., Mouritsen, O. G., and Zuckermann, M. J.** The effects of anaesthetics on the dynamic heterogeneity of lipid membranes. *Chem. Phys. Lipids* 65, 205, 1993.
39. **Sabra, M. C., Jørgensen, K., and Mouritsen, O. G.** Calorimetric and theoretical studies of the effects of lindane on lipid bilayers with different acyl-chain length. *Biochim. Biophys. Acta* 1233, 89, 1995.
40. **Mouritsen, O. G. and Bloom, M.** Mattress model of lipid-protein interactions in membranes. *Biophys. J.* 46, 141, 1984.
41. **Mouritsen, O. G. and Bloom, M.** Models of lipid-protein interactions in membranes. *Annu. Rev. Biophys. Biomol. Struct.* 22, 147, 1993.
42. **Fattal, D. R. and Ben-Shaul, A.** A molecular model for the lipid-protein interaction in membranes: the role of hydrophobic mismatch. *Biophys. J.* 65, 1795, 1993.
43. **Sperotto, M. M. and Mouritsen, O. G.** Monte Carlo simulation studies of lipid order parameter profiles near integral membrane proteins. *Biophys. J.* 59, 261, 1991.
44. **Sperotto, M. M. and Mouritsen, O. G.** Mean-field and Monte Carlo simulation studies of the lateral distribution of proteins in membranes. *Eur. Biophys. J.* 19, 157, 1991.
45. **Zhang, Z., Sperotto, M. M., Zuckermann, M. J., and Mouritsen, O. G.** A microscopic model for lipid/protein bilayers with critical mixing. *Biochim. Biophys. Acta* 1147, 154, 1993.
46. **Mouritsen, O. G., Sperotto, M. M., Risbo, J., Zhang, Z., and Zuckermann, M. J.** Computational approach to lipid-protein interactions in membranes. In *Advances in Comparative Biology* (Villar, H.O., Ed.), JAI Press, Greenwich, CN (in press).

47. **Helfrich, W.** Elastic properties of lipid bilayers: theory and possible experiments. *Z. Naturforsch.* 28c, 693, 1973.
48. **Mouritsen, O. G.** *Computer Studies of Phase Transitions and Critical Phenomena.* Springer-Verlag, Heidelberg, 1984.
49. **Zhang, Z., Mouritsen, O. G., and Zuckermann, M. J.** Detecting phase equilibria in models of thermotropic and lyotropic liquid crystals. *Mod. Phys. Lett. B* 7, 217, 1993.
50. **Ferrenberg, A. M. and Swendsen, R. H.** New Monte Carlo technique of studying phase transitions. *Phys. Rev. Lett.* 61, 2635, 1988.
51. **Lee, J. and Kosterlitz, J. M.** Finite-size scaling and Monte Carlo simulations of first order phase transitions. *Phys. Rev. B* 43, 3265, 1991.
52. **Helfrich, W.** Steric interaction of fluid membranes in multilayer systems. *Z. Naturforsch.* 33a, 305, 1978.
53. **Allen, M. P. and Tildesley, D. J.** *Computer Simulation of Liquids.* Oxford Science Publications, Oxford, 1987.
54. **Morrow, M. R., Whitehead, J. P., and Lu, D.** Chain-length dependence of lipid bilayer properties near the liquid crystal to gel phase transition. *Biophys. J.* 63, 18, 1992.
55. **Chapman, D., Williams, R. M., and Ladbrooke, B. D.** Physical studies of phospholipids. *Chem. Phys. Lipids* 1, 445, 1967.
56. **Albon, N. and Sturtevant, J. M.** Nature of the gel to liquid crystal transition of synthetic phosphatidylcholines. *Proc. Natl. Acad. Sci. U.S.A.* 75, 2258, 1978.
57. **Corvera, E., Laradji, M., and Zuckermann, M. J.** Application of finite-size scaling to the Pink model for lipid bilayers. *Phys. Rev. E* 47, 696, 1993.
58. **Ipsen, J. H., Jørgensen, K., and Mouritsen, O. G.** Density fluctuations in saturated phospholipid bilayer increase the acyl-chain length decreases. *Biophys. J.* 58, 1099, 1990.
59. **Mouritsen, O. G. and Zuckermann, M. J.** Softening of lipid bilayers. *Eur. Biophys. J.* 12, 75, 1985.
60. **Mouritsen, O. G. and Jørgensen, K.** Dynamic lipid-bilayer heterogeneity: a mesoscopic vehicle for membrane function? *BioEssays* 14, 129, 1992.
61. **Mouritsen, O. G. and Biltonen, R. L.** Protein-lipid interactions and membrane heterogeneity. In *Protein-Lipid Interactions. New Comprehensive Biochemistry* (Watts, A., Ed.), Vol. 25, Elsevier, Amsterdam, 1993, 1.
62. **Knoll, W., Schmidt, G., and Sackmann, E.** Critical demixing in fluid bilayers of phospholipid mixtures. A neutron diffraction study. *J. Chem. Phys.* 79, 3439, 1983.
63. **Huang, J., Swanson, J. E., Dibble, A. R. G., Hinderliter, A. K., and Feigenson, G. W.** Nonideal mixing of phosphatidylserine and phosphatidylcholine in the fluid lamellar phase. *Biophys. J.* 64, 413, 1993.
64. **Huang, J. and Feigenson, G. W.** Monte Carlo simulation of lipid mixtures: finding phase separation. *Biophys. J.* 65, 1788, 1993.
65. **Thewalt, J. and Bloom, M.** Phosphatidylcholine:cholesterol phase diagrams. *Biophys. J.* 63, 1176, 1993.
66. **Almeida, P. F. F., Vaz, W. L. C., and Thompson, T. E.** Lateral diffusion in liquid phases of dimyristoylphosphatidylcholine/cholesterol lipid bilayers. A free volume analysis. *Biochemistry* 31, 6739, 1992.
67. **Almeida, P. F. F., Vaz, W. L. C., and Thompson, T. E.** Percolation and diffusion in three-component lipid bilayers: effect of cholesterol on an equimolar mixture of two phosphatidylcholines. *Biophys. J.* 64, 399, 1993.
68. **Ipsen, J. H., Karlström, G., Mouritsen, O. G., Wennerström, H., and Zuckermann, M. J.** Phase equilibria in the phosphatidylcholine-cholesterol system. *Biochim. Biophys. Acta* 905, 162, 1987.
69. **Corvera, E., Mouritsen, O. G., Singer, M. A., and Zuckermann, M. J.** The permeability and the effect of acyl chain length for phospholipid bilayers containing cholesterol: theory and experiment. *Biochim. Biophys. Acta* 1107, 261, 1992.
70. **Michels, B., Fazel, N., and Cerf, R.** Enhanced fluctuations in small phospholipid bilayer vesicles containing cholesterol. *Eur. Biophys. J.* 17, 187, 1989.
71. **Mouritsen, O. G. and Sperotto, M. M.** Thermodynamics of lipid-protein interactions in lipid membranes: the hydrophobic matching condition. In *Thermodynamics of Membrane Receptors and Channels* (Jackson, M., Ed.), CRC Press, Boca Raton, FL, 1993, 127.
72. **Morrow, M. R., Huschilt, J. C., and Davis, J. H.** Simultaneous modeling of phase and calorimetric behavior in an amphiphilic peptide/phospholipid model membrane. *Biochemistry* 24, 5396, 1985.
73. **Morrow, M. R. and Davis, J. H.** Differential scanning calorimetry and ^{2}H NMR studies of the phase behavior of gramicidin-phosphatidylcholine mixtures. *Biochemistry* 27, 2024, 1988.
74. **Sperotto, M. M. and Mouritsen, O. G.** Lipid selectivity of integral membrane proteins in two-component lipid bilayers. Bacteriorhodopsin in DLPC-DSPC mixtures (in preparation, 1995).
75. **Piknova, B., Perochon, E., and Tocanne, J.-F.** Hydrophobic mismatch and long-range protein-lipid interactions in bacteriorhodopsin/phosphatidylcholine vesicles. *Eur. J. Biochem.* 218, 385, 1993.
76. **Rubin, E., Miller, K. W., and Roth, S. H., Eds.** Molecular and Cellular Mechanisms of Alcohols and Anaesthetics. *Ann. N.Y. Acad. Sci.* Vol. 625, 1991.
77. **Singer, M. A. and Jain, M. K.** Interaction of four local anesthetics with phospholipid bilayer membranes: permeability effects and possible mechanisms. *Can. J. Biochem.* 58, 815, 1984.
78. **van Osdol, W., Ye, Q., Johnson, M. L., and Biltonen, R. L.** The effects of the anesthetic dibucaine on the kinetics of the gel-liquid crystalline transition of dipalmitoylphosphatidylcholine multilamellar vesicles. *Biophys. J.* 63, 1011, 1992.

79. **Mountcastle, D. B., Biltonen, R. L., and Halsey, M. J.** Effect of anesthetics and pressure on the thermotropic behavior of multilamellar dipalmitoylphosphatidylcholine liposomes. *Proc. Natl. Acad. Sci. U.S.A.* 75, 4906, 1978.

80. **Antunes-Madeira, M. C. and Madeira, V. M. C.** Partitioning of lindane in synthetic and native membranes. *Biochim. Biophys. Acta* 820, 165, 1985.

81. **Mouritsen, O. G., Jørgensen, K., and Hønger, T.** Permeability of lipid bilayers near the phase transition. In *Permeability and Stability of Lipid Bilayers.* (Disalvo, E. A. and Simon, S. A., Eds.), CRC Press, Boca Raton, FL 1995, p. 137.

82. **Cruzeiro-Hansson, L. and Mouritsen, O. G.** Passive ion permeability of lipid membranes modelled via lipid-domain interfacial area. *Biochim. Biophys. Acta* 944, 63, 1988.

83. **Pink, D. A. and Hamboyan, H.** Effect of integral proteins upon bilayer permeability to ions. *Eur. Biophys. J.* 18, 245, 1990.

84. **Glaus, U.** Monte Carlo simulation of self-avoiding surfaces in three dimensions. *Phys. Rev. Lett.* 56, 1996, 1986.

85. **Sackmann, E.** Molecular and global structure and dynamics of membranes and lipid bilayers. *Can. J. Phys.* 68, 999, 1990.

86. **Helfrich, W.** Effect of thermal undulations on the rigidity of fluid membranes and interfaces. *J. Phys. (Paris)* 46, 1236, 1985.

87. **Polyakov, A.** Fine structure of strings. *Nucl. Phys.* B268, 406, 1986.

88. **Peliti, L. and Leibler, S.** Effects of thermal fluctuations on systems with small surface tension. *Phys. Rev. Lett.* 54, 1690, 1985.

89. **Kleinert, H.** Thermal softening of curvature elasticity in membranes. *Phys. Lett.* A114, 263, 1986.

90. **Förster, D.** On the scale dependence due to thermal fluctuations of the elastic properties of membranes. *Phys. Lett.* A114, 115, 1986.

91. **Ichinose, S.** Background-field approach to extended string theory. *Phys. Lett.* B209, 461, 1988.

92. **Renken, R. L. and Kogut, J. B.** The crumpling transition in presence of quantum theory. *Nucl. Phys.* B354, 328, 1991.

93. **Baillie, C. F. and Johnston, D. A.** Weak self-avoidance and crumpling in random surfaces with extrinsic curvature. *Phys. Lett.* B273, 380, 1991.

94. **Armbjørn, J., Jurkiewicz, J., Varsted, S., Irback, A., and Peterson, B.** Critical properties of the dynamical random surface with extrinsic curvature. *Phys. Lett.* B275, 295, 1992.

95. **Münkel, C. and Heermann, D. W.** The crumpling transition of dynamically triangulated random surfaces (preprint, 1993).

96. **Boal, D. and Rao, M.** Scaling behaviour of fluid membranes in three dimensions. *Phys. Rev. A* 45, R6947, 1992.

97. **Kroll, D. M. and Gompper, G.** The conformation of fluid membranes: Monte Carlo simulations. *Science* 255, 968, 1992.

98. **Kroll, D. M. and Gompper, G.** Scaling behaviour of randomly triangulated self-avoiding surfaces. *Phys. Rev. A* 46, 3119, 1992.

99. **Ipsen, J. H. and Jeppesen, C.** The persistence length in a self-avoiding random-surface model. *J. Phys. I. (Paris),* (in press, 1995).

100. **Gompper, G. and Kroll, D. M.** Shape of inflated vesicles. *Phys. Rev. A* 46, 7466, 1992.

101. **Baumgärtner, A.** Inflated vesicles: a lattice model. *Physica A* 190, 63, 1992.

102. **Dammann, B., Fogedby, H. C., Ipsen, J. H., and Jeppesen, C.** Monte Carlo study of the inflation-deflation transition in a fluid membrane. *J. Phys. I. (Paris)* 4, 1139, 1994.

103. **Banavar, J. R., Maritan, A., and Stella, A.** Geometry, topology, and universality of random surfaces. *Science* 252, 825, 1991.

104. **Karowski, M. and Thun, H. J.** Phase structure of systems of self-avoiding surfaces. *Phys. Rev. Lett.* 54, 2556, 1985.

105. **David, F.** O(n) gauge models and self-avoiding random surfaces in three dimensions. *Europhys. Lett.* 9, 575, 1989.

106. **Höfsass, T. and Kleinert, H.** Gaussian curvature in an Ising model of microemulsions. *J. Chem. Phys.* 86, 3565, 1987.

107. **Vannimenus, J.** Phase transitions for polymers on fractal lattices. *Physica* 38D, 351, 1989.

108. **Jeppesen, C. and Ipsen, J. H.** Scaling properties of self-avoiding surfaces with free topology. *Europhys. Lett.* 22, 713, 1993.

Chapter 5

Molecular Theory of Acyl Chain Packing in Lipid and Lipid-Protein Membranes

Deborah R. Fattal and Avinoam Ben-Shaul

CONTENTS

I. INTRODUCTION

In this chapter we describe a molecular level theory of lipid packing and chain statistics in membrane bilayers and demonstrate some of its recent biophysical applications.[1,2] The central quantity in this "mean field" theory is the singlet probability of lipid tail conformations. The singlet probability distribution function is derived by minimizing the system free energy subject to the relevant packing constraints on lipid conformational statistics. As will be shown in the next section, these packing constraints can be expressed in a simple mathematical form, based on one reasonable and commonly accepted physical assumption — namely, that the hydrophobic score of the lipid membrane (in its fluid state) is uniformly packed by chain segments at liquid-like density.[3-5] This assumption implies packing constraints which depend on the aggregation geometry, that is, on the curvature of the hydrocarbon-water interface and on the average area per headgroup, as measured at this interface. This yields a mathematically simple expression for the singlet probability distribution, with the aid of which one can calculate any desired conformational property and, in the mean field approximation, any thermodynamic property of interest.

Previous applications of this theory include amphiphile chain packing statistics (e.g., bond orientational order parameters and segment spatial distributions) in micelles, bilayers, and monolayers;[1,2] curvature and stretching elasticity of pure and mixed membranes;[6] and a simple model for lipid protein interaction that will be described in Section 4.[7] Recently, the theory has been extended to monolayers of grafted polymers ("brushes").[8]

Being a mean field theory, the approach described here is obviously approximate as far as thermodynamic properties are concerned. As has been and will be demonstrated (Section 3), its predictions with regard to "single chain" conformational properties, which depend only on the singlet distribution, show good agreement with both experiments and large-scale computer simulations. As a molecular level theory it is certainly more detailed than phenomenological, e.g., continuum theories of the lipid membrane or models based on geometric packing considerations.[3-5] On the other hand, it is less detailed (yet much simpler to implement) than many-molecule computer simulations, such as molecular dynamics (MD) or Monte Carlo (MC) calculations.

In MD simulations of a lipid membrane one solves the classical equations of motion governing the dynamics of all the atoms of all the constituent lipid molecules, as well as those of the surrounding water molecules (see, e.g., References 9 to 13). To avoid finite size effects the simulated membrane should include at least several hundred lipids and a similar number of water molecules. The computation times are enormous even on the fastest present-day computers,[9] and the corresponding physical times sampled are rather short (a few hundred picoseconds). Another, inevitable difficulty associated with these

0-8493-4731-9/96/$0.00+$.50
© 1996 by CRC Press, Inc.

calculations are the uncertainties involved with the multitude of intermolecular potentials required for such calculations, especially those governing the interfacial (aqueous) regions of the membrane.[13] The long equilibration times and the uncertainties in interaction potentials can sometimes lead to computational artifacts but, most importantly, considerably limit the number of systems and conditions (e.g., membrane shapes) which can be faithfully modeled. Nevertheless, with the fast advent of computers and computing algorithms, MD simulations are rapidly becoming more common and more reliable and will certainly play an increasing central role in providing information on the dynamics and structure of complex many-molecule systems, including lipid and lipid-protein membranes.

The uncertainties in intermolecular potentials also affect the applicability of MC simulations. In these simulations, unlike in MD, one does not solve the equations of motions, but, rather, tries to sample as many equilibrium configurations of the system in question as possible (see, e.g., References 14 to 16). Various algorithms have been developed to efficiently "move" from one configuration to another. A major difficulty in MC studies of lipid membranes is related to the high, liquid-like density of these systems and the polymeric character of the lipid chains. These facts impose severe limitations on the MC sampling procedure, which in some cases may result in "unexplored" phase space regions. Yet, as with the MD simulations, increasingly sophisticated MC procedures are constantly being developed, and it can be anticipated that many important (especially structural) properties of lipid membranes will be investigated using these methods.

In mean field theories of the lipid membrane one treats, usually in great detail, the conformational properties of one "central" molecule, but the effects of neighboring molecules are treated approximately. They are assumed to provide a "mean field" for the "motion" of the central molecule. In general, the mean field appears as a variational parameter (or parameters) in the singlet probability distribution of the central chain and its numerical evaluation involves a solution of "self-consistency" equations. For instance, in the theory presented in the next section, the mean field acting on the central lipid molecule is represented by the "lateral pressure profile" exerted on this chain by its neighbors. (As the average area per molecule decreases, the lateral pressure increases, and the lipid chains are stretched further along the membrane normal.) The self-consistency equations represent the packing constraints on the lipid chains which, as mentioned above, reflect the assumption of a uniform, liquid-like, hydrophobic core.

Several mean field theories of chain packing statistics in bilayers have been formulated to account for structural and thermodynamic properties of lipid membranes. The first theory of this kind was proposed by Marelja,[17,18] primarily in order to analyze the fluid-solid ("gel-liquid crystalline") transition in lipid bilayers. This theory, formulated in the spirit of the Maier-Saupe theory of the isotropic to nematic phase transition in liquid crystals,[19] assumes that interchain interactions are governed by the anisotropy of the interaction potential between neighboring chain segments. In this respect it is different from the theory described below in which excluded volume (packing) interactions are the dominant ones. These interactions also play the decisive role in the lattice theory originally presented by Dill and Flory[20] and then further developed by Dill and Stigter.[21] Gruen's theory,[22,23] developed at about the same time, is in many respects similar to ours,[24-32] as has been discussed in detail elsewhere.[30] Another mean field approach, extensively applied to analyze lipid membrane properties, has been developed by the Wageningen group, based on the mean field theory of polymeric systems originally proposed by the Scheutjens and the Fleer (see, e.g.,References 33 to 36). Unlike our approach, this theory is based on the assumption that the lipid chain segments occupy the sites of a given underlying lattice.

Most of the mean field theories mentioned above share several common features and differ in some other (often subtle) respects. Our intention in this chapter is not to compare the different theories but, rather, to describe one consistent approach, its possible applications, and some of its achievements as well as possible deficiencies. To this end, after introducing the basic theoretical concepts in Section 2, we will describe three different applications of biological relevance in Sections 3 to 5 and conclude with some preliminary results relevant to the application of the theory to vesicle fusion in Section 6.

II. MOLECULAR THEORY OF MEMBRANE STRUCTURE

In this section we describe a rather simple statistical thermodynamic molecular theory of lipid chain packing in membranes in their fluid state. As usual, we treat the membrane as a two-dimensional (2D) film, composed of a central hydrophobic region comprising the hydrocarbon chains ("tails") of the lipid molecules and two interfacial regions containing the lipid polar headgroups (Figure 1). We shall focus on the conformational properties of the lipid chains. The interactions involving the polar heads will only

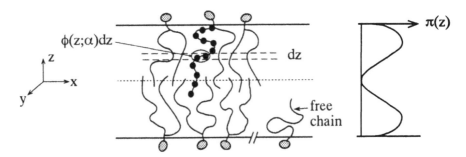

Figure 1 Schematic illustration of a (planar) lipid bilayer and the quantities appearing in the derivation of the singlet probability of chain conformations (see Equation 5): $\phi(z;\alpha)dz$ denotes the number of chain segments which, for a chain in conformation α, fall within the shell $z, z + dz$. The 'free chain' is a (hypothetical) chain with no neighbors around it. The lateral pressure profile $\pi(z)$ schematically illustrated in the figure accounts for the pressure exerted on a given (stretched) chain by its neighbors in the bilayer.

be treated in an approximate, "phenomenological" fashion. The main reason for that is the highly specific nature of these interactions which depend sensitively on the size, charge (e.g., ionic vs. zwitterionic), and chemical composition of the headgroups. On the other hand, the theoretical treatment of the hydrophobic region of the membrane is in many respects simpler and more general. This is because the hydrophobic core is, to a very good approximation, uniformly packed with lipid chain segments at hydrocarbon liquid-like density.[3-5] This basic fact is valid and thus simplifies considerably the theoretical description, even when the core is composed of different types of acyl chains (e.g., saturated and partially unsaturated or long and short chains).

The membrane free energy can be expressed as a sum of three terms,

$$F = F_t + F_s + F_h = 2N(f_t + f_s + f_h) \tag{1}$$

representing, respectively, the contributions of the hydrocarbon tails, the surface free energy corresponding to the hydrocarbon-water interface, and the (solution-mediated) interaction free energy between the headgroups. f_i ($= t,s,h$) denotes the corresponding free energies, per molecule, with $2N$ denoting the number of molecules in the membrane (N per monolayer, on average). All three terms depend on the lipid composition of the two monolayers comprising the membrane bilayer and the ambient solution conditions, as well as on the membrane area and curvature. The interplay between these terms dictates the equilibrium geometry of the membrane (area and curvature), fluctuations around the equilibrium state, and all microscopic (e.g., chain conformations) and thermodynamic (e.g., elastic) properties of the membrane. We shall first consider each term separately, devoting most of the discussion to F_t.

B. CHAIN FREE ENERGY

The formalism outlined below can be applied to a bilayer membrane of arbitrary lipid composition, whether symmetric or nonsymmetric with respect to the two monolayers, as well as to an arbitrary membrane thickness and/or curvature. Also, the theory can be and has been applied to other amphiphilic aggregates such as micelles (see also Section V), as well as to systems containing hydrophobic solutes (see Section IV). Yet, to introduce the basic concepts involved and to derive the main working equations and expressions, let us consider the simplest possible membrane: namely, a planar and symmetric bilayer composed of a single lipid component. To further simplify the description, let us assume that the lipids are single (saturated) chain amphiphiles of the form $\mathcal{P}-(CH_2)_{n-1}-CH_3$, with $\mathcal{P}$ symbolizing the polar headgroup. The generalization to doubly tailed lipids, including nonsaturated lipid chains, is straightforward, as described in Section III.

Our goal is to derive an expression for $P(\alpha)$, the probability of finding the lipid hydrocarbon chain in conformation α. In general, $P(\alpha)$ is fully specified by the coordinates of all the atoms constituting the tail. In practice, one can employ rather accurate but simpler schemes like the rotational isomeric state (RIS) model of acyl chains,[37] whereby α is specified by the trans/gauche sequence of the CH_2-CH_2 bonds along the chain and by the overall orientation of the chain (specified by three Euler angles) relative to some arbitrary fixed system of coordinates (see Figure 1). For a lipid of the form $\mathcal{P}-(CH_2)_{n-1}-CH_3$ the number of trans/gauche sequences is 3^{n-1} (treating the $P-(CH_2)_1$ bond as a CH_2-CH_2 bond). In the numerical examples presented in the following sections, the calculations involve generation of all possible sequences, each multiplied by several dozens overall chain orientations as well as several lateral displacements of the headgroup relative to the hydrocarbon-water interface.[31]

Knowing the equilibrium $P(\alpha)$ we can calculate any desirable "single-chain" conformational property. Of particular interest are those properties which can be either measured experimentally or calculated by detailed computer simulations. In addition to the physical significance of these properties, their calculation provides a test for the expression which we will soon derive for $P(\alpha)$. The most commonly measured or calculated conformational properties are the bond orientational order parameters and the segment spatial distributions to be defined later. Furthermore, $P(\alpha)$ can be used to calculate various thermodynamic properties of interest, such as the free energy per chain, f_t, and related quantities such as the curvature elasticity moduli of the membrane.[6,38-41]

The free energy per chain is given, in terms of $P(\alpha)$, as

$$f_t = \sum_\alpha P(\alpha)\epsilon(\alpha) + kT \sum_\alpha P(\alpha) \ln P(\alpha) \tag{2}$$

where k is Boltzmann's constant, T is the temperature, and $\epsilon(\alpha)$ is the internal (trans/gauche) energy of a chain in conformation α. More specifically, $\epsilon(\alpha) = n_g(\alpha)e_g + n_t(\alpha)e_t$, where $n_g(\alpha)$ and $n_t(\alpha)$ are the numbers of gauche and trans conformers along the chain, with e_g and e_t representing their respective energies. One usually sets $e_t = 0$, implying that $e_g \simeq 500$ cal/mol.

The first term in Equation 2 is the energetic contribution to the chain free energy, while the second is the conformational entropy contribution. Both depend on the curvature of the membrane, its thickness, and its chemical composition.

Equation 2 representing the chain free energy as a function of the singlet probability distribution (spd) of chain conformations, is also the key expression in the variational procedure for deriving the equilibrium expression for $P(\alpha)$. We derive the desired (equilibrium) spd by minimization of f_t with respect to $\{P(\alpha)\}$, subject to whichever constraints $P(\alpha)$ must fulfill. Except for the trivial normalization condition ($\sum_\alpha P(\alpha) = 1$) the only additional constraint which we impose on $P(\alpha)$ results from one simple and common assumption. Namely, we assume that the liquid-like hydrophobic core is uniformly packed by chain segments. The mathematical expression of this constraint is

$$\int ds\sigma(s) \sum_\alpha P(\alpha;s)\psi(r;\alpha,s) = \rho(r) \quad \text{(all } r\text{)} \tag{3}$$

which for a symmetric planar bilayer reduces to

$$\sum_\alpha P(\alpha)[\phi(z;\alpha) + \phi(-z;\alpha)] = a\rho \quad \text{(all } z\text{)} \tag{4}$$

The quantities appear in Equations 3 and 4 are as follows: $P(\alpha;s)$ denotes the spd corresponding to chains originating from point s of the hydrocarbon-water interface; for simplicity s may be regarded as the headgroup position. $\sigma(s)$ is the lateral density of headgroups at the interface; i.e., $\sigma(s)ds$ is the number of chains originating from an area element ds at the interface. It should be noted that the s integration in Equation 3 includes both interfaces. The quantity $\psi(r;\alpha,s)dr$ denotes the number of segments of a chain in conformation α, originating at s, which fall within a small volume element dr (around r) of the hydrophobic core. $\rho(r)$ is the average segment density at r which, for a "compact" core, is constant: $\rho(r) = \rho = 1/v$, where v is the average volume per chain segment in the hydrophobic core.

In passing from Equation 3 to Equation 4 we have specifically considered a symmetric planar single-component bilayer. For this system we have $\sigma(s) = constant = 1/a$, where a is the average cross-sectional area per chain, measured at the hydrocarbon-water interface. This important structural characteristic of the membrane is usually referred to as "the area per headgroup." Also, for the simple bilayer, $P(\alpha;s) = P(\alpha)$ is independent of s. We now choose a coordinate system whose origin is at the bilayer midplane, with its z axis pointing towards the "upper" interface. Clearly, for a chain with headgroup coordinates $6s = x,y = 0,0$ the quantity $\psi(r,\alpha s) = \psi(r,\alpha,0)$ is only a function of z. Then the left-hand side of Equation

3 can be integrated over x and y to obtain Equation 4, in which $\phi(z;\alpha)dz = [\int \psi(\mathbf{r};\alpha)dxdy]dz$ is simply the number of segments of an α-chain falling within the shell $z, z + dz$ of the hydrophobic core (see Figure 1). The two terms within the square brackets in Equation 4 represent the contribution to the segment density in z due to chains anchored to the "upper" and "lower" interfaces, respectively. Of course Equation 4 could be immediately written down for the symmetric bilayer. We have emphasized here that it is a special case of the more general form Equation 3, which will be of use in Section IV.

To derive $P(\alpha)$ we now minimize Equation 2 subject to Equation 4 and obtain

$$P(\alpha) = \frac{1}{q}\exp\left[-\beta\epsilon(\alpha) - \beta\int \pi(z)\phi(z;\alpha)dz\right] \qquad (5)$$

with

$$q = \sum_\alpha \exp\left[-\beta\epsilon(\alpha) - \beta\int \pi(z)\phi(z;\alpha)dz\right] \qquad (6)$$

representing the conformational partition function of the chain, and $\beta \equiv 1/kT$. The quantities $\{\pi(z)\}$ in these equations are the Lagrange multipliers conjugated to the packing constraints Equation 4. They have the dimensions and the physical significance of "lateral pressures", as discussed in detail elsewhere[1,2,24,25] (see Figure 1).

The numerical values of the $\{\pi(z)\}$ terms are determined by the self-consistency equations resulting from substitution of Equation 5 into the packing constraint (Equation 4). This results in a set of nonlinear algebraic equations which can be solved quite simply and efficiently for any chain model.

Substituting Equation 5 into Equation 2 we obtain

$$f_t = -kT \ln q - a\rho \int \pi(z)\,dz \qquad (7)$$

Thus, after evaluating the $\pi(z)$ we can calculate any chain conformational property derivable from $P(\alpha)$ through Equation 5 or any thermodynamic property derivable from f_t using Equation 7. Clearly, all the results depend sensitively, through $\pi(z)$, on the value of the area per headgroup, a.

The results of Equations 5 to 7, corresponding to the symmetric, planar, single-component bilayer, can be easily generalized to more complex systems. Let us briefly mention a few cases of interest.

Curved bilayers — In this case, instead of a constant cross-sectional area per chain, a, we have $a = a(c_1, c_2, z)$, where c_1 and c_2 denote the local interfacial curvatures $c_i = 1/R_i$ are the principal curvatures; R_i denotes the radius of curvature). In this case the packing constraints (Equation 4) should be replaced by

$$\chi_E \sum_\alpha P_E(\alpha)\phi(z;\alpha) + \chi_I \sum_\beta P_I(\beta)\phi(z;\beta) = \rho a(z) = a(0)\left[1 + \left(c_1 + c_2\right)z + c_1 c_2 z^2\right] \qquad (8)$$

with $P_E(\alpha)$ and $P_I(\beta)$ representing the spd's of chains originating at the "external" (E) and "internal" (I) monolayers comprising the bilayer, e.g., of a vesicle. (Clearly, for a spherical vesicle of radius $R, c_1 = c_2 = 1/R$.) The quantities χ_E and χ_I are the "mole fractions" of lipids in the two monolayers; $\chi_e = N_E/N$ and $\chi_I = N_I/N$ ($\chi_E + \chi_I = 1$, $N_E + N_I = N$), with N_E and N_I denoting the number of chains originating from the E and I interfaces, respectively.

The free energy of the system in this case is given by

$$F_t = N_E \sum_\alpha P_E(\alpha)\left[\epsilon(\alpha) + kT \ln P_E(\alpha)\right] + N_I \sum_\beta P_I(\beta)\left[\epsilon(\beta) + kT \ln P_I(\beta)\right] \qquad (9)$$

Minimization of Equation 9 with respect to Equation 8 yields for $P_E(\alpha) = P_E(\alpha; a, c_1, c_2)$ and $P_I(\beta) = P_I(\beta; a, c_1, c_2)$ expressions similar to Equation 5, except that now $\pi(z)$ and $q_E(q_I)$ depend not only on a but also on c_1 and c_2.

Micelles — Here again, all we need to do is to account for the curvature dependence of $a(z)$. For a cylindrical micelle, for instance, as will be discussed in Section V, instead of Equation 4 we have

$$\sum_\alpha P(\alpha)\phi(\pi;\alpha) = \rho a(R)r / R \tag{10}$$

where R is the radius of the micelle and r is the radial distance from the cylinder axis. Here again $P(\alpha)$ has the same form as Equation 5, but now the pressure profile $\{\pi(r)\}$ depends on the micellar radius R. For micelles composed of chains of a given length n the average area per chain at the interface, $a(R)$, is uniquely determined by R through the geometric packing condition $a(R) = 2v/R$, where v is the chain volume. For simple alkyl chains, $-(CH_2)_{n-1}-CH_3$, $v \cong (n + 1)v$, where $v \simeq 27$ Å^3 is the specific volume of a CH_2 group.

Two- (or more) component systems — Consider for instance a planar symmetric bilayer composed of two types of lipid chains, A and B. Then, instead of Equation 4 we write

$$X_A \sum_\alpha P_A(\alpha)\left[\phi_A(z,\alpha) + \phi_A(-z;\alpha)\right] + X_B \sum_\beta P_B(\beta)\left[\phi_B(z,\beta) + \phi_B(-z,\beta)\right] = a\rho \tag{11}$$

where X_A and $X_B = 1 - X_A$ denotes the mole fractions of the two types of chains. The free energy of this system is given by

$$F_t = N\left(X_A f_{t,A} + X_B f_{t,B}\right) \tag{12}$$

with $f_{t,A}$ and $f_{t,B}$ defined according to Equation 2. Again, minimization of Equation 12 subject to Equation 11 yields $P_A(\alpha)$ and $P_B(\beta)$ of the Equation 5 form. The same $\pi(z)$ appears in both expressions.

Nonuniform membranes — The presence of a hydrophobic solute in the membrane — say, an integral protein, — "breaks" the translational symmetry of the planar bilayer. Thus, $P(\alpha,s)$ will depend on the headgroup position s, as measured, for instance, with respect to the position of the protein. In this more general case, we have

$$F_t = \int ds\sigma(s)f_t(s) \tag{13}$$

where $f_t(s)$ is the local free energy of a chain originating at s. The relevant packing constraint is now Equation 3. Minimization of Equation 13 with respect to Equation 3 yields[7]

$$P(\alpha;s) = \frac{1}{q(s)}\exp\left[-\beta\epsilon(\alpha) - \beta\int dr\lambda(r)\psi(r;\alpha,s)\right] \tag{14}$$

with the $\lambda(r)$ corresponding to the Lagrange parameters conjugated to the packing constraints (Equation 3). In this case, due to the lower symmetry of the system, the calculations are considerably more involved (chains must be generated and classified for different points s), but are feasible, as illustrated in Section 4 for a model of lipid protein interaction.[7]

B. ADDING F_s AND F_h

The conformational free energy f_t, as given by Equations 2 and 7, decreases as the average cross-sectional area per chain, a, increases. This follows simply from the fact that as a increases the lateral dimensions of the chains also increase, allowing for more conformational freedom. More precisely, the energetic contribution to the free energy $\langle\epsilon_t\rangle = \Sigma P(\alpha)\epsilon(\alpha)$ increases with a since the average number of gauche bonds increases. However, the increase in conformational entropy (chain flexibility) overcompensates for the increase in $\langle\epsilon_t\rangle$, resulting in a net decrease of f_t. Thus, the conformational free energy corresponds effectively to a repulsive interaction between chains. This implies a positive lateral pressure

$$\Pi_t = -\frac{\partial f_t}{\partial a} = -\rho \int \pi(z)\,dz > 0 \qquad (15)$$

which tends to expand the bilayer.

The interaction between the headgroups, whether electrostatic or steric, is generally also repulsive;[1-3] i.e.,

$$\Pi_h = -\frac{\partial f_h}{\partial a} > 0 \qquad (16)$$

where $f_h = F_h/2N$ is the average interaction free energy per headgroup.

The interfacial free energy F_s provides the force which opposes both interheadgroup and interchain repulsions. The origin of this force is the "hydrophobic interaction", resulting from the increased hydrocarbon-water contact area upon increasing α. $f_s = F_s/2N$ is usually expressed as a simple surface energy $f_s = \gamma a$, with γ (often taken as $\gamma = 50$ dyn/cm = 0.12 kT/Å^2 at T = 300 K) denoting the effective surface tension.[3] With this representation of f_s one has

$$\Pi_s = -\frac{\partial f_s}{\partial a} = -\gamma < 0 \qquad (17)$$

The equilibrium area per headgroup, a_{eq}, is determined by the balance of the three forces, that is,

$$\Pi_t + \Pi_h + \Pi_s = 0 \qquad (18)$$

Figure 2 shows the three contributions, f_t, f_s, and f_h, to the average free energy per molecule

$$f = f_t + f_s + f_h \qquad (19)$$

as a function of the area per molecule in a planar bilayer. The chain contribution $f_t(a)$ was calculated using Equation 6 for three values of chain length ($n = 12,14,16$). The hydrocarbon tails correspond to simple alkyl chains, $-(CH_2)_{n-1}-CH_3$, modeled using the rotational isomeric state model. For f_s we have used $f_s = \gamma a$ with $\gamma = 0.12$ kT/Å^2. The headgroup contribution is represented here by the simple (and common) form[3,5,7]

$$f_h = C/a \qquad (20)$$

where C is a phenomenological constant. For the calculations shown in Figure 2 it has been chosen to yield $a_{eq} = 32$ Å^2 for the $n = 14$ chains ($C = 48$ kT).

It should be stressed that Equation 20 is a highly approximate representation of f_h. Several authors have suggested more elaborate expressions for f_h, based on detained models for electrostatic and/or steric repulsions (see, e.g., References 31 and 42 to 49). Unfortunately, these expressions are usually system specific and contain some poorly known molecular parameters. Thus, in some respects it is more reasonable to Equation 20 (or alternative phenomenological expressions; see Section 5) and treat C as a semiempirical parameter, as we did in Figure 2.

A rather common approximation in phenomenological treatments of amphiphile self-assembly is to set f_t to be a constant, independent of chain length and structure as well as of aggregation geometry. Thus, in bilayers, for example, f_t is assumed to be independent of chain length and of the area per headgroup a. In this case, using Equation 19 for f for Equation 20 for f_h, and setting $f_s = \gamma a$, one finds that $a_{eq} = (C/\gamma)^{1/2}$, independent of chain length. Experiments suggest that in lipid bilayers a_{eq} is indeed nearly independent of chain length.[3,50] In the calculations shown in Figure 2 the chain term, f_t, is explicitly included and as can be seen $f_t(a)$ depends sensitively on a. Yet the value of a_{eq} varies only weakly with n. In fact, based on simple scaling arguments it can be shown that $a_{eq} \sim n^\alpha$ with $\alpha \leq 1/3$, explaining the very slow increase of a_{eq} with n.[2,31] Furthermore, our calculations suggest that chain repulsion is actually stronger than headgroup repulsion, i.e., $\Pi_t > \Pi_h$. Thus, the slow variation of a_{eq} with n should not be regarded as justification for the approximation $f_t = constant$. As noted above, in this approximation, also known as

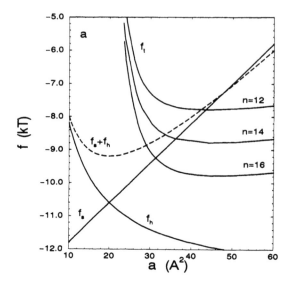

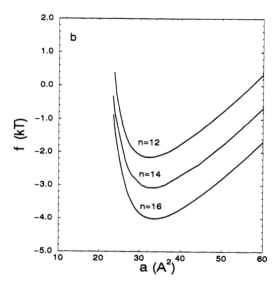

Figure 2 (a) The variation of the chain (f_t), headgroup (f_h) and surface (f_s) contributions to the average free energy per molecule, as a function of the average cross sectional area per chain (headgroup), a (see Equation 19). f_h is calculated using Equation 20 with $C = 48$ kT. $f_s = \gamma a$ with $\gamma = 0.12$ kT/Å². f_t is calculated using the meanfield theory for C_{12}, C_{14}, and C_{16} chains. (b) The sum of the three contributions above revealing that a_{eq} (~32 Å²) increases slowly with chain length. (From Ben-Shaul, A., *Handbook of Physics of Biological Systems*, Vol. 1, Lipowsky, R. and Sackman, E., Eds., Elsevier Science, Amsterdam, 1995, Chap. 7. With permission.)

"the hydrocarbon droplet assumption",[3-5] the chain conformational properties are assumed to be independent of the hydrophobic core geometry. Considering the fact that the dimensions of the hydrophobic region are comparable to the hydrocarbon chain length, there is no *a priori* justification for this assumption.

From the variation $\delta f(a) = f(a) - f(a_{eq})$ of $f(a)$ around the equilibrium area, $a = a_{eq}$, one can evaluate the area compressibility modulus κ_a defined by[38]

$$\frac{1}{a_{eq}}\delta f(a) = \frac{1}{2}\kappa_a\left(\frac{\delta a}{a_{eq}}\right)^2 \tag{21}$$

From the data in Figure 2 it follows that $\kappa_a \sim 0.2$ kT/Å^2. One can also calculate $\delta f(a,c_1,c_2) = f(a,c_1,c_2) - f(a_{eq},c_1^{eq},c_2^{eq})$ and thus derive explicit expressions and numerical values for curvature elasticity moduli. The relevant formalism and its applications have been described in detail elsewhere.[2,6] The calculation of membrane elastic moduli is a particular *thermodynamic* application of the theory outlined in this section. Other thermodynamic applications are described in Sections 4 and 5. Clearly, these calculations are only approximate, since the free energy is calculated using the *singlet* probability distribution function $P(\alpha)$ rather than the multichain distribution $P(\alpha_1,...,\alpha_N)$. It should be stressed that these "mean field" calculations of the free energies and related thermodynamic functions are approximate even if $P(\alpha)$ were the exact singlet distribution

$$P(\alpha) = \sum_{\alpha_2,...,\alpha_N} P\left(\alpha_1 = \alpha, \alpha_2, \alpha_3,...,\alpha_N\right) \tag{22}$$

The variational derivation of the singlet probability distribution (Equation 5) started indeed with the mean field free energy expression (Equation 2). However, an alternative derivation of Equation 5 is possible, based on expansion of the many chain configurational partition functions, lending further theoretical support to the accuracy of this form for the singlet probability distribution.[25]

In the next section we briefly describe the application of $P(\alpha)$ for calculating single-chain conformational properties and compare the results obtained to experimental and computer simulation data.

III. CHAIN CONFORMATIONAL PROPERTIES

One of the most familiar characteristics of conformational chain statistics in membranes is the bond orientational order profile of the C–H bonds along the lipid hydrocarbon tails.[5,51-55] The orientational order parameters are commonly measured by nuclear magnetic resonance (NMR) methods, using selective (or nonselective) deuteration of the chains. Specifically, the measured quantity is the orientational order parameter of the C_k–H (C_k–D) bond for $-(CH_2)_{n-1}-CH_3$ chains, $k = 1,...,n$), defined as

$$S_k = \left\langle P_2(\cos\theta_k)\right\rangle = \sum_\alpha P(\alpha)\left[3\cos^2\theta_k(\alpha) - 1\right]/2 \tag{23}$$

where $P_2(x) = (3x^2 - 1)/2$ is the second Legendre polynomial and $\theta_k(\alpha)$ denotes, for a chain conformation α, the angle between the kth bond and the membrane normal (the "director"). The C–H order parameters can be related to the skeletal order parameters, $\tilde{S}_k$, corresponding to the vectors $r_{k-1,k+1}$ connecting carbons $k - 1$ and $k + 1$ of the chain: $\tilde{S}_k = -2S_k$ for all k except for the terminal methyl group ($-CH_3$), for which $\tilde{S}_n = -3S_n$.[55]

The orientational order parameter profiles provide a measure of chain flexibility, reflecting the "fluidity" of the hydrophobic core. In the perfectly ordered state of the membrane, when all lipid chains are in their all-*trans* conformation, with the chain axis along the membrane normal, one has $S_k = -0.5(\theta_k = \pi/2)$ or $\tilde{S}_k = 1$, for $k = 1,...,n - 1$. In the opposite limit, where bond orientations are random, $S_k = \tilde{S}_k = 0$.

Typically, bond order parameter profiles of (saturated) lipid chains in planar bilayers are characterized by a roughly constant value of S_k for about the first half of the chain (the "plateau region"), followed by a monotonic decrease of S_k towards the chain terminus. The last chain segments, those which reach and possibly cross ("interdigitate" through) the bilayer midplane, are characterized by $S_k \sim 0$, indicating nearly random bond orientations. This behavior also indicates "high fluidity" in the central part of the hydrophobic core, as compared to the regions bordering the interface, where chain orientational ordering is relatively large. The magnitude of S_k in the plateau region increases as the membrane thickness d increases or, equivalently, as the average cross-sectional area per chain, a, decreases. This behavior is to be expected, since as d increases, the hydrocarbon chains must be further stretched out, resulting in a higher degree of chain ordering along the membrane normal. Note that this trend is a direct consequence of the tight (fluid-like) packing condition of the chains within the hydrophobic core. In other words, the packing constraints rather than, say, the relative trans/gauche energy of the chains are the important determinant of chain ordering in membranes. These qualitative trends have been quantitatively analyzed and confirmed by molecular level calculations based on Equation 5.[28]

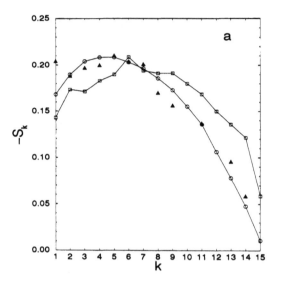

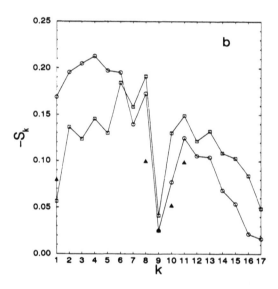

Figure 3 Orientational order parameter profiles of the C–H bonds along the palmitoyl (a) and oleoyl (b) chains of DMPC (adapted from Reference 31). ▲ — experimental results,[51] □ — molecular dynamics calculations,[9] ○ — meanfield theory. (From Fattal, D. R. and Ben-Shaul, A., *Biophys. J.*, 67, 983, 1994. With permission.)

Orientational bond order profiles calculated using Equation 5 for $P(\alpha)$ have been calculated for various sytems, including "pure" (i.e., single-component) and mixed micelles and membranes. For single-chain amphiphiles the results have been compared, whenever possible, to experimental data and computer simulation results, showing generally very good agreement.[1,2] Similarly, good agreement has been obtained with respect to other conformational properties such as segment spatial distributions and conformational energy or entropy per chain. In Figure 3 we show two sets of bond order parameter profiles, corresponding to the two hydrocarbon chains of palmitoyl-oleoyl-phosphatidylcholine (POPC).[31] Here the palmitoyl chain is a saturated $-(CH_2)_{14}-CH_3$ alkyl chain whereas the C_{17}(17-carbon) oleoyl chain contains one *cis* double bond, between carbons 8 and 9: $-(CH_2)_7-(CH=CH)-(CH_2)_7-CH_3$. The two chains are connected through the glycerol backbone, which is also connected to the switterionic phosphatidyl-choline headgroup. The results shown in Figure 3 correspond to POPC molecules packed in a bilayer of (hydrophobic core) thickness $d = 30.0$ Å at T = 300 K. Simple packing considerations (see below) imply that this thickness corresponds to an average cross-sectional area per headgroup of $a = 60.5$ Å^2, which is also the average cross-sectional area of the lipid tail (composed of one oleoyl and one palmitoyl chain).

The POPC bilayer has been chosen primarily in order to compare the predictions of the mean field theory (Equation 5) with those of a most comprehensive molecular dynamics (MD) simulation of the same system.[9] The MD results as well as (partial) experimental results for the POPC bilayer[51] are also shown in Figure 3. The agreement between these sets of results is quite satisfactory considering the complexity of the system modeled. The typical "plateau" region of the saturated chain is reproduced, as well as the very distinctive drop in the orientational order parameter at the double bond region of the oleoyl chain. Differences in the S_k values appear mainly for the first few C–H bonds of the oleoyl chain. This is probably due to the approximate mean field treatment of the glycerol backbone of the lipid from which the two chains emanate. In fact, in the mean field calculation of the POPC bilayer the hydrophobic core of the membrane has been modeled as an equimolar mixture of palmitoyl and oleoyl chains satisfying a packing constraint of the form in Equation 11. In other words, the connectivity of the oleoyl and palmitoyl chains through the glycerol backbone has not been explicitly taken into account. This approximation is consistent with the assumption that the interactions between chains originating from the same headgroup are no different from those originating from different headgroups. The reason for this approximation is "technical". As noted in the previous section, the calculation of $P(\alpha)$ involves the generation of all possible conformations of the chain considered. Thus, for a pure aggregate composed, say, of saturated $-(CH_2)_n-CH_3$ chains, the number of possible (RIS) conformations is of the order of 3^n. For each conformation generated one counts and classifies the $\phi(z;\alpha)$ terms into groups of degenerate conformations corresponding to their distribution within the hydrophobic core. For a mixture of $-(CH_2)_n-CH_3$ and $-(CH_2)_m-CH_3$ chains this is done separately for each type of chain. The number of conformations enumerated is thus $3^n + 3^m$. On the other hand, the total number of conformations corresponding to two chains originating from the same headgroup is $3^n \cdot 3^m$. This, of course, implies an enormous numerical effort. Furthermore, there is no real reason for treating the correlations between two chains belonging to the same headgroup more accurately than those originating from different headgroups. After all, the chains comprising the hydrophobic region are tightly packed and correlated. The mean field theory treats these correlations only indirectly, through the packing constraints, Equations 4 or 11.

We note that despite the inherent approximations involved in the mean field analysis its predictions compare well with those derived from MD simulations. The difference in computation time between the two approaches is enormous. Obviously, MD simulations provide much more detailed information, including information on dynamic properties, which the mean field equilibrium theory cannot treat at all. Yet, even with the best interaction potentials known and the fastest computers available, the number of systems which can be studied in detail by MD methods to date is limited, and even those are followed over relatively short periods of time. On the other hand, the mean field approach described above, though approximate, can be easily applied to a very wide range of systems (e.g., different lipid compositions) and a wide range of conditions (e.g., membranes of different curvatures). Thus, as noted already in Section I, while the quality of large-scale computer simulations is rapidly growing, there are many systems and properties (e.g., curvature elastic moduli) which can only be studied by approximate mean-field theories.

Another measurable structural characteristic of lipid membranes is the distribution of different chain segments across the bilayer hydrophobic core. Figure 4 shows the distribution (number of segments per unit length) of terminal (CH_3) groups, double-bonded carbons ($-CH=$), and the sum of all methylene ($-CH_2-$) segments comprising the two hydrophobic tails of dioleoyl phosphatidylcholine (DMPC). The lipid tail of these molecules is composed of two $-(CH_2)_7-(CH\triangleq CH)-(CH_2)_7-CH_3$ chains. One set of curves represents experimental results obtained by X-ray and neutron scattering.[56,57] The other represents the predictions of the mean field theory.[31] The thickness of the hydrophobic region, defined as the average distance between the lipid carbonyl groups on opposite interfaces of the membrane, is $d = 32$ Å. The agreement between the measured and calculated results is quite satisfactory, at least with respect to the peaks of the various segment distributions. The main difference appears in the width of the CH_2 distribution, showing wider "wings" of the experimental results. Yet it should be noted that no attempts were made to fit the calculated results to experiments. The only input into the calculation was the thickness of the membrane, allowing, as in other calculations of this kind, for small fluctuations of the lipid headgroup around the hydrocarbon-water interface. Further details of this and other calculations are discussed in Reference 31.

The only input parameter in the mean field calculations is the hydrophobic thickness of the membrane, d. It enters into the packing constraints (see, e.g., Equation 4) through the average cross-sectional area

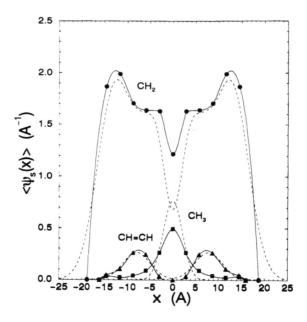

Figure 4 The distribution of lipid chain segments across a POPC bilayer (adapted from Reference 31). ●, ■, and ▲ represent the results calculated by the mean-field theory for, the (sum of all) CH_2 segments, CH_3 groups, and CH=CH groups, respectively. The dashed lines are the experimental results.[57] (From Fattal, D. R. and Ben-Shaul, A., *Biophys. J.*, 67, 983, 1994. With permission.)

per chain, a. The relationship between d and a follows immediately from the assumption that the hydrophobic core is uniform and liquid like. Namely, using v to denote the volume occupied by the hydrocarbon tail, it follows immediately that $a = 2v/d$. The tail volume can be calculated by adding the specific volumes of the various segments comprising the chain, as measured for bulk liquid hydrocarbons, e.g., $v(CH_2) \cong 27$ Å^3, $v(CH_3) \cong 54$ Å^3, and $v(-CH=) \cong 21.5$ Å^3 (Reference 31, see also Reference 58). It should be noted, however, that these values are only used to convert from d to a. The quantity which actually enters the calculation is the hydrophobic thickness, d, rather than the area a. Only the relative values (not the absolute ones), e.g., $v(CH_3/v(CH_2)$, are important for implementing the packing constraints. For further details see, e.g., Reference 31.

We conclude this section with another comparison between experimental results and mean field theory calculations. Figure 5 shows the lateral (in-plane) fluctuations of the CH_2 segments of DPPC (dipalmitoyl phosphatidylcholine) tails, packed in a planar bilayer with four different areas per chain, ranging from $\alpha = 25.5$ to 31.3 Å^2 (the corresponding areas per headgroup are twice these values). More explicitly, the figure shows σ_{xy}, the root-mean-square (rms) deviations, in the xy plane (parallel to the membrane), of carbons $k = 1$ to 15 along the palmitoyl chain $-(CH_2)_{14}-CH_3$. The rms deviation of carbon k is calculated using

$$\left(\sigma_{xy}^k\right)^2 = \sum_\alpha P(\alpha)\left\{\left[x_k(\alpha) - \langle x_k \rangle\right]^2 + \left[y_k(\alpha) - \langle y_k \rangle\right]^2\right\} = \langle x_k^2 \rangle + \langle y_k^2 \rangle \tag{24}$$

with $x_k(\alpha)$ denoting the x coordinate of the k segment when the chain is in conformation α. Using $\langle x_0 \rangle = \langle y_0 \rangle = 0$ to denote the headgroup position, then for a fluid membrane $\langle x_k \rangle = \langle y_k \rangle = 0$ for all chain segments k. σ_{xy}^k is thus a measure of the lateral fluctuations of the kth segment around the membrane normal (originating from the headgroup). We note, as expected, that as k increases (i.e., towards the chain terminus), σ_{xy}^k also increases, indicating a higher degree of chain flexibility and lateral mobility. Furthermore, and again as expected, the amplitude of the fluctuations increase as the average cross-sectional area per chain, a, increases. All these findings are consistent with those inferred from the behavior of the profile of orientational bond order parameters.

The σ_{xy}^k have been measured for DPPC chains by incoherent quasi-elastic neutron scattering.[59] These measurements suggest, for $a = 29.6$ Å^2, that σ_{xy}^k varies nearly linearly from ~0.6 Å for $k = 1$ to σ_{xy} ~7 Å for $k = 15$. The calculated results shown in Figure 5 are in good agreement with these findings.

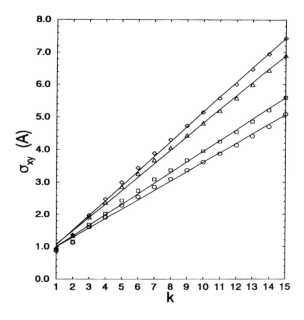

Figure 5 The lateral fluctuations (root mean square, in-plane, deviations) of the carbon atoms (k = 1,...,15) of DPPC calculated by the mean field theory (adapted from Reference 31). ○, □, △, and ◇ correspond to membranes in which the average cross sectional area per chain is a = 25.5, 26.6, 29.6, and 31.3 Å², respectively. The straight lines are linear fits. The calculated results show good agreement with experiments.[59] (From Fattal, D. R. and Ben-Shaul, A., *Biophys. J.*, 67, 983, 1994. With permission.)

We have shown in this section that the predictions of the mean field theory outlined in Section 2 show generally good agreement with observed experimental and computer simulation data. Good qualitative and often quantitative agreement has also been found with respect to other measurable properties, such as curvature elastic constants of mixed monolayers and bilayers.[6] Encouraged by these confirmations of the simple mean field theory we turn, in the following sections, to describe some of its applications to more complex phenomena.

IV. LIPID-PROTEIN INTERACTION

The presence of a hydrophobic "solute", such as the hydrophobic part of an integral protein, modifies the conformational properties of the lipids around it. In general, this "perturbation" increases the free energy of the surrounding lipids, so that when two or more hydrophobic solutes are in close proximity to each other the lipid-mediated interaction between them is attractive, thus favoring solute aggregation. The driving force for this aggregation is the tendency to minimize the contact area, and hence the extent of lipid perturbation, between the hydrophobic solutes and their surrounding lipid chains.

Various theoretical models have been proposed to describe and calculate the effects of an integral protein, usually treated as a rigid hydrophobic perturbation, on the lipid environment.[7,18,60-72] Some of those are continuum theories, based on treating the lipid bilayer as an elastic sheet of finite thickness. Several other statistical thermodynamic theories invoke Landau-type expansions of the free energy in terms of some order parameter, e.g., the "hydrophobic mismatch" (the difference between the protein and bilayer hydrophobic thickness[71]), addressing such issues as the effect of proteins on the solid/fluid ("gel-liquid crystal") transition temperature of the membrane. Very few models have considered the lipid-protein interaction on a molecular level. One such model, based on the molecular theory presented in Section II, will be outlined in this section.[7] Before turning to a more detailed description of this model it should be stressed that, as is often the case, the different theoretical approaches to the very complex issue of lipid-protein interaction should be regarded as complimentary rather than contradictive. For instance, the continuum elastic theories,[62-65] which are inherently valid for perturbations which are large on a molecular scale, are useful for understanding long-range elastic interactions between inclusions. Simple phenomenological approaches, such as the "mattress mode",[71] are useful for understanding the qualitative trends on membrane phase transitions as a function of the hydrophobic mismatch. Molecular

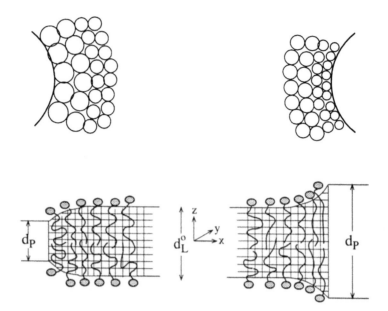

Figure 6 Schematics of the lipid-protein interaction model described in the text (adapted from Reference 7). Bottom panel is a "side view" of the bilayer, depicting the protein as a rigid wall of thickness d_P, either larger (right, "positive mismatch") or smaller (left, "negative mismatch") than the unperturbed bilayer thickness d_L^0. The chains in the vicinity of the protein are either stretched (when $d_P > d_L^0$) or compressed (when $d_P < d_L^0$) in order to bridge over the hydrophobic mismatch. The top panel is a "top view" of the membrane illustrating the corresponding changes in the average cross sectional area per chain as a function of the distance from the protein. (From Fattal, D. R. and Ben-Shaul, A., *Biophys. J.*, 65, 1795, 1993. With permission.)

models of the type presented below can provide numerical estimates on the lipid-protein interaction free energy, the range of perturbation of the lipid environment, and the origin of its dependence on the hydrophobic mismatch. Finally, it should be noted that all types of models mentioned above, including the one described below, expose only one aspect of the very complex and diverse phenomenon of lipid-protein interaction. More explicitly, by treating the protein as a rigid hydrophobic solute, we ignore not only the details of its complex structure but also the various electrostatic and steric interactions prevailing both in the core and in the interfacial regions.

Let d_L^0 denote the thickness of the lipid hydrophobic core. It is commonly assumed that when a rigid protein (or other inclusion) of hydrophobic thickness $d_P \sim d_L^0$ is incorporated into the membrane, the flexible lipid chains around it will adjust their length so as to shield the protein from direct contact with the surrounding water (see Figure 6). Using $d_L(x)$ to denote the bilayer thickness at a distance x from the protein, this assumption implies that $d_L(0) = d_P$. The variation of $d_L(x)$, between d_P at $x = 0$ are d_L^0 at $x \to \infty$, will be modeled as

$$d_L(x) = d_L^0 + \left(d_P - d_L^0\right)\exp(-x/\xi) \qquad (25)$$

with ξ measuring the range (or the "coherence length") of the perturbation. The model treats ξ as a variational parameter whose value is determined by minimization of the total perturbation free energy. The exponential variation of the membrane thickness profile (Equation 25) has been derived by some of the Landau-type theories of lipid-protein interaction.[67,68] Yet in the present model it should only be regarded as a convenient parametrization of $d_L(x)$. In fact, some of the continuum elastic theories of lipid-protein interaction predict more complicated, including nonmonotonic, functional forms for $d_L(x)$.[63-65]

In the model illustrated in Figure 6 the protein is treated as a rigid cylinder embedded in the membrane. The diameter, D, of the cylinder cross section is assumed to be much larger than the average lateral dimension of the lipid chains, i.e., $D \gg a^{1/2}$, where a is the average cross-sectional area per chain. Accordingly, to the lipids in its periphery the protein appears as a planar wall. Free energy calculations have been performed assuming that the protein wall is flat and is extending normally and symmetrically around the bilayer midplane.[7] Other geometries, e.g., a conical inclusion, can be treated similarly.

Assuming that the protein wall is parallel to the zy plane (z is the direction normal to the membrane plane), the lipid-protein interaction free energy per unit length of the protein perimeter (along the y direction) is given by

$$\Delta F = 2\int dx\left[\sigma(x)f(x) - \sigma^0 f^0\right] \tag{26}$$

where $f(x)$ is the local free energy per molecule at distance x from the protein and $\sigma(x)dxdy$ is the number of molecules originating from a small area element $dxdy$ of one of the two membrane interfaces. (More precisely, $dxdy$ is the projection of this area element onto the bilayer midplane.) σ^0 and f^0 are the corresponding quantities for the unperturbed membrane; that is, $\sigma^0 = \sigma(x)$ and $f^0 = f(x)$ as $x \to \infty$. The factor 2 in front of the integral accounts for the two leaflets of the bilayer.

In the planar bilayer $\sigma^0 = 1/a^0$, where $a^0 = 2v/d_L^0$ is the average area per chain; v denotes the volume of the hydrophobic tail. Assuming, as we did throughout this chapter, that the hydrophobic core is uniform and liquid-like, we have $\sigma(x) = 2v/d_L(x)$. Note, however, that except for the planar bilayer $\sigma(x) \neq 1/a(x)$, where $a(x)$ is the average local interface area per chain in the vicinity of the protein. This latter quantity is given by

$$a(x) = \frac{2v}{d_L(x)}\left\{1 + \frac{1}{4}\left(\frac{\partial d_L(x)}{\partial x}\right)^2\right\}^{1/2} \tag{27}$$

As in Section 2, the free energy per molecule can be expressed as a sun of tail, surface, and headgroup contributions:

$$f(x) = f_t(x) + f_s(x) + f_h(x) \tag{28}$$

The tail free energy and the corresponding spd are given By equations 13 and 14, respectively, with $s \to x$ and $\mathbf{r} \to x,z$. The numerical calculation of $f_t(x)$ is considerably more complex than in the planar bilayer case, since $\lambda = \lambda(x,z)$ varies along both x and z, whereas for the planar bilayer $\lambda \to \pi(z)$ is only a function of z. Nevertheless, the calculations are feasible and representative numerical results will be shown below. In these calculations the surface free energy is modeled as $f_s = \gamma a(x)$, with $\gamma = 0.12$ KT/Å^2, and $f_h(x)$ is represented by the simple form $f_h(x) = C/a(x)$, with $a(x)$ given by Equation 27. The parameter C was chosen such that for a planar bilayer composed of $C_{14}(\mathcal{P}\text{-}(CH_2)_{13}\text{-}CH_3)$ lipids the equilibrium area per chain is $a_0 \cong 32$ Å^2 (see Figure 2).

The perturbation free energy ΔF of the C_{14} bilayer is shown in Figure 7 for three choices of the headgroup interaction parameter: $C = 48$ kT, 12 kT, and 0. First we note that $\Delta F \neq 0$ even for the case where no hydrophobic mismatch occurs, when $d_P = d_L^0$. In this case there is no contribution to ΔF from the surface ($\Delta F_s = 0$) or headgroup ($\Delta F_h = 0$) terms; only $\Delta F_t > 0$. Even though there is no change in the average chain length, the presence of the impenetrable protein wall reduces the conformational freedom of nearby chains, resulting in excess chain orientational ordering and non-negligible positive contribution to F_t. These notions are confirmed by explicit calculations of bond orientational order parameter profiles, showing increased $\langle S_k \rangle$ values for chains near the protein as compared to those away from it.[7] The chain conformational calculations also show a finite (though small) average tilt angle of the chains (away from the wall). It should be noted that the first molecular model of lipid-protein interaction, which was proposed by Marčelja, has been formulated for the $d_P = d_L^0$ case.[18] In Marčelja's model, like in the one presented here, $\Delta F_t > 0$ due to the loss of lipid chain conformational freedom in vicinity of the protein.[18]

When $d_P > d_L^0$ the lipid tails are stretched beyond their length in the unperturbed membrane, resulting in $\Delta F_t > 0$. In parallel, the average area per headgroup decreases[26] and consequently $\Delta F_s < 0$. The opposite behavior characterizes the case $d_P < d_L^0$. The contribution of headgroup repulsion (ΔF_{h_-} to ΔF is, at least according to the model described, small compared to ΔF_s and ΔF_t. Thus, since as $d_P - d_L^0$ increases ΔF_t increases whereas ΔF_s decreases, the minimum of $\Delta F = \Delta F_t + \Delta F_s + \Delta F_h \simeq \Delta F_t + \Delta F_s$ is generally around $d_P - d_L^0 = 0$. However as seen in Figure 6, the minimum of ΔF shifts to a negative $d_P - d_L^0$ value when the strength of headgroup repulsion increases. In other words, negative hydrophobic mismatch can in fact relieve some of the lipid-protein interaction free energy when headgroup repulsion is strong. Similarly,

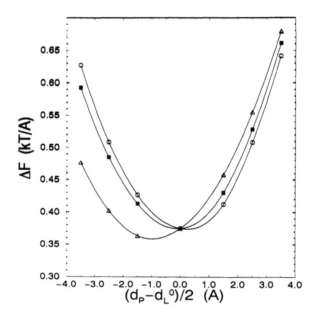

Figure 7 The lipid-protein interaction free energy (per unit length of perimeter length) for a bilayer of C_{14} chains, as a function of the hydrophobic mismatch (adapted from Reference 7). $\circ$, ■, and $\triangle$ correspond to three different choices of the headgroup repulsion strength, C = 0, 12, 48 kT, respectively. (From Fattal, D. R. and Ben-Shaul, A., *Biophys. J.*, 65, 1795, 1993. With permission.)

positive mismatch can reduce ΔF (compared to the case $d_P = d_L^0$) in the case of strong chain repulsion. This effect has recently been predicted by Safran and Dan using a continuous elastic theory for the effect of hydrophobic inclusions on membrane properties.[64,65] Its origin, according to their analysis, is the nonzero spontaneous curvature of the monolayers comprising the bilayer. To understand the effect it is worthwhile to elaborate on the role of spontaneous curvature in lipid bilayers.

Consider one of the two monolayers comprising a lipid bilayer and assume it is planar. The three forces, headgroup repulsion, surface tension, and chain repulsion, balance each other at some equilibrium area per chain, a_{eq}. These forces also exert moments which may prefer a finite "spontaneous" curvature for the monolayer. The curvature may be either positive (the hydrocarbon-water interface convex towards the water), negative, or zero. Large moments of headgroup repulsion will tend to induce positive spontaneous curvature. Large moments of chain repulsion will act in the opposite manner. When two monolayers are brought into contact to form a planar bilayer, both are "frustrated" energetically since their curvature is not the optimal (spontaneous) one. Yet the planar bilayer geometry usually involves the least curvature energy cost for the two monolayers. Now suppose that headgroup repulsion is strong enough to favor positive spontaneous curvature for the monolayer. If $d_P < d_L^0$ then the lipids around the protein wall are packed with positive spontaneous curvature (see Figure 5), thus relieving some of the frustration energy associated with the formation of the planar bilayer. The case C = 48 kT in Figure 6 corresponds to strong headgroup repulsion and hence positive spontaneous curvature. Indeed, we note that for this case the minimum in ΔF takes place at a negative value of $d_P - d_L^0$. Similarly, stronger chain repulsion would shift the minimum towards positive $d_P - d_L^0$ values.

Various other structural and thermodynamic characteristics of the lipid-protein bilayer can be derived from the model described in this section. One result of particular interest is the spatial range of the perturbation, ξ. The perturbation of lipid order by the protein wall extends to ~ 3ξ. The calculations show that ξ ~ 5 Å. Since, typically, the lateral dimension of a lipid chain is $a^{1/2} \simeq$ 5 to 6 Å, it follows that the range of perturbation corresponds to just a few molecular diameters.

V. THE VESICLE–MICELLE TRANSITION

Most lipid molecules in aqueous solution self-assemble spontaneously into extended 2D bilayers. The spontaneous (equilibrium, minimal free energy) curvature of bilayers is generally zero; i.e., they tend to

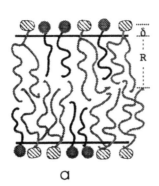 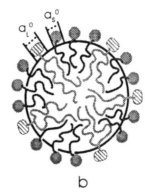

a   b

Figure 8 Schematic illustration of a mixed lipid surfactant bilayer (a) and a mixed micelle (b) (adapted from Reference 32). a_L^0 and a_S^0 are the bare headgroup areas of the two amphiphilic components, δ is the distance from the plane of headgroup repulsion to the hydrocarbon-water interface. (From Fattal, D. R., Andelman, D., and Ben-Shaul, A., *Langmuir*, submitted.)

be planar. To avoid the excess free energy associated with the exposure of their edges to water, the bilayers often close on themselves to from vesicles, at least in dilute solutions.[73] At higher lipid concentrations they may organize into multilamellar structures.[73,74] Other, nonlipid amphiphiles, which in dilute solution self-assemble into high-curvature aggregates such as cylindrical micelles, usually organize into multilamellar phases at higher concentration. These phases are stabilized by interaggregate interactions which overcome the intrinsic preference of the molecules to pack in highly curved aggregates.

Surfactant molecules, such as octylglucoside or bile salts, form micelles in dilute solution, reflecting their high spontaneous curvature.[75,76] Recall that any point of a curved surface can be characterized by two local principal curvatures, c_1 and c_2, with $R_i = 1/c_i$ ($i = 1,2$) denoting the corresponding radius of curvature. Thus, for example, the hydrocarbon-water interface of a spherical micelle of radius R is characterized everywhere by $c_1 = c_2 = 1/R$, with $R \leq l$, where l is the length of the fully extended amphiphile tail. Similarly, in cylindrical micelles (except at the hemispherical caps), $c_1 = 1/R_1$ whereas $c_2 = 0$, R_1 denoting the radius of the cylinder cross section ($R_1 \leq l$) and $R_2 \to \infty$ denoting the radius of the cylinder axis. In planar bilayers $c_1 = c_2 = 0$, and in spherical vesicles of radius R, $c_1 = c_2 = 1/R$, with $R \gg l$.

Consider now a dilute binary aqueous solution of lipids whose spontaneous aggregation geometry is planar ($c_1 = c_2 \simeq 0$) and surfactants which in dilute solution prefer organization in, say, cylindrical micelles ($c_1 \cong 1/l$, $c_2 = 0$). Let $\chi = \chi_L = N_L/(N_L + N_S) = N_L/N$ denote the mole fraction of lipids and $\chi_S = 1 - \chi$ the mole fraction of surfactants in solution, N_L and N_S denoting the number (or concentration) of lipid and surfactant molecules, respectively. (The mole fractions involve only the amphiphilic components, not the solvent; $\chi_S + \chi_L = 1$.) In the limits $\chi = 1$ and $\chi = 0$ the amphiphiles form lipid vesicles ($c_1 = c_2 \simeq 0$) and surfactant micelles, respectively. When a small amount of surfactant molecules is added to a system composed of lipid vesicles, a certain fraction of them (typically very small, corresponding to the cmc [critical micellar concentration][1,3-5]) are dispersed as monomers in the solution; the rest are incorporated ("solubilized") into the vesicles. As in ordinary binary mixtures, the thermodynamic driving force for the incorporation of the surfactant into the lipid bilayer is the mixing entropy, which overcounts the tendency of the surfactant and lipid molecules to pack, separately, according to their energetically preferred aggregation geometries. Similarly, upon adding small amounts of lipids to a surfactant-rich system they will be solubilized in the surfactant micelles (hardly any of them will be present as monomers, due to the extremely low cmc of lipids). Schematic illustrations of a mixed lipid-surfactant bilayer and a mixed cylindrical micelle are shown in Figure 8.

In ordinary binary molecular solutions, say of A and B molecules, phase separation of an A-rich and a B-rich phase can take place, provided the effective A-B interaction $w = w_{AB} - (w_{AA} + w_{BB})/2$ is repulsive, i.e., $w > 0$; $w_{IJ}(I = A,B)$ denotes the interaction energy between I and J molecules (integrated over distances and orientations or, in lattice models, between neighboring molecules). More precisely, phase separation occurs below a certain critical temperature T_c (proportional to w) and only over a certain range of (intermediate) compositions, which broadens as T decreases farther from T_c.

An analogous scenario can, and usually does, happen in aqueous solutions of lipids and surfactants. The analogue of w in these systems is the difference in the packing (free) energy of surfactants and lipids in a mixed system, compared to their packing in separate aggregates. The coexisting phases in lipid-surfactant solutions, if and when phase separation takes place, are vesicles with amphiphile composition x_v (v = vesicle) and micelles with composition x_m (m = micelle), such that $x_m < x_v$. The separated phases appear in different regions of space (vesicles and micelles floating in the aqueous solution) and are characterized by very different symmetries: nearly planar lipid-rich bilayer vesicles vs. elongated (or sometimes globular) surfactant-rich micelles.

This qualitative thermodynamic scenario does indeed take place in many lipid-surfactant systems[77-81] and is of considerable biological importance, e.g., for membrane reconstitution.[76] What typically happens is that a lipid vesicle can take up surfactant molecules up to a limit corresponding to a lipid content x_v. Beyond that limit the vesicles break into micelles with lipid content x_m. In lecithin-bile salt and lecithin-octylglucoside mixtures the compositions of coexisting vesicles and micelles are $x_v \sim 1/2$ and $x_m \sim 1/4$.

A simple qualitative theoretical model of the vesicle–micelle transition has recently been formulated by Andelman, Kozlov, and Helfrich.[83] These authors have expressed the free energy of both the (mixed) bilayer and micelle as a sum of a curvature energy term and a mixing entropy term. Explicitly, the average (Helmholtz) free energy per molecule in each of the two aggregation geometries is written as

$$\psi(x) = \frac{1}{2}\kappa\left[c_1 + c_2 - c_0(x)\right]^2 + kT\left[x \ln x + (1-x)\ln(1-x)\right] \qquad (29)$$

with x denoting the lipid mole fraction in the aggregate. (Actually in Reference 83 x denotes the area fraction of lipids, measured at the hydrocarbon-water interface. This difference is irrelevant for the present discussion and for understanding this phenomenon.) The second term in Equation 29 is an ideal mixing entropy contribution. The first term is the common, Helfrich form of the bending free energy of a membrane.[38] κ denotes the curvature (splay) elastic modulus and $c_0(x)$ is the spontaneous curvature of an aggregate with composition x. The spontaneous curvature has been assumed to vary linearly with x, say from $c_0(x = 0) = 1/R$, corresponding to a cylindrical surfactant micelle of radius R, to $c_0(x = 1) = 0$, corresponding to a planar lipid bilayer (or a very large vesicle). κ was treated as a constant, independent of x or aggregation geometry. Then $f(x)$ is calculated for the vesicle ($c_1 = c_2 \simeq 0$) and for the cylindrical micelle ($c_1 = 1/R, c_2 = 0$), and by equating the chemical potentials of the lipid and surfactant in the two geometries (common tangent construction) a general expression can be derived for the lipid composition of the vesicle and micelle at the transition.

This simple and elegant model can predict some interesting qualitative trends, e.g., the dependence of the coexisting compositions on $c_0(x)$, T and κ. Yet it must be remembered that the bending energy term in Equation 29 is valid only for small deviations, mainly of planar films, around the equilibrium curvature. Vesicles and cylindrical micelles correspond to very different equilibrium curvatures. It is highly unlikely that the harmonic (quadratic) form of the bending energy, with the same κ for all geometries, can faithfully describe the geometry dependence of the amphiphile packing free energy. Molecular-level calculations of the type described in previous sections confirm this question. Furthermore, molecular calculations of κ show that it depends sensitively on x.[6,84] (For instance, the bending rigidity of lipid membranes decreases rapidly upon adding short-chain surfactants to the bilayer.) Interestingly enough, recent calculations of this kind show that in some mixtures $c_0(x)$ varies nearly linearly with x over a wide range of composition.[84]

An obvious alternative to the first term in Equation 29 is to calculate the packing free energy of mixed lipid-surfactant bilayers and micelles using the molecular mean field theory described in Sections 2 to 4. Thus, instead of Equation 29 one writes

$$\psi_g(x) = xf_g^L(x) + (1-x)f_g^S(x) + kT\left[x \ln x + (1-x)\ln(1-x)\right] \qquad (30)$$

with $f_g^L(x)$ denoting the packing free energy per lipid molecule in a mixed aggregate of geometry g (g = vesicle, micelle) and composition x. Then, plotting $\psi_g(x)$ vs. x for the two aggregation geometries one can evaluate the coexisting compositions, x_v and x_m, using common tangent construction.

The results of one calculation of this kind, corresponding to a given set of molecular parameters (see below), are shown in Figure 9. Also marked on the figure is the composition $x_v = 0.47$, below which the

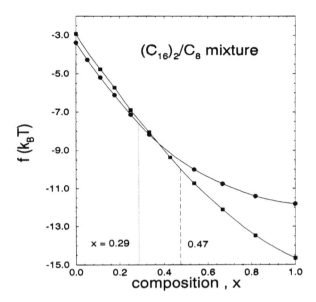

Figure 9 The average free energy per molecule in a mixed $(C_{16})_2/C_8$ lipid surfactant bilayer (■) and cylindrical micelle (●) as a function of lipid mole fraction (adapted from Reference 32). The compositions of the bilayer (x = 0.47) and micelle (x = 0.29) at the vesicle-micelle transitions are evaluated by common tangent construction (corresponding to equating the chemical potentials of each component in both aggregation geometries). (From Fattal, D. R., Andelman, D., and Ben-Shaul, A., *Langmuir*, submitted.)

vesicle is unstable, and the composition $x_m = 0.29$, corresponding to the micelles initially formed when the vesicles break. Conversely, x_m is the maximal lipid content in a cylindrical micelle, beyond which vesicles of composition x_v begin to form.

The results shown in Figure 9 as well as several additional cases are discussed in more detail elsewhere.[32] Here we shall only mention the basic assumptions. The system considered is a mixture of saturated double-chain lipids $\mathcal{P}_L$–$[(CH_2)_{15}$–$CH_3]_2$ and short single-chain surfactants $\mathcal{P}_S$–$(CH_2)_7$–CH_3, with $\mathcal{P}_L$ and $\mathcal{P}_S$ denoting the lipid and surfactant headgroups, respectively. As in previous sections, the free energy of the mixed bilayer, and the mixed cylinder, has been expressed as a sum of tail (f_t), surface ($f_s = \gamma a$), and headgroup contributions. The headgroup contribution to $xf_g^L(x) + (1 - x)f_g^S(x)$ has been modeled as a steric repulsion free energy:[44]

$$f_h^g = -kT \ln\left(1 - a_h / \overline{a}^g\right) \tag{31}$$

Here $\overline{a}^g = \overline{a}^g(x)$ is the average area per headgroup in aggregates of geometry $g(g = v,m)$. This area is measured at the plane of headgroup repulsion, assumed to be located at distance δ from the hydrocarbon-water interface. For a planar bilayer (large vesicle) $\overline{a}^v = a^v$, where a^v is the area per headgroup at the interface; for a cylinder micelle of radius R, $\overline{a}^m = a^m (1 + \delta/R)$. The quantity $a_h = a_h(x) = xa_h^L + (1 - x)a_h^S$ is the average bare (hard-core) headgroup area per molecule, at the plane of headgroup interactions. a_h^L and a_h^S denote, respectively, the bare headgroup areas per lipid and per surfactant molecule.

The surface contribution to the free energy is modeled as

$$f_s^g(x) = x\gamma\left(a^g - a_L^h\right) + (1 - x)\gamma\left(a^g - a_s^h\right) = \gamma\left(a^g a_h\right) \tag{32}$$

The chain conformational contributions were calculated using the mean field theory for mixed systems, as outlined in Section 2.

The numerical values used for the specific calculation shown in Figure 9 were $\gamma = 0.12$ kT/Å, $a_h^L = 42$ Å², $a_h^S = 50$ Å², and $\delta = 1.1$ Å. The choice of a_h^L ensures that the equilibrium average cross-sectional area per molecule in a pure lipid bilayer is 68 Å², as commonly found for lecithin bilayers.[80] The numerical values of a_h^S and δ (which in Reference 78 were treated as parameters controlling the spontaneous

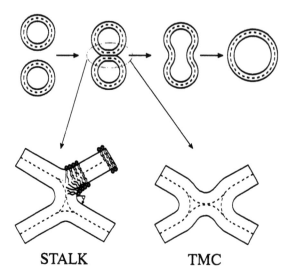

STALK   TMC

Figure 10 Schematic illustration of the several stages in the fusion of vesicle bilayers (top). The bottom figure shows in more detail two of the proposed structural intermediates along the fusion pathway.85,86

curvature of the surfactant) ensure that the optimal packing geometry of the C_8 surfactant molecules in dilute solution is a cylindrical micelle, with an average area per surfactant headgroup $a \simeq 55$ Å^2.

The free energy per molecule has been calculated as a function of composition x for both geometries, and the values of the amphiphile in the vesicles and the micelles at the transition ($x^v = 0.47$ and $x^m = 0.29$) were determined by common tangent construction. These values for x^v and v^m are in the range observed experimentally.[79-82] The molecular parameters used ($a_h^L, a_h^S, \gamma, \delta$) are all very reasonable and, moreover, have been adjusted so as to ensure micelle formation at $x \to 0$ and vesicle formation at $x \to 1$. Nevertheless, it must be mentioned that the uncertainties involved in choosing these parameters are considerable. Different choices of, say, a_h^L and δ can lead to substantial shifts in the values inferred for x_v and x_m. The semiempirical adjustment of such parameters by referring to limiting (i.e., the pure) cases seems, at present, to be the most plausible procedure. Notwithstanding these reservations, the model described in this section does account for the basic interactions and trends characteristic of the vesicle–micelle transition. Other structural transitions, such as from lamellar to inverted hexagonal or cubic phases, can possibly be accounted for using a similar approach.

VI. SUMMARY

In some cases, especially in systems of low symmetry such as the lipid-protein membrane, the molecular mean field theory described in this chapter requires some nontrivial calculations. Nevertheless, the computational effort involved is still substantially less than that required in large-scale computer simulations. Various other systems and processes in membrane biophysics can be studied and analyzed on a molecular level using this approach. These include, for example, the thermodynamic stability of mixed vesicles, pore (and other defect) formation in membranes, and the phase transition from lamellar to inverted hexagonal phases. Let us conclude this section by mentioning some very preliminary results concerning an issue of considerable biological relevance: the fusion of lipid vesicles.

Several authors have proposed phenomenological models for the mechanism and the structural intermediates involved in the process of membrane fusion[85-87] after the initial adhesion process.[88,90] One of the suggested pathways, the "modified Stalk mechanism" suggested by Siegel,[45] is schematically illustrated in Figure 10. Siegel has also estimated the excess free energies associated with the formation of the structural intermediates, using the continuum theory for membrane curvature and stretching elasticity.[38]

We have recently performed calculations of the kind described in Sections 2 and 3 for the excess free energy of the "Stalk" intermediate (Figure 10) for a pure lipid membrane composed of C_{14} chains. The results obtained are $\Delta F \geq 100$ kT. These numbers are in surprisingly good agreement with those obtained using the continuum theory. The agreement is surprising because the structural intermediates involve variations of packing geometry extending over only a few molecular diameters. However, additional calculations are called for before this good agreement can be confirmed.

ACKNOWLEDGMENT

We would like to thank E. Sackmann, W. M. Gelbart, I. Szleifer, D. Lichtenberg, Y. Barenholz, D. Andelman, Y. Talmon, S. Safaran, M. Kozlov, and D. Siegel for helpful discussions and comments on the topics outlined in this chapter. The financial support of the National Science Foundation administered by the Israel Academy of Science and Humanities and the Yeshaya Horowitz Association is gratefully acknowledged. The Fritz Haber Research Center, where this research was done, is supported by the Minerva Gesellschaft für die Forschung, mbH, Munich, Germany.

REFERENCES

1. **Ben-Shaul, A. and Gelbart, W. M.,** Statistical thermodynamics of amphiphile self-assembly: structure and phase transitions in micellar solutions, in: *Micelles, Membranes, Microemulsions and Monolayers,* Gelbart, W. M., Ben-Shaul, A., and Roux, D., Eds., Springer-Verlag, New York, 1994, chap. 1.
2. **Ben-Shaul, A.,** Molecular theory of chain packing, elasticity and lipid protein interaction in lipid bilayers, in: *Handbook of Physics of Biological Systems,* Vol. 1, Lipowsky, R. and Sackmann, E., Eds., Elsevier Science, Amsterdam, 1995, chap. 7.
3. **Tanford, C.,** *The Hydrophobic Effect,* 2nd ed., Wiley-Interscience, New York, 1980.
4. **Israelachvili, J. N.,** *Intermolecular and Surface Forces,* Academic Press, London, 1985.
5. **Wennerström, H. and Lindman, B.,** Micelles, physical chemistry of surfactant association, *Phys. Rep.,* 52, 1, 1979.
6. **Szleifer, I., Kramer, D., Ben-Shaul, A., Gelbart, W. M., and Safran, S. A.,** Molecular theory of curvature elasticity in surfactant films, *J. Chem. Phys.,* 92, 6800, 1990.
7. **Fattal, D. R. and Ben-Shaul, A.,** A molecular model for lipid-protein interaction in membranes: the role of hydrophobic mismatch, *Biophys. J.,* 65, 1795, 1993.
8. **Carignano, M. A. and Szleifer, I.,** Statistical theory of grafted polymer layers, *J. Chem. Phys.,* 98, 5006, 1993.
9. **Heller, H., Schaefer, M., and Schulten, K.,** Molecular dynamics simulation of a bilayer of 200 lipids in the gel and in the liquid crystal phases, *J. Phys. Chem.,* 97, 8343, 1993.
10. **van der Ploeg, P. and Berendsen, H. J. C.,** Molecular dynamics of a bilayer membrane, *Mol. Phys.,* 49, 233, 1983; Egberts, E. and Berendsen, H. J. C., Molecular dynamics simulation of a smectic liquid crystal with atomic detail, *J. Phys. Chem.,* 89, 3718, 1988.
11. **Biswas, A. and Schurman, B. L.,** Molecular dynamics simulation of a dense model bilayer of chain molecules with fixed head groups, *J. Chem. Phys.,* 95, 5377, 1991.
12. **Marrink, S. J., Berkowitz, M., and Berendsen, H. J. C.,** Molecular dynamics of a membrane/water interface: the ordering of water and its relation to the hydration force, *Langmuir,* 9, 3122, 1993.
13. **Raghavan, K., Reddy, M. R., and Berkowitz, M. L.,** A molecular dynamics study of the structure and dynamics of water between dilauroylphosphatidylethanolamine bilayers, *Langmuir,* 8, 233, 1992.
14. **Levine, Y. K., Kolinski, A., and Skolnick, J.,** Lattice dynamics study of a Langmuir monolayer of monounsaturated fatty acids, *J. Chem. Phys.,* 98, 7581, 1993; Levine, Y. K., Monte Carlo dynamics study of cis and trans unsaturated hydrocarbon chains, *Mol. Phys.,* 78, 619, 1993.
15. **Scott, H. L. and Cherng, S. L.,** Monte Carlo studies of phospholipid lamellae. Effects of proteins, cholesterol, bilayer curvature, and lateral mobility on order parameters, *Biochim. Biophys. Acta,* 510, 209, 1978.
16. **Ipsen, J. H., Jørgensen, K., and Mouritsen, O. G.,** Density fluctuations in saturated phospholipid bilayers increase as the acyl-chain length decreases, *Biophys. J.,* 58, 1099, 1990.
17. **Marelja, S.,** Chain ordering in liquid crystals. II. Structure of bilayer membranes, *Biochim. Biophys. Acta,* 367, 165, 1974.
18. **Marelja, S.,** Lipid-mediated protein interaction in membranes, *Biochim. Biophys. Acta,* 455, 1, 1976.
19. **See e.g., de Gennes, P. G. and Prost, J.,** *The Physics of Liquid Crystals,* 2nd ed., Clarendon Press, Oxford, 1993.
20. **Dill, K. A. and Flory, P.,** Interphases of chain molecules: monolayers and lipid bilayer membranes, *Proc. Natl. Acad. Sci. U.S.A.,* 77, 3115, 1980.
21. **Dill, K. A. and Stigter, D.,** Lateral interactions among phosphatidylcholine and phosphatidylethanolamine head groups in phospholipid monolayers and bilayers, *Biochem. J.,* 27, 3446, 1988.
22. **Gruen, D. W. R.,** A model for the chains in amphiphilic aggregates. I. Comparison with a molecular dynamics simulation of a bilayer, *J. Phys. Chem.,* 89, 146, 1985.
23. **Gruen, D. W. R.,** The standard picture of ionic micelles, *Prog. Colloid Polymer Sci.,* 70, 6, 1985.
24. **Ben-Shaul, A., Szleifer, I., and Gelbart, W. M.,** Statistical thermodynamics of amphiphile chains in micelles, *Proc. Natl. Acad. Sci. U.S.A.,* 81, 4601, 1984.
25. **Ben-Shaul, A., Szleifer, I., and Gelbart, W. M.,** Chain organization and thermodynamics in micelles and bilayers. I. Theory, *J. Chem. Phys.,* 83, 3597, 1985.

26. **Szleifer, I., Ben-Shaul, A., and Gelbart, W. M.,** Chain organization and thermodynamics in micelles and bilayers. II. Model calculations, *J. Chem. Phys.,* 83, 3612, 1985.

27. **Szleifer, I., Ben-Shaul, A., and Gelbart, W. M.,** Statistical thermodynamics of molecular organization in mixed micelles and bilayers, *J. Chem. Phys.,* 86, 7094, 1987.

28. **Szleifer, I., Ben-Shaul, A., and Gelbart, W. M.,** Chain statistics in micelles: effects of surface roughness and internal energy, *J. Chem. Phys.,* 85, 5345, 1986.

29. **Szleifer, I., Ben-Shaul, A., and Gelbart, W. M.,** Chain packing statistics and thermodynamics of amphiphilic monolayers, *J. Phys. Chem.,* 94, 5081, 1990.

30. **Ben-Shaul, A. and Gelbart, W. M.,** Alkyl chain packing in micelles and bilayers, *Annu. Rev. Phys. Chem.,* 36, 179, 1985.

31. **Fattal, D. R. and Ben-Shaul, A.,** Mean-field calculations of chain packing and conformational statistics in lipid bilayers: comparison with experiments and molecular dynamics studies, *Biophys. J.,* 67, 983, 1994.

32. **Fattal, D. R., Andelman, D., and Ben-Shaul, A.,** The vesicle-micelle transition in mixed lipid-surfactant systems: a molecular model, *Langmuir,* 11, 1154, 1995.

33. **Scheutjens, J. M. H. M. and Fleer, G. J.,** Statistical theory of the adsorption of interacting chain molecules. I. Partition function, segment density distribution, and adsorption isotherms, *J. Phys. Chem.,* 83, 1619, 1979.

34. **Leermakers, F. A. M. and Scheutjens, J. M. H. M.,** Statistical thermodynamics of association colloids. I. Lipid bilayer membranes, *J. Chem. Phys.,* 89, 3264, 1988.

35. **Barneveld, P. A., Hesselink, D. E., Leermakers, F. A. M., Lyklema, J., and Scheutjens, J. M. H. M.,** Bending moduli and spontaneous curvature. II. Bilayers and monolayers of pure and mixed ionic surfactants, *Langmuir,* 10, 1084, 1994.

36. **Meijer, L. A., Leermakers, F. A. M., and Nelson, A.,** Modelling of the electrolyte ion-phospholipid layer interaction, *Langmuir,* 10, 1199, 1994.

37. **Flory, P. J.,** *Statistical Mechanics of Chain Molecules,* Wiley-Interscience, New York, 1969.

38. **Helfrich, W.,** Elastic properties of lipid bilayers: theory and possible experiments, *Z. Naturforsch.,* 28c, 693, 1973; **Helfrich, W.,** Blocked lipid exchange in lipid bilayers and its possible influence on the shape of vesicles, *Z. Naturforsch.,* 29c, 510, 1974.

39. **Evans, E. A. and Skalak, R.,** Mechanics and thermodynamics of biomembranes, in: *Critical Reviews in Bioengineering,* CRC Press, Boca Raton, FL, 1979, p. 181.

40. **Petrov, A. G. and Bivas, I.,** Elastic and flexoelectric aspects of out-of-plane fluctuations in biological and model membranes, *Prog. Surf. Sci.,* 16, 389, 1984.

41. **Lasic, D. D.,** The mechanism of vesicle formation, *Biochem. J.,* 256, 1, 1988.

42. **Stigter, D., Mingins, J., and Dill, K. A.,** Phospholipid interactions in model membrane systems. II. Theory, *Biophys. J.,* 61, 1616, 1992.

43. **Andelman, D.,** Electrostatic properties of membranes, in: *Handbook of Physics of Biological Systems,* Vol. 1, Lipowsky, R. and Sackmann, E., Eds., Elsevier Science, Amsterdam, 1995.

44. **Puvvada, S. and Blankschtein, D.,** Theoretical and experimental investigations of micellar properties of aqueous solutions containing binary mixtures of nonionic surfactants, *J. Phys. Chem.,* 96, 5579, 1992; Nagarajan, R., Molecular theory for mixed micelles, *Langmuir,* 1, 331, 1985.

45. **Cevc, G.,** Membrane electrostatics, *Biochim. Biophys. Acta,* 1031, 311, 1990.

46. **Ennis, J.,** Spontaneous curvature of surfactant films, *J. Phys. Chem.,* 97, 663, 1992.

47. **Winterhalter, M. and Helfrich, W.,** Bending elasticity of electrically charged bilayers: coupled monolayers, neutral surfaces, and balancing stresses, *J. Chem. Phys.,* 96, 327, 1992.

48. **Mitchell, D. J. and Ninham, B. W.,** Curvature elasticity of charged membranes, *Langmuir,* 5, 1121, 1989.

49. **Lekkerkerker, H. N. W.,** Contribution of the electric double layer to the curvature elasticity of charged amphiphilic monolayers, *Physica,* A159, 319, 1989.

50. **Lewis, B. A. and Engelman, D. M.,** Lipid bilayer thickness varies linearly with acyl chain length in fluid phosphatidylcholine vesicles, *J. Mol. Biol.,* 166, 211, 1983.

51. **Seelig, J. and Waespe-Šarčević, N.,** Molecular order in cis and trans unsaturated phospholipid bilayers, *Biochem. J.,* 17, 3310, 1978.

52. **Seelig, J. and Seelig, A.,** Lipid conformation in model membranes and biological membranes, *Q. Rev. Biophys.,* 13, 19, 1980.

53. **Bloom, M., Evans, E., and Mouritsen, O. G.,** Physical properties of the fluid lipid-bilayer component of cell membranes: a perspective, *Q. Rev. Biophys.,* 24, 293, 1991.

54. **Nagle, J. F.,** Area/lipid of bilayers from NMR, *Biophys. J.,* 64, 1476, 1993.

55. **Edholm, O.,** Order parameters in hydrocarbon chains, *Chem. Phys.,* 64, 259, 1982.

56. **Wiener, M. C. and White, S. H.,** Structure of a fluid dioleoylphosphatidylcholine bilayer determined by joint refinement of X-ray and neutron diffraction data. II. Distribution and packing of terminal methyl groups, *Biophys. J.,* 61, 428, 1992.

57. **Wiener, M. C. and White, S. H.,** Structure of a fluid dioleoylphosphatidylcholine bilayer determined by joint refinement of X-ray and neutron diffraction data. III. Complete structure, *Biophys. J.,* 61, 434, 1992.

58. **Small, D. M.,** *The Physical Chemistry of Lipids,* Plenum Press, New York, 1986.

59. König, S., Pfeiffer, W., Bayerl, T., Richter, D., and Sackmann, E., Molecular dynamics of lipid bilayers studied by incoherent quasi-elastic neutron scattering, *J. Phys. II (Paris)*, 2, 1589, 1992.

60. Abney, J. R. and Owicki, J. C., Theories of protein-lipid and protein-protein interactions in membranes, in: *Progress in Protein-Lipid Interactions*, Watts, A. and de Pont, J. J. H. M., Eds., Elsevier, Amsterdam, 1985, 1.

61. Mouritsen, O. G. and Bloom, M., Models of lipid-protein interactions in membranes, *Annu. Rev. Biophys. Biomol. Struct.*, 22, 145, 1993.

62. Huang, H. W., Deformation free energy of bilayer membrane and its effect on gramicidin channel lifetime, *Biophys. J.*, 50, 1061, 1986.

63. Helfrich, P. and Jakobsson, E., Calculation of deformation energies and conformations in lipid membranes containing gramicidin channels, *Biophys. J.*, 57, 1075, 1990.

64. Dan, N., Pincus, P., and Safran, S. A., Membrane-induced interactions between inclusions, *Langmuir*, 9, 2768, 1993.

65. Dan, N. and Safran, S. A., *J. Israel Chem. Soc.*, submitted, 1995.

66. Caillé, A., Pink, D., De Verteuil, F., and Zuckermann, M. J., Theoretical models for quasi-dimensional mesomorphic monolayers and membrane bilayers, *Can. J. Phys.*, 58, 581, 1980.

67. Owicki, J. C. and McConnel, H. M., Theory of protein-lipid and protein-protein interactions in bilayer membranes, *Proc. Natl. Acad. Sci. U.S.A.*, 76, 4750, 1979.

68. Jähnig, F., Critical effects from lipid-protein interaction in membranes, *Biophys. J.*, 36, 329, 1981.

69. Jähnig, F., Vogel, H., and Best, L., Unifying description of the effect of membrane proteins on lipid order. Verification for the melittin/dimyristoylphosphatidylcholine system, *Biochemistry*, 21, 6790, 1982.

70. Scott, H. L. and Coe, T. J., A theoretical study of lipid-protein interactions in bilayers, *Biophys. J.*, 42, 219, 1983.

71. Mouritsen, O. G. and Bloom, M., Mattress model of lipid-protein interactions in membranes, *Biophys. J.*, 46, 141, 1984.

72. Sperotto, M. M. and Mouritsen, O. G., Dependence of lipid membrane phase transition temperature on the mismatch of protein and lipid hydrophobic thickness, *Eur. Biophys. J.*, 16, 1, 1988.

73. Lasic, D. D., *Liposomes: From Physics to Applications*, Elsevier Science, Amsterdam, 1993.

74. Gelbart, W. M., Ben-Shaul, A., and Roux, D., Eds., *Micelles, Membranes, Microemulsions and Monolayers*, Springer-Verlag, New York, 1994.

75. Rosevear, P., VanAken, T., Baxter, J., and Ferguson-Miller, S., Alkyl glycoside detergents: a simpler synthesis and their effects on kinetic and physical properties of cytochrome c oxidease, *Biochemistry*, 19, 4108, 1980; VanAken, T., Foxall-VanAken, S., Casetleman, S., and Ferguson-Miller, S., Alkyl glycoside detergents: synthesis and applications to the study of membrane proteins, *Methods Enzymol.*, 125, 27, 1986.

76. Jennis, R. B., *Biomembranes: Molecular Structure and Function*, Springer-Verlag, New York, 1989.

77. Vinson, P. K., Talmon, Y., and Walter, W., Vesicle-micelle transition of phosphatidylcholine and octyl glucoside elucidated by cryo-transmission electron microscopy, *Biophys. J.*, 56, 669, 1989.

78. Shurtenberger, P., Mazer, N., and Känzig, W., Micelle to vesicle transition in aqueous solutions of bile salt and lecithin, *J. Phys. Chem.*, 89, 1042, 1985.

79. Almog, S., Litman, B. J., Wimley, W., Cohen, J., Wachtel, E., Barenholtz, E. J., Ben-Shaul, A., and Lichtenberg, D., State of aggregation and phase transformations in mixtures of phosphatidylcholine and octylglucoside, *Biochemistry*, 29, 4582, 1990.

80. Ollivon, M., Eidelman, O., Blumenthal, R., and Walter, A., Micelle-vesicle transition of egg phosphatidylcholine and octyl glucoside, *Biochemistry*, 27, 1695, 1988; See also; Eidelman, O., Blumenthal, R., and Walter, A., Composition of octyl glucoside-phosphatidylcholine mixed micelles, *Biochemistry*, 27, 2839, 1988.

81. Edwards, K., Gustafsson, J., Almgren, M., and Karlsson, G., Solubilization of lecithin vesicles by a cationic surfactant: intermediate structures in the vesicle-micelle transition observed by cryo-transmission electron microscopy, *J. Colloid. Interface Sci.*, 161, 299, 1993.

82. Paternostre, M. T., Roux, M., and Rigaud, J. L., Mechanisms of membrane protein insertion into liposomes during reconstitution procedures involving the use of detergents. I. Solubilization of large unilamellar liposomes (prepared by reverse-phase evaporation) by Triton X-100, octyl glucoside, and sodium cholate, *Biochemistry*, 27, 2668, 1988.

83. Andelman, D., Kozlov, M. M., and Helfrich, W., Phase transitions between vesicles and micelles driven by competing curvatures, *Europhys. Lett.*, 25, 231, 1994.

84. May, S. and Ben-Shaul, A., Spontaneous curvature of mixed lipid vesicles, in preparation.

85. Siegel, D. P., Energetics of intermediates in membrane fusion: comparison of stalk and inverted micellar intermediate mechanisms, *Biophys. J.*, 65, 2124, 1993.

86. Kozlov, M. M., Leikin, S. L., Chernomordik, V., Markin, V. S., and Chizmadzhev, Y. A., Stalk mechanism of vesicle fusion. Intermixing of aqueous contents, *Eur. Biophys. J.*, 17, 121, 1989.

87. Verkleij, A. J., VanEchteld, C. J. A., Gerritsen, W. J., Cullis, P. R., and DeKruijff, B., The lipidic particle as an intermediate structure in membrane fusion processes and bilayer to hexagonal H_{II} transitions, *Biochim. Biophys. Acta*, 600, 620, 1980.

88. Leckband, D. E., Helm, C. A., and Israelachvili, J., Role of calcium in the adhesion and fusion of bilayers, *Biochemistry*, 32, 1127, 1993.

89. Rand, R. P. and Parsegian, V. A., Hydration forces between phospholipid bilayers, *Biochim. Biophys. Acta*, 988, 351, 1989.

Chapter 6

On the Mechanisms of the Lamellar → Hexagonal H$_{II}$ Phase Transition and the Biological Significance of the H$_{II}$ Propensity

Paavo K. J. Kinnunen

CONTENTS

I. SUMMARY

Both eukaryote and prokaryote membranes invariably contain considerable amounts of lipids which, under appropriate conditions, undergo a transition from the lamellar (L) bilayer to the inverted hexagonal (H$_{II}$) phase. The driving force for the L → H$_{II}$ transition can be governed qualitatively by alterations in the effective molecular geometries of such lipids in the lamellar and hexagonal phases and, more quantitatively, in terms of the respective spontaneous curvatures. The roles of these H$_{II}$ phase-forming lipids in biomembranes have remained enigmatic. Several functions of membranes appear to be connected to the H$_{II}$ propensity, such as fusion of bilayers, activation of protein kinase C by lipid surfaces, and the growth of microorganisms.

I have previously suggested that H$_{II}$ phase-forming lipids may adopt the so-called extended conformation.[1] In this conformation the hydrocarbon chains of a lipid are not aligned in parallel but extend in opposite directions from the headgroup. The extended conformation would manifest itself, for instance, in the contact site between two adhering bilayers, so that the acyl chains are embedded in the opposing leaflets of the contacting membranes while the headgroup is retained in the interface. Importantly, the extended conformation allows the delineation of molecular-level mechanisms for the lamellar → H$_{II}$ (as well as the lamellar → inverted cubic phase) transition and for membrane fusion via a common intermediate structure. Notably, both of these processes may now proceed without the exposure of the hydrophobic parts of the participating lipids to water as well as without the formation of intramembrane lipidic particles. The extended conformation also allows for a novel mechanism for the membrane attachment of proteins such as protein kinase C. The anchorage of peripheral proteins via extended lipids may also explain the significance of membranes with H$_{II}$ propensity as a general and ancient growth-supporting signal in cells. It is further suggested that this mechanism is functional also in eukaryotes and that it maintains in part the constitutive growth of cancer cells.

II. INTRODUCTION

Intense research on the physicochemical properties of lipids constituting biological membranes has provided a wealth of information from both *in vitro* as well as *in vivo* systems. Likewise, elucidation of the biochemistry of the participation of lipids in signaling mechanisms controlling cellular growth, for instance, is progressing at a rapid pace. Yet, while lipids are best characterized in terms of their physical properties,[2,3] the understanding of the significance of these properties to the functions of biological

0-8493-4731-9/96/$0.00+$.50
© 1996 by CRC Press, Inc.

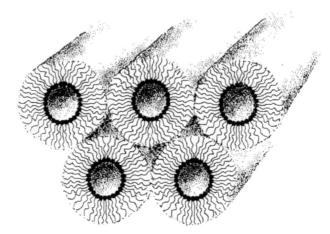

Figure 1 Schematic structure of the inverted hexagonal H_{II} phase. Dimensions of the lipids as well as the H_{II} lattice are arbitrary. It is reasonable to assume that the water-filled cylinders would actually be deformed to the extent allowed by the intrinsic curvature of the lipids so as to allow for maximal reduction of the voids between the tubes (not shown).

Table 1 Common Lipids of Biomembranes Capable of Forming the H_{II} Phase

Lipid	Refs.
Phosphatidylethanolamine (PE)	5–10
Plasmalogen PE	11
Cardiolipin	12,13
Monogalactosidyl diglyceride	14

membranes is limited.[4] In keeping with this, the roles of different lipid structures in biomembranes are poorly understood. This brief review focuses on the possible significance of lipids capable of forming the so-called hexagonal H_{II} phase, schematically illustrated in Figure 1. A characteristic of this phase is that lipids arrange to form long, water-filled tubes, with the headgroups in contact with water and the hydrophobic parts pointing towards the perimeter. The H_{II} phase is formed by several lipids, as exemplified by the compounds listed in Table 1. In addition, the H_{II} phase has been demonstrated for certain lipid mixtures (see Table 2). To date, perhaps the most intensively studied H_{II} phase-forming lipid is phosphatidylethanolamine (PE), abundant in both pro- and eukaryote membranes. Another well-known example is provided by cardiolipin (CL), which forms the H_{II} phase isothermally upon protonation or in the presence cations. The reason for the enrichment of PE in, for instance, the inner, cytoplasmic leaflet of eukaryote plasma membranes remains unknown.[15] Likewise, the role(s) of cardiolipin has (have) remained elusive.[16] Importantly, a large number of compounds, drugs, detergents, solvents, and metabolites have been shown to influence the lamellar $\rightarrow H_{II}$ phase transition, as compiled in Table 2. The rationale for their effects is currently understood in terms of molecular geometries and the respective spontaneous curvatures at free energy minima. The ability of lipids such as PE to form the H_{II} phase may be further influenced by peroxidation of their acyl chain double bonds.[43]

Several excellent reviews have been published focusing on the various aspects of the H_{II} phase.[44-54] As will be discussed in more detail below, the abilities to (1) form or promote the formation of the H_{II} phase and (2) undergo or promote fusion are intimately related. Another interesting, emerging concept is (3) the correlation between the H_{II} propensity and the activation of protein kinase C by lipids. Finally, (4) optimal growth of bacteria requires a certain degree of H_{II} propensity. Possible molecular mechanisms linking the above are presented. More specifically, the mechanistic feasibility of lipids in the so-called extended conformation[1] for processes 1 to 4, as well as for (5) the involvement of lipids in maintaining constitutive growth of cancer cells is outlined. There are alternative views regarding the transition mechanism and fusion;[44,55] for these the reader is referred to the analyses of Siegel.[56-58]

Table 2   Examples of Additives Either Promoting or
Inhibiting the H_{II} Phase Formation

Compound	Effect on H_{II}	Refs.
Hexachlorophene	Promotion	17
Anisodamine	Promotion	18
Antiviral peptides	Inhibition	19
Trehalose	Promotion	20
Detergents	Inhibition	21
Alkanes	Promotion	22–26
Tetradecanol	Promotion	26
Squalene	Promotion	27
Dolichols	Promotion	28
Retinoids	Promotion	29
Acylcarnitine	Inhibition	30
Phosphatidylcholine	Inhibition	31, 32
Lysophosphatidylcholine	Inhibition	21
Digalactosyl diglyceride	Inhibition	33
Triacylglycerols[a]	Promotion	34
Diacylglycerols	Promotion	21, 24, 34–36
Monoacylglycerols	Promotion	34
Cholesterol	Promotion	37, 38
Cholesteryl esters	Promotion	39
Cholesteryl sulfate	Inhibition	40
Fatty acids	Promotion	41, 42

[a]   Potency of acylglycerols decreases in the sequence tri > di > mono.

III. PHYSICAL BASIS OF THE H_{II} PHASE

In contrast to our present lack of molecular-level understanding of the biological significance of the H_{II} phase-forming lipids, the basic underlying physical principles governing the structure of this phase are reasonably well established. This is due to intense research involving techniques such as differential scanning colorimetry (DSC),[6,59] time-resolved X-ray diffraction,[51,60-62] temperature-jump freeze-fracture electron microscopy (EM),[63] electron spin resonance (ESR),[64] fluorescence,[65-67] infrared (IR) spectroscopy,[68] and nuclear magnetic resonance (NMR).[44,49,69-71] A major breakthrough was the elucidation of the importance of molecular geometries and shapes in determining the three-dimensional organization of lipids to yield various types of membrane structures.[44,72-74] In this context it is mandatory to emphasize the fact that it is the *effective* molecular shape which matters. Accordingly, characterization of the properties of a lipid in this respect requires not only knowledge about its chemical structure, determining only the van der Waals volume and shape of the molecule, but also information on its hydration shell, conformation, conformational dynamics, and intermolecular interactions. In other words, the effective geometry of the lipid could be represented as a soft and diffuse surface determined by the amplitudes and frequencies of the thermal motion of the molecule, by fields due to its charges and dipoles, by interactions with its neighboring lipids (e.g., hydrogen bonding), and by the hydration shell of its headgroup. The above is complemented by information on membrane free volume distribution[75] and its dependence on factors such as hydration.[76]

Israelachvili et al.[74] introduced a parameter P describing the molecular geometry and defined as

$$P = V/(a \cdot l_c)$$

where V = the effective volume of the hydrophobic part of the molecule, a = the limiting surface area of its hydrophilic part, and l_c = the length of its hydrophobic part.

Parameter P thus emphasizes the relative sizes of the different parts of the molecule, and its values for lipids forming bilayers and H_{II} phases are $1/2 < P < 1$ and >1, respectively. For H_{II}-forming lipids the headgroup is small compared to the hydrophobic part of the molecule. This view is readily compatible with the known properties of lipids adopting the H_{II} phase. The headgroup of PE is poorly hydrated, which results in a small effective size in comparison to phosphatidylcholine (PC) with its large hydration shell.

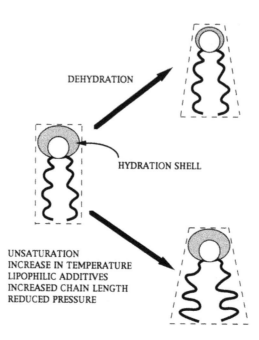

Figure 2 Examples of factors influencing the effective molecular shapes of lipids undergoing the lamellar → H$_{II}$ transition. Similar schema should be applicable to the tetraacylphospholipid, cardiolipin.

Likewise, introduction of *cis* double bonds into the acyl chains favors the H$_{II}$ phase. This is schematically illustrated in Figure 2. The balance determining between lamellar or H$_{II}$ organization can also be influenced by certain additives.

While the original considerations on the correlation between the molecular shapes and the structure of lipid phases employed terms such as inverted cone and wedge-shaped, it is obvious that more delicate features of molecular dynamics have to be introduced to obtain the effective shape. In keeping with this, the acyl chain region of H$_{II}$ phase-forming lipids have been recently suggested to be more accurately described as a "soft balloon".[77] The parameters describing the effective molecular geometries have been suggested to be additive in lipid mixtures.[78] In accordance, the effective shapes have been recently shown to accurately predict the influence of a constant molar proportion of different amphiphiles on the lamellar → hexagonal H$_{II}$ phase transition temperature (T$_H$) of palmitoyl-oleoyl PE (POPE).[79] More specifically, these authors demonstrated a linear correlation between the headgroup volume of several dioleoyl lipids and their effects on the transition midpoint of POPE.

The relatively simple considerations of effective molecular shapes provide a useful, intuitive way of analyzing qualitatively the organization and phase behavior of lipid assemblies.[44,72-74,80,81] Yet, as pointed out by Gruner et al., a more quantitative analysis of the H$_{II}$ phase can be performed by taking into account the bending energy and the spontaneous curvature of the membrane in question.[82-84] Membrane lipids have been stated to be under dual-solvent stress.[26] Accordingly, both water activity[85-89] (influencing the extent of hydration), as well as the presence of hydrophobic additives such as alkanes (Table 2), determine the free energy minimum of the system with the respective intrinsic curvature and, thus, the possibility for the formation of H$_{II}$ phase and related structures.

IV. MECHANISM(S) OF THE LAMELLAR → H$_{II}$ TRANSITION

The exact molecular details of the lamellar to H$_{II}$ (L$_\beta$/L$_\alpha$ → H$_{II}$) transition have been shown considerable interest, as reflected in the large number of articles available on this subject.[50-52,56,60-62,90] At present, very little remains to be added in order to review the data collected, and thus only the major conclusions are briefly summarized here. Compared with the gel → liquid crystalline (main) transition of lipids such as DPPC, the enthalpy for the transition into the H$_{II}$ phase is much less.[59] In general, the lamellar → H$_{II}$ phase transition is easily comprehended qualitatively in terms of changes in the effective molecular shapes. Accordingly, this transition can be promoted by decreasing the effective headgroup size by dehydration

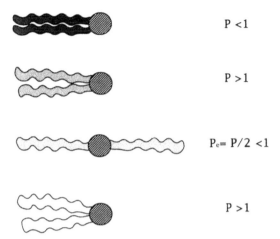

$P < 1$

$P > 1$

$P_e = P/2 < 1$

$P > 1$

Figure 3 Effective molecular geometries and conformations of a phospholipid such as phosphatidylethanolamine upon increasing degree of thermal excitation during the lamellar → H_{II} transition. Significance of the geometric parameters is discussed in the text. (Modified from Kinnunen, P. K. J., *Chem. Phys. Lipids*, 63, 251, 1992. With permission.)

(Figure 2) or deionization. Promotion of the H_{II} phase formation by acidic lipids in the presence of divalent cations, most prominently Ca^{2+}, is likely to reflect both of these processes. In addition, the H_{II} phase is promoted by factors influencing the hydrocarbon region of the lipids, i.e., decrease in pressure, increase in temperature, acyl chain unsaturation, increase in acyl chain length, and the presence of proper lipophilic additives, surfactants, solvents, or metabolites. Chaotropic agents such as KSCN and urea have been shown to produce a significant increase in the T_H of PE.[86] This effect is readily compatible with the suggestion that the affinity of water to the PE headgroup should decrease upon entering the H_{II} phase.[59] Accordingly, we may envisage the effect of chaotropic substances to be due to a diminished affinity of water to the bulk aqueous phase, which thus allows PE to maintain the hydration water at higher temperatures, counteracting the thermal increase in the *trans → gauche* isomerization and the corresponding increase in the effective volume V of the acyl chain region.[68] Sensitivity of the H_{II} phase for hydration and water activity is of great interest in light of the recent increasing recognition of the importance of osmotic forces in the regulation of cellular functions.

Proposals on the mechanism of this transition generally involve inverted intramembrane micelles as intermediates.[55-57] However, it was also pointed out that the transition may well proceed via several alternative pathways.[91] Recent exploitation of more precise experimental techniques, such as temperature-jump cryo electron microscopy[63] and real-time X-ray diffraction,[51,52,60] has not provided evidence for the involvement of intramembrane inverted micelles as transition intermediates. Lack of appearance of intramembrane particles (IMPs) in the course of the transition[60,62,63,91] has been assumed to be due to their short lifetime and highly transient nature during this process.

I have recently suggested an alternative mechanism involving a lipid conformation nominated as extended. In this conformation the acyl chains of the lipid in question are pointed in opposite directions from the headgroup, instead of the usual conformation wherein acyl chains are aligned in parallel (Figure 3). The formation of the extended conformation requires acyl chain reversal to take place, driven by conformational dynamics.[92-94] Evidence for the extended conformation of phospholipids is provided by high-resolution NMR.[93,94] In a small proportion of the total phospholipid molecules of normal and inverted micelles the antiperiplanar conformation of the C(2)-C(3) glycerol bond is evident and has been shown on the NMR time scale to be in rapid exchange with the predominant *gauche* conformations of the torsion angle Θ_4. Evidence for the extended conformation of triacylglycerol has been obtained by the lipid monolayer technique.[95]

The extended conformation should be favored under specific conditions as follows. Upon increasing temperature, for instance, the degree of *trans → gauche* isomerization as well as the rates and amplitudes of chain motion of diacylphospholipids increase, so as to increase the effective size of the lipid acyl chains. Accordingly, the geometric parameter P → 1 and the pressure within the center of the membrane

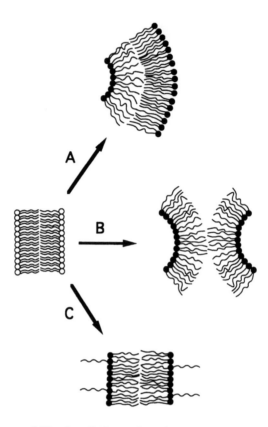

Figure 4 Illustration of the possibilities for relieving steric strain due to, for instance, thermal excitation of lipids pushing the system towards the H$_{II}$ phase. See text for details.

increases. Because of this the free energy of the system begins to increase due to bending stress as the intrinsic curvature for a single monolayer should increase. Accordingly, maintaining the lipid bilayer at a free energy minimum would require a change in membrane curvature, as illustrated in Figure 4. For an isolated bilayer, such as is present in large unilamellar vesicles, this is not possible, as increasing the curvature of one monolayer would require a decrease in the curvature of the other leaflet (Figure 4A). On the other hand, a simultaneous increase in the curvature of the two leaflets is also not possible (Figure 4B). Therefore, the planar configuration would be favored in spite of the augmented packing pressure in the hydrocarbon region of the membrane. However, the planarity requires part of the lipid molecules to adopt the extended conformation, with one of the acyl chains sticking out into the aqueous phase. Accordingly, packing pressure in the membrane interior is relieved (Figure 4C). Obviously, the increase in free energy of the system due to the exposure of the acyl chains to water should be compensated by free energy gain due to avoidance of membrane bending. A membrane in a state with a high degree of H$_{II}$ formation propensity will hereafter in this text be referred to as being in L$_e$ phase. The formation of this phase and, accordingly, changes in the surface properties of the membrane should start several degrees below the onset of the transition.

   In terms of the geometric parameter P the above can be analyzed as follows (Figure 3). The effective volume V$_e$ of the hydrocarbon region of a lipid in the extended conformation is defined as

$$V_e = V/2$$

Thus, for a lipid in the extended conformation, P$_e$ is defined as

$$P_e = V/(2al_c)$$

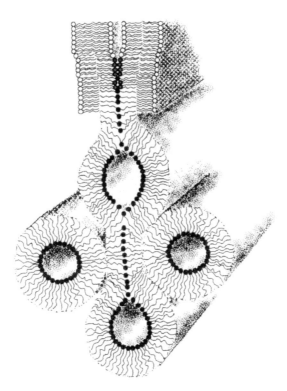

Figure 5 Molecular mechanism of lamellar → H_{II} transition via an intermediate lipid organization consisting of molecules in the extended conformation. For a thermal transition the progression corresponds, from top to bottom, to an increasing temperature. (Modified from Kinnunen, P. K. J., *Chem. Phys. Lipids*, 63, 251, 1992. With permission.)

Due to the additivity[78] of the values for P for composite membranes (or lipids adopting different conformations) it follows that for a lamellar membrane containing lipids in the extended conformation we have

$$[nP_e + (N - n)P_a]/N < 1$$

where N = the total number of lipids and n = the number of lipids in the extended conformation. In other words the *system* value of P can be maintained at <1 when a sufficient number of lipids adopt the extended conformation. Upon further increase in temperature, however, the system of adhered bilayers collapses. The volume of the acyl chains increases further so that also $P_e \to 1$ and the system value for P exceeds unity. This necessitates reorganization of the assembly into the H_{II} phase, concomitantly with the realignment of the acyl chains. Importantly, for isolated unilamellar vesicles no formation of the H_{II} phase can be detected.[96] Instead, for closely packed lamellae such as those in multilamellar liposomes the possibility of removing the extended, external chain from contact with water can be achieved by insertion into the opposing monolayer. For a stack of membranes a possibility for the intermediate structure is shown in Figure 5.

Dehydration of the lipid surface upon approaching T_H has been reported.[97] The extended conformation allows for an easy progression into the H_{II} phase, without the involvement of inverted micelles (Figure 6). At the contact site the lipids are at a rapid conformational and positional equilibrium (Figure 7). This situation should result in lipid mixing between the contacting leaflets. The flattened tubes demonstrated by Hui et al.[91] and the X-ray data[60] are compatible with the depicted $L \to H_{II}$ transition mechanism. Notably, in this mechanism the driving force is provided essentially by the same general principles as in the previous suggestions, i.e., changes in the effective shapes and spontaneous curvatures of the molecules undergoing the transition.

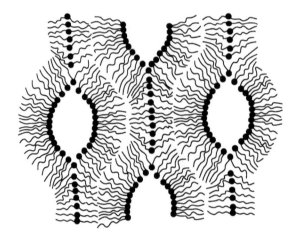

Figure 6 Possible arrangement of lipids in a stack of membranes undergoing the lamellar $\rightarrow H_{II}$ transition.

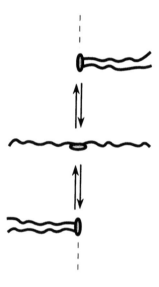

Figure 7 Kinetic equilibrium between aligned-extended-aligned conformations for a lipid with two acyl chains, allowing lipid mixing between the adhering leaflets.

The isotropic signal seen by ^{31}P-NMR is generally taken as diagnostic for the H_{II} phase and lipid intramembrane particles.[9,98-100] Importantly, Thayer and Kohler have pointed out that such spectra may also arise due to proper headgroup conformations and rapid rotation of the molecule about its long axis while retaining the bilayer configuration, i.e., in the absence of the formation of the H_{II} phase.[69] The isotropic ^{31}P-NMR signal is also seen for microsomal membranes.[101-103] However, this signal is not present for extracted lipids dispersed in water but requires the presence of proteins.[102]

As was already pointed out above, the extended conformation requires the so-called chain reversal process to take place. Due to conformational restrictions imposed by the glycerophospholipid structure it is readily conceivable that it is the sn-2 acyl chain which would be more amenable to extension out from the bilayer. This property should be greatly augmented by unsaturation and thus may explain the specificity of positional distribution of saturated and unsaturated fatty acids at the sn-1 and sn-2 carbons, respectively, of the glycerol backbone. Furthermore, the formation of the L_e phase as a fundamental property of biological membranes would also explain the reasoning behind selecting lipids with *two* acyl chains as major constituents.

V. FUSION OF LIPID BILAYERS

Parallels between H_{II} phase formation and lipid bilayer fusion are at present firmly established, and these two processes are known to be closely related.[57,58,104-111,155] The various H_{II} phase-forming, inverted cone-shaped lipids themselves promote fusion of lipid bilayers. Likewise, compounds promoting the formation of the H_{II} phase, such as diacylglycerol and hexadecane, enhance fusion.[35,111] These two processes have been postulated to proceed via a common pathway involving inverted micelles as intermediates. This view is supported by the appearance of intramembrane lipidic particles (IMPs) under conditions where fusion or lamellar $\rightarrow H_{II}$ phase transition is occurring. In addition to electron microscopy data, evidence supporting the involvement of IMPs is derived from ^{31}P-NMR studies which report an isotropic signal under conditions promoting fusion as well as in the so-called isotropic lipid phases present below the actual onset of the lamellar $\rightarrow H_{II}$ transition. Yet direct causality has been questioned[96,100,112,113] and evidence demonstrating lack of involvement of IMPs in fusion has been published.[60,63,91,96,100,112,113] As was pointed out above, the isotropic ^{31}P-NMR signal may also arise in lipids retaining the bilayer configuration.[69]

Several distinct stages can be distinguished in membrane fusion. First, contact between the bilayers is required.[114] In order to allow for fusion, partial dehydration of the lipids at the contact site should take place. In keeping with this, vesicle-vesicle contacts are promoted by lipids such as PE as well as dehydration of acidic phospholipids by divalent cations. Contact between membranes may be reversible and not result in fusion. Several processes may take place during the contact. Lentz et al. recently showed that the transbilayer migration of phosphatidylglycerol (PG) is enhanced by vesicle-vesicle associations and occurs in the absence of formation of IMPs.[115] Prior to fusion, i.e., merging of the aqueous contents of two contacting vesicles, an intermediate state, "hemifusion", has been postulated to occur.[116-118] In this state, rapid mixing of the outer leaflet lipids takes place without coalescence of the aqueous contents of the vesicles. Importantly, measured as a function of temperature, the rate of bilayer fusion is maximal several degrees below T_H, i.e., in the L_e phase, and slows down upon further approaching the transition temperature.[109]

In order to result in fusion, some kind of destabilization of the bilayer structure must take place. The nature of the intermediate phase preceeding fusion has remained ambiguous. I have recently described a molecular mechanism of membrane fusion which proceeds without the formation of inverted micelles.[1] This mechanism is based on the extended lipid conformation delineated above, and it thus is readily related to the properties of lipids forming the H_{II} phase. In other words, compounds promoting fusion should either form or promote the formation of the H_{II} phase. Likewise, conditions promoting fusion should also favor the formation of the H_{II} phase. The essentials of this model are schematically depicted in Figure 8, illustrating a sequence of membrane-level molecular reorganization processes (from A to F). In the following these stages are described in detail.

Phase A: The opposing bilayers are separated by an aqueous layer. The contributions from repulsive forces vary depending on the composition of the membranes and the water phase.[119]

Phase B: The two bilayers at the contact site are now to a large extent dehydrated and adhere. The headgroups of the participating phospholipids are marked with hatched lines. Both step A and step B are generally included in membrane fusion models.

Phase C: Due to the absence of water between the opposing bilayers there is no energy barrier due to hydrophobic effects, and an acyl chain in a phospholipid can now, due to conformational fluctuations, reverse its direction and extend into the opposing membrane while the headgroup remains in the interface. This extended conformation should be present, for instance, in the isotropic phase existing between the L_α and H_{II} phases for dioleoyl PE.[109] Notably, the extended conformation would also allow for lipid mixing between the outermost monolayers of the opposing membranes without fusion occurring. The postulated mechanism of lipid exchange is closely related to that of lipid transfer, (Figure 9). It has been demonstrated that there is a lipid-concentration-dependent exchange process at higher concentrations of liquid crystalline vesicles which is mediated by transient vesicle-vesicle associations.[120,121] This mode of lipid transfer is enhanced in the presence of phosphatidylethanolamine.[122] Conformational flexibility of the acyl chains would be required for the adoption of the extended conformation during fusion. This process should be more effective for unsaturated lipids, in accordance with results from studies on the fusion of cells as well as vesicles.[123,124] In electron microscopy images the appearance of this intermediary structure should be indistinguishable from that of phase B.

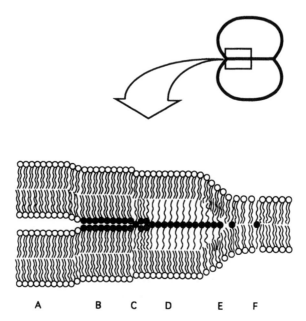

Figure 8 Merging of two contacting bilayers in the course of fusion by a mechanism involving extended lipids. See text for details. (From Kinnunen, P. K. J., *Chem. Phys. Lipids*, 63, 251, 1992. With permission.)

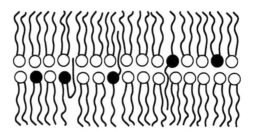

Figure 9 Contact-enhanced lipid exchange between two bilayers via an extended lipid conformation.

Phase D: For illustrative purposes a membrane domain is shown, composed solely of phospholipids in the extended conformation.

Phase E: The acyl chains of the phospholipids in the inner leaflets of the opposing membranes now interdigitate with those of the lipids in the middle of the assembly and in the extended conformation. This results in the formation of the normal bilayer arrangement, possibly retaining some extended phospholipids.

Phase F: A bilayer diaphragm now separates the aqueous spaces of the vesicles. For weakly hydrating lipids the extended conformation could be accommodated also in the planar diaphragm, as tentatively illustrated.

For further details the reader is referred to the original publication.[1] We may conclude that the molecular requirements for lipid bilayer fusion as well as for the formation of the H_{II} phase are readily compatible with the extended lipid conformation. The proposed mechanism of fusion also alleviates the exposure of the hydrophobic chains of the lipids to water and does not require the formation of intermediate structures such as inverted micelles within the membranes.

VI. PERIPHERAL MEMBRANE ANCHORAGE OF PROTEINS BY LIPIDS IN THE EXTENDED CONFORMATION

Membrane proteins are commonly divided into two distinctly different types. Integral proteins contain at least one membrane-spanning, hydrophobic domain, which mediates the lipid-protein interactions and

Figure 10 Membrane association of a peripheral protein by an extended lipid anchor.

keeps the protein firmly in the bilayer. Instead, peripheral proteins are associated with the lipid surface. In contrast to the integral proteins, the membrane association and detachment of peripheral proteins can be rapidly controlled so as to cause changes in their function. Accordingly, a thorough understanding of mechanisms responsible for and controlling peripheral interactions between proteins and lipids is warranted. Well-known examples of peripheral membrane proteins are cytochrome c,[125,126] protein kinase C (PKC),[127] and CTP:phosphocholine cytidyl transferase.[128] The lipid requirements for the membrane association and activity of these proteins have been elucidated in considerable depth. This topic has been recently summarized in more detail and, importantly, certain unifying principles can be identified.[129]

Due to its central role in cellular signal transmission cascades the lipid-PKC interaction has been intensively explored. These studies have provided clear evidence that the activation of PKC can be accomplished by H_{II} phase-forming lipids. However, it has been pointed out by Epand that it is the *propensity* for the H_{II} phase formation and *not* the actual formation of the H_{II} phase which is important.[130,131] Based on the available data we have proposed a novel mechanism for the peripheral membrane association of proteins, schematically illustrated in Figure 10. In brief, a peripheral protein may become associated with the surface via an extended lipid anchor.[129] Due to considerations arising from lipid packing, provided above, this mode of binding would be promoted by a propensity for H_{II} formation; i.e., this anchorage via extended lipids to surfaces would in part be controlled by the L_e phase, while specific lipid headgroup-protein interactions are likely to be additionally involved. The above mechanism requires a hydrophobic cavity or groove in the protein. It is noteworthy that this mode of surface association involves hydrophobic forces. Accordingly, it avoids the problem of electrostatic interactions being disrupted by the ions present in physiological fluids.

VII. L_e PHASE AS A GENERAL GROWTH PROMOTER

Studies on the growth of microorganisms revealed an adaptation of the lipid composition of their membranes due to alterations in the supplementation of nutrients as well as changes in the growth temperature. These data were interpreted in terms of "homeoviscous adaptation".[132] In brief, according to this view, cells tend to maintain constant microviscosity of their membranes. Recent measurements also support this hypothesis.[133,134] Yet there is also evidence to indicate that homeoviscous adaptation is an oversimplification and that the situation is more complex.[135,136] Systematic studies on microorganisms under a range of conditions[14,33,137-144] have revealed that, depending on growth conditions, their membrane compositions are varied so that optimum growth is observed when the temperature is *above* the gel $\rightarrow$ liquid crystalline transition as well as approximately 10° *below* the $L_\alpha \rightarrow H_{II}$ transition. The relatively good correlation between the molecular geometries of membrane lipids[145] and their phase behavior has led to the conclusion that microorganisms such as *Escherichia coli* and *Acholeplasma laidlawii* maintain their membrane lipid composition such that they are lamellar and fluid, yet have a certain degree of propensity for H_{II} formation, i.e., are in the L_e phase. These general conclusions appear to be applicable to a range of conditions and allow one to predict both qualitative as well as quantitative changes in membrane lipid composition due to changes in temperature or due to introduction of exogenous lipophilic compounds.[139,140] Most recently it was demonstrated using an *E. coli* mutant incapable of PE synthesis that this strain is capable of growing in the presense of metal cations, compensating under these conditions for the absence of PE by having the required H_{II} propensity due to cardiolipin.[142-144] These findings suggest that membranes with the proper fluidity and, importantly, with H_{II} phase propensity function as a general growth signal in bacteria by associating and activating a proper set of proteins to the membranes by extended lipid anchorage. The proteins attached by this mechanism to bacterial membranes would (1) regulate the chemical composition of the membrane so as to maintain the L_e phase with H_{II} propensity, (2) initiate and drive the replication of DNA as well as other cellular structures, and (3) in general, support mechanisms maintaining maximal growth of the microorganism.

It is worth noting that biomembranes with H_{II} propensity and thus the capability of controlling the activity of membrane proteins would be ideally suited for regulatory purposes. First of all, as long as the membrane is below T_H the temperature range for the L_e phase is relatively broad. This would reduce the temperature sensitivity of the organism and allow for the accommodation of the synthetic pathways to the prevailing conditions. After proper accommodating changes have taken place in the membrane (lipid) composition, a growth-supporting extent of the L_e phase should be reestablished, together with the presence of specific lipid structures. Regarding the latter, cells may further add to the specificity of regulation by the extended anchorage by involving specific lipid-protein interactions, as exemplified by the activation of PKC by phosphatidyl-L-serine.[146]

Basically, cells can be understood to be self-organizing, supramolecular assemblies capable of replication, representing the living state of matter. As with all biomolecular assemblies, there is an intimate coupling between the chemical composition, physical state, organization, and function. Accordingly, although the organization of a living cell is vastly complex, there must be a limited number of configurations of certain cellular elements controlling the fate and behavior of the cell. We may then envisage the functional change of a cell from the resting to the mitotic state as a transition involving specific alterations in chemical composition, organization, and physical state. This would also apply for the transition into the malignant state. In keeping with the range of oncogenes functioning at different levels in cellular signal transmission controlling growth, it is clear that the transforming signal may originate at different cellular locations, such as the plasma membrane and the nucleus. Ultimately, however, the transforming signal must convert the entire cell to the malignant, constitutively growing state. This scheme is illustrated in Figure 11 and is concordant with "epigenetic" transformation of cultured cells, taking place in the absence of a mutation.[147]

Taking into account the conserved general principles of function of nucleic acids and proteins in prokaryotes and eukaryotes, it is suggested that the L_e phase would function as a general growth signal also in the latter type of cells. In this context it is revealing to compare the behavior of bacteria with that of malignant cells. Bacteria are reminiscent of cancer cells in the sense that they appear to strive for a maximal growth rate. Accordingly, taking into account the role of eukaryote plasma membrane as a site normally involved in growth signal initiation and transmission, we should find lipids with H_{II} propensity to be enriched in the plasma membrane of cancer cells. It is postulated that, analogously to prokaryotes, also in the membranes of transformed eukaryote cells extended lipid anchorage provides a constitutive growth signal, resulting in the membrane attachment and activation of proteins maintaining cellular growth.

High contents of lipids promoting the formation of the H_{II} phase have been demonstrated in several types of cancer cells. More specifically, plasma membranes of cancer cells have been shown to contain large amounts of neutral lipids, acylglycerols, and cholesteryl esters.[148,149] In the plasma membranes of *ras*-transformed cells the content of diacylglycerol has been observed to be as high as 10 mol% of the total lipid.[150] Strongly isotropic NMR signals are observed, and because of the limited solubility of acylglycerols and cholesteryl esters in phospholipid bilayers part of these neutral lipids are likely to exist embedded in the center of the membrane lipid bilayer as intramembrane lipid droplets.[148] The amount of isotropically tumbling triacylglycerol has been recently shown to increase 25-fold in activated T-lymphocytes.[151]

The excessive presence of L_e-phase lipids in the plasma membrane should also greatly facilitate the fusion of intracellular lipid vesicles to this membrane. Such poorly controlled fusion should result in an aberrant expression of various types of proteins as well as lipids on the cell surface. Likewise, one would expect excessive accumulation of membrane material on the surface, which in turn would cause abnormal

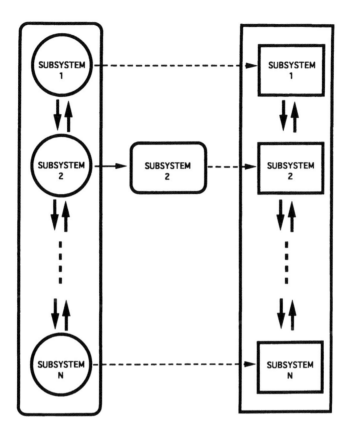

Figure 11 Transformation of a cell represented as a system transition. Accordingly, transition of a cellular structure such as the plasma membrane into a state signaling constitutive growth may induce a transition of other cellular mechanisms to the same state.

cellular morphology. The excess membrane would subsequently be shed from the plasma membrane of transformed cells.[152]

As has been pointed out previously, the development of membrane-bounded organelles in eukaryotes allows for a rather independent control of their lipid composition as well as their phase state. Continuing this line of argument further, the L_e phase should provide, regardless of the lipid structures constituting it, a general growth signal also in the different organelle membranes. Accordingly, in nuclear membranes this should result in replication and in endoplasmic reticulum it should cause an enhanced synthesis of lipids, for instance. In mitochondria, on the other hand, the extended anchorage should be used to activate processes generating ATP, independently from the cell cycle. In the plasma membrane constitutive growth signaling would be maintained.

Finally, I believe that mere metabolic compartmentalization was not the major reason for the development of cell organelles in eukaryotes. Instead, we may look upon the evolution of organelles as the development of separate, supramolecular membrane assemblies for which their protein and lipid compositions and phase state can be controlled sufficiently independently so as to allow for the differential control of the various metabolic activities in a cell. The need for such separate membrane structures within a cell originated due to the limited possibilities of organizing functionally and compositionally differentiated domains in a single, fluid bilayer such as in prokaryotes. Although distinct, laterally segregated membrane domains are likely to be found also in bacteria, the phase state of its membrane is likely to be altered more or less *de toto*. It follows that, unlike in eukaryotes, the set of proteins activated by the L_e phase also should be almost solely controlled by gene expression.

ACKNOWLEDGMENTS

I wish to thank Drs. Jukka Y. A. Lehtonen, Marjatta Rytömaa, Anu Koiv, and Pekka Mustonen for rewarding brainstorming discussions on the topics of this article. They create the enthusiastic atmosphere of our laboratory, making it a pleasure to work there. Financial support was provided by the Finnish State Medical Research Council, the Sigrid Juselius Foundation, and University of Helsinki Biotechnology Fund.

REFERENCES

1. **Kinnunen, P. K. J.,** Fusion of lipid bilayers: a model involving mechanistic connection to H_{II} phase forming lipids, *Chem. Phys. Lipids,* 63, 251, 1992.

2. **Kinnunen, P. K. J. and Laggner, P.,** Eds., Phospholipid phase transitions, *Chem. Phys. Lipids,* 57, 109–498, 1991.

3. **Slater, J. L. and Huang, C.,** Interdigitated bilayer membranes, *Prog. Lipid Res.,* 27, 325, 1988.

4. **Kinnunen, P. K. J.,** On the principles of functional ordering in biological membranes, *Chem. Phys. Lipids,* 57, 375, 1991.

5. **Tilcock, C. P. S. and Cullis, P. R.,** The polymorphic phase behaviour and miscibility properties of synthetic phosphatidylethanolamines, *Biochim. Biophys. Acta,* 684, 212, 1982.

6. **Epand, R. M.,** High sensitivity differential scanning calorimetry of the bilayer to hexagonal phase transitions of diacylphosphatidylethanolamines, *Chem. Phys. Lipids,* 36, 387, 1985.

7. **Lewis, R. N. A. H. and McElhaney, R. N.,** Calorimetric and spectroscopic studies of the polymorphic phase behaviour of a homologous series of n-saturated 1,2-diacyl phosphatidylethanolamines, *Biophys. J.,* 64, 1081, 1993.

8. **Leventis, R., Fuller, N., Rand, R. P., Yeagle, P. L., Sen, A., Zuckermann, M. J., and Silvius, J. R.,** Molecular organization and stability of hydrated dispersions of headgroup-modified phosphatidylethanolamine analogues, *Biochemistry,* 30, 7212, 1991.

9. **Boggs, J. M., Stamp, D., Hughes, D. W., and Deber, C. M.,** Influence of ether linkage on the lamellar to hexagonal transition of ethanolamine phospholipids, *Biochemistry,* 20, 5728, 1981.

10. **Harlos, K. and Eibl, H.,** Hexagonal phase in phospholipids with saturated chains: ethanolamines and phosphatidic acids, *Biochemistry,* 20, 2888, 1981.

11. **Lohner, K., Hermetter, A., and Paltauf, F.,** Phase behaviour of ethanolamine plasmalogen, *Chem. Phys. Lipids,* 34, 163, 1984.

12. **Powell, G. L. and Marsh, D.,** Polymorphic phase behaviour of cardiolipin derivatives by ^{31}P NMR and X-ray diffraction, *Biochemistry,* 24, 2902, 1985.

13. **Seddon, J. M., Kaye, R. D., and Marsh, D.,** Induction of the lamellar-inverted hexagonal phase transition in cardiolipin by protons and monovalent cations, *Biochim. Biophys. Acta,* 734, 347, 1983.

14. **Lindblom, G., Brentel, M., Sjölund, M., and Wieslander, Å.,** Phase equilibria of membrane lipids from *Acholeplasma laidlawii:* importance of a single lipid forming non-lamellar phases, *Biochemistry,* 25, 7502, 1986.

15. **Zachowski, A.,** Phospholipids in animal eucaryotic membranes: transverse asymmetry and movement, *Biochem. J.,* 294, 1, 1993.

16. **Hoch, F. L.,** Cardiolipins and biomembrane function, *Biochim. Biophys. Acta,* 1113, 71, 1992.

17. **Epand, R. M., Epand, R. F., Leon, B. T.-C., Menger, F. M., and Kuo, J. F.,** Evidence for the regulation of the activity of protein kinase C through changes in membrane properties, *Biosci. Rep.,* 11, 59, 1991.

18. **Hwang, F., Wu, Y., and Wen, D.,** Anisodamine induces the hexagonal phase in phospholipid liposomes, *Biochim. Biophys. Acta,* 860, 713, 1986.

19. **Epand, R. M.,** Virus replication inhibitory peptide inhibits the conversion of phospholipid bilayers to the hexagonal phase, *Biosci. Rep.,* 6, 647, 1986.

20. **Wistrom, C. A., Rand, R. P., Crowe, L. M., Spargo, B. J., and Crowe, J. H.,** Direct transition of dioleoylphosphatidylethanolamine from lamellar gel to inverted hexagonal phase caused by trehalose, *Biochim. Biophys. Acta,* 984, 238, 1989.

21. **Madden, T. D. and Cullis, P. R.,** Stabilization of bilayer structure for unsaturated phosphatidylethanolamines by detergents, *Biochim. Biophys. Acta,* 684, 149, 1982.

22. **Epand, R. M.,** Diacylglycerols, lysolecithin, or hydrocarbons alter the bilayer to hexagonal phase transition temperature of phosphatidylethanolamines, *Biochemistry,* 24, 7092, 1985.

22a. **Sjölund, M., Lindblom, G., Rilfors, L., and Arvidson, G.,** Hydrophobic molecules in lecithin-water systems. I. Formation of reversed hexagonal phases at high and low water contents, *Biophys. J.,* 52, 145, 1987.

23. **Sjölund, M., Rilfors, L., and Lindblom, G.,** Reversed hexagonal phase formation in lecithin-alkane-water systems with different acyl chain unsaturation and alkane length, *Biochemistry,* 28, 1323, 1989.

24. **Siegel, D. P., Banschbach, J., and Yeagle, P. L.,** Stabilization of H_{II} phase by low levels of diglycerides and alkanes: an NMR, calorimetric, and X-ray diffraction study, *Biochemistry,* 28, 5010, 1989.

25. **Lafleur, M., Cullis, P. R., Fine, B., and Bloom, M.,** Comparison of the orientational order of lipid chains in the L_{α} and H_{II} phases, *Biochemistry,* 29, 8325, 1990.

26. **Rand, R. P., Fuller, N. L., Gruner, S. M., and Parsegian, V. A.,** Membrane curvature, lipid segregation, and structural transitions for phospholipids under dual-solvent stress, *Biochemistry,* 29, 76, 1990.

27. **Lohner, K., Degovics, G., Laggner, P., Gnamusch, E., and Paltauf, F.,** Squalene promotes the formation of non-bilayer structures in phospholipid model membranes, *Biochim. Biophys. Acta,* 1152, 69, 1993.

28. **Valtersson, C., van Duyn, G., Chojnacki, T., de Kruijff, B., and Dallner, G.,** The influence of dolichol, dolichol esters, and dolichyl phosphate on phospholipid polymorphism and fluidity in model membranes, *J. Biol. Chem.,* 260, 2742, 1985.

29. **Ortiz, A., Aranda, F. J., Villalaín, J., and Gómez-Fernández, J. C.,** Influence of retinoids on phosphatidylethanolamine lipid polymorphism, *Biochim. Biophys. Acta,* 1112, 226, 1992.

30. **Epand, R. M., Robinson, K. S., Andrews, M. E., and Epand, R. F.,** Dependence of the bilayer to hexagonal phase transition on amphiphile chain length, *Biochemistry,* 28, 9398, 1989.

31. **Tate, M. W. and Gruner, S. M.,** Lipid polymorphism of mixtures of dioleoylphosphatidylethanolamine and saturated and monounsaturated phosphatidylcholines of various chain lengths, *Biochemistry,* 26, 231, 1987.

32. **Epand, R. M. and Bottega, R.,** Determination of the phase behaviour of phosphatidylethanolamine admixed with other lipids and the effects of calcium chloride: implications for protein kinase C regulation, *Biochim. Biophys. Acta,* 944, 144, 1988.

33. **Wieslander, Å., Christiansson, A., Rilfors, L., and Lindblom, G.,** Lipid bilayer stability in membranes. Regulation of lipid composition in *Acholeplasma laidlawii* as governed by molecular shape, *Biochemistry,* 19, 3650, 1980.

34. **Epand, R. M., Epand, R. F., and Lancaster, C. R. D.,** Modulation of the bilayer to hexagonal phase transition of phosphatidylethanolamines by acylglycerols, *Biochim. Biophys. Acta,* 945, 161, 1988.

35. **Siegel, D. P., Banschbach, J., Alford, D., Ellens, H., Lis, L. J., Quinn, P. J., Yeagle, P. L., and Bentz, J.,** Physiological levels of diacylglycerols in phospholipid membranes induce membrane fusion and stabilize inverted phases, *Biochemistry,* 28, 3703, 1989.

36. **Das, S. and Rand, R. P.,** Modification by diacylglycerol of the structure and interactions of various phospholipid bilayers, *Biochemistry,* 25, 2882, 1985.

37. **Gallay, J. and de Kruijff, B.,** Correlation between molecular shape and hexagonal H_{II} phase promoting ability of sterols, *FEBS Lett.,* 143, 133, 1982.

38. **Cheetham, J. J., Wachtel, E., Bach, D., and Epand, R. M.,** Role of the stereochemistry of the hydroxyl group of cholesterol and the formation of nonbilayer structures in phosphatidylethanolamine, *Biochemistry,* 28, 8928, 1989.

39. **Tilcock, C. P. S., Hope, M. J., and Cullis, P. R.,** Influence of cholesterol esters of varying unsaturation on the polymorphic phase preferences of egg phosphatidylethanolamine, *Chem. Phys. Lipids,* 35, 363, 1984.

40. **Cheetham, J. J., Epand, R. M., Andrews, M., and Flanagan, T. D.,** Cholesterol sulfate inhibits the fusion of Sendai virus to biological and model membranes, *J. Biol. Chem.,* 265, 12404, 1990.

41. **Marsh, D. and Seddon, J. M.,** Gel-to-inverted hexagonal (L_β-H_{II}) phase transitions in phosphatidylethanolamines and fatty acid-phosphatidylcholine mixtures, demonstrated by ^{31}P-NMR spectroscopy and X-ray diffraction, *Biochim. Biophys. Acta,* 690, 117, 1982.

42. **Epand, R. M., Epand, R. F., Ahmed, N., and Chen, R.,** Promotion of hexagonal phase formation and lipid mixing by fatty acids with varying degree of unsaturation, *Chem. Phys. Lipids,* 57, 75, 1991.

43. **van Duijn, G., Verkleij, A. J., and de Kruijff, B.,** Influence of phospholipid peroxidation on the phase behaviour of phosphatidylcholine and phosphatidylethanolamine in aqueous dispersions, *Biochemistry,* 23, 4969, 1984.

44. **Cullis, P. R. and de Kruijff, B.,** Lipid polymorphism and the functional roles of lipids in biological membranes, *Biochim. Biophys. Acta,* 559, 399, 1979.

45. **Gruner, S. M., Cullis, P. R., Hope, M. J., and Tilcock, C. P. S.,** Lipid polymorphism: the molecular basis of nonbilayer phases, *Annu. Rev. Biophys. Biophys. Chem.,* 14, 211, 1985.

46. **Boggs, J. M.,** Lipid intermolecular hydrogen bonding: influence on structural organization and membrane function, *Biochim. Biophys. Acta,* 906, 353, 1987.

47. **Yeagle, P. L.,** Lipid regulation of cell membrane structure and function, *FASEB J.,* 3, 1833, 1989.

48. **Lindblom, G. and Rilfors, L.,** Cubic phases and isotropic structures formed by membrane lipids: possible biological relevance, *Biochim. Biophys. Acta,* 988, 221, 1989.

49. **Lafleur, M., Bloom, M., and Cullis, P. R.,** Lipid polymorphism and hydrocarbon chain order, *Biochem. Cell. Biol.,* 68, 1, 1990.

50. **Seddon, J. M.,** Structure of the inverted hexagonal (H_{II}) phase, and non-lamellar phase transitions of lipids, *Biochim. Biophys. Acta,* 1031, 1, 1990.

51. **Laggner, P. and Kriechbaum, M.,** Phospholipid phase transitions: kinetics and structural mechanisms, *Chem. Phys. Lipids,* 57, 121, 1991.

52. **Tate, M. W., Eikenberry, E. F., Turner, D. C., Shyamsunder, E., and Gruner, S. M.,** Nonbilayer phases of membrane lipids, *Chem. Phys. Lipids,* 57, 147, 1991.

53. **Tournois, H. and de Kruijff, B.,** Polymorphic phospholipid phase transitions as tools to understand peptide-lipid interactions, *Chem. Phys. Lipids,* 57, 327, 1991.

54. **Lohner, K.,** Effects of small organic molecules on phospholipid phase transitions, *Chem. Phys. Lipids,* 57, 341, 1991.

55. **Verkleij, A. J., van Echteld, C. J. A., Gerritsen, W. J., Cullis, P. R., and de Kruijff, B.,** The lipidic particle as an intermediate structure in membrane fusion processes and bilayer to hexagonal H_{II} phase transitions, *Biochim. Biophys. Acta,* 600, 620, 1980.

56. **Siegel, D. P.,** Inverted micellar intermediates and the transitions between lamellar, cubic, and inverted hexagonal lipid phases. I. Mechanism of the $L_\alpha \leftrightarrow H_{II}$ phase transitions, *Biophys. J.,* 49, 1155, 1986.

57. **Siegel, D. P.,** Inverted micellar intermediates and the transitions between lamellar, cubic, and inverted hexagonal lipid phases. II. Implications for membrane-membrane interactions and membrane fusion, *Biophys. J.,* 49, 1171, 1986.

58. **Siegel, D. P.,** Membrane-membrane interactions via intermediates in lamellar-to-inverted hexagonal phase transitions, in *Cell Fusion,* Sowers, A. E., Ed., Plenum Press, New York, 1987, 181.

59. **Seddon, J. M., Cevc, G., and Marsh, D.,** Calorimetric studies of the gel-fluid (L_β-L_α) and lamellar-inverted hexagonal (L_α-H_{II}) phase transitions in dialkyl- and diacylphosphatidylethanolamines, *Biochemistry,* 22, 1280, 1983.

60. **Caffrey, M.,** Kinetics and mechanism of the lamellar gel/lamellar liquid crystal and lamellar/inverted hexagonal phase transition in phosphatidylethanolamine: a real-time X-ray diffraction study using synchrotron radiation, *Biochemistry,* 24, 4826, 1985.

61. **Caffrey, M., Magin, R. L., Hummel, B., and Zhang, J.,** Kinetics of the lamellar and hexagonal phase transitions in phosphatidylethanolamine. Time-resolved X-ray diffraction study using microwave-induced temperature jump, *Biophys. J.,* 58, 21, 1990.

62. **Tate, M. W., Shyamsunder, E., Gruner, S. M., and D'Amico, K. L.,** Kinetics of the lamellar-inverse hexagonal phase transition determined by time-resolved X-ray diffraction, *Biochemistry,* 31, 1081, 1992.

63. **Siegel, D. P., Green, W. J., and Talmon, Y.,** The mechanism of lamellar-to-inverted hexagonal phase transitions: a study using temperature-jump cryo-electron microscopy, *Biophys. J.,* 66, 402, 1994.

64. **Hardman, P. D.,** Spin-label characterization of the lamellar-to-hexagonal (H_{II}) phase transition in phosphatidylethanolamine aqueous dispersions, *Eur. J. Biochem.,* 124, 95, 1982.

65. **Nieva, J.-L., Castresana, J., and Alonso, A.,** The lamellar to hexagonal phase transition in phosphatidylethanolamine liposomes: a fluorescence anisotropy study, *Biochem. Biophys. Res. Commun.,* 168, 987, 1990.

66. **Epand, R. M. and Leon, B. T.-C.,** Hexagonal phase forming propensity detected in phospholipid bilayers with fluorescent probes, *Biochemistry,* 31, 1550, 1992.

67. **Epand, R. M.,** Detection of hexagonal phase forming propensity in phospholipid bilayers, *Biophys. J.,* 64, 290, 1993.

68. **Mantsch, H. H., Martin, A., and Cameron, D. G.,** Characterization by infrared spectroscopy of the bilayer to nonbilayer phase transition of phosphatidylethanolamines, *Biochemistry,* 20, 31318, 1981.

69. **Thayer, A. M. and Kohler, S. J.,** Phosphorus-31 nuclear magnetic resonance spectra characteristic of hexagonal and isotropic phases generated from phosphatidylethanolamine in the bilayer phase, *Biochemistry,* 20, 6831, 1981.

70. **Gagné, J., Stamatatos, L., Diacovo, T., Hui, S. W., Yeagle, P. L., and Silvius, J. R.,** Physical properties and surface interactions of bilayer membranes containing N-methylated phosphatidylethanolamines, *Biochemistry,* 24, 4400, 1985.

71. **Veiro, J. A., Khalifah, R. G., and Rowe, E. S.,** P-31 nuclear magnetic resonance studies of the appearance of an isotropic component in dielaidoylphosphatidylethanolamine, *Biophys. J.,* 57, 637, 1990.

72. **Israelachvili, J. N., Mitchell, D. J., and Ninham, B. W.,** Theory of self-assembly of hydrocarbon amphiphiles into micelles and bilayers, *J. Chem. Soc. Faraday Trans. 2,* 72, 1525, 1976.

73. **Israelachvili, J. N., Mitchell, D. J., and Ninham, B. W.,** Theory of self-assembly of lipid bilayers and vesicles, *Biochim. Biophys. Acta,* 470, 185, 1977.

74. **Israelachvili, J. N., Marcelja, S., and Horn, R. G.,** Physical principles of membrane organization, *Q. Rev. Biophys.,* 13, 121, 1980.

75. **Xiang, T.,** A computer simulation of free-volume distributions and related structural properties in a model lipid bilayer, *Biophys. J.,* 65, 1108, 1993.

76. **Lehtonen, J. Y. A. and Kinnunen, P. K. J.,** Changes in the lipid dynamics of liposomal membranes induced by poly(ethylene glycol). Free volume alterations revealed by inter- and intramolecular excimer forming phospholipid analogs, *Biophys. J.,* 66, 1981, 1994.

77. **Fenske, D. B., Jarrell, H. C., Guo, Y., and Hui, S. W.,** Effect of unsaturated phosphatidylethanolamine on the chain order profile of bilayers at the onset of the hexagonal phase transition. A ^{2}H NMR study, *Biochemistry,* 29, 11222, 1990.

78. **Kumar, V. V.,** Complementary molecular shapes and additivity of the packing parameter of lipids, *Proc. Natl. Acad. Sci. U.S.A.,* 88, 444, 1991.

79. **Lee, Y.-C., Taraschi, T. F., and Janes, N.,** Support for the shape concept of lipid structure based on a headgroup volume approach, *Biophys. J.,* 65, 1429, 1993.

80. **Hui, S. W. and Sen, A.,** Effects of lipid packing on polymorphic phase behaviour and membrane properties, *Proc. Natl. Acad. Sci. U.S.A.,* 86, 5825, 1989.

81. **Hui, S. W.,** Lipid molecular shape and high curvature structures, *Biophys. J.,* 65, 1361, 1993.

82. **Gruner, S. M.,** Intrinsic curvature hypothesis for biomembrane lipid composition: a role for nonbilayer lipids, *Proc. Natl. Acad. Sci. U.S.A.,* 82, 3665, 1985.

83. **Tate, M. W. and Gruner, S. M.,** Lipid polymorphism of mixtures of dioleoylphosphatidylethanolamine and saturated and monounsaturated phosphatidylcholines of various chain lengths, *Biochemistry,* 26, 231, 1987.

84. **Gruner, S. M.,** Stability of lyotropic phases with curved interfaces, *J. Phys. Chem.,* 93, 7562, 1989.

85. **Yeagle, P. L. and Sen, A.,** Hydration and the lamellar to hexagonal II phase transition of phosphatidylethanolamine, *Biochemistry,* 25, 7518, 1986.

86. **Yeagle, P. L. and Sen, A.,** Hydration and the lamellar to hexagonal II phase transition of phosphatidylethanolamine, *Biochemistry,* 25, 7518, 1986.

87. **Epand, R. M.,** Hydrogen bonding and the thermotropic transitions of phosphatidylethanolamines, *Chem. Phys. Lipids,* 52, 227, 1990.

88. **Webb, M. S., Hui, S. W., and Steponkus, P. L.,** Dehydration-induced lamellar-to-hexagonal-II phase transitions in DOPE/DOPC mixtures, *Biochim. Biophys. Acta,* 1145, 93, 1993.

89. **Gawrisch, K., Parsegian, V. A., Hajduk, D. A., Tate, M. W., Gruner, S. M., Fuller, N. L., and Rand, R. P.,** Energetics of a hexagonal-lamellar-hexagonal phase transition sequence in dioleoylphosphatidylethanolamine membranes, *Biochemistry,* 31, 2856, 1992.

90. **Kirk, G. L., Gruner, S. M., and Stein, D. L.,** A thermodynamic model of the lamellar to inverse hexagonal phase transition of lipid membrane-water systems, *Biochemistry,* 23, 1093, 1984.

91. **Hui, S. W., Stewart, T. P., and Boni, L. T.,** The nature of lipidic particles and their roles in polymorphic transitions, *Chem. Phys. Lipids,* 33, 113, 1983.

92. **Ben-Shaul, A., Szleifer, I., and Gelbart, W. M.,** Statistical thermodynamics of amphiphile chains in micelles, *Proc. Natl. Acad. Sci. U.S.A.,* 81, 4601, 1984.

93. **Hauser, H., Guyer, W., Pascher, I., Skrabal, P., and Sundell, S.,** Polar group conformation of phosphatidylcholine. Effect of solvent and aggregation, *Biochemistry,* 19, 366, 1980.

94. **Hauser, H., Pascher, I., and Sundell, S.,** Preferred conformation and dynamics of the glycerol backbone in phospholipids. An NMR and X-ray single-crystal analysis, *Biochemistry,* 27, 9166, 1988.

95. **Fahey, D. A. and Small, D. M.,** Surface properties of 1,2-dipalmitoyl-3-acyl-*sn*-glycerols, *Biochemistry,* 25, 4468, 1986.

96. **Allen, T. M., Hong, K., and Papahadjopoulos, D.,** Membrane contact, fusion, and hexagonal (H_{II}) transition in phosphatidylethanolamine liposomes, *Biochemistry,* 29, 2976, 1990.

97. **Katsaras, J., Jeffrey, K. R., Yang, D. S.-C., and Epand, R. M.,** Direct evidence for the partial dehydration of phosphatidylethanolamine bilayers on approaching the hexagonal phase, *Biochemistry,* 32, 10700, 1993.

98. **Hui, S. W., Stewart, T. P., Yeagle, P. L., and Albert, A. D.,** Bilayer to non-bilayer transition in mixtures of phosphatidylethanolamine and phosphatidylcholine: implications for membrane properties, *Arch. Biochem. Biophys.,* 207, 227, 1981.

99. **Boni, L. T. and Hui, S. W.,** Polymorphic phase behaviour of dilinoleoylphosphatidylethanolamine and palmitoyloleoylphosphatidylcholine mixtures. Structural changes between hexagonal, cubic and bilayer phases, *Biochim. Biophys. Acta,* 731, 177, 1983.

100. **Verkleij, A. J.,** Lipidic intramembranous particles, *Biochim. Biophys. Acta,* 779, 43, 1984.

101. **Stier, A., Finch, S. A. E., and Bösterling, B.,** Non-lamellar structure in rabbit liver microsomal membranes. A ^{31}P-NMR study, *FEBS Lett.,* 91, 109, 1978.

102. **de Kruijff, B., van den Besselaar, A. M. H. P., Cullis, P. R., van den Bosch, H., and van Deenen, L. L. M.,** Evidence for isotropic motion of phospholipids in liver microsomal membranes. A ^{31}P NMR study, *Biochim. Biophys. Acta,* 514, 1, 1978.

103. **de Kruijff, B., Rietveld, A., and Cullis, P. R.,** ^{31}P NMR studies on membrane phospholipids in microsomes, rat liver slices and intact perfused rat liver, *Biochim. Biophys. Acta,* 600, 343, 1980.

104. **Verkleij, A. J., Mombers, C., Gerritsen, W. J., Leunissen-Bijvelt, L., and Cullis, P. R.,** Fusion of phospholipid vesicles in association with the appearance of lipidic particles as visualized by freeze fracturing, *Biochim. Biophys. Acta,* 555, 358, 1979.

105. **Hoekstra, D. and Martin, O. C.,** Transbilayer redistribution of phosphatidylethanolamine during fusion of phospholipid vesicles. Dependence on fusion rate, lipid phase separation, and formation of nonbilayer structures, *Biochemistry,* 21, 6097, 1982.

106. **Ellens, H., Bentz, J., and Szoka, F. C.,** pH-induced destabilization of phosphatidylethanolamine-containing liposomes: role of bilayer contact, *Biochemistry,* 23, 1532, 1984.

107. **Bentz, J., Ellens, H., Lai, M.-Z., and Szoka, F. C., Jr.,** On the correlation between H_{II} phase and the contact-induced destabilization of phosphatidylethanolamine-containing membranes, *Proc. Natl. Acad. Sci. U.S.A.,* 82, 5742, 1985.

108. **Ellens, H., Bentz, J., and Szoka, F. C.,** Destabilization of phosphatidylethanolamine liposomes at the hexagonal phase transition temperature, *Biochemistry,* 25, 285, 1986.

109. **Ellens, H., Bentz, J., and Szoka, F. C.,** Fusion of phosphatidylethanolamine-containing liposomes and mechanism of L_α-H_{II} phase transition, *Biochemistry,* 25, 4141, 1986.

110. **Ellens, H., Siegel, D. P., Alford, D., Yeagle, P., Boni, L., Lis, L., Quinn, P. J., and Bentz, J.,** Membrane fusion and inverted phases, *Biochemistry,* 28, 3692, 1989.

111. **Walter, A., Yeagle, P. L., and Siegel, D. P.,** Diacylglycerol and hexadecane increase divalent cation-induced lipid mixing rates between phosphatidylserine large unilamellar vesicles, *Biophys. J.,* 66, 366, 1994.

112. **Bearer, E. L., Düzgünes, N., Friend, D. S., and Papahadjopoulos, D.,** Fusion of phospholipid vesicles arrested by quick-freezing. The question of lipidic particles as intermediates in membrane fusion, *Biochim. Biophys. Acta,* 693, 93, 1982.

113. **Papahadjopoulos, D., Nir, S., and Düzgünes, N.,** Molecular mechanisms of calcium-induced membrane fusion, *J. Bioenerg. Biomembr.,* 22, 157, 1990.

114. **Rand, R. P.**, Interacting phospholipid bilayers: measured forces and induced structural changes, *Annu. Rev. Biophys. Bioeng.*, 10, 277, 1981.

115. **Lentz, B. R., Whitt, N. A., Alford, D. R., Burgess, S. W., Yeates, J. C., and Nir, S.**, The kinetic mechanism of cation-catalyzed phosphatidylglycerol transbilayer migration implies close contact between vesicles as an intermediate state, *Biochemistry*, 28, 4575, 1989.

116. **Day, E. P., Kwok, A. Y. W., Hark, S. K., Ho, J. T., Vail, W. J., Bentz, J., and Nir, S.**, Reversibility of sodium-induced aggregation of sonicated phosphatidylserine vesicles, *Proc. Natl. Acad. Sci. U.S.A.*, 77, 4026, 1980.

117. **Wilschut, J., Düzgünes, N., and Papahadjopoulos, D.**, Calcium/magnesium specificity in membrane fusion: kinetics of aggregation and fusion of phosphatidylserine vesicles and the role of bilayer curvature, *Biochemistry*, 20, 3126, 1981.

118. **Wilschut, J., Düzgünes, N., Hong, K., Hoekstra, D., and Papahadjopoulos, D.**, Retention of aqueous contents during divalent cation-induced fusion of phospholipid vesicles, *Biochim. Biophys. Acta*, 734, 309, 1983.

119. **Rand, R. P. and Parsegian, V. A.**, Hydration forces between phospholipid bilayers, *Biochim. Biophys. Acta*, 988, 351, 1989.

120. **Jones, J. D. and Thompson, T. E.**, Spontaneous phosphatidylcholine transfer by collision between vesicles at high lipid concentration, *Biochemistry*, 28, 129, 1989.

121. **Jones, J. D. and Thompson, T. E.**, Mechanism of spontaneous, concentration-dependent phospholipid transfer between bilayers, *Biochemistry*, 29, 1593, 1990.

122. **Wimley, W. C. and Thompson, T. E.**, Phosphatidylethanolamine enhances the concentration-dependent exchange of phospholipids between bilayers, *Biochemistry*, 30, 4200, 1991.

123. **Zaks, W. J. and Creutz, C. E.**, Evaluation of the annexins as potential mediators of membrane fusion in exocytosis, *J. Bioenerg. Biomembr.*, 22, 97, 1990.

124. **Roos, D. S. and Choppin, P. W.**, Biochemical studies on cell fusion. II. Control of fusion response by lipid alteration, *J. Cell Biol.*, 101, 1591, 1985.

125. **Rytömaa, M., Mustonen, P., and Kinnunen, P. K. J.**, Reversible, non-ionic, and pH-dependent association of cytochrome c with cardiolipin-phosphatidylcholine liposomes, *J. Biol. Chem..*, 267, 22243, 1992.

126. **Rytömaa, M. and Kinnunen, P. K. J.**, Evidence for two distinct acidic phospholipid-binding sites in cytochrome c, *J. Biol. Chem.*, 269, 1770, 1994.

127. **Newton, A. C.**, Interaction of proteins with lipid headgroups: lessons from protein kinase C, *Annu. Rev. Biophys. Biomol. Struct.*, 22, 1, 1993.

128. **Jamil, H., Hatch, G. M., and Vance, D. E.**, Evidence that binding of CTP:phosphocholine cytidyltransferase to membranes is modulated by the ratio of bilayer- to nonbilayer-forming lipids, *Biochem. J.*, 291, 419, 1993.

129. **Kinnunen, P. K. J., Koiv, A., Rytömaa, M., Lehtonen, J. Y. A., and Mustonen, P.**, Lipid dynamics and peripheral interactions of proteins with membrane surfaces, *Chem. Phys. Lipids*, 73, 181, 1994.

130. **Epand, R. M.**, The relationship between the effects of drugs on bilayer stability and on protein kinase C activity, *Chem. Biol. Interactions*, 63, 239, 1987.

131. **Epand, R. M.**, Relationship of phospholipid hexagonal phases to biological phenomena, *Biochem. Cell Biol.*, 68, 17, 1990.

132. **Sinensky, M.**, Homeoviscous adaptation: a homeostatic process that regulates the viscosity of membrane lipids in *Escherichia coli*, *Proc. Natl. Acad. Sci. U.S.A.*, 71, 522, 1974.

133. **Monck, M. A., Bloom, M., Lafleur, M., Lewis, R. N. A. H., McElhaney, R. N., and Cullis, P. R.**, Influence of lipid composition on the orientational order in *Acholeplasma laidlawii* strain B membranes: a deuterium NMR study, *Biochemistry*, 31, 10037, 1992.

134. **Moore, D. J. and Mendelsohn, R.**, Adaptation to altered growth temperatures in *Acholeplasma laidlawii* B: Fourier transform infrared studies of acyl chain conformational order in live cells, *Biochemistry*, 33, 4080, 1994.

135. **Zaritsky, A., Parola, A. H., Abdah, M., and Masalha, H.**, Homeoviscous adaptation, growth rate and morphogenesis in bacteria, *Biophys. J.*, 48, 337, 1985.

136. **Parola, A. H., Ibdah, M., Gill, D., and Zaritsky, A.**, Deviation from homeoviscous adaptation in *Escherichia coli* membranes, *Biophys. J.*, 57, 621, 1990.

137. **Rilfors, L., Lindblom, G., Wieslander, Å., and Christiansson, A.**, Lipid bilayer stability in biological membrane, *Biomembranes*, 12, 205, 1884.

138. **Christiansson, A., Eriksson, L. E. G., Westman, J., Demel, R., and Wieslander, Å.**, Involvement of surface potential in regulation of polar membrane lipids in *Acholeplasma laidlawii*, *J. Biol. Chem.*, 260, 3984, 1985.

139. **Wieslander, Å., Rilfors, L., and Lindblom, G.**, Metabolic changes of membrane lipid composition in *Acholeplasma laidlawii* by hydrocarbons, alcohols, and detergents: arguments for effects on lipid packing, *Biochemistry*, 25, 7511, 1986.

140. **Lindblom, G., Hauksson, J. B., Rilfors, L., Bergenståhl, B., Wieslander, Å., and Eriksson, P.-O.**, Membrane lipid regulation in *Acholeplasma laidlawii* grown with saturated fatty acids, *J. Biol. Chem.*, 268, 16198, 1993.

141. **Elhaney, R. N.**, The influence of membrane lipid composition and physical properties of membrane structure and function in *Acholeplasma laidlawii*, *Crit. Rev. Microbiol.*, 17, 1, 1989.

142. **De Chavigny, A., Heacock, P. N., and Dowhan, W.**, Sequence and inactivation of the *pss* gene of *Escherichia coli*. Phosphatidylethanolamine may not be essential for cell viability, *J. Biol. Chem.*, 266, 5323, 1991.

143. **Rietveld, A. G., Killian, J. A., Dowhan, W., and de Kruijff, B.**, Polymorphic regulation of membrane phospholipid composition in *Escherichia coli*, *J. Biol. Chem.*, 268, 12427, 1993.

144. **Killian, J. A., Koorengevel, M. C., Bouwstra, J. A., Gooris, G., Dowhan, W., and de Kruijff, B.**, Effect of divalent cations on lipid organization of cardiolipin isolated from *Escherichia coli* strain AH930, *Biochim. Biophys. Acta*, 1189, 225, 1994.

145. **Goldfine, H.**, Bacterial membranes and lipid packing theory, *J. Lipid Res.*, 25, 1501, 1984.

146. **Lee, M.-H. and Bell, R. M.**, Phospholipid functional groups involved in protein kinase C activation, phorbol ester binding, and binding to mixed micelles, *J. Biol. Chem.*, 264, 14797, 1989.

147. **Chow, M., Yao, A., and Rubin, H.**, Cellular epigenetics: topochronology of progressive "spontaneous" transformation of cells under growth constraint, *Proc. Natl. Acad. Sci. U.S.A.*, 91, 599, 1994.

148. **Mountford, C. E., Wright, L. C., Holmes, K. T., Mackinnon, W. B., Gregory, P., and Fox, R. M.**, High-resolution proton nuclear magnetic resonance analysis of metastatic cancer cells, *Science*, 226, 1415, 1984.

149. **Mountford, C. E. and Wright, L. C.**, Organization of lipids in the plasma membranes of malignant and stimulated cells: a new model, *Trends Biochem. Sci.*, 13, 172, 1988.

150. **Wolfman, A. and Macara, I. G.**, Elevated levels of diacylglycerol and decreased phorbol ester sensitivity in *ras*-transformed fibroblasts, *Nature*, 325, 359, 1987.

151. **Dingley, A. J., King, N. J. C., and King, G. F.**, An NMR investigation of the changes in plasma membrane triglyceride and phospholipid precursors during the activation of T-lymphocytes, *Biochemistry*, 31, 9098, 1992.

152. **Black, P. H.**, Shedding from the surface of normal and cancer cells, *Adv. Cancer Res.*, 32, 75, 1980.

Chapter 7

Computational Molecular Models of Lipid Bilayers Containing Mixed-Chain Saturated and Monounsaturated Acyl Chains

Ching-hsien Huang and Shusen Li

CONTENTS

I. INTRODUCTION

Liposomes, or aqueous dispersions of enclosed multilamellar lipid vesicles, are well known to be the result of the remarkable self-assembly capabilities exhibited by amphipathic lipid molecules in excess water. Many important properties of liposomes, such as the permeabilities of ions and small molecules, including water molecules, often mimic closely those of biological membranes, in which amphipathic lipids are a major component.[1-3] Consequently, the aqueous dispersion of lipid vesicles prepared from various amphipathic lipids such as diacylphospholipids and glycolipids has become a particularly attractive and challenging model system for biological membranes. The systematic and in-depth study of liposomes prepared from various phospholipids is relevant and important because the results obtained may provide insights into the structure-property relationships of these amphipathic lipids in cell membranes.

In the literature, there is a large body of experimental data describing the physical and transport properties of liposomes or lipid vesicles prepared from various phospholipids. However, most of them have not been correlated with the molecular structures of the underlying lipids. In fact, diacylphospholipids, in excess water, may exhibit polymorphic phase behavior upon heating.[4,5] The detailed molecular structures of most phospholipids in the various phases, particularly the liquid-crystalline state, are not known at an accurate atomic level in terms of atomic coordinates such as torsion angles. Although the crystal structures of more than 30 amphipathic lipids have been reported,[6] the repertoire of amphipathic lipids derived from biological membranes is so bewildering that, at the present time, the crystal structures of only a very tiny fraction of membrane lipids are known. For instance, 1-palmitoyl-2-oleoyl-phosphatidylcholine is the most abundant unsaturated membrane lipid found in animal cells; information on the crystal structure of this lipid is completely lacking. Nevertheless, the crystal structures of those lipids that are already known from X-ray diffraction can be used as important references for molecular modeling calculations.

Recently, the rapid increase in computing power of desktop and laboratory class computers together with the decrease in the costs of computers and software packages has enabled many biochemistry laboratories to use computer-based molecular modeling as a research technique. This technique offers the possibility of simulating the unknown structure of phospholipids based on information about the crystal structure of model lipid molecules.[7-9] In this communication, we describe how the structures of mixed-chain saturated phospholipids and monoenoic phospholipids can be simulated using the crystal structure data of dimyristoyl phosphatidylcholine and the oleate chain of cholesteryl oleate. Once the monomeric structure is simulated, higher orders of aggregated structures are successively built up, leading eventually to a large three-dimensional system which resembles the structure of the multilamellar lipid vesicles. In addition, the structures of various kinked hydrocarbon chains are simulated, since the water transport across the lipid bilayer and the phospholipid-cholesterol interaction in the bilayer have been suggested to depend on the kinked structure of the lipid's acyl chain.[10,11]

II. BASIC EQUIPMENT, SOFTWARE, AND COMPUTATION

Computer graphics were drawn with the software package HyperChem™ (Autodesk, Inc., Sausalito, CA) performed on a 486 platform.

The software MM2 (Version 85) and MM3 (Version 92), originally developed by Allinger and co-workers[12,13] and supplied by Quantum Chemistry Program Exchange (Department of Chemistry, Indiana University, Bloomington), were run on an IBM RS/6000 computer workstation to perform the molecular mechanics (MM) calculations for molecular modeling.

Allinger's MM calculations were performed *in vacuo,* including various potential energy functions. The sum of these functions is the steric energy (E_s). For the MM2 program, $E_s = E_{st} + E_b + E_{tor} + E_{dip} + E_{VDW} + E_{sb}$, where E_{st} is the stretching energy of bonds, E_b is the energy associated with the angle bending, E_{tor} is the torsional energy of bonds, E_{dip} is the energy of bond-dipole interactions, E_{VDW} is the energy of van der Waals interactions, and E_{sb} is the energy associated with the coupling between bond stretching and bond bending. Some of these terms, such as E_{st}, E_b, and E_{VDW}, are modified in the MM3 program;[13] moreover, in the MM3 program additional terms such as torsion-stretching and bending-bending interactions are included in the expression of E_s used in the molecular mechanics force field. Consequently, the value of E_s calculated from the MM3 program is in general greater than that from the MM2 program. The structure parameters and some thermodynamic properties obtained with the MM3 program are, in general, improved considerably over those calculated from the MM2 program.[13]

Of the various potential energy functions included in the steric energy, the van der Waals term, E_{VDW}, plays the most important role in determining the structure and steric energy of the diglyceride moiety of lipid molecules. The van der Waals interaction energy for the nonbonded $C(sp^3)$-$C(sp^3)$ pair calculated by MM2 and MM3 programs, respectively, can be used as an example, which is illustrated in Figure 1. The plot shows that using the MM2 program the optimal van der Waals attractive interaction (E_{VDW} = –0.0514 kcal/mol) between the nonbonded C-C pair occurs at the separation distance of 3.80 Å, corresponding to the sum of the van der Waals radii. Using the MM3 program, however, the sum of van der Waals radii between two nonbonded carbons is 4.08 Å and the minimal energy is –0.0302 kcal/mol. The separation distances corresponding to the zero van der Waals interaction are 3.34 Å (calculated by the MM2 program) and 3.58 Å (calculated by the MM3 program), as shown in Figure 1.

Using three-dimensional periodic boundary conditions, the simulated energy-minimized structure of a tetrameric lipid assembly packed in a rectangular box surrounded by 26 replicas of image boxes and its steric energy can be obtained. These computations were performed on a 486 platform using the MM⁺ program included in the software package HyperChem™.

In this communication, various phospholipids lacking headgroups are used exclusively for our MM calculations. Hence, the headgroup-headgroup interaction and the interaction of the headgroup with the diglyceride moiety of the lipid molecule are not considered in our MM calculations. Thus, the results obtained are assumed for lipids with the same headgroup, and only relative comparisons between lipids with the same headgroup should be made.

III. STRUCTURE CONSTRUCTION OF THE LIPID BILAYER COMPRISED OF IDENTICAL-CHAIN C(14):C(14)PC

The global structure of the diglyceride moiety of C(14):C(14)PC in the lipid bilayer at temperatures below the crystalline-to-gel ($L_c \rightarrow L'_\beta$) phase is assumed to adopt the "h"-shaped geometry. In this state, the lipid

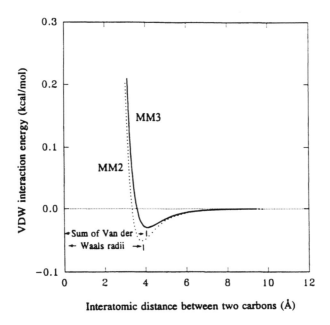

Figure 1 The van der Waals (VDW) interaction energy of a pair of nonbonded C(sp³)-C(sp³) atoms as a function of the interatomic distance. The curve was calculated using Allinger's MM2 (dotted line) or MM3 program (solid line).

Table 1 Structure Parameters of the Diglyceride Moiety of Phosphatidylcholines

C(X):C(Y)PC	E_s	θ_1	θ_3	β_1	β_2	β_3	β_4	β_5	γ_1	δ_2	δ_1	Δ	δ_1'	δ_2'
C(14):C(14)PC (A)	23.60	58	–178	82	172	–81	45	171	–177					
C(14):C(14)PC (B)	19.83	168	166	120	179	–134	67	180	102					
C(14):C(14)PC (O)	81.35	180	180	120	180	–60	–60	180	180					
C(14):C(14)PC (M)	19.53	178	180	104	176	–62	–56	–178	–178					
C(18):C(10)PC	22.79	178	180	105	177	–62	–56	–178	–178					
Oleate chain, crystal	—	—	—	—	—	—	—	—	—	70	123	1	128	174
C(16):C(18:1Δ⁹)PC	22.78	–178	–179	103	177	–62	–57	–177	178	66	134	0	154	177

Note: The fatty acids ester-linked at the *sn*-1 and *sn*-2 positions are presented as C(X) and C(Y) in the shorthand notation of phosphatidylcholine or C(X):C(Y)PC; Δ^9 denotes the unsaturated fatty acid with a *cis* double bond at the position of carbon number 9 starting from the carboxyl end. E_s denotes the steric energy in kcal/mol. θ_n, β_n, γ_n, δ_n, Δ, and δ_n' are all torsion angles in degrees.[8] Four molecular species of C(14):C(14)PC are specified: (A) and (B) refer to the crystal structures A and B;[6] (O) and (M) refer to the crude and energy-minimized structures discussed in the text.

molecule in the bilayer is further assumed to be highly immobilized, and the lipid's chain-chain interaction is maximal. Based on these basic assumptions, together with the fact that the ester bond has a partial double-bond character owing to the resonance hybrid effect,[11] the torsion angles of the various bonds in the diglyceride moiety of C(14):C(14)PC can be estimated,[8] as shown in Table 1. The convention for naming the various torsion angles as well as the crude skeletal structure of the diglyceride moiety of C(14):C(14)PC are illustrated in Figure 2A. A computer-graphics representation of the same moiety using the space-filling model is shown in Figure 2B.

The estimated atomic coordinates (e.g., torsion angles) for C(14):C(14)PC, given in Table 1, are used as the input data in constructing the refined structure of monomeric C(14):C(14)PC by MM2 force field calculations. The refinement involves basically the computer search for the optimal structure of C(14):C(14)PC based on the minimal molecular energy. This is accomplished by the systematic adjustment of the structural parameters, such as the bond length, the bond angle, and the van der Waals distance between two nonbonded atoms, for all atoms in the diglyceride moiety of C(14):C(14)PC by repeated automatic cycles of the Newton-Raphson minimization technique. Specifically, after each cycle, the sum

176

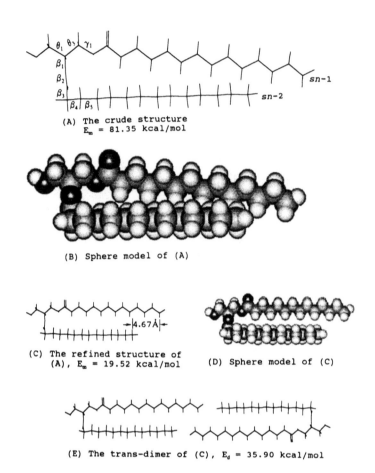

(A) The crude structure
E_m = 81.35 kcal/mol

(B) Sphere model of (A)

(C) The refined structure of (A), E_m = 19.52 kcal/mol

(D) Sphere model of (C)

(E) The trans-dimer of (C), E_d = 35.90 kcal/mol

Figure 2 Simulated molecular structures of the diglyceride moiety of C(14):C(14)PC. (A) The crude skeletal structure of C(14):C(14)PC. This is constructed based on the estimated torsion angles, shown in Table 1, derived from the basic assumptions discussed in the text. The convention for naming the various torsion angles (β_n, γ_n, and θ_n) is also indicated. (B) The same crude structure presented in the form of a space-filling model. (C) The refined structure of A. The molecule has been subjected to energy minimization using the MM2 program. (D) Same as C but presented in the form of the space-filling model. (E) The transbilayer dimer of C(14):C(14)PC. The nonbonded van der Waals contact distance separating the two terminal methyl groups along the long molecular axis is 4.16 Å. However, a nonbonded distance of 3.79 Å is observed between the two terminal methyl groups of the *sn*-1 acyl chains in the transbilayer dimer. The nonbonded interatomic distances between C(2), C(4), C(6), C(8), or C(10) of the *sn*-1 acyl chain and C(6), C(8), C(10), C(12), or C(14) of *sn*-2 acyl chain within each molecule are 3.93, 3.94, 3.94, 3.93, and 3.85 Å, respectively. This energy-minimized transbilayer dimer has a steric energy of 35.90 kcal/mol obtained with the MM2 program.

of the various potential energy functions used in the molecular mechanics force field (e.g., bond lengths, bond angles, etc.) is calculated as the steric energy for the lipid molecule with a set of structural parameters. These cycles of self-adjusted computation, however, come to a halt as the steric energy reaches the minimal value. The structure of the diglyceride moiety of C(14):C(14)PC with the minimal steric energy simulated by the MM2 program is considered to be the equilibrium structure or the energy-minimized structure in this communication. Figures 2C and D illustrate the skeletal and space-filling representations, respectively, of the energy minimized C(14):C(14)PC molecule. In comparing Figures 2A and C, it is evident that the global conformation of the diglyceride moiety of C(14):C(14)PC, after energy minimization, is roughly similar to that of the crude one, having a characteristic "h"-shaped geometry. However, the long chain axes of the two acyl chains, shown in Figure 2D, are nearly parallel. In contrast, the same two chains shown in Figure 2B are nonparallel; obviously, the chain-chain separation distance in the crude lipid structure is sterically improbable. This comparison serves to demonstrate that, via the coupled rotations of the torsion angles in the lipid molecule, the overall net result of the energy minimization by force field computation is to relieve appreciably the molecular overcrowding or torsional strains in some parts of the lipid molecule and, simultaneously and cooperatively, to increase optimally the nonbonded van der Waals favorable interactions elsewhere in the same lipid molecule. The steric energy of the resulting structure (E_m = 19.52 kcal/mol) is thus considerably smaller than that (E_m = 81.35 kcal/mol) of the initial structure. Values of some of the torsion angles for the various parts of the C(14):C(14)PC molecule, before and after the energy minimization, are summarized in Table 1.

Two structural features for the energy-minimized C(14):C(14)PC molecule are worthy of special note. (1) The two long chain axes of the sn-1 and sn-2 acyl chains run nearly parallel with each other, whereas the zigzag planes of the two acyl chains are oriented perpendicular to each other. This latter feature of mutually perpendicular orientation has been detected experimentally.[14] If the zigzag plane of the sn-1 acyl chain is placed in the plane of the paper and the direction of the sn-1 acyl chain is chosen to run from the left (carboxyl end) to the right (methyl end), then the zigzag plane of the sn-2 acyl chain is oriented perpendicularly to the plane of the paper, extending slightly out of the paper plane toward the viewer. The spatial separation distances between C(1), C(5), and C(9) of the sn-1 acyl chain and the zigzag plane of the sn-2 acyl chain can be estimated from the MM2 computation to be 4.46, 4.66, and 4.69 Å, respectively. (2) The sn-1 acyl chain is ester-linked to glycerol backbone carbon one, C_1. The linked segment and the sn-1 acyl chain are fully extended in the crystalline state; hence, the effective chain length of the sn-1 acyl chain extending from glycerol backbone carbon two (C_2) is longer than the actual sn-1 acyl chain length by a fully extended segment of C_2-C_1-O-C(1), where C(1) is carbon one or the carbonyl carbon of the sn-1 acyl chain. The effective chain length of the sn-2 acyl chain, however, is shortened by 1.2 Å due to the sharp bend at C(2), as shown in Figure 2. The difference in the effective chain length between the sn-1 and sn-2 acyl chains is 4.67 Å (Figure 2C). This effective chain length difference, abbreviated as ΔC, corresponds to the separation distance along the long molecular axis between the two terminal methyl groups in the sn-1 and sn-2 acyl chains. This ΔC value of 4.67 Å or 3.68 C-C bond lengths is for C(14):C(14)PC packed in the crystalline state. For C(14):C(14)PC packed in the gel-state bilayer, the ΔC value is known experimentally to be 1.8 Å or 1.5 C-C bond lengths,[15] suggesting that the sn-1 acyl chain in the gel state is not fully extended.

Once the energy-minimized structure of monomeric C(14):C(14)PC is constructed, the next step is to build a dimeric model following a well-defined packing motif. For C(14):C(14)PC, a transbilayer dimer with a partially interdigitated packing motif is simulated as shown in Figure 2E. In this packing motif, the long sn-1 acyl chain and the short sn-2 acyl chain of one energy-minimized C(14):C(14)PC molecule are aligned in a common plane with the sn-2 and sn-1 acyl chains, respectively, of another energy-minimized C(14):C(14)PC molecule. This packing motif mimics the overall packing geometry of two lipid molecules residing on the two opposing leaflets of the lipid bilayer in the crystalline or gel phase.

The constructed transbilayer dimer is then subjected to energy minimization using Allinger's MM2 program. The refined structure bears a striking resemblance to the original structure, shown in Figure 2E, with a steric energy (E_d^P) of 35.90 kcal/mol. Here the superscript P in the steric energy term denotes the partially interdigitated packing motif and the subscript d refers to two lipid molecules packed in a transbilayer dimer. The stabilization energy of the transbilayer dimer contributed by its constituent monomer, ΔE_d^P, can be calculated from the relation $\Delta E_d^P = (E_d^P - 2E_m)/2 = -1.57$ kcal/mol. This value by itself is not of any great significance, however, except when it is used on a relative basis. This point will become clear in a later paragraph.

The transbilayer dimer discussed above cannot be regarded as a simple bilayer model, since lateral lipid-lipid interaction is totally lacking. Consequently, higher levels of lipid assembly must be considered. The next higher level of lipid organization is the tetramer in which the transbilayer dimer is an assembly unit. There are many ways in which two transbilayer dimers may interact, giving rise to a wide variety of overall geometry of the tetramers. Here we are considering only two types of tetrameric architecture: the front-to-back (F-B) tetramer and the up-and-down (U-D) tetramer. Figure 3A graphically depicts the geometric relationship between two transbilayer dimers in a F-B tetramer. In this packing motif, two sn-1 and two sn-2 acyl chains in each dimeric unit form a matching face. As one dimeric unit is superimposed over the other, the matching faces are in direct contact, thus allowing van der Waals interactions to occur between the two contiguous faces. The U-D tetramer is illustrated in Figure 3B. In this arrangement, two energy-minimized dimers, lying on a common plane, are aligned side by side. In this packing motif, the dimer-dimer contact surface is limited to the lateral chain-chain interaction.

The F-B and U-D tetramers illustrated in Figure 3 can be subjected to energy minimization using Allinger's MM2 program. Each of the resulting tetramers is placed in a rectangular box (45.62 × 29.48 × 56.10 Å) using the MM+ program of HyperChem™. Three-dimensional periodic boundaries are then

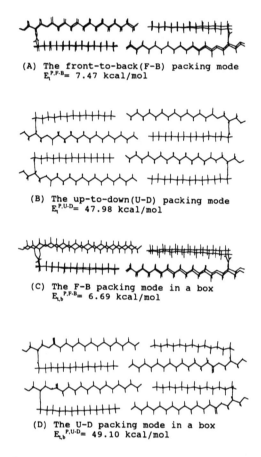

(A) The front-to-back(F-B) packing mode
$E_t^{P,F-B} = 7.47$ kcal/mol

(B) The up-to-down(U-D) packing mode
$E_t^{P,U-D} = 47.98$ kcal/mol

(C) The F-B packing mode in a box
$E_{t,b}^{P,F-B} = 6.69$ kcal/mol

(D) The U-D packing mode in a box
$E_{t,b}^{P,U-D} = 49.10$ kcal/mol

Figure 3 Packing motifs of tetrameric C(14):C(14)PC. (A) Two transbilayer dimers are superimposed in the eclipsed position to form the front-to-back (F-B) packing motif, minimized by the MM2 program. (B) Two transbilayer dimers, lying in the same plane, are aligned side-by-side to form the up-and-down (U-D) packing motif, refined with the MM2 program. (C) The refined structure of A after subjecting it to three-dimensional periodic boundary conditions using the MM+ program. (D) The refined structure of B after subjecting it to three-dimensional periodic boundary conditions using the MM+ program.

employed to simulate an assembly of $3^3 - 1$ boxes surrounding each of the original boxed tetramers using the software package of HyperChem™. Consequently, each tetramer is periodically replicated in all directions; this is analogous to the periodic repeat of a unit cell observed in single crystals. This large three-dimensional system with a total of 27 boxes is subsequently energy-minimized. The refined structures of the F-B and U-D tetramers for C(14):C(14)PC packed in the central box surrounded by 26 identical image boxes are presented in Figures 3C and D, respectively. The averaged sum of these two types of multiple tetramers is taken to mimic, to a first approximation, the orthorhombic closed-packed structures of phospholipids assembled in the lipid bilayer in the crystalline phase.

The steric energies of the monomer ($E_{m,b}$) and the F-B ($E_{t,b}^{P,F-B}$) and U-D ($E_{t,b}^{P,U-D}$) tetramers for C(14):C(14)PC packed in the central box of 27 boxes, as calculated by the MM+ program of HyperChem™ are presented in Table 2. The stabilization energies of the F-B and U-D tetramers contributed by the monomer, $\Delta E_{t,b}^{P,F-B} = (E_{t,b}^{P,F-B} - 4E_{m,b})/4$ and $\Delta E_{t,b}^{P,U-D} = (E_{t,b}^{P,U-D} - 4E_{m,b})/4$, give rise to a value of -13.02 kcal/mol as the overall averaged stabilized energy of the tetramer contributed by its constituent monomer $[\Delta E_{ave,b}^{P} = \frac{1}{2}(\Delta E_{t,b}^{P,F-B} + \Delta E_{t,b}^{P,U-D})]$, which is also given in Table 2. It should be emphasized that this value (-13.02) by itself should not be taken as the true absolute value of the overall averaged stabilized energy for multiple C(14):C(14)PC tetramers packed in the partially interdigitated bilayer, since it depends on the molecular mechanics program (MM+, MM2, or MM3) used. However, the calculated value can be used for comparative purposes on a relative basis, provided that the same program of force field computation is applied to a series of lipids with the same molecular weight and the same packing motif or to the same lipid species self-assembled in different packing motifs. For instance, the overall averaged

Table 2 Stabilization Energy of Various Diglyceride Moieties of Phosphatidylcholines

C(X):C(Y)PC	$E_{m,b}$	$E_{t,b}^{P,F-B}$	$E_{t,b}^{P,U-D}$	$E_{t,b}^{M,F-B}$	$E_{t,b}^{M,U-D}$	$-\Delta E_{ave,b}^{P}$	$-\Delta E_{ave,b}^{M}$
C(14):C(14)PC	19.99	6.69	49.10	0.26	66.34	13.02	11.66
C(18):C(10)PC	23.24	6.61	45.09	−7.06	48.27	16.78	18.09
C(16):C(18:1Δ^9)PC	23.24	17.17	56.97	24.24	87.95	13.97	9.22

Note: The energy terms used in various columns are expressed in kcal/mol. The superscripts P, M, F-B, and U-D denote the partially interdigitated, the mixed interdigitated, the front-to-back, and the up-and-down motifs, respectively. The subscripts m, t, and b denote the monomer, tetramer, and boxed tetramer, respectively; $\Delta E_{ave,b} = 1/2(\Delta E_{t,b}^{F-B} + \Delta E_{t,b}^{U-D})$.

stabilization energy for tetrameric C(14):C(14)PC packed in the mixed interdigitated motif, $\Delta E_{ave,b}^{M}$, is calculated by the MM$^+$ program to be −11.66 kcal/mol (Table 2). The mixed interdigitated packing motif will be discussed in the next section. On a relative basis, the overall average stabilization energy is thus more negative for tetrameric C(14):C(14)PC packed in the partially interdigitated assembly, implying that the partially interdigitated bilayer of C(14):C(14)PC is more stable at equilibrium than the mixed interdigitated bilayer comprised of the same lipid species of C(14):C(14)PC. Our computational results thus suggest that C(14):C(14)PC tends to self-assemble preferably into the partially interdigitated bilayer at $T < T_m$, where T_m is the gel-to-liquid-crystalline phase transition temperature.

IV. STRUCTURE CONSTRUCTION OF THE LIPID BILAYER COMPRISED OF MIXED-CHAIN C(18):C(10)PC

The energy-minimized structure of identical-chain C(14):C(14)PC, obtained by molecular modeling calculations as illustrated in Figures 2C and D, has provided a framework on which to initiate the construction of energy-minimized structures for other mixed-chain phosphatidylcholines.[8] In this section, we shall consider the construction of C(18):C(10)PC structure as a monomer and higher levels of aggregates. Mixed-chain C(18):C(10)PC is a highly asymmetric lipid species in which the effective chain length of the *sn*-1 acyl chain is about twice that of the *sn*-2 acyl chain. Consequently, the packing geometry of C(18):C(10)PC molecules in the gel-state bilayer is markedly different from that of identical-chain C(14):C(14)PC, being assembled in a mixed interdigitated type of packing motif.[16-18]

The initial crude structure of monomeric C(18):C(10)PC can be readily constructed as follows. Four methylene units, each unit being linked by the *trans* C-C bond, are added first to the *sn*-1 acyl chain of the energy-minimized structure of C(14):C(14)PC. Then four methylene units are deleted from the *sn*-2 acyl chain of the same molecule. The resulting structure of C(18):C(10)PC, shown in Figure 4A, is subsequently subjected to energy minimization using Allinger's MM2 program. The refined C(18):C(10)PC structure in the monomeric form is illustrated in Figure 4B. Some of the torsion angles of the C-C bonds exhibited by the refined structure are given in Table 1. Overall, the global geometries of the two structures, Figures 4A and B, are virtually identical. The steric energies, however, are distinctly different, as indicated in Figures 4A and B.

The immediately higher level of structure is the sum of two monomers arranged in a defined transbilayer architecture. The structure of the transbilayer dimer of C(18):C(10)PC is derived initially by arranging the two monomers according to the mixed interdigitated packing motif, and the model is then refined by the MM approach. In this packing motif, the long *sn*-1 octadecanoyl chain spans the bilayer hydrocarbon core with its terminal methyl group positioned at the hydrocarbon/water interface, and the short *sn*-2 decanoyl chain packs end-to-end with a second *sn*-2 decanoyl chain from another C(18):C(10)PC molecule located in the opposing bilayer leaflet. This transbilayer dimer, after energy minimization, is illustrated in Figure 4C. Using Allinger's MM2 program, the steric energies for the transbilayer dimer (E_d^M) and the monomer (E_m) are calculated to be 29.05 and 22.79 kcal/mol, respectively, as indicated in Figure 4.

In general, the intermolecular interaction between two monomers within a dimer is intimately related to the stability of the dimer. If the geometry of the dimer is such that extensive attractive interactions can occur between the two constituent monomers, then this dimer can be expected to be stable at equilibrium. This general concept can be applied to lipid systems. Hence, a fundamental aspect of intermolecular interaction between two lipid molecules within the isolated transbilayer dimer is the affinity of the two

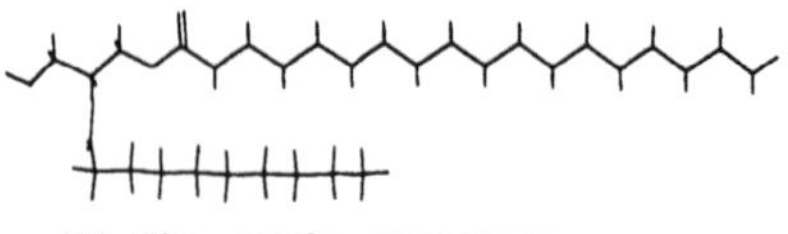

(A) The crude structure
E_m = 24.54 kcal/mol

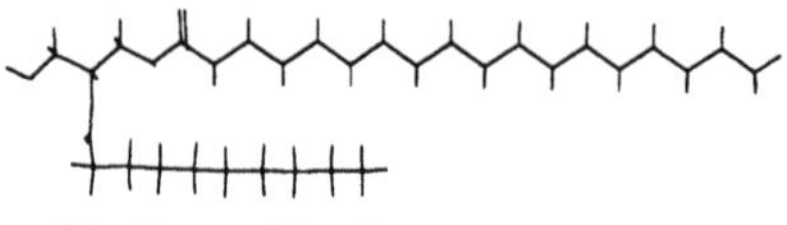

(B) The refined structure of (A)
E_m = 22.79 kcal/mol

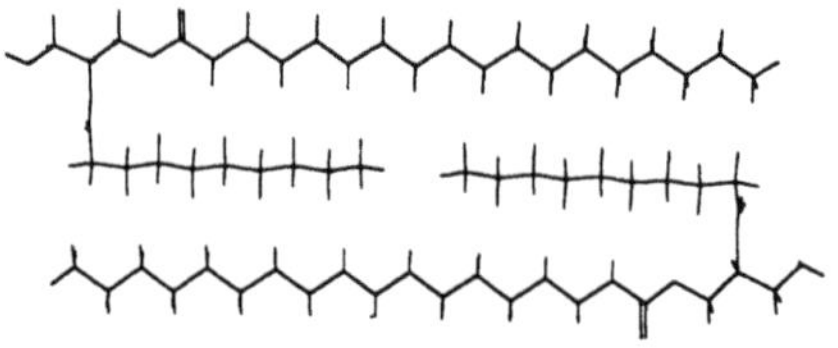

(C) The mixed trans-dimer of (B)
E_d = 29.05 kcal/mol

Figure 4 Simulated molecular structures of the diglyceride moiety of C(18):C(10)PC. (A) The crude structure originated from the energy-minimized structure of C(14):C(14)PC. (B) The energy-minimized structure of A. (C) A refined transbilayer dimer packed according to the mixed interdigitated motif.

for each other, which is a measure of the stabilization energy of the transbilayer dimer contributed by its constituent monomer. The magnitude of the stabilization energy contributed by each monomer for dimeric C(18):C(10)PC packed according to the mixed interdigitated motif can be calculated from the relation $\Delta E_d^M = (E_d^M - 2E_m)/2 = -8.27$ kcal/mol. If the intramolecular interaction within the monomer of C(18):C(10)PC is taken to be the same as that of C(14):C(14)PC, then the value of ΔE_d^M has to be modified slightly by substituting the E_m value of C(14):C(14)PC for that of C(18):C(10)PC in the relation just given. The resulting value of the modified ΔE_d^M is –5.00 kcal/mol. This value of –5.00 kcal/mol is more negative than the corresponding value of –1.57 kcal/mol calculated for dimeric C(14):C(14)PC with a partially interdigitated packing motif (Figure 2E). This 3.2-fold difference in stabilization energy indicates that the chain-chain contact interactions between two highly asymmetric C(18):C(10)PC molecules within the isolated mixed interdigitated dimer are significantly greater than those between two identical-chain C(14):C(14)PC molecules within the isolated partially interdigitated dimer. The rationalization can be understood by comparing the MM refined structural models for the two types of transbilayer dimer. In the case of a mixed interdigitated dimer, each of the two sn-2 acyl chains that extend from C(2) to the terminal C(10) at the methyl end is shown in Figure 4C to be in closest van der Waals contact with the two neighboring sn-1 acyl chains. In contrast, the sn-2 acyl chain of C(14):C(14)PC in the partially interdigitated dimer can make the similar closest van der Waals contact with only one neighboring sn-1 acyl chain (Figure 2E). The architecture of the mixed interdigitated C(18):C(10)PC dimer produces a more stable aggregated structure with a larger negative value of stabilization energy because of the extensive interchain interactions between the two constituent monomers.

The above-described energy-minimized structure of dimeric C(18):C(10)PC provides a building block for the construction of C(18):C(10)PC participating in higher order structures in which lateral lipid-lipid interactions in the opposing two leaflets are included. Figure 5A shows the refined structure of the juxtaposition of two such transbilayer dimers called the F-B tetramer. This refined F-B tetramer can be placed in a box and then subjected to periodic boundaries, leading to a large assembly of 27 identical boxes arranged in the three-dimensional space. The energy-minimized structure of tetrameric

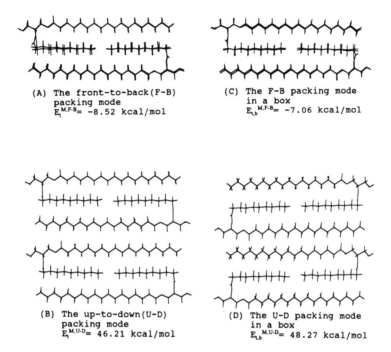

(A) The front-to-back(F-B) packing mode
$E_t^{M,F-B}= -8.52$ kcal/mol

(C) The F-B packing mode in a box
$E_{t,b}^{M,F-B}= -7.06$ kcal/mol

(B) The up-to-down(U-D) packing mode
$E_t^{M,U-D}= 46.21$ kcal/mol

(D) The U-D packing mode in a box
$E_{t,b}^{M,U-D}= 48.27$ kcal/mol

Figure 5 Packing motifs of tetrameric C(18):C(10)PC. (A) Refined structure of the F-B tetramer. (B) Refined structure of the U-D tetramer. (C) The refined structure of A after subjecting it to the three-dimensional periodic boundary conditions. (D) The refined structure of B after subjecting it to the three-dimensional periodic boundary conditions.

C(18):C(10)PC with the mixed interdigitated packing motif assembled in the central box of 27 identical boxes is illustrated in Figure 5C. Moreover, the refined structure of the U-D tetramer, in which two transbilayer dimers are aligned side-by-side, and the structure of the U-D tetramer packed in the central box of 27 boxes are also simulated by molecular mechanics calculations. These structures are presented in Figures 5B and D. The stabilization energies of these structures are given in Table 2. For comparisons, the transbilayer dimer of C(18):C(10)PC packed according to the partially interdigitated motif and the tetrameric C(18):C(10)PC constructed on the basis of the partially interdigitated motif and assembled in the central box of a large system of 27 boxes have also been modeled; their steric and stabilization energies are summarized in Table 2.

It is interesting to examine the values of the overall averaged stabilization energy for boxed tetramers of C(18):C(10)PC with the partially interdigitated and the mixed interdigitated packing motifs, $\Delta E_{ave,b}^P$ and $\Delta E_{ave,b}^M$, as shown in Table 2. Results from MM calculations clearly indicate that the magnitude of the negative overall averaged stabilization energy for the mixed interdigitated tetramer ($-\Delta E_{ave,b}^M$) is larger, suggesting that highly asymmetric C(18):C(10)PC molecules tend to self-assemble preferably into the mixed interdigitated bilayer over the partially interdigitated bilayer at $T < T_m$. The opposite situation, however, is observed for boxed C(14):C(14)PC with which the $\Delta E_{ave,b}^P$ is more negative (Table 2). Consequently, it is energetically more favorable for C(14):C(14)PC to pack into a partially interdigitated type of motif at $T < T_m$. These calculated results are indeed consistent with the mixing behavior of binary lipid mixtures composed of C(18):C(10)PC and C(14):C(14)PC.[19]

V. STRUCTURE CONSTRUCTION OF THE LIPID BILAYER COMPRISED OF C(16):C(18:1Δ⁹)PC

Thus far we have been concerned with the molecular modeling of saturated phosphatidylcholines as monomers and higher orders of structures by molecular mechanics approaches. Since biological membranes contain primarily mixed-chain phospholipids in which the *sn*-2 acyl chain is usually an unsaturated hydrocarbon chain with various numbers and positions of the *cis* double bond (Δ), the construction of refined structural models for an unsaturated phosphatidylcholine packed in various aggregated forms is thus biologically relevant. 1-Palmitoyl-2-oleoyl-phosphatidylcholine (POPC), C(16):C(18:1Δ⁹)PC, is the most abundant phospholipid found in animal cell membranes. In this section, we choose POPC as an example of biological lipids for the construction of the bilayer model by the computer graphics-supported MM calculations. The X-ray single-crystal structure of POPC has yet to be established; however, the torsion angles of cholesteryl oleate in single crystals are available.[20] The structure and steric energy of POPC in the crystalline state can thus be initially modeled using the single-crystal data of cholesteryl oleate and the atomic coordinates of the refined structure of C(14):C(14)PC. The crude structure can then be improved by force field refinement using the MM2 program.[9] It should be stated that the strategy involved in constructing the equilibrium structure for POPC by computer modeling calculations described in this section may be easily extended to other homologous phospholipids with monoeroic acyl chains.

Prior to the presentation of the simulated structure of POPC, it is pertinent to mention that as a *cis* carbon-carbon double bond (Δ) is introduced into a long all-*trans* hydrocarbon chain, the two carbon-carbon single bonds adjacent to the Δ-bond become highly flexible. The relationships between the steric energy and the torsion angles for rotations about C(6)-C(7) in *n*-dodecane and C(5)-C(6) in *cis*-dodecene-6, obtained with the MM3 program, are presented in Figures 6A and B, respectively. The steric energy difference between the *trans* (torsion angle = ±180°) and *gauche* (torsion angle = ±65°) conformations in *n*-dodecane is 0.89 kcal/mol, and the energy barrier for transformation of the *trans* into the *gauche* conformation by rotation is 3.36 kcal/mol (Figure 6A). In contrast, the steric energy diagram for rotation about a single C-C bond adjacent to the Δ-bond in *cis*-dodecene-6 is very different, having two very broad wells centered at torsion angles of ±110° (Figure 6B). Moreover, the torsional barrier between the two wells is only 1.94 kcal/mol. Because of the broad wells, the C-C bond next to the Δ-bond can readily adopt a wide variety of torsion angles. The hydrocarbon chain with a Δ-bond can thus be thought of as existing in a wide range of conformations, and each has an energy that is nearly the same as the other.[9]

For monounsaturated mixed-chain phospholipids, a great variation in the conformation of the unsaturated *sn*-2 acyl chain with very similar low energy can be expected to exist at $T < T_m$. Taking POPC as an example, it has been proposed that 16 low-energy conformations may exist normally at $T < T_m$, although these conformations contain some redundancy.[9] Unfortunately, these low-energy conformations have not been confirmed due to the complete lack of X-ray diffraction data on unsaturated phospholipids. Nevertheless, the structural data of an oleate chain obtained at 123 K from single crystals of cholesteryl oleate are available.[20] These crystal data indicate that the oleate chain is nearly linear; moreover, it is packed with antiparallel orientation next to an adjacent oleate chain from another cholesteryl oleate. Hence, the spatial packing arrangement is relevant to the acyl chain packing motif of unsaturated diacylphospholipids in the lipid bilayer. The crystal structure, however, represents only a unique low-energy conformation of the oleate chain at low temperature. The refined structural model for POPC, to be described in the following paragraph, is computationally constructed on the basis of the single-crystal structure of cholesteryl oleate; consequently, it should be regarded as one of many possible low-energy conformations that are likely to interconvert rapidly at $T < T_m$.

The molecular conformation of the oleate chain established by X-ray crystallographic data of cholesteryl oleate is used as the starting geometry for the *sn*-2 acyl chain of POPC. Specifically, the torsion angles of the *sn*-2 acyl chain of POPC starting from C(3) to the methyl terminus of the chain are taken from the crystal data (Table 1), and all other torsion angles of the lipid molecule are those, shown in Table 1, exhibited by the energy-minimized structure of C(14):C(14)PC illustrated in Figure 2C, except that the *sn*-1 acyl chain is extended by two methylene units in the *trans* form. The assembled initial structure of POPC as represented by the skeletal drawing is given in Figure 7A. Here the chain axis of the lower segment of the *sn*-2 acyl chain is seen to lie in between the *sn*-1 acyl chain and the upper segment of the *sn*-2 acyl chain. Moreover, the zigzag plane of the lower segment of the *sn*-2 acyl chain is virtually perpendicular to the zigzag plane of the *sn*-1 acyl chain. The nonbonded interatomic distance between the olefinic C(10) atom in the *sn*-2 acyl chain and the C(6) atom of the *sn*-1 acyl chain amounts to 1.96 Å, indicating that the initially constructed crude structure is sterically improbable. The assembled structure is improved drastically after force field refinement, as graphically depicted in Figure 7B. In this refined structure, the two acyl chains are in closest van der Waals contact. For instance, the nonbonded interatomic distance between the olefinic C(10) atom of the *sn*-2 acyl chain and C(6) of the *sn*-1 acyl chain has now increased from 1.96 to 3.62 Å, a favorable van der Waals interaction distance. Some of the torsion angles of C-C bonds in the refined *sn*-2 acyl chain are summarized in Table 1. In this table, the torsion angles of the C-C single bonds surrounding the Δ-bond are designated as δ_n and δ'_n, denoting the

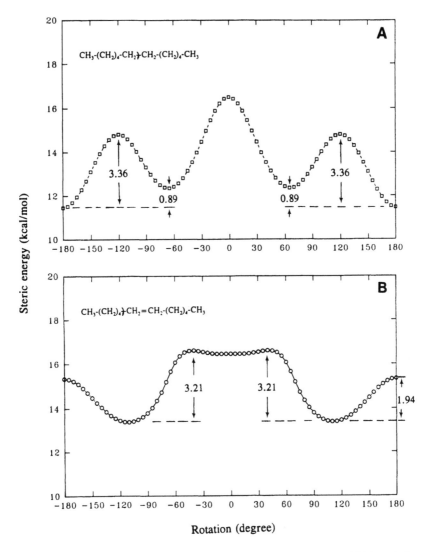

Figure 6 Potential energy as a function of torsion angles for rotations about C(6)-C(7) in *n*-dodecane (A) and C(5)-C(6) in *cis*-dodecene-6 (B). Both profiles were obtained with the MM3 program.

*n*th carbon-carbon bond preceding and succeeding the Δ-bond, respectively. Most interestingly, the energy-minimized structure established by using atomic coordinates of cholesteryl oleate as shown in Figure 7B, has a kink motif around the Δ-bond that is essentially identical to one of the 16 motifs proposed for POPC based on molecular mechanics calculations.[9]

The transbilayer dimer and the two boxed tetrameric POPC, the F-B and U-D tetramers, each in a large system of 27 boxes, are presented in Figures 7C, D, and E. In these higher orders of structure, the transbilayer is packed according to the partially interdigitated motif; this is supported by the larger negative value of the overall averaged stabilization energy calculated for this motif (Table 2). In addition, the global conformation of POPC in these higher orders of structure is seen to be basically similar to that of monomeric POPC shown in Figure 7B, indicating that the most favorable conformational state of POPC in bilayers at low temperatures can be approximated by the energy-minimized structure of monomeric POPC, which has the following structural characteristics. (1) The *sn*-1 acyl chain is aligned in a straight *trans* conformation. (2) The torsion angles of the *sn*-2 acyl chain starting from C(3) are all *trans* except for δ_2, δ_1, Δ, and δ_1'. This sequence $g^+S^+\Delta S^+$, with the torsion angles δ_2, δ_1, Δ, and δ_1' produces a kink (called the $g^+S^+\Delta S^+$ or type IIIb kink[9]) near the middle of the *sn*-2 acyl chain. Here S refers to the *skew* (120 ± 30°) bond. Because of the $g^+S^+\Delta S^+$ kink and the sharp bend at the C(2) position, the total

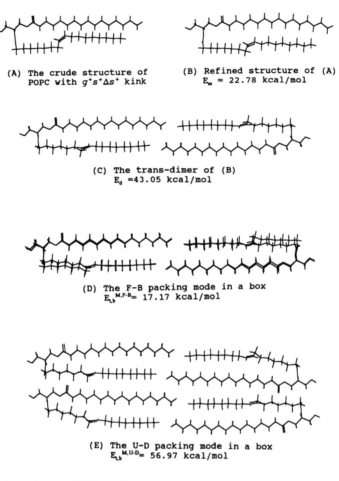

(A) The crude structure of POPC with $g^+s^+\Delta s^+$ kink

(B) Refined structure of (A) $E_m = 22.78$ kcal/mol

(C) The trans-dimer of (B) $E_d = 43.05$ kcal/mol

(D) The F-B packing mode in a box $E_{t,b}^{M,F\text{-}B} = 17.17$ kcal/mol

(E) The U-D packing mode in a box $E_{t,b}^{M,U\text{-}D} = 56.97$ kcal/mol

Figure 7 Simulated structures of POPC. (A) The assembled structure of POPC in which the *sn*-2 acyl chain starting from C(3) has the same conformations as the oleate chain of crystal cholesteryl oleate established by X-ray crystallography and the rest of the molecule has the same conformation as that of C(14):C(14)PC shown in Figure 2C. (B) The refined structure of A using the MM2 program. (C) The transbilayer dimer. (D) The F-B tetramer after subjecting it to three-dimensional periodic boundary conditions. (E) The U-D tetramer after subjecting it to three-dimensional periodic boundary conditions.

numbers of *trans* C-C bonds in the upper and lower segments of the oleate chain in POPC are 4 and 7, respectively. (3) The axes of the *sn*-1 acyl chain and the upper and lower segments of *sn*-2 acyl chains are coplanar; furthermore, they are oriented parallel to each other within the molecule, enabling the favorable van der Waals interactions to occur intramolecularly. Energetically, this parallel packing also favors the lateral interactions between two transbilayer dimers in the tetramer.

VI. COUPLED ISOMERIZATIONS OF CARBON-CARBON BONDS IN LIPID ACYL CHAINS — KINKS AND JOGS

The equilibrium molecular structures of monomeric C(14):C(14)PC, C(18):C(10)PC, and C(16):C(18:1Δ⁹)PC constructed by molecular mechanics calculations, as discussed earlier, all possess a fully extended *sn*-1 acyl chain. In addition, for saturated phosphatidylcholines, the *sn*-2 acyl chain also has an all-*trans* acyl chain beyond C(3). Even for unsaturated C(16):C(18:1Δ⁹)PC, the *sn*-2 acyl chain has two linear segments separated by the Δ-bond. Each segment has a defined length of consecutive C-C bonds in *trans* form. These various structures are simulated based on X-ray single-crystal data; hence, they correspond to the monomeric structures packed in the lipid bilayer in the crystalline state. In this state, it is expected that the long polymethylene chain of each lipid species adopts the minimum potential

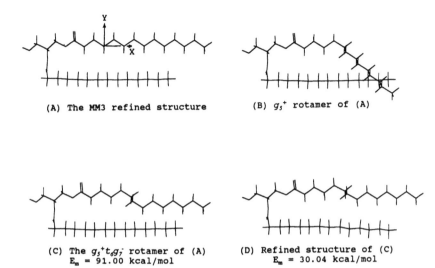

(A) The MM3 refined structure

(B) g_5^+ rotamer of (A)

(C) The $g_5^+ t_6 g_7^-$ rotamer of (A)
$E_m = 91.00$ kcal/mol

(D) Refined structure of (C)
$E_m = 30.04$ kcal/mol

Figure 8 C(14):C(14)PC with single and coupled *trans → gauche* rotational isomerizations. (A) Refined structure of C(14):C(14)PC with an all-*trans* sn-1 acyl chain. The refinement was carried out with the MM3 program. (B) The sn-1 acyl chain has an isolated *gauche*(+) bond located between C(5) and C(6) atoms. (C) A typical 2g1 kink is introduced into the sn-1 acyl chain between the C(5) and C(8) atoms. (D) The energy-minimized structure of C.

energy conformation which corresponds to the fully extended all-*trans* configuration. Upon heating, however, there is an increase in rotation about the C-C bonds in the long hydrocarbon chain, resulting in the formation of *gauche* rotamers. There are two *gauche* conformations, namely a *gauche*(+) with a torsion angle of 65° and a *gauche*(–) with a torsion angle of –65°. Energetically, *gauche* conformations are 0.89 kcal/mol higher than the *trans* conformation and the energy barrier between *trans* and *gauche* rotamers is 3.36 kcal/mol for rotation about the C(6)-C(7) bond in *n*-dodecane (Figure 6A).

Figure 8A illustrates the energy-minimized structure of C(14):C(14)PC in which the sn-1 acyl chain and the sn-2 acyl chain starting from C(3) are in the *trans* form. The zigzag plane of the sn-1 acyl chain is defined as lying on the X-Y plane. The origin of the Cartesian coordinate system is placed at C(4). The positive direction of the X-axis runs from C(4) toward the methyl terminus of the chain, and the positive Y-axis of the coordinate system runs away from the sn-2 acyl chain, as indicated in Figure 8A. Now, as a single *gauche*(+) bond is introduced at the C(5)-C(6) bond in the sn-1 acyl chain, the chain is observed to make a pronounced bend of 120° at C(6), as shown in Figure 8B. By viewing along the chain axis, the lower linear segment starting from C(7) is seen to bend downward and extends away from the viewer along the negative Z-direction. It is evident that the lower part of the bent segment and the last four C-C bonds of the sn-2 acyl chain are in closest proximity — closer than the sum of their van der Waals radii, leading to repulsion between them. The bent conformation caused by the single *gauche* rotation is thus improbable. However, by elevating the temperature, the conformational disordering process of *trans →* *gauche* isomerizations can be coupled. The occurrence of coupled isomerizations will not cause the lipid chain to bend sharply. Many types of coupled isomerizations are known to exist.[21-23] Two of them, kinks and jogs, have been widely speculated to occur progressively in the lipid bilayer comprised of phospholipids with saturated chains as the temperature is heated gradually above the crystalline state toward the liquid-crystalline state. In the liquid-crystalline state, the kinks and jogs are probably formed most frequently.

A. THE 2g1 KINKS
Of the various coupled isomerizations, the sequential *gauche*(±)-*trans-gauche*(∓) bonds, known as the 2g1 kinks, have been best characterized by Pechhold and Blasenbrey.[22,23] The occurrence of a 2g1 kink in a lipid acyl chain converts the fully extended chain into two segments. These two segments are colinear with each other; they run perpendicular to the bilayer surface. Geometrically, the hydrocarbon chain is kinked in the shape of a crankshaft, as shown in Figure 8C. In this example, a g^+ bond between C(5) and C(6) is followed by a t bond between C(6) and C(7), which is then sequentially followed by a g^- bond between C(7) and C(8) in the sn-1 acyl chain of the energy-minimized structure of C(14):C(14)PC.

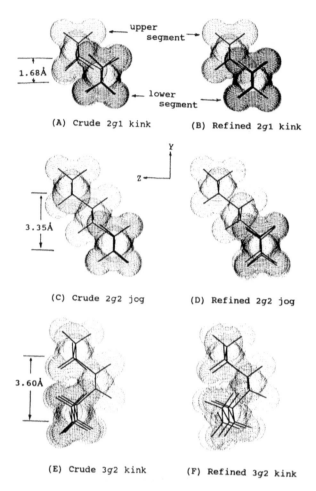

Figure 9 The computer graphics-generated views of the kinked *sn*-1 acyl chains in C(14):C(14)PC as seen from the chain methyl terminus projected on the Y-Z plane. (A) The 2*g*1 kink. (B) The refined 2*g*1 kink. (C) The 2*g*2 jog. (D) The refined 2*g*2 jog. (E) The 3*g*2 kink. (F) The refined 3*g*2 kink. The displacement of the lower segment along the negative Y-axis is indicated in A, C, and E.

Clearly, the lower linear segment of the *sn*-1 acyl chain is shifted downward on the X-Y plane. Actually, the lower linear segment is diagonally shifted; it is related, on the Y-Z plane, to the upper linear segment by a 1.68 Å translation along the negative Y-axis (Figure 9A) and a 1.29 Å translation along the negative Z-axis. In the graphics display shown in Figure 8C, the geometric arrangements of C(8), C(10), and C(12) methylene units of the *sn*-1 acyl chain are in slightly repulsive contact distances with the methylene units in the C(10)–C(14) segment of the *sn*-2 acyl chain. After energy minimization, however, the altered torsion angles of the sequence $tg_7^+ \ tg_9^- \ t$ (175.5°, 65.1°, −179.6°, −71.5°, 178.6°) in the *sn*-1 acyl chain appear to be exquisitely tailored to bring the two acyl chains into the optimal van der Waals contact (Figure 8D). Moreover, the steric energy is decreased significantly from 91.0 kcal/mol (Figure 8C) to 30.04 kcal/mol (Figure 8D) using Allinger's MM3 program.

The above-discussed (*sn*-1)$g_5^+ \ tg_7^-$ isomer serves as an example to illustrate the structural characteristics of a 2*g*1 kink in the lipid acyl chain. The overall effect of a 2*g*1 kink is to shorten the acyl chain by 1.27 Å and to shift the chain diagonally. However, the general orientation of the chain is conserved, being perpendicular to the bilayer surface. The steric energies (E$_s$) of various kinks along the *sn*-1 and *sn*-2 acyl chains have been calculated using Allinger's MM2 and MM3 programs. In general, the values of E$_s$ are relatively constant in each chain. The results are given in Table 3. Rotamers with 2*g*1 kinks in the *sn*-2 acyl chain always have higher values of E$_s$. The last four kinks in the *sn*-1 acyl chain have E$_s$ values which are smaller than the averaged E$_s$ value according to MM3 calculations, indicating that the 2*g*1 kink is

Table 3 The Steric Energies of C(14):C(14)PC with Various Kinks Calculated by MM2 and MM3 Programs

Kink position	Along the *sn*-1 acyl chain				Along the *sn*-2 acyl chain			
	E_s^{MM2}	ΔE_s^{MM2}	E_s^{MM3}	ΔE_s^{MM3}	E_s^{MM2}	ΔE_s^{MM2}	E_s^{MM3}	ΔE_s^{MM3}
$g_{\overline{10}}^{} t_{11} g_{12}^{+}$	21.85	2.32	29.96	2.04	23.35	3.82	30.85	2.93
$g_{10}^{+} t_{11} g_{\overline{12}}^{}$	21.86	2.33	29.94	2.02	22.87	3.34	30.55	2.63
$g_{\overline{9}}^{} t_{10} g_{11}^{+}$	21.74	2.21	30.01	2.09	22.62	3.09	30.56	2.64
$g_{9}^{+} t_{10} g_{\overline{11}}^{}$	21.46	1.93	29.90	1.98	22.74	3.21	30.57	2.65
$g_{\overline{8}}^{} t_{9} g_{10}^{+}$	22.41	2.88	30.33	2.41	23.28	3.85	30.93	3.01
$g_{8}^{+} t_{9} g_{\overline{10}}^{}$	22.35	2.82	30.25	2.33	22.96	3.43	30.76	2.86
$g_{\overline{7}}^{} t_{8} g_{9}^{+}$	22.07	2.54	30.24	2.32	22.79	3.26	30.67	2.75
$g_{7}^{+} t_{8} g_{\overline{9}}^{}$	21.89	2.36	30.05	2.13	22.68	3.15	30.59	2.67
$g_{\overline{6}}^{} t_{7} g_{8}^{+}$	22.10	2.57	30.34	2.42	23.25	3.72	31.00	3.08
$g_{6}^{+} t_{7} g_{\overline{8}}^{}$	21.94	2.41	30.19	2.21	23.24	3.71	30.94	3.02
$g_{\overline{5}}^{} t_{6} g_{7}^{+}$	22.13	2.60	30.32	2.39	22.96	3.43	30.86	2.93
$g_{5}^{+} t_{6} g_{\overline{7}}^{}$	21.90	2.37	30.04	2.12	22.57	3.04	30.76	2.84
$g_{\overline{4}}^{} t_{5} g_{6}^{+}$	22.24	2.71	30.45	2.53	24.03	4.50	31.12	3.21
$g_{4}^{+} t_{5} g_{\overline{6}}^{}$	22.06	2.53	30.18	2.26	23.75	4.22	31.27	3.35
$g_{\overline{3}}^{} t_{4} g_{5}^{+}$	22.08	2.55	30.42	2.50	22.08	2.55	29.95	2.03
$g_{3}^{+} t_{4} g_{\overline{5}}^{}$	22.81	2.28	30.06	2.14	22.83	3.30	31.53	3.61
$g_{\overline{2}}^{} t_{3} g_{4}^{+}$	21.13	2.60	30.59	2.67	—	—	—	—
$g_{2}^{+} t_{3} g_{\overline{4}}^{}$	22.83	2.30	30.24	2.32	—	—	—	—
$g_{\overline{1}}^{} t_{2} g_{3}^{+}$	21.49	1.96	30.89	2.97	—	—	—	—
$g_{1}^{+} t_{2} g_{\overline{3}}^{}$	21.45	1.92	30.48	2.56	—	—	—	—
Average	21.94	2.41	30.24	2.32	23.01	3.48	30.81	2.89

Note: Energy unit is kcal/mol; ΔE is the energy difference between the C(14):C(14)PC with a kink and the same molecule with an all-*trans* conformation.

formed more readily at the methyl end of the *sn*-1 acyl chain. How relevant are the relative E_s values observed for the various $2g1$ kinks in the *sn*-1 and *sn*-2 acyl chains to the effective chain length difference (ΔC) between the *sn*-1 and *sn*-2 acyl chain in C(14):C(14)PC? Earlier we mentioned that the effective chain length difference (ΔC) between two acyl chains in C(14):C(14)PC is decreased from 3.68 to 1.55 C-C bond lengths as the bilayer is transformed from the crystalline to the gel phase. This indicates that the *sn*-1 acyl chain is less fully extended in the gel state. This, of course, is consistent with the calculated results indicating that the $2g1$ kink is more readily formed in the *sn*-1 acyl chain.

Because of the diagonal displacement of the chain segment caused by the $2g1$ kink, a special property of the $2g1$ kink is its ability to form a hydrophobic pocket. Small molecules, including water molecules, thus may be accommodated in a snug fit by the hydrophobic pocket within the closely packed hydrocarbon interior of a bilayer. Interestingly, the migration of the sequences $g^+ t g^-$ and $g^- t g^+$ along a saturated lipid acyl chain has been proposed to act as a mobile site in the bilayer interior to carry a water molecule across the lipid bilayer.[10] However, it should be noted that the odd carbon and the even carbon in a lipid acyl chain occupy notably different geometric positions on the X-Y plane. As shown in Figure 8, the lower linear segment of the *sn*-1 acyl chain displays a downward shift on the X-Y plane, but it also bends away from the viewer, if a $g_5^+ t g_7^-$ kink is introduced into the *sn*-1 acyl chain. In contrast, the lower linear segment of the *sn*-1 acyl chain bends upward on the X-Y plane, and it also extends toward the viewer, if a $g_5^+ t g_7^-$ kink is formed. In other words, the diagonal displacement observed in the lower segment of the *sn*-1 acyl chain caused by the $g_{odd}^+ t g_{odd}^-$ kink is shifted in the opposite direction from that exhibited by the $g_{even}^+ t g_{even}^-$ kink. However, if the $g_8^- t g_{10}^+$ kink is introduced into the *sn*-1 acyl chain, the lower segment of the *sn*-1 acyl chain will display the same diagonal shift as the one exhibited by the $g_5^+ t g_7^-$ kink. For a kink to migrate along a saturated acyl chain with a fixed width of the hydrophobic pocket, the kink must, therefore, alternate repeatedly between the $g_{odd}^+ t g_{odd}^-$ (or $g_{even}^+ t g_{even}^-$) and the $g_{even}^- t g_{even}^+$ (or $g_{odd}^- t g_{odd}^+$) sequences along the acyl chain, as indicated in Figure 10. These sequential changes in the kink position along the lipid acyl chain provide structural insight into the coupled rotational isomerizations that may be involved in the process of water molecule transport in the bilayer membrane.

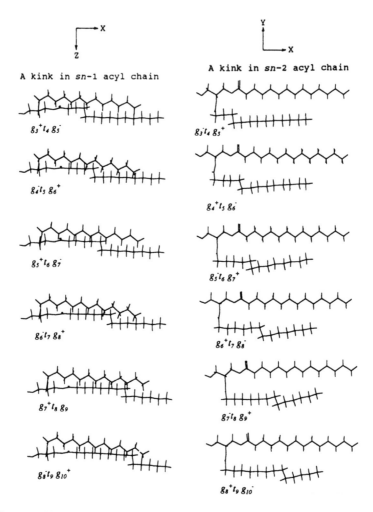

Figure 10 Different positions of a 2*g*1 kink along the *sn*-1 (left) and *sn*-2 (right) acyl chains in a C(14):C(14)PC molecule. These two series of computer graphics-generated structures illustrate two possible ways that a hydrophobic pocket created by the 2*g*1 kink can move across the bilayer interior.

B. THE 2*g*2 JOGS

A second class of coupled *trans* → *gauche* isomerizations which result in a diagonal displacement of the acyl chain but can keep the displaced chain segment perpendicular to the bilayer surface are 2*g*2 jogs. This is accomplished by the occurrence of a consecutive sequence of $g^{\pm}tttg^{\mp}$ moieties in the hydrocarbon chain. Here, a *gauche* bond and its mirror-image *gauche* bond are separated by three consecutive *trans* bonds. The introduction of a 2*g*2 jog into an all-*trans* hydrocarbon chain shortens the chain by about two C-C bond lengths along the chain axis, or 2.54 Å. It should be stated that the nomenclature of the kinks and jogs for saturated hydrocarbon chains follows the ngm rule convention, where *g* is the *gauche* bond, n corresponds to the number of *gauche* bonds, and m is the number of C-C bond lengths shortened by the coupled isomerizations.

Figure 11A shows a $g_5^{+} tttg_7^{-}$ jog in the *sn*-1 acyl chain of C(14):C(14)PC. The g^{+} (torsion angle = +65°) and g^{-} (torsion angle = –65°) bonds are introduced into the C(5)-C(6) and C(9)-C(10) positions. The *sn*-1 acyl chain is seen to be comprised of two parallel linear segments: an upper segment consisting of C(1)–C(6) methylene units and a lower segment consisting of C(9)–C(14) methylene units. Relative to the geometric position of the upper segment, the lower segment has shifted diagonally toward the negative Y-axis by 3.35 Å and toward the negative Z-axis by 2.53 Å. This is most clearly demonstrated by viewing end-on the *sn*-1 acyl chain from the chain methyl terminus up the chain axis as illustrated in Figure 9C. The diagonal shift with ΔY and ΔZ values of –3.35 and –2.53 Å, respectively, is changed somewhat after

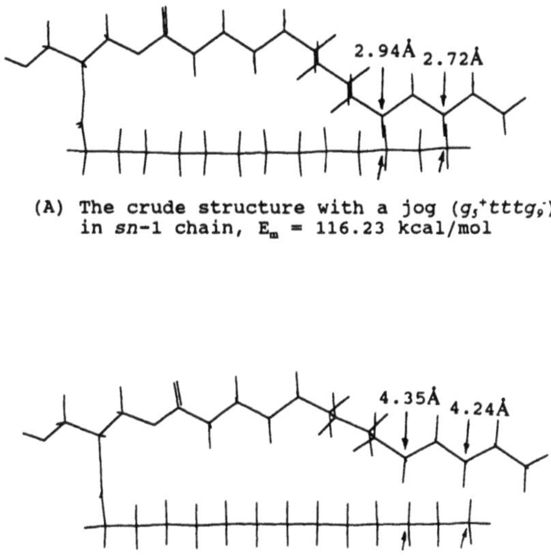

(A) The crude structure with a jog ($g_5^+tttg_9^-$)
in sn-1 chain, E_m = 116.23 kcal/mol

(B) The refined structure of (A)
E_m = 29.93 kcal/mol

Figure 11 A 2g2 jog in the sn-1 acyl chain of C(14):C(14)PC before (A) and after (B) the energy minimization using the MM3 program. The g^+ and g^- of the sequence g^+ttttg^- are introduced at the C(5)-C(6) and C(9)-C(10) bonds, respectively.

energy minimization (ΔY = −3.47 Å, ΔZ = −2.47 Å). The most striking feature of this diagonally shifted sn-1 acyl chain is the change in the orientation of the contact surface with the sn-2 acyl chain. On the X-Y plane (Figures 11A and B), the upper segment is directly at the top of the zigzag surface of the sn-2 acyl chain. In contrast, the lower segment packs against the back side of the sn-2 acyl chain and forms complementary van der Waals interactions with a different face of the sn-2 acyl chain. Prior to force field refinement, the nonbonded interatomic distance between C(10) and C(12) in the lower segment of the sn-1 acyl chain and C(12) and C(14) in the sn-2 acyl chain are 2.94 and 2.72 Å, respectively (Figure 11A), indicating repulsive van der Waals interactions between the two acyl chains near the terminal ends of the chains. After energy minimization by Allinger's MM3 program, however, the nonbonded interatomic distances between the same two pairs of carbon atoms are increased from 2.94 and 2.72 Å to 4.35 and 4.24 Å, respectively, indicating that there are attractive van der Waals interactions between them (Figure 11B). Accompanying this energy minimization, the torsion angles of the sequence $g_5^+ tttg_9^-$ are changed to 64.7°, 176.5°, 177.8°, −168.7°, and −59.7° and the steric energy (E_s) is decreased significantly from 116.23 to 29.93 kcal/mol. It is interesting to note that the steric energy of the refined C(14):C(14)PC with an $(sn-1)g_5^+ tttg_9^-$ jog, 29.93 kcal/mol, is virtually identical to the E_s value, 30.04 kcal/mol, calculated for refined C(14):C(14)PC with an $(sn-1)g_5^+ tg_7^-$ kink.

After energy minimizations, the various steric energies of C(14):C(14)PC molecules containing a 2g2 jog at different positions along the sn-1 as well as the sn-2 acyl chain are presented in Table 4. The average E_s values obtained with the MM3 program for C(14):C(14)PC molecules with a 2g2 jog in the sn-1 and the sn-2 acyl chains are 30.83 and 31.85 kcal/mol, respectively, indicating that the formation of a jog is, on the average, slightly more favorable in the sn-1 acyl chain. Also, the E_s values of the last four jogs near the sn-1 methyl terminus are smaller than the average E_s value for all jogs, suggesting a higher probability of having jogs near the methyl terminus. The same phenomenon is also observed for the last four 2g1 kinks in the sn-1 acyl chain (Table 3). Taken together, these data reflect that the coupled trans → gauche isomerizations are, in general, more likely to occur near the methyl terminus.

C. THE 3g2 KINKS

Next, we shall proceed to discuss briefly the effect of the sequence $g^\pm tg^\pm tg^\pm$, or the 3g2 kinks, on the structure of a saturated acyl chain. First, let us consider an isolated long hydrocarbon chain with its all-trans zigzag plane lying on the paper (or X-Y) plane. The occurrence of a sequence $g^+tg^+tg^+$ near the center of the chain causes the chain to kink in the shape of a crankshaft. The resulting two segments run

Table 4 The Steric Energies of C(14):C(14)PC with Various $2g2$ Jogs Calculated by MM2 and MM3 Programs

Jog position	Along the sn-1 acyl chain				Along the sn-2 acyl chain			
	E_s^{MM2}	ΔE_s^{MM2}	E_s^{MM3}	ΔE_s^{MM3}	E_s^{MM2}	ΔE_s^{MM2}	E_s^{MM3}	ΔE_s^{MM3}
$g_{\bar{8}}^+ ttt\, g_{12}^+$	22.46	2.93	30.33	2.41	22.87	3.34	30.50	2.58
$g_8^+ ttt\, g_{\bar{12}}$	22.43	2.90	30.25	2.33	25.07	5.54	31.68	3.76
$g_{\bar{7}}^+ ttt\, g_{11}^+$	21.61	2.08	30.35	2.43	25.80	6.27	32.06	4.14
$g_7^+ ttt\, g_{\bar{11}}$	21.81	2.28	29.97	2.05	24.14	4.16	31.40	3.48
$g_{\bar{6}}^+ ttt\, g_{10}^+$	23.79	4.26	31.08	3.16	22.81	3.28	30.37	2.45
$g_6^+ ttt\, g_{\bar{10}}$	23.61	4.08	30.91	2.99	26.62	7.09	32.52	4.60
$g_{\bar{5}}^+ ttt\, g_9^+$	21.86	2.33	30.60	2.68	27.27	7.74	32.86	4.94
$g_5^+ ttt\, g_{\bar{9}}$	21.85	2.32	29.93	2.01	24.12	4.59	31.71	3.79
$g_{\bar{4}}^+ ttt\, g_8^+$	25.49	5.96	31.97	4.05	22.23	2.70	29.97	2.05
$g_4^+ ttt\, g_{\bar{8}}$	25.19	5.66	31.73	3.81	28.17	8.64	33.31	5.39
$g_{\bar{3}}^+ ttt\, g_7^+$	21.89	2.36	30.87	2.95	28.30	8.77	33.23	5.31
$g_3^+ ttt\, g_{\bar{7}}$	21.83	2.30	29.95	2.03	24.42	4.89	32.61	4.69
$g_{\bar{2}}^+ ttt\, g_6^+$	26.86	7.33	32.69	4.77	—	—	—	—
$g_2^+ ttt\, g_{\bar{6}}$	21.15	1.62	32.32	4.40	—	—	—	—
$g_{\bar{1}}^+ ttt\, g_5^+$	21.05	1.52	29.86	1.94	—	—	—	—
$g_1^+ ttt\, g_{\bar{5}}$	21.75	2.22	30.47	2.55	—	—	—	—
Average	23.10	3.57	30.83	2.91	25.15	5.62	31.85	3.93

Note: Energy unit is kcal/mol; ΔE is the energy difference between the C(14):C(14)PC with a jog and the same molecule with an all-*trans* conformation.

parallel with each other along the long chain axis; moreover, the zigzag planes of the two segments lie in the same X-Y plane. The lower segment, however, is displaced from the upper segment in a nearly parallel manner by 3.60 Å in the negative Y-direction (Figure 9E).

When the same $3g2$ kink is introduced into the sn-1 acyl chain of C(14):C(14)PC with the first g^+ bond ($+65°$) located at the C(5)–C(6) bond, the sequence $g_5^+\, tg_7^+\, tg_9^+$ shifts the lower linear segment of the sn-1 acyl chain toward the adjacent sn-2 acyl chain, resulting in a sterically forbidden situation (Figure 12A). In this case, the van der Waals radii of carbon atoms C(9) to C(14) in the sn-1 acyl chain undergo severe steric collisions with those of carbon atoms C(10) to C(14) in the sn-2 acyl chain; thus, the steric energy (E_s) of this improbable structure is 2966.38 kcal/mol. After energy minimization by Allinger's MM3 program, the torsion angles of the sequence $g_5^+\, tg_7^+\, tg_9^+$ in the sn-1 acyl chain change to 67.4°, 172.0°, 68.2°, 179.1°, and 71.0°, and the E_s value of the refined lipid structure is drastically reduced to 30.57 kcal/mol. The refined lipid structure with a modified $3g2$ kink is graphically illustrated in Figure 12B. It is evident that on the X-Y plane the sn-2 acyl chain is bent downward somewhat. Most importantly, the lower segment of the sn-1 acyl chain is now poised to undergo favorable van der Waals contact interactions with the sn-2 acyl chain. The nonbonded interatomic pair distances between C(10), C(11), and C(12) in the sn-1 acyl chain and C(12), C(13), and C(14) in the sn-2 acyl chain are 4.24, 4.37, and 4.26 Å, respectively. This energy-minimized structure does serve as an example to demonstrate that phospholipid molecules with refined $3g2$ kinks can exist sterically. The steric energy of the refined $3g2$ kink-containing phospholipid, however, is slightly higher than that exhibited by the same lipid molecule with either a refined $2g1$ kink or a refined $2g2$ jog at the corresponding chain position in the same acyl chain.

D. THE Δ-CONTAINING KINKS

In regard to the process of coupled *trans* → *gauche* rotational isomerizations, we have already seen in this section several examples of kinked acyl chains in saturated phosphatidylcholine molecules which are most likely to occur at temperatures above the crystalline → gel phase transition temperature. For phospholipids isolated from biological membranes, the fatty acid linked by an ester bond to C_2 of the glycerol backbone is often found to be an unsaturated sn-2 acyl chain containing one or more *cis* carbon-carbon double bonds. As discussed earlier, the introduction of a *cis* double bond (Δ) into a hydrocarbon chain enables the chain to adopt a large number of conformations with very similar steric energies.[9] Moreover, the gel → liquid-crystalline phase transition temperature (T_m) is drastically reduced for bilayers comprised of unsaturated *cis* bonds, and it is not clear whether these bilayers can exist in the

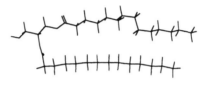

(A) The crude structure of $3g2$
($g_5{}^+tg_7{}^+tg_9{}^+$) kink

(B) The refined srtucture of (A)
E_m = 30.57 kcal/mol

Figure 12 A $3g2$ kink in the *sn*-1 acyl chain of C(14):C(14)PC before (A) and after (B) the energy minimization using the MM3 program. The first, second, and third g^+ of the sequence $g^+tg^+tg^+$ are introduced at the C(5)-C(6), C(7)-C(8), and C(9)-C(10) bonds, respectively.

crystalline phase (L_c) at a reasonable temperature.[24] Nevertheless, the possible existence of kinking of the *sn*-2 acyl chain in monoenoic phospholipids at T < T_m was implicated early on.[11,25] The long-awaited experimental identification of such Δ-containing kinks with crankshaft-like topology, however, has yet to be established; this is largely due to the fact that biological phospholipids with unsaturated *sn*-2 acyl chains are notoriously difficult to crystallize, and hence no X-ray crystallographic data are available. Recently, computer-based molecular mechanics calculations have been applied to simulate the Δ-containing kinks in phospholipids.[9] These computational results indicate that for POPC alone 16 types of Δ-containing kinks can occur theoretically below the T_m. Most interestingly, one of the 16 kinks calculated for POPC is indistinguishable from the energy-minimized kink derived experimentally from the cholesteryl oleate as discussed earlier (Figure 7B). Now we will not consider all 16 types of Δ-containing kinks; instead, we select the type Ia kink with sequence $\Delta s^- g^-$ as a representative Δ-containing kink and discuss its simulated structure.

When a carbon-carbon single bond in the center of an isolated hydrocarbon chain is substituted by a *cis* carbon-carbon double bond (Δ) such as *cis*-dodecene-6, it can be shown by molecular graphics that the *cis* double bond introduces a bend in the molecule with an obtuse angle of 104.7° (Figure 13A). This *cis* isomer is in fact under steric strain because of overcrowding of two methylene groups on the same side of the Δ-bond. After energy minimization, the δ_1' value is adjusted slightly from −180° to −177.9° to relieve the steric strain; consequently, the angle between the upper and lower segments is 135.9°, as shown in Figure 13B. If the s^- (torsion angle = −140°) and g^- (−65°) bonds are subsequently introduced into the chain at the δ_1' and δ_2' positions, respectively, followed by energy minimization, the molecular shape of *cis*-dodecene-6 is kinked in the shape of a crankshaft (Figure 13C). The δ_1' and δ_2' values, after energy minimization, are adjusted to −122.1° and −69.4°, respectively, in the kinked *cis*-dodecene-6 rotamer.

After the kink-forming sequence $\Delta s^- g^-$ with a set of (δ_1' = −140, δ_2' = −65°) torsion angles is introduced into the *sn*-2 acyl chain of POPC, the lower segment of the *sn*-2 acyl chain starting at C(11) bends downward in the negative Y-direction and it also extends away from the viewer, as graphically illustrated in Figure 13D. The energy minimization is then performed by the MM3 program. The torsion angles of δ_2, δ_1, Δ, δ_1' and δ_2' are seen to change from 180, −180, 0, −140, and −65°) to (175.2, −160.6, −3.2, −132.7, and −71.4°, respectively. The chain axis of the lower segment now runs parallel to that of the *sn*-1 acyl chain; moreover, their zigzag planes are coplanar (Figure 13E). The most striking feature of the refined structure is that the C(11), C(13), C(15), and C(17) atoms of the *sn*-2 acyl chain fit optimally into the

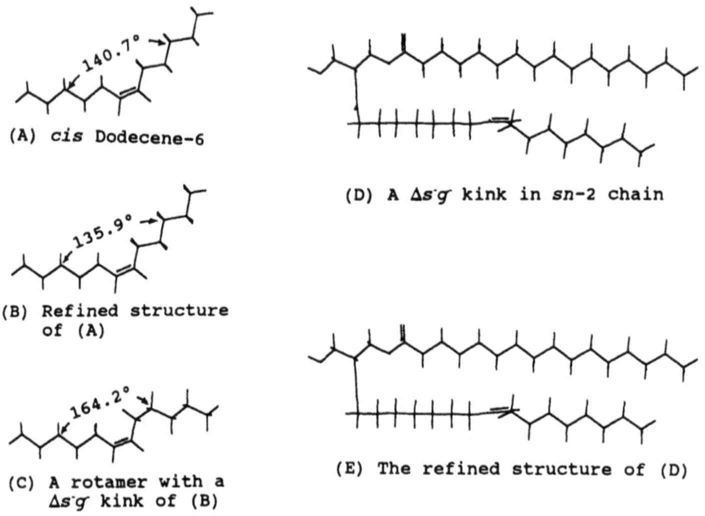

(A) *cis* Dodecene-6

(B) Refined structure of (A)

(C) A rotamer with a Δs⁻g⁻ kink of (B)

(D) A Δs⁻g⁻ kink in *sn*-2 chain

(E) The refined structure of (D)

Figure 13 Computer graphics-generated Δ-containing kinks. (A) A single *cis* double bond introduces a bend in a long hydrocarbon chain as demonstrated by *cis*-dodecene-6. (B) The energy-minimized structure of A. (C) A rotamer of B, with a Δs⁻g⁻ kink in which torsion angles of −140° and −65° are introduced into δ_1' and δ_2', respectively. (D) A Δs⁻g⁻ kink in the *sn*-2 acyl chain of POPC. (E) The energy-minimized structure of D. All energy minimizations used in the various figures were carried out with the MM3 program.

centers of the V-shaped grooves formed by the C(6)–C(8), C(8)–C(10), C(10)–C(12), and C(12)–C(14) atoms, respectively, of the neighboring *sn*-1 acyl chain, with an average nonbonded interatomic distance of 4.03 Å. The steric complementarity of the van der Waals contact surfaces makes a major contribution to the overall stability of POPC with the type Ia kink. The steric energy of the refined structure is 35.61 kcal/mol as obtained with the MM3 program or 22.75 kcal/mol as calculated by the MM2 program.

It should be noted that the molecular structure of POPC with the type Ia kink, shown in Figure 13E, is clearly distinct from that of the same molecule with the type IIIb kink, as illustrated in Figure 7B. For instance, the orientations of the lower segment of the *sn*-2 acyl chain in the two rotamers differ by about 90°. The steric energies (22.8 kcal/mol) calculated by the MM2 program, however, are identical. The identical value of E_s implies that POPC with different kink motifs can interconvert readily. In fact, POPC with the type Ia kink in the bilayer has the proper geometric features to form a hydrophobic pocket which can accommodate nicely the two angular methyl groups of the cholesterol molecule protruding from the β-surface of the steroid nucleus.[11] Here we further indicate that the type Ia kink and the type IIIb kink are interconvertible, suggesting that the hydrophobic pocket generated by the Δ-containing kinks is adjustable. It is thus tempting to hypothesize that the adjustable hydrophobic pocket generated by the Δ-containing kinks may offer a possible site within the bilayer by which the methyl groups of farnesylated or geranylgeranylated membrane proteins can fit complementarily like fingers fit into a glove. Hence, the Δ-containing kink may be regarded as an important mechanical device with extraordinary flexibility that may regulate the stability and functions of some membrane proteins.

VII. SUMMARY AND CONCLUSIONS

Phospholipids are major constituents of biological membranes. The hydrophobic moiety of phospholipid is the diglyceride, which consists of a glycerol backbone and two long fatty acyl chains. Due to the wide diversity of the fatty acyl chain in terms of chain length and degree of unsaturation, the number of possible structures of a given class of phospholipids is enormous. In this communication, we report the general strategy used in our laboratory to construct the structures of monomeric and aggregated forms of mixed-chain saturated and monounsaturated diglyceride moieties of phosphatidylcholines. These structures are built up initially from a subset of atomic coordinates of saturated identical-chain dimyristoylphosphatidyl-choline, C(14):C(14)PC, and the oleate chain of cholesteryl oleate using the computer-based molecular modeling approach. Specifically, the molecular mechanics programs (MM⁺, MM2, and MM3) originally developed by Allinger are applied to calculate the energy-minimized structure and steric energy of the lipid under study, using an IBM RS/6000 computer workstation. The atomic coordinates of the energy-minimized structure are then transferred to HyperChem™ operated on a 486 platform, from which the atomic coordinates can be excised and hence the molecular graphics of lipid molecules can be visualized and printed out.

An important observation from this molecular modeling study is the occurrence of coupled *trans* → *gauche* isomerizations of C-C bonds in both saturated and monounsaturated acyl chains, which transforms the lipid acyl chain into a kinked structure while maintaining the overall chain direction perpendicular to the bilayer surface. Consequently, an assembly of relatively ordered lipid chains can exist in the bilayer interior at temperatures above the crystalline-to-gel phase transition temperature. Moreover, the kinked monoenoic hydrocarbon chain can result in an adjustable hydrophobic pocket in the hydrocarbon interior of the bilayer membrane, implying that such kinks may be extremely valuable for forming hydrophobic sites to accommodate methyl groups from other membrane components in the bilayer interior and hence for promoting the stability and function of biological membranes in general.

ACKNOWLEDGMENTS

This research was supported, in part, by U.S. Public Health Service Grant GM-17452 from the National Institute of General Sciences, National Institutes of Health, U.S. Department of Health and Human Services. The secretarial assistance of Ms. Linda Saunders is gratefully acknowledged.

REFERENCES

1. Bangham, A. D., Standish, M. M., and Watkins, J. C., Diffusion of univalent ions across the lamellae of swollen phospholipids. *J. Mol. Biol.*, 13, 238, 1965.
2. Bangham, A. D., Membrane models with phospholipids, *Prog. Biophys. Mol. Biol.*, 18, 29, 1968.
3. Cohen, B. E., The permeability of liposomes to nonelectrolytes, *J. Memb. Biol.*, 20, 205, 1975.
4. Chapman, D., Phase transitions and fluidity characteristics of lipids and membranes, *Q. Rev. Biophys.*, 8, 185, 1975.
5. Chen, S. C., Sturtevant, J. M., and Gaffney, B. J., Scanning calorimetric evidence for a third phase transition in phosphatidylcholine bilayer, *Proc. Natl. Acad. Sci. U.S.A.*, 77, 5060, 1980.
6. Pascher, I., Lundmark, M., Nyholm, P.-G., and Sundell, S., Crystal structures of membrane lipids, *Biochim. Biophys. Acta*, 1113, 339, 1992.
7. Vanderkooi, G., Multibilayer structure of dimyristoylphosphatidylcholine dihydrate as determined by energy minimization, *Biochemistry*, 30, 10760, 1991.
8. Li, S., Wang, Z.-q., Lin, H.-n., and Huang, C., Energy-minimized structures and packing states of a homologous series of mixed-chain phosphatidylcholines: a molecular mechanics study on the diglyceride moieties, *Biophys. J.*, 65, 1415, 1993.
9. Li, S., Lin, H.-n., Wang, Z.-q., and Huang, C., Identification and characterization of kink motifs in 1-palmitoyl-2-oleoyl-phosphatidylcholines. A molecular mechanics study, *Biophys. J.*, 66, 2005, 1994.
10. Träuble, H., The movement of molecules across lipid membranes: a molecular theory, *J. Membr. Biol.*, 4, 193, 1971.
11. Huang, C., A structural model for the cholesterol-phosphatidylcholine complexes in bilayer membranes, *Lipids*, 12, 348, 1977.
12. Allinger, N. L., Conformational analysis. 130. MM2. A hydrocarbon force field utilizing V_1 and V_2 torsional terms, *J. Am. Chem. Soc.*, 99, 8127, 1977.
13. Allinger, N. L., Yuh, Y. H., and Lii, J.-H., Molecular mechanics. The MM3 force field for hydrocarbons, *J. Am. Chem. Soc.*, 111, 8551, 1989.
14. Lewis, R. N. A. H., and McElhaney, R. N., Structures of the subgel phases of *n*-saturated diacyl phosphatidylcholine bilayer: FTIR spectroscopic studies of $^{13}C=O$ and ^2H-labelled lipids, *Biophys. J.*, 61, 63, 1992.
15. Zaccai, G., Büldt, G., Seelig, A., and Seelig, J., Neutron diffraction studies on phosphatidylcholine model membranes, *J. Mol. Biol.*, 134, 693, 1979.
16. McIntosh, T. J., Simon, S. A., Ellington, J. C., and Porter, N. A., New structural model for mixed-chain phosphatidylcholine bilayers, *Biochemistry*, 23, 4038, 1984.
17. Hui, S. W., Mason, J. T., and Huang, C., Acyl chain interdigitation in saturated mixed-chain phosphatidylcholine bilayer dispersions, *Biochemistry*, 23, 5570, 1984.
18. Mattai, J., Scripada, P. K., and Shipley, G. G., Mixed-chain phosphatidylcholine bilayer structure and properties, *Biochemistry*, 26, 3287, 1987.
19. Huang, C., Mixed-chain phospholipids and interdigitated bilayer systems, *Klin. Wochenschr.*, 68, 149, 1990.
20. Gao, Q., and Craven, B. M., Conformation of the oleate chains in crystals of cholesteryl oleate at 123 K, *J. Lipid Res.*, 27, 1214, 1986.

21. Seelig, A., and Seelig, J., The dynamic structure of fatty acyl chains in a phospholipid bilayer measured by deuterium magnetic resonance, *Biochemistry*, 13, 4839, 1974.

22. Pechhold, W., and Blasenbrey, S., Kooperative Rotationsisomerie in Polymeren. I. Schweltztheorie und Kinkenkonzehtrationen, *Kolloid Z.Z. Polym.*, 216/217, 235, 1967.

23. Pechhold, W. and Blasenbrey, S., Molekülbewegung in Polymeren. III. Teil: Mikrostructure und mechanische Eigenschaften, *Kolloid Z.Z. Polym.*, 241, 955, 1970.

24. Huang, C., Structure and properties of self-assembled phospholipids in excess water, in *Phospholipid-Binding Antibodies*, Harris, E. N., Exner, T., Hughes, G. R. V., and Asherson, R.A., Eds., CRC Press, Boca Raton, FL, 1991, chap. 1.

25. Lagaly, G., Weiss, A., and Stuke, E., Effect of double-bonds on bimolecular films in membrane models, *Biochim. Biophys. Acta*, 470, 331, 1977.

Chapter 8

Properties of Mixed-Chain-Length Phospholipids and Their Relationship to Bilayer Structure

Jeffrey T. Mason

CONTENTS

I. INTRODUCTION

Liposome is a generic term used to describe a variety of bilayer structures formed by many natural or synthetic phospholipids and glycolipids when these amphiphiles are dispersed by various means in an aqueous solution. Liposomes generally fall into one of three categories based upon their gross morphology. These categories are multilamellar vesicles, small unilamellar vesicles, and large unilamellar vesicles. Model membrane systems, such as liposomes, were originally developed as a means to simplify the study of the structural and functional properties of complex biological membranes. Indeed, in this capacity model membrane systems have made a tremendous contribution to our understanding of biomembrane structure and function.[1] At the same time, this historical origin has imposed a constraint on the types of membrane structures that have been studied. Biological membranes consist predominately of phospholipids whose two constituent acyl (or alkyl) chains are approximately equal in length (symmetric). Model bilayers formed from such phospholipids adopt an organization where each phospholipid headgroup spans two hydrocarbon chains and the hydrocarbon core of the bilayer contains a well-defined bilayer center.[2] Thus, the bilayer consists of two discrete leaflets that are, in many respects, functionally independent.

However, biological membranes also contain classes of lipids that frequently possess two constituent hydrocarbon chains that are highly dissimilar in length. Such membrane lipids include sphingomyelins, neutral glycosphingolipids, gangliosides, some classes of phosphatidylethanolamines (PEs), lysophospholipids, and certain lipid autocoids, such as platelet-activating factor. Despite this fact, biochemical and biophysical studies on the effects of lipid chain-length asymmetry on the structure and dynamics of bilayers have only recently been undertaken. In this chapter, work on the effects of an asymmetry in the length of the two constituent hydrocarbon chains on the bilayer properties of mixed-chain-length phospholipids will be described.

0-8493-4731-9/96/$0.00+$.50
© 1996 by CRC Press, Inc.

One of our original motivations for studying the effects of chain-length asymmetry in phospholipids was the observation by Mabrey and Sturtevant[3] that binary mixtures of saturated symmetric-chain-length phosphatidylcholines (PCs) differing in chain length by six carbons are monotectic and show lateral phase separation in the gel phase. Mixed-chain-length phospholipids represent an interesting variation on the above studies, since the acyl chains are forced to mix by virtue of their attachment to the same glycerol backbone. I will present evidence from a variety of physical techniques that the bilayers' response to this dilemma is to compensate for the phospholipid chain-length asymmetry by adopting a gel-phase arrangement where the acyl chains interdigitate across the bilayer center. Depending upon the magnitude of chain-length asymmetry, the mixed-chain-length phospholipids can adopt a noninterdigitated, a partially interdigitated, a mixed-interdigitated, or a fully interdigitated bilayer gel-phase organization. The magnitude of the chain-length asymmetry will also be shown to determine the morphology of the phospholipid assemblies at temperatures above the chain-melting phase-transition temperature of the phospholipids.

In the past 10 years, model membranes have begun to move out of the biological realm and find applications in such diverse fields as synthetic photosynthesis,[4] magnetic film technology,[5] semiconductor fabrication[6] and molecular electronics,[7] chemical assays,[8] nanosystems,[9] chemical synthesis,[10] and agriculture.[11] In many of these applications the physical properties of artificial bilayers modeled after symmetric-chain-length phospholipids have proven somewhat restrictive. It would clearly be advantageous to be able to modify this fundamental membrane structure in ways that would optimize the properties of the bilayer for the target application. As will be evident from the following sections, considerable manipulation of the structural and dynamic properties of the bilayer phase can be achieved through the use of mixed-chain-length phospholipids. Potential applications of bilayers formed from mixed-chain-length phospholipids to areas of liposome technology will also be discussed.

II. MOLECULAR STRUCTURE OF PHOSPHOLIPIDS

The interpretation of the results of our studies with mixed-chain-length phospholipids has relied heavily on a detailed knowledge of the preferred molecular conformation of phospholipids. Over the past 15 years, considerable effort has been directed toward the study of phospholipid conformation.[12] These studies have revealed several important phospholipid conformational features that appear to be conserved irrespective of the temperature, the solvent environment, or the morphology of the phospholipid assembly. In this section, I will briefly review these findings and emphasize those aspects of phospholipid conformation that are essential to the understanding of the behavior of mixed-chain-length phospholipids, specifically the phospholipid interface and acyl chain regions.

A. INTERFACE REGION

X-ray diffraction data are available on single crystals of 1,2-dilauroyl-*rac*-glycero-3-phosphoethanolamine[13] and 1,2-dimyristoyl-*sn*-glycero-3-phosphocholine dihydrate.[14] A model showing the molecular conformation of 1-stearoyl-2-lauroyl-*sn*-glycero-3-phosphocholine [C(18)C(12)PC]* based upon the crystal data of Pearson and Pascher[14] is shown in Figure 1. The interface region of the phospholipid molecule consists of the glycerol backbone and is also generally considered to include the ester linkages and the α-methylenes of the *sn*-1 and *sn*-2 acyl chains. The acylglycerol ester linkage can have two resonance forms; this imparts significant double-bond character to the ester bond. Since rotation about the axis of a double bond is highly restricted, the five atoms of the acylglycerol ester linkage are essentially coplanar,[15] as shown in Figure 1. Numerous studies have confirmed the structural rigidity of the interface region in hydrated phospholipid assemblies.[16-18] For example, [2]H-nuclear magnetic resonance (NMR) spectroscopic studies[18] identified two quadrapole splittings for specifically deuterated [C3-^{2}H$_2$]-glycerol moieties introduced into a variety of hydrated phospholipids. By employing stereospecifically deuterated 1,2-dipalmitoyl-*sn*-[C3-(*R*)-^{2}H]-glycero-3-phosphocholine, it was shown that these two signals originate from each of the two magnetically inequivalent deuterons and not from slowly exchanging conformations.[19] These results are consistent with a rigid conformation for the phospholipid interface region, even in the liquid-crystalline bilayer phase.

---

* The notation C(n_1)C(n_2)PC(PE) is used to denote symmetric-chain-length and mixed-chain-length PCs and PEs where n_1 and n_2 are the number of carbon units in the *sn*-1 and *sn*-2 acyl chains, respectively, and 18:1 denotes oleic acid.

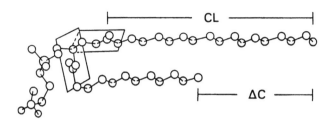

Figure 1   Model revealing the molecular conformation of C(18)C(12)PC based upon the crystal structure data of Pearson and Pascher.[14] ΔC is the effective chain-length difference between the two acyl chains, and CL is the effective length of the longer of the two acyl chains.

B. ACYL CHAIN REGION

In the phospholipid crystal, the *sn*-1 acyl chain and the carbon atoms of the glycerol moiety form an all-*trans* linear chain oriented almost parallel to the long axis of the molecule. The initial segment of the *sn*-2 acyl chain extends out from the glycerol backbone at a right angle and runs parallel to the lamellar surface. This conformation results in the planes of the two acylglycerol ester linkages being oriented almost perpendicular to each other as shown in Figure 1. The *sn*-2 acyl chain bends by about 90° at the C2-C3 bond of the acyl chain, which allows the remainder of the *sn*-2 acyl chain to run parallel to the *sn*-1 chain.[14] Because of the unique conformation of the initial segment of the *sn*-2 acyl chain, the terminal ends of the two chains display an axial displacement. In the phospholipid crystal this displacement equals three methylene units, or 3.7 Å.

A number of techniques, including Raman spectroscopy,[20] ^{2}H-NMR spectroscopy,[21] and neutron diffraction analysis,[22] have demonstrated that the conformational inequivalence of the two phospholipid acyl chains that is seen in the crystal persists in hydrated phospholipid phases. Neutron diffraction studies[22] on C(16)C(16)PC specifically deuterated at the α-methylene sites of the acyl chains have confirmed that the bent conformation of the initial segment of the *sn*-2 acyl chain is present in both the gel and liquid-crystalline bilayer phases. However, in these hydrated phases the chain inequivalence is reduced and the terminal ends of the chains are out of step by just 1.5 carbon-carbon chain lengths, or 1.8 Å. ^{2}H-NMR spectroscopy[21,23] has been employed to demonstrate the existence of chain inequivalence in a wide variety of phospholipids, and it can now be stated with confidence that the conformational inequivalence of the *sn*-1 and *sn*-2 acyl chains described above is a general feature of phospholipids.

On the basis of the phospholipid conformational properties described above, the effective difference in chain length between the *sn*-1 and *sn*-2 acyl chains of mixed-chain-length phospholipids can be quantitatively defined by the parameter ΔC, which is given by the expression $\Delta C = |n_1 - n_2 + 1.5|$. Here ΔC is the effective chain-length difference (in units of carbon-carbon bond lengths), n_1 and n_2 are the number of carbon atoms in the *sn*-1 and *sn*-2 acyl chains, respectively, and the quantity 1.5 represents the contribution to ΔC that arises from the conformational inequivalence of the two chains in the bilayer gel phase.[24] Another useful quantitative parameter is the effective length of the longer of the two acyl chains in a given mixed-chain-length phospholipid. This parameter is designated CL (in units of carbon-carbon bond lengths) and is assigned the larger of the two values $n_1 - 1$ or $n_2 - 2.5$. These two parameters are depicted in Figure 1. The effective chain-length difference, expressed as a fraction of the overall length of the hydrocarbon region of the mixed-chain-length phospholipid, is given by the expression $\Delta C/CL$, which is designated the chain inequivalence parameter.[24] As I will show, the chain inequivalence parameter is fundamental to the analysis of the behavior of mixed-chain-length phospholipids.

III. SINGLE-COMPONENT MIXED-CHAIN-LENGTH PHOSPHOLIPID BILAYERS

A. SATURATED MIXED-CHAIN-LENGTH PHOSPHATIDYLCHOLINES

1. Overview

The effect of chain-length asymmetry on the properties of mixed-chain-length PC bilayers has been systematically investigated by studying the properties of a series of synthetic PCs where the *sn*-1 acyl chain is fixed at 18 carbons (stearic acid) while the length of the *sn*-2 acyl chain is shortened by two methylene units for each successive member of the series. Our notation (see footnote in Section II.A) for this series is as follows: C(18)C(18)PC, C(18)C(16)PC, C(18)C(14)PC, ..., C(18)C(2)PC, C(18)C(0)PC. Methods for the synthesis of isomerically pure mixed-chain phospholipids have been developed.[25-28] The principal technique used to study the properties of these mixed-chain-length PCs has been differential scanning calorimetry (DSC).[29] This technique allows for the determination of the thermodynamic parameters associated with the phase transitions of the phospholipids, such as their transition tempera-tures and their transition enthalpy (ΔH) and entropy (ΔS) changes. DSC can also be used to determine the kinetics associated with phase changes and to detect phase metastability. The thermodynamic parameters associated with the phase transitions exhibited by the mixed-chain-length PC bilayer series are listed in Table 1, as are those for other mixed-chain-length phospholipids discussed below.

The thermotropic properties of the mixed-chain-length phospholipids can be conveniently analyzed by plotting the magnitude of the entropy change associated with the chain-melting phase transition (gel $\leftrightarrow$ liquid-crystalline or gel $\rightarrow$ micellar) of the phospholipids as a function of their chain inequivalence parameters ($\Delta C/CL$). This entropy change reflects, principally, the strength of the lateral chain-chain interactions and the degree of intrachain conformational order in the bilayer gel phase. To compare phospholipids with different molecular weights, the transition entropy change must be normalized prior to plotting. A reasonable estimate for the maximum transition entropy change for a given phospholipid would be the fusion entropy of the corresponding free fatty acids. This entropy is given by the relation-ship[30] $\Delta S_{exp} = (1.9 \text{ cal} \cdot \text{mol}^{-1} \cdot \text{K}^{-1}) \times (n_1 + n_2)$. The plot of $\Delta S/\Delta S_{exp}$ vs. $\Delta C/CL$ is shown in Figure 2. The thermodynamic parameters of the mixed-chain-length PCs used to derive the plot shown in Figure 2 are those given in Table 1. Code numbers or letters relating the mixed-chain-length PCs to their correspond-ing points in the phase diagram of Figure 2 are also listed in Table 1. The plot is divided into four regions corresponding either to different phospholipid gel-phase acyl chain packing conformations or to different morphologies for the chain-melted phase of the PCs. These four regions will be described individually.

2. Region I

The phospholipids in Region I are organized into lamellar structures at all temperatures and adopt either a noninterdigitated or a partially interdigitated acyl chain packing conformation in the bilayer gel phase. The PCs in Region I include C(18)C(18)PC, C(18)C(16)PC, and C(18)C(14)PC. As shown in Figures 3A and 3C, the main gel $\leftrightarrow$ liquid-crystalline phase transition occurs at 55.1°C for C(18)C(18)PC and at 29.8°C for C(18)C(14)PC. These phase transitions are reversible upon cooling (Figure 3B) and are independent of the thermal history of the sample. Figure 2 shows that the normalized transition entropy change for the main gel $\leftrightarrow$ liquid-crystalline phase transition of C(18)C(18)PC, C(18)C(16)PC, and C(18)C(14)PC decreases linearly with an increase in the chain inequivalence parameter. This linear relationship suggests that the acyl chain conformational order and the lateral chain-chain interactions of these PCs in the bilayer gel phase are perturbed proportionally by a progressive increase in the chain-length asymmetry of the PCs.

Raman spectroscopic measurements were carried out on the mixed-chain-length PCs to determine the validity of this interpretation.[31] Temperature profiles for a series of mixed-chain-length PCs from C(18)C(10)PC to C(18)C(18)PC derived from the peak-height intensity ratio of $I(2936 \text{ cm}^{-1})/I(2884 \text{ cm}^{-1})$ are shown in Figure 4. The magnitude of this spectral index is related to the lateral chain-chain packing disorder within the bilayers; the larger ratio for C(18)C(14)PC suggests that the C(18)C(14)PC molecules exhibit weaker interchain interactions in the bilayer gel phase relative to the other PCs in the series. The conformational disorder of the acyl chains in the bilayer gel phase for this PC series was compared based on their Raman spectral intensity ratio of $I(1088 \text{ cm}^{-1})/I(1065 \text{ cm}^{-1})$. This plot is shown in Figure 5. The magnitude of this spectral index is proportional to the *gauche/trans* ratio along the phospholipid acyl chains. The data conform to a bell-shaped curve with a maximum conformational disorder for the C(18)C(14)PC gel-phase bilayer.[29,31] The relatively weak interchain interactions observed for C(18)C(14)PC in the gel phase can thus be attributed to the presence of significant numbers of *gauche* conformers along the acyl chains of the C(18)C(14)PC molecules.

X-ray diffraction analysis was used to gain a better understanding of the gel-phase packing associa-tions of the C(18)C(14)PC bilayer.[32,33] Upon cooling, the C(18)C(14)PC liquid-crystalline phase trans-formed to a rippled $P_{\beta'}$ gel phase, which in turn transformed to an $L_{\beta'}$ gel phase at lower temperatures. The analysis of the diffraction data indicated that the phospholipid molecules were packed in the bilayer gel phase such that each phosphocholine headgroup spanned two acyl chains. Within the $L_{\beta'}$ phase, the chain segments containing the first 10 or 11 methylene units were packed with rotational disorder in a two-dimensional hexagonal lattice and were the source of the symmetric 4.2 Å reflection seen in the

Table 1 Thermodynamic Parameters of Mixed-Chain-Length Phospholipid Dispersions as Determined by Differential Scanning Calorimetry (DSC)

Phospholipid	Code[a]	$\Delta C/CL$	Subtransition T_s	ΔH	ΔS	Pretransition T_p	ΔH	ΔS	Main Transition T_m	ΔH	ΔS	Ref.[b]
C(18)C(18)PC-C(18)C(0)PC Series												
C(18)C(18)PC	A	0.088	32.1[c]	4.81	15.8	51.0	1.21	3.73	55.1	10.2	31.1	u
C(18)C(16)PC	B	0.206	33.2[d]	5.12	16.7	30.1	0.71	2.31	44.1	7.6	24.0	u
C(18)C(14)PC	C	0.324	23.4[d]	6.40	21.6	19.1	1.33	4.55	29.8	5.62	18.6	u
C(18)C(12)PC	D	0.441	18.5[e]	14.7	50.4				17.9	7.74	26.6	u
C(18)C(10)PC	E	0.559							19.1	9.90	33.9	u
C(18)C(8)PC	F	0.676							0.2[f]	2.51	9.19	u
C(18)C(6)PC	G	0.794							8.2[f]	3.33	11.8	u
C(18)C(4)PC	H	0.912							14.0[f]	4.14	14.4	u
C(18)C(2)PC	I	1.03	13.9[c]	2.9	10.1	17.1	2.61	8.96	18.6[f]	2.60	8.92	u
C(18)C(0)PC	J	1.15							26.0[f]	6.92	23.1	u
Other Mixed-Chain-Length PCs[g]												
C(16)C(18)PC	1	0.030							48.7	8.15	25.3	u
C(13)C(15)PC	2	0.040							25.5	6.0	20.1	49
C(14)C(18)PC	3	0.161							37.9	7.83	25.2	u
C(15)C(13)PC	4	0.250							18.8	5.3	18.2	49
C(11)C(19)PC	5	0.394							17.3	4.4	15.1	49
C(10)C(18)PC	6	0.419							10.9	6.04	21.3	u
C(18)C(11)PC	7	0.491							21.4	9.12	31.0	u
C(18)C(9)PC	8	0.609							11.1	8.59	30.3	u
C(19)C(9)PC	9	0.639							13.2	8.3	29.0	85
Unsaturated Mixed-Chain-Length PCs												
C(16)C(18:1)PC	10	0.013							–0.8	5.4	19.8	63
C(18:1)C(16)PC	11	0.192							–7.9	4.6	17.4	63
C(18)C(18:1)PC	12	0.106							8.3	5.4	19.2	63
C(18:1)C(18)PC	13	0.072							10.9	6.7	23.6	63
Sphingomyelins												
C(16)SPM	14	0.167							41.3	6.8	21.6	66
C(18)SPM	15	0.265							45	7.0	22.0	67
C(24)SPM	16	0.457				40	4.35	13.9	47.5	6.25	19.5	70
C(18)C(18)PE-C(18)C(10)PE Series												
C(18)C(18)PE	—	0.088	74.2[e]	22.2	63.9				72.9	10.9	31.5	25
C(18)C(16)PE	—	0.206	65.8[e]	14.6	43.1				64.4	9.21	27.3	25
C(18)C(14)PE	—	0.324	59[e]	13.2	39.8				53.5	6.82	20.8	25
C(18)C(12)PE	—	0.441	50.8[e]	14.2	43.9	13.9	2.61	9.06	36.9	4.21	13.6	25
C(18)C(10)PE	—	0.559	39.2[e]	12.2	39.1				21.1	9.20	31.1	25

Note: Subtransition refers to any phase transition where the low-temperature phase is an L_c crystalline phase. Pretransition refers to any phase transition where both the low- and high-temperature phases are gel phases. Main transition refers to the chain-melting phase transition of the phospholipid. The thermodynamic parameters for the subtransition were determined from samples that had been incubated at 0°C [–20°C for C(18)C(2)PC] for at least 1 month. The listed thermodynamic parameters were derived from heating DSC scans; other details of the DSC measurements are given in the indicated references. T_s, T_p, and T_m are the sub-, pre-, and main phase-transition temperatures (in °C), respectively, of the phospholipid dispersions. ΔH is the phase transition enthalpy change (in kcal/mol) and ΔS is the phase transition entropy change (in cal · mol^{-1} · K^{-1}) for the phase transitions. ΔS is calculated by the formula $\Delta S = \Delta H/T$, where T is the relevant phase-transition temperature. The chain inequivalence parameter ($\Delta C/CL$) is calculated as described in the text.

[a] Code refers to the number or letter for the corresponding plotted point in Figure 2 or Figure 8.

[b] Reference is the source publication for the indicated thermodynamic parameters; u = unpublished data from DSC runs performed for this review conducted as described.[25]

[c] $L_c \rightarrow L_{\beta'}$ subtransition.

[d] $L_c \rightarrow P_{\beta'}$ subtransition.

[e] $L_c \rightarrow L_\alpha$ subtransition.

[f] Gel $\rightarrow$ micellar phase transition; the other main phase transitions are gel $\leftrightarrow$ liquid-crystalline phase transitions.

[g] Thermodynamic parameters are given for the main phase transition only.

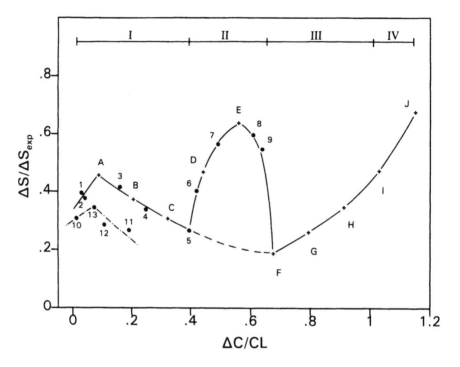

Figure 2 Plot of the normalized transition entropy change versus the chain inequivalence parameter for the mixed-chain-length PCs listed in Table 1. The numbers or letters by the data points refer to the corresponding entries in Table 1. The trend indicated by the dotted and dashed line is for the unsaturated mixed-chain-length PCs. The plot is divided into four regions corresponding to either different phospholipid gel-phase chain packing conformations or different morphologies for the chain-melted phase of the phospholipids as described in the text. The dashed line represents the predicted shape of the profile in the absence of the mixed-interdigitated packing conformation.

wide-angle diffraction region. The last seven or eight methylene units of the sn-1 stearoyl chains (the ΔC region) were extended across the bilayer center to form a partially interdigitated bilayer packing. In the partially interdigitated bilayer phase (designated $L_{\beta'}^{p}$), the longer chain of a phospholipid molecule on one side of the bilayer packs end-to-end with the shorter chain of another phospholipid in the opposing bilayer leaflet.[29] However, the interdigitation of the acyl chains in the C(18)C(14)PC gel phase has been shown to be more dynamic than static in nature.[31,32] This packing association is depicted in Figure 6C.

In summary, the above findings indicate that the mixed-chain-length PCs in Region I form gel-phase bilayers where the phospholipid acyl chains are progressively interdigitated as the chain-length asymmetry is increased for $\Delta C/CL$ values greater than ~0.1. However, due to the intrachain conformational disorder and weak interchain packing associations of the partially interdigitated bilayer center, the magnitude of the normalized transition entropy change for the main gel $\leftrightarrow$ liquid-crystalline phase transition of the PCs decreases linearly with an increase in the chain inequivalence parameter. Symmetric-chain-length phospholipids, or mixed-chain-length phospholipids with chain inequivalence parameters below 0.1, form a noninterdigitated gel-phase packing conformation.

For the mixed-chain-length PCs in Region I that have been studied, the partially interdigitated $L_{\beta'}^{p}$ gel phase has been shown to be metastable and will convert to a crystalline phase upon incubation at low temperatures.[34] Figure 3A shows the thermogram of a suspension of C(18)C(18)PC liposomes that have been incubated at 0°C for 1 month. Two low-temperature phase transitions occur at 32.1°C and 51.0°C, in addition to the gel $\leftrightarrow$ liquid-crystalline phase transition at 55.1°C.[29] The transition at 32.1°C is absent upon cooling (Figure 3B) or immediately reheating (Figure 3C) the sample. The phase transition at 32.1°C is thus assigned to the $L_{c} \rightarrow L_{\beta'}$ subtransition, which corresponds to the conversion from a poorly hydrated crystalline phase with orthorhombic chain packing to the $L_{\beta'}$ hydrated gel phase with distorted hexagonal or orthorhombic chain packing.[35] The phase transition at 51.0°C is reversible and independent of the thermal history of the sample. This phase transition is assigned to the $L_{\beta'} \leftrightarrow P_{\beta'}$ pretransition, where the $P_{\beta'}$ gel phase is characterized by hexagonal chain packing and the presence of a periodic ripple

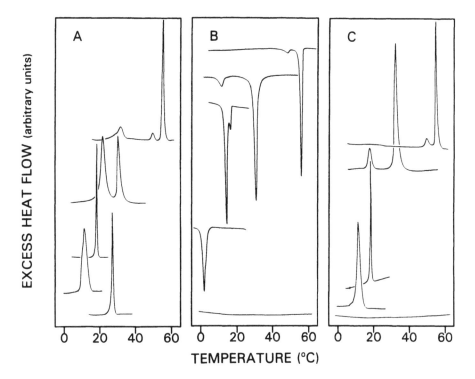

Figure 3 Heat flow versus temperature profiles for selected members of the C(18)C(18)PC through C(18)C(0)PC mixed-chain-length PC series. The scans in each panel are, from top to bottom, for aqueous dispersions (15 mg/ml in 50 m*M* KCl) of C(18)C(18)PC, C(18)C(14)PC, C(18)C(10)PC, C(18)C(6)PC, and C(18)C(0)PC. Differential scanning calorimetry was performed as described.[25] (A) Heating scans of samples that were held at 0°C for 1 month. (B) Cooling scans conducted immediately after the heating scans in A. (C) Heating scans conducted immediately after the cooling scans in B. The individual profiles have been arbitrarily enlarged in the direction of the heat flow axis to enhance their detail and are not drawn to scale.

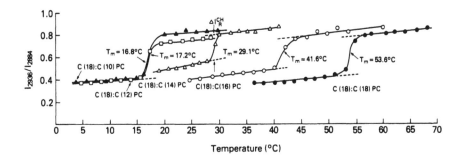

Figure 4 Temperature profiles for the mixed-chain-length PC series C(18)C(10)PC through C(18)C(18)PC derived from the Raman spectral $I(2936 \text{ cm}^{-1})/I(2884 \text{ cm}^{-1})$ peak-height intensity ratio. (From Huang, C., Mason, J. T., and Levin, I. W., *Biochemistry*, 22, 2775, 1983. With permission.)

structure in the bilayer. Thus, the overall phase behavior of C(18)C(18)PC can be described by the relationships given below:

$$L_c \underset{\text{slow}}{\xleftarrow{\hspace{0.5cm}}} L_{\beta'} \longleftrightarrow P_{\beta'} \longleftrightarrow L_\alpha$$

Figure 3A shows the thermogram of a suspension of C(18)C(14)PC liposomes that has been held at 0°C for 1 month. This thermogram shows the presence of only one low-temperature phase transition, which occurs at 23.4°C. This phase transition is not reversible, but instead it is replaced by a lower enthalpy phase transition, which occurs at 13°C upon cooling (Figure 3B) or at 19.1°C upon immediately

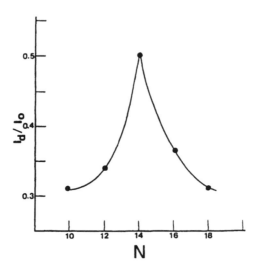

Figure 5 The relative Raman peak-height ratio (I_d/I_o) between the 1088 cm⁻¹ and 1065 cm⁻¹ C-C stretching bands recorded 10°C below the phospholipid T_m versus the carbon number (N) of the *sn*-2 acyl chain of the C(18)C(10)PC through C(18)C(18)PC series. (From Huang, C. and Mason, J. T., *Biochim. Biophys. Acta*, 864, 423, 1986. With permission.)

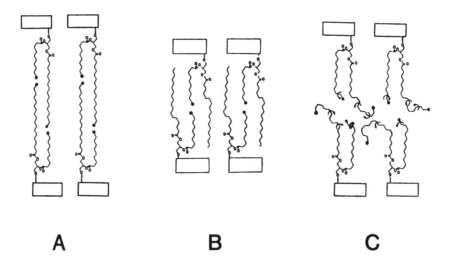

Figure 6 Acyl chain packing models for the mixed-chain-length phospholipid bilayers. (A) Static partially interdigitated chain packing characteristic of the L_c phase of the Region I and Region II phospholipids, such as C(18)C(14)PC and C(18)C(12)PC. (B) Mixed-interdigitated gel-phase packing characteristic of the phospholipids in Region II, such as C(18)C(10)PC. (C) Partially (dynamic) interdigitated gel-phase packing characteristic of the phospholipids in Region I, such as C(18)C(14)PC. (From Hui, S. W., Mason, J. T., and Huang, C., *Biochemistry*, 23, 5570, 1984. With permission.)

reheating (Figure 3C) the sample.[29] Based upon X-ray diffraction analysis of the mixed-chain-length PC bilayers,[34,36] the phase transition seen after prolonged incubation of the C(18)C(14)PC sample is assigned to the $L_c \rightarrow P_{\beta'}$ subtransition, and the phase transition seen upon cooling or immediately reheating the sample is assigned to the $L_{\beta'} \leftrightarrow P_{\beta'}$ pretransition.[34,36] Thus, the $L_{\beta'}$ phase is absent in annealed samples of C(18)C(14)PC. C(18)C(16)PC bilayers display a pattern of thermotropic behavior similar to that of C(18)C(14)PC. Thus, the overall phase behavior of C(18)C(16)PC and C(18)C(14)PC can be described by the relationships given below:

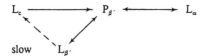

The crystalline bilayer phases formed by the mixed-chain-length PCs in Region I all reveal the presence of two wide-angle reflections at about 4.5 and 3.9 Å.[34] This implies a similar acyl chain packing mode for all of the PCs in which the acyl chains are stabilized by occupying specific lattice sites. It can be inferred[37] that the acyl chains of the phospholipid molecules adopt a partially interdigitated chain packing in order to maximize van der Waals contacts among the chains. Thus, the mixed-chain-length PCs of Region I form crystalline phase bilayers (designated L_c) where the phospholipid acyl chains are progressively interdigitated as the chain-length asymmetry is increased. However, unlike the gel phase, the partial interdigitation in the crystalline phase is static, as depicted in Figure 6A.

The properties of the C(18)C(18)PC, C(18)C(16)PC, and C(18)C(14)PC bilayers described above are representative of the thermotropic behavior of phospholipids whose chain inequivalence parameters place them in Region I. Some of the interesting trends in thermotropic behavior exhibited by the mixed-chain-length phospholipids in Region I are described below.

1. The difference between the main phase-transition temperature (T_m) and the subtransition temperature (T_s) for the mixed-chain-length PCs decreases as the chain inequivalence parameter increases. Based upon the trends exhibited by T_m and T_s for the mixed-chain-length PCs, Wang et al.[38] predicted that these two phase transitions would become coincident for a $\Delta C/CL$ value of 0.29. Indeed, this is observed for bilayers of C(12)C(18)PC.[38] For mixed-chain-length PCs with $\Delta C/CL$ values greater than 0.29, the crystalline phase is more stable than the $P_{\beta'}$ gel phase at all temperatures. For these phospholipids, an $L_c \rightarrow L_\alpha$ subtransition is observed, which occurs at a higher temperature than the $P_{\beta'} \leftrightarrow L_\alpha$ main phase transition. Bilayers composed of C(11)C(17)PC exhibit this property.[38] A direct conversion of the L_c phase to the L_α phase has also been observed for annealed samples of symmetric-chain-length PCs such as C(13)C(13)PC.[39] Thus, as the chain-length asymmetry is progressively increased, the L_c phase first becomes more stable than the $L_{\beta'}$ gel phase and then more stable than the $P_{\beta'}$ gel phase.

2. In general, the rate of conversion of the $L_{\beta'}$ gel phase to the L_c crystalline phase increases as the chain-length asymmetry is increased. For bilayers composed of C(13)C(21)PC, the rate of conversion is sufficiently fast that a single apparent $P_{\beta'} \rightarrow L_c$ exothermic transition is observed upon cooling.[40] However, it has been argued that the formation of the L_c phase always proceeds by way of the $L_{\beta'}$ gel phase.[34]

3. The value of ΔS for the subtransition increases with increasing values of the chain inequivalence parameter due to the corresponding progressive increase in the disorder of the $P_{\beta'}$ gel phase. Accordingly, this progressive increase in ΔS for the subtransition is balanced by a concomitant decrease in the value of ΔS for the $P_{\beta'} \leftrightarrow L_\alpha$ main phase transition. As a result, the value of ΔS for the combined $L_c \rightarrow L_\alpha$ phase transition is essentially independent of $\Delta C/CL$. This total entropy change, normalized to the number of methylene groups per molecule, is $\Delta S \sim 1.7$ cal · mol^{-1} · K^{-1}.[34]

4. By studying the thermotropic behavior of a large number of mixed-chain-length PCs, Huang and co-workers[38,40-42] have established the range for Region I at $\Delta C/CL \sim 0.09$–0.41.

Finally, in Figure 2 it can be seen that the normalized transition entropy change decreases slightly for mixed-chain-length PCs whose values of $\Delta C/CL$ are less than that for C(18)C(18)PC. For these phospholipids, such as C(16)C(18)PC, the value of $\Delta C \sim 0.5$, which means that the terminal methyl groups of the phospholipid acyl chains are separated by a distance of only one half of a carbon-carbon bond length. Since the terminal methyl group occupies about twice the effective volume of a methylene unit,[37,43] the close proximity of the terminal methyls can be expected to greatly distort the chain packing at the center of the noninterdigitated bilayer. Alternatively, it has been suggested by Lin et al.[40] that the acyl chain with the shorter effective length will form a 2gl kink to shorten the chain so that its terminal methyl group will pack next to a methylene unit. In either case, the resulting distortion in the gel-phase chain packing will decrease the normalized transition entropy change for these bilayers relative to that of C(18)C(18)PC.

3. Region II

The phospholipids in Region II are organized into lamellar structures at all temperatures and adopt a mixed-interdigitated acyl chain packing conformation in the bilayer gel phase. The phospholipids in Region II include C(18)C(12)PC and C(18)C(10)PC. For these phospholipids the chain inequivalence parameter is close to 0.5, which means that one hydrocarbon chain of these lipid molecules in the fully extended all-*trans* conformation is about twice the length of the other.

Figure 3A shows the thermogram of a liposomal suspension of C(18)C(10)PC that was incubated at 0°C for 1 month. A single cooperative phase transition is observed in the heating scan, which is reversible upon immediately reheating (Figure 3C) the sample. This transition, which occurs at 19.1°C, is assigned to the main gel $\leftrightarrow$ liquid-crystalline phase transition of C(18)C(10)PC.[24] The cooling scan (Figure 3B) reveals two exothermic phase transitions at 17.7 and 14.6°C. The assignment for these phase transitions will be discussed below. As is evident from Figure 2, the values of the normalized transition entropy change for the gel $\leftrightarrow$ liquid-crystalline phase transitions of C(18)C(12)PC and C(18)C(10)PC are significantly larger than that for C(18)C(14)PC. This suggests that molecules of C(18)C(12)PC and C(18)C(10)PC are packed in the bilayer gel phase in a manner that is quite different from the partially interdigitated packing associations exhibited by the phospholipids in Region I.

As discussed in the previous section, Raman spectroscopic measurements were conducted on a series of mixed-chain-length PCs from C(18)C(10)PC to C(18)C(18)PC.[31] These measurements indicate that the strength of the lateral chain-chain interactions and the intramolecular conformational order of the acyl chains in the C(18)C(12)PC and C(18)C(10)PC gel-phase bilayers are significantly greater than those of C(18)C(14)PC and, in fact, are similar to those of the C(18)C(18)PC gel-phase bilayer. The symmetric-chain-length C(18)C(18)PC molecules are known to adopt a noninterdigitated gel-phase bilayer packing with a nearly all-*trans* conformation of the stearoyl chains.[44,45] Thus, despite their significant chain-length asymmetries, the acyl chains of the phospholipids in Region II must also adopt a nearly all-*trans* packing conformation in the bilayer gel phase.

X-ray diffraction analysis was employed to determine the gel-phase packing associations of the C(18)C(10)PC bilayer.[32-34] Based upon mensuration measurements,[32] the bilayer thickness for C(18)C(10)PC in the gel phase at 10°C was determined to be 33 Å, the average area per C(18)C(10)PC headgroup at the bilayer-water interface was found to be 63 Å^2, and the cross-sectional area of the acyl chains was determined to be 20.4 Å^2. The latter two values indicate that the phosphocholine headgroup area at the bilayer surface encompasses three acyl chains. A bilayer thickness of 33–34 Å was also determined from the electron density profile of C(18)C(10)PC at 13°C.[33] A bilayer thickness of 33–34 Å is consistent with the length of an all-*trans* stearic acid chain (27 Å) plus the overall thickness of the phosphocholine headgroups on both sides of the bilayer (7.3 Å).[12] Finally, a sharp symmetric wide-angle reflection at 4.1–4.2 Å was observed for the C(18)C(10)PC gel-phase bilayer, which indicates that the acyl chains of the phospholipid molecules are not tilted with respect to the bilayer normal. Similar measurements were obtained for C(18)C(12)PC bilayer dispersions.[32]

The packing model for the C(18)C(10)PC and C(18)C(12)PC gel-phase bilayers that is consistent with the above observations is depicted in Figure 6B. In this packing model the long stearoyl chains, in an all-*trans* conformation, are fully interdigitated across the entire hydrocarbon width of the bilayer, while the short acyl chains from lipid molecules in the opposing bilayer leaflets pack end-to-end. This has been designated a mixed-interdigitated bilayer packing[32] and given the symbol L$_\beta^M$. The mixed-interdigitated gel-phase packing is characteristic of the phospholipids in Region II.[32,34,46] Support for the mixed-interdigitated packing is provided by freeze-fracture electron microscopy studies of the C(18)C(10)PC gel-phase bilayer.[32] Freeze-fracture electron micrographs of symmetric-chain-length bilayers such as C(18)C(18)PC display smooth fracture planes due to the weak intermolecular interactions at the bilayer center. In contrast, the freeze-fracture planes of C(18)C(10)PC lamellae are interrupted repeatedly by up-and-down steps. This discontinuous fracture plane is consistent with the mixed-interdigitated bilayer packing, which has no preferred weak bonding plane.

The barotropic behavior of C(18)C(10)PC dispersions in D$_2$O was studied by high-pressure Fourier transform infrared spectroscopy[47] to determine the orientation of the acyl chain planes in the mixed-interdigitated gel-phase packing lattice. At pressures above 5.5 kbar at room temperature, the bilayers form a mixed-interdigitated chain packing where the acyl chain dynamics are similar to those exhibited by C(18)C(10)PC bilayers in the gel phase at low temperatures. These studies indicate that the planes of the linear all-*trans* acyl chains are oriented, both intra- and intermolecularly, nearly perpendicular to each other within their two-dimensional packing lattice.

Huang and co-workers[48,49] have studied the properties of a large number of PCs with chain inequivalence parameters near 0.5. These studies have revealed that the range for Region II is $\Delta C/CL \sim 0.42\text{--}0.67$. This narrow range is a consequence of the stringent geometric requirements for the formation of a stable mixed-interdigitated gel-phase bilayer packing. As shown in Figure 2, the normalized transition entropy changes of the PCs in Region II conform to a bell-shaped curve with a maximum at $\Delta C/CL \sim 0.57$, which corresponds to C(18)C(10)PC.[49] For C(18)C(10)PC, the sum of the effective lengths of two opposing decanoyl chains packed end-to-end plus the van der Waals contact distance between the two opposing terminal methyl groups (about 1.7 carbon-carbon bond lengths)[43] matches exactly the effective chain length of the stearoyl chains. This permits the terminal methyl group of the stearoyl chains to pack at the level of the phospholipid interface region at the bilayer surface.

Decreasing the value of $\Delta C/CL$ from 0.57 to 0.42 corresponds to mixed-chain-length PCs with progressively diminished chain-length asymmetry [C(18)C(12)PC is an example]. Consequently, a progressively greater number of $2gl$ kinks must be introduced into the shorter acyl chains of these PCs to allow the terminal methyl groups of the longer chains to pack at the surface of the mixed-interdigitated bilayer. In contrast, $\Delta C/CL$ values from 0.57 to 0.67 correspond to mixed-chain-length PCs with progressively enhanced chain-length asymmetry [C(18)C(9)PC is an example]. In this case, a progressively greater number of $2gl$ kinks must be introduced into the longer chains of these PCs to prevent the terminal methyl groups of these chains from protruding out into the aqueous solvent. The lateral chain-chain packing interactions and intrachain conformational order in the mixed-interdigitated gel-phase bilayer are perturbed by these kinks. The bell-shaped curve that describes the variation in the normalized transition entropy change with the value of $\Delta C/CL$ for the mixed-chain-length PCs in Region II is a consequence of this effect.

The packing characteristics discussed above also have consequences for the metastability of the mixed-interdigitated gel phase. To date, a crystalline phase for fully hydrated bilayers of C(18)C(10)PC has not been identified.[34] However, the suboptimally packed mixed-interdigitated gel-phase bilayers of the rest of the Region II phospholipids are metastable and will convert to a crystalline phase after prolonged incubation at low temperatures.[34,40,50] In the crystalline phase the acyl chains adopt a partially interdigitated packing conformation with two acyl chains per lipid headgroup.[34,37] Consequently, the crystalline-phase packing conformations of the Region II PCs form a continuous series with the partially interdigitated crystalline-phase packing conformations of the Region I PCs. The L_c phase is more stable than the L_β^M phase at all temperatures; thus, a single $L_c \rightarrow L_\alpha$ phase transition is observed for these phospholipids, which occurs at a higher temperature than the corresponding $L_\beta^M \leftrightarrow L_\alpha$ phase transition.[34,40,50] The thermotropic behavior of C(18)C(10)PC can be described by the relationship

$$L_\beta^M \longleftrightarrow L_\alpha$$

The thermotropic behavior of the rest of the Region II mixed-chain-length PCs that have been characterized can be described by the following relationships:

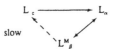

As noted above, two exothermic phase transitions are seen by DSC in cooling scans of C(18)C(10)PC dispersions that have been incubated at low temperatures. This is a general property of Region II phospholipids.[25,51] The appearance of the two peaks depends upon the thermal history of the sample. Freshly prepared samples of these PCs, such as C(18)C(10)PC, exhibit only the low-temperature exotherm, which occurs at 14.6°C for C(18)C(10)PC. However, if the samples are held at –20°C for 24 h, all subsequent cooling scans of the samples will show both of the exothermic phase transitions. It has been suggested[29,34] that this behavior results from the formation of an intermediate between the L_α and the L_β^M phases, perhaps the $L_{\beta'}^P$ phase. However, other studies have concluded that this behavior is not associated with any change in the packing conformation of the lipid acyl chains but may arise instead from an alteration in the size or structure of microdomains present in the liquid-crystalline bilayer phase. In this case, both exothermic peaks represent $L_\alpha \rightarrow L_\beta^M$ phase transitions, but they arise from different liquid-crystalline phase microdomains.[51] Clearly, more studies are required to resolve these issues.

In the liquid-crystalline phase, fully hydrated C(18)C(10)PC forms a lamellar organization with two acyl chains per lipid headgroup.[32-34] The bilayer thickness of the C(18)C(10)PC liquid-crystalline phase is less than that of C(14)C(14)PC in the liquid-crystalline phase. This indicates that there is some degree of partial interdigitation of the lipid acyl chains in the C(18)C(10)PC liquid-crystalline bilayer. The partially interdigitated liquid-crystalline phase has been given the designation[46] L_α^p. This finding is supported by magic-angle NMR studies of the C(18)C(10)PC liquid-crystalline phase.[52] Although the overall motional behavior of the acyl chains was clearly liquid-like, both [1]H- and ^{13}C-spin-lattice relaxation times indicated that there were unusual restrictions on the segmental reorientation of the C(18)C(10)PC acyl chains at megahertz frequencies as compared to C(16)C(16)PC in the liquid-crystalline phase. More studies are clearly required to determine if dynamic acyl chain interdigitation in the liquid-crystalline bilayer phase is a general property of the PCs in Region II.

4. Region III

The PCs in Region III include C(18)C(8)PC, C(18)C(6)PC, and C(18)C(4)PC. These phospholipids are the least studied members of the C(18)C(18)PC through C(18)C(0)PC series due, in part, to their low transition temperatures and complex metastable behavior.[29,53] ^{31}P-NMR spectra of the above three phospholipids confirm that they are organized into lamellar structures at temperatures below their respective main phase-transition temperatures. Results from Raman spectroscopy (C. Huang and J. T. Mason, unpublished work) and calorimetry[29,53] suggest that the phospholipids in Region III adopt a partially interdigitated acyl chain conformation in the bilayer gel phase. However, preliminary results from ^{31}P-NMR spectroscopic studies of the phospholipids suggest that these phases are metastable and may convert to other, more stable gel phases and also to lamellar crystalline phases after prolonged incubation of the dispersions at low temperatures.

Upon heating, the lamellar phase of C(18)C(8)PC, C(18)C(6)PC, and C(18)C(4)PC will transform to an "isotropic-viscous" phase. Dispersions of phospholipids in this phase are optically transparent but highly viscous and, at high enough concentrations, can have the consistency of an agar. ^{31}P-NMR spectra of this phase yield sharp, isotropic phosphorous resonances consistent with a cubic, rhombic, or micellar phase.[54] Freeze-fracture electron micrographs of the isotropic-viscous phase appear to reveal the presence of both individual micelles and long tubular structures, perhaps cylindrical micelles. Thus, the main phase transition of the phospholipids in Region III is most likely a lamellar → micellar phase transition.

The thermogram of a preparation of C(18)C(6)PC held at 0°C for 1 month is shown in Figure 3A. This heating scan shows two highly overlapped endothermic transition peaks centered at 8.2 and 11.2°C. When the sample was immediately reheated (Figure 3C), only the endotherm at 8.2°C was seen. On cooling (Figure 3B), a slightly asymmetric exothermic transition peak was seen at 1.4°C. This thermotropic behavior is representative of the other mixed-chain-length PCs in Region III. In contrast to the phospholipids in Region I, the normalized transition entropy change for the gel → micellar phase transition of the phospholipids in Region III increases linearly with an increase in the chain inequivalence parameter. This observation suggests that once the depth of interdigitation has reached a critical level, further increases in chain-length asymmetry give rise to progressively more stable partially interdigitated gel-phase chain packings. This phenomenon will be discussed in greater detail in the Summary (Section III.E).

In summary, the mixed-chain-length phospholipids in Region III most likely adopt a partially interdigitated gel-phase packing conformation and organize into cylindrical micelles at temperatures above the chain-melting phase-transition temperature.

5. Region IV

The PCs in Region IV are characterized by their association into discrete, nonspherical micelles above a critical temperature and into extended fully interdigitated lamellar sheets below this critical temperature. The mixed-chain-length PCs C(18)C(0)PC (1-stearoyl-2-lyso-sn-glycero-3-phosphocholine) and C(18)C(2)PC are members of this group. The thermotropic properties of C(18)C(0)PC will be described in detail.

Lysophospholipids may be regarded, structurally, as an extreme case of a mixed-chain-length phospholipid where the entire sn-2 acyl chain has been replaced by a hydrogen atom. The thermotropic behavior of the lysophosphatidylcholine (lysoPC) C(18)C(0)PC has been studied extensively.[55-57] After prolonged incubation at 0°C, C(18)C(0)PC dispersions display a sharp, cooperative lamellar → micellar phase transition at 26°C when examined by calorimetry, as shown in Figure 3A. This phase transition is more cooperative and has a larger transition entropy change per acyl chain than the gel ↔ liquid-crystalline phase transition of C(18)C(18)PC.[55] However, as shown in Figure 3B and 3C, the lamellar → micellar phase transition is irreversible without prolonged incubation of the sample at temperatures near 0°C.

The kinetics of the micellar → lamellar phase transition of C(18)C(0)PC has been studied by a variety of techniques.[56] The rate of lamellar formation from micelles was shown to require significant supercooling of the micelles and to have an extremely large negative temperature coefficient. A lag time always preceded the conversion of the micelles to the lamellar phase. Raman spectroscopy[55] has shown that the C(18)C(0)PC acyl chains are more tightly packed in micelles at supercooling temperatures than at temperatures above 26°C. In addition, the formation enthalpy, measured by calorimetry, for the micellar → lamellar conversion at temperatures near 0°C was shown to decrease with decreasing incubation temperature. Based upon these findings, it was proposed that the micellar → lamellar phase transition of C(18)C(0)PC observed at supercooling temperatures is best described by a two-dimensional nucleation and growth process.[56] At supercooling temperatures, the micelles associate into large aggregates. The micelles within the aggregates are under stress due to the distortion of the micellar shape that results from the ordering of the lysoPC acyl chains at supercooling temperatures. Thus, at numerous sites within the micellar aggregates, closely apposed micelles will convert into lamellar nuclei. Those nuclei that are smaller than a critical size will collapse and re-form the micellar structure. Those nuclei that are larger than a critical size will grow to form larger lamellar organizations, which, in turn, will coalesce with other such structures within the aggregate. This process will continue until the whole aggregate has been converted to the thermodynamically stable lamellar organization. Finally, extended lamellar sheets will be formed by the annealing of numerous smaller lamellar aggregates.

Raman spectroscopy of the C(18)C(0)PC lamellar phase reveals that the lysoPC acyl chains are packed in a hexagonal lamellar lattice whose chain-chain lateral interactions and intrachain conformational order are greater than those of the acyl chains of C(18)C(18)PC in the bilayer gel phase, which is consistent with the calorimetry results described above.[55] Subsequently, X-ray diffraction was employed to determine the structure of the C(18)C(0)PC lamellar phase.[57] The electron density profile of the C(18)C(0)PC lamellar phase yields a phosphate-to-phosphate bilayer thickness of 35 Å and indicates that the electron density trough at the bilayer center seen in the profiles of noninterdigitated bilayers is absent. From mensuration measurements, the average area per lysoPC headgroup at the bilayer-water interface was found to be 45.5 Å^2, and the cross-sectional area of the acyl chains was determined to be 19.5 Å^2. These two values indicate that the lysoPC headgroup area at the bilayer surface encompasses two acyl chains.[57] The 35 Å bilayer thickness is equal to the sum of the length of an all-*trans* stearic acid chain (27 Å) plus the overall thickness of the two phospholipid headgroups (7.3 Å). These results are consistent with a packing model where the stearoyl chains extend across the entire hydrocarbon width of the bilayer and interact laterally with the stearoyl chains from the opposing bilayer leaflet. The terminal ends of the stearoyl chains pack underneath the interface region of the lysoPCs, which results in an organization where each lysoPC headgroup spans two acyl chains. The observation of a sharp, symmetric, high-angle X-ray reflection at 4.11 Å indicates that the acyl chains are not tilted with respect to the bilayer normal.[57] This packing is referred to as a fully interdigitated bilayer conformation (given the symbol L_β^F), which is shown in Figure 7D. The greater order of the C(18)C(0)PC gel lamellar phase relative to that of C(18)C(18)PC most likely results from the lack of a distorting bilayer center in the fully interdigitated C(18)C(0)PC packing conformation and the resulting stronger van der Waals contacts among the acyl chains.

In contrast to the highly ordered bilayer interior of the C(18)C(0)PC lamellar phase, the lysoPC headgroups display a significant degree of dynamic disorder, even at temperatures as low as −20°C.[58] This has been explained by the fact that the C2 atom of the glycerol moiety is not anchored to the bilayer interior by way of an *sn*-2 acyl chain in lysophospholipids. Consequently, the C1–C2 glycerol bond is able to rotate freely, which imparts an increased range of dynamic motion to the lysophospholipid headgroup. This model was used to explain the smaller chemical shift anisotropy of lysophospholipids relative to that of symmetric-chain-length phospholipids.[58]

Extensive studies on the thermotropic behavior of aqueous suspensions of C(18)C(2)PC have also been conducted.[29,59] This phospholipid is of interest because it is an analogue of platelet-activating factor, a lipid autocoid, and is structurally related to several lipids that have anti-tumor or anti-human immunodeficiency virus activity. This highly asymmetric phospholipid differs from C(18)C(0)PC in that the C(18)C(2)PC fully interdigitated gel phase is metastable and will convert to a lamellar crystalline phase

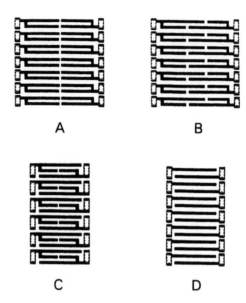

Figure 7 Summary of the noninterdigitated and interdigitated phospholipid packing models. (A) Noninterdigitated chain packing. (B) Partially interdigitated chain packing. (C) Mixed-interdigitated chain packing. (D) Fully interdigitated chain packing.

after prolonged incubation at low temperatures. In the L_c phase, the C(18)C(2)PC molecules are packed in a fully interdigitated orthorhombic lattice with minimally hydrated headgroups.[59] When heated, the C(18)C(2)PC L_c phase transforms to a fully interdigitated gel phase at 13.9°C. The gel lamellar phase transforms to the micellar phase at 18.6°C; however, a small thermal pretransition (at 17.1°C) immediately precedes the lamellar → micellar phase transition.[29] The different thermotropic behavior of C(18)C(2)PC relative to that of C(18)C(0)PC is believed to be due to the presence of the short sn-2 acetyl moiety in C(18)C(2)PC.[29] This short chain segment can anchor the glycerol backbone to the bilayer hydrocarbon interior at low temperatures, which allows the formation of the L_c phase. At higher temperatures, however, the dynamic motions of the acetyl moiety perturb the chain packing in the bilayer interior, which leads first to the $L_c \to L_\beta^F$ phase transition, followed by the thermal pretransition at 17.1°C.

B. SATURATED MIXED-CHAIN-LENGTH PHOSPHATIDYLETHANOLAMINES

PEs differ from PCs in that PEs can engage in headgroup-headgroup hydrogen bonding interactions with adjacent PE molecules in the bilayer.[60] The strong hydrogen-bonding propensity of PEs accounts for most of the differences in thermotropic behavior seen between symmetric-chain-length PCs and PEs. Thus, it was of interest to determine if the ability of mixed-chain-length PEs to engage in strong intermolecular headgroup interactions would alter the interdigitated acyl chain packing conformations that were observed for the corresponding mixed-chain-length PCs.

The thermotropic behavior of the mixed-chain-length PE series C(18)C(18)PE, C(18)C(16)PE, C(18)C(14)PE, C(18)C(12)PE, and C(18)C(10)PE has been investigated by calorimetry and ^{31}P-NMR spectroscopy.[25] The thermodynamic parameters of these PEs are listed in Table 1, and the normalized transition entropy changes for their gel ↔ liquid-crystalline phase transitions are plotted against the corresponding values for the chain inequivalence parameter in Figure 8. Unhydrated samples (lipids dispersed in water at 0°C) of all of these PEs exhibited a lamellar crystalline → liquid-crystalline phase transition whose transition temperature (T_H) increased in proportion to the molecular weight of the lipid. ^{31}P-NMR spectroscopy of these preparations confirmed that the PEs form a poorly hydrated crystalline phase at temperatures below T_H and a liquid-crystalline bilayer phase at temperatures above T_H. Thus, the lamellar $L_c \to L_\alpha$ phase transition corresponds to the combined hydration and acyl chain melting of a poorly hydrated crystalline phase.

When the PE dispersions were cooled from temperatures above T_H and immediately reheated, reversible gel ↔ liquid-crystalline phase transitions were observed whose transition temperatures (T_L) were less than the corresponding values of T_H. For dispersions of C(18)C(18)PE, C(18)C(16)PE, and C(18)C(14)PE, the ^{31}P-NMR spectra at $T > T_L$ were consistent with a liquid-crystalline bilayer phase

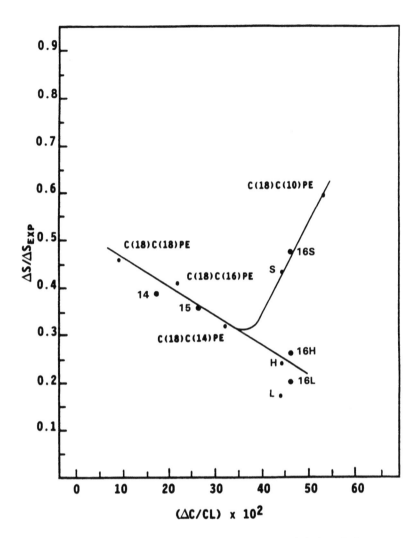

Figure 8 Plot of the normalized transition entropy change versus the chain inequivalence parameter for the sphingomyelins and the C(18)C(18)PE through C(18)C(10)PE series. The thermodynamic parameters of these phospholipids are given in Table 1, as are the number codes for the sphingomyelins. For hydrated samples of C(18)C(12)PE, values are indicated for the low-temperature (L) and the high-temperature (H) phase transitions and the sum (S) of these two phase transitions. For hydrated samples of C(24)SPM, values are indicated for the low-temperature (16L) and high-temperature (16H) phase transitions and the sum (16S) of these two phase transitions. (Modified from Mason, J. T. and Stephenson, F. A., *Biochemistry*, 29, 590, 1990. With permission.)

while the spectra at $T < T_L$ were indicative of a hydrated gel lamellar phase. As shown in Figure 8, the normalized transition entropy changes for these three PEs decrease linearly with an increase in $\Delta C/CL$, the same trend that was observed for the corresponding PCs.[24] In fact, the points for the corresponding mixed-chain-length PEs and PCs are virtually superimposable. Thus, C(18)C(18)PE adopts a noninterdigitated gel-phase bilayer packing, while C(18)C(16)PE and C(18)C(14)PE pack with partial (dynamic) interdigitation of their acyl chains across the bilayer center in the gel phase.

Bilayers of C(18)C(10)PE yielded thermodynamic parameters that were virtually identical to those of C(18)C(10)PC, including the presence of a double-peak exothermic transition profile in descending-temperature DSC scans of the PE. ^{31}P-NMR spectra of C(18)C(10)PE and C(18)C(10)PC in the gel phase were identical; both yielded bilayer line shapes with chemical shift anisotropies of ~82 ppm. Gel-phase ^{31}P-NMR bilayer line shapes with chemical shift anisotropies above 80 ppm are diagnostic of the mixed-interdigitated chain packing conformation.[61] The fact that the transition temperatures of C(18)C(10)PC, C(18)C(10)PE, and C(18)C(10)PA (phosphatidic acid; J.T. Mason, unpublished data) are all virtually identical indicates that the increased headgroup spacing in the mixed-interdigitated gel phase largely

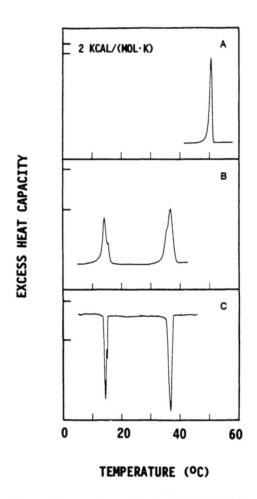

Figure 9 Excess heat capacity versus temperature profiles for C(18)C(12)PE dispersions. (A) Heating scan of an unhydrated preparation. (B) Heating scan of a hydrated preparation that was heated to 55°C, rapidly cooled to 0°C, and then immediately scanned. (C) Cooling scan conducted immediately after the heating scan of B. (From Mason, J. T. and Stephenson, F. A., *Biochemistry*, 29, 590, 1990. With permission.)

serves to negate intermolecular headgroup interactions; thus, the thermotropic behavior of the phospholipids is dictated, almost exclusively, by the hydrocarbon chain packing associations.

Bilayers of C(18)C(12)PE exhibited a thermotropic behavior that differed from that of C(18)C(12)PC. As shown in Figure 9A, unhydrated samples of C(18)C(12)PE exhibit a lamellar crystalline → liquid-crystalline phase transition at 50.8°C. When this sample was immediately rescanned from a temperature above 55°C, a pair of reversible phase transitions were observed at about 14 and 37°C (Figures 9B and 9C). ^{31}P-NMR spectra of the three phases bordering the two phase transitions indicate that the low-temperature transition corresponds to a mixed-interdigitated ↔ partially interdigitated gel-phase transition, while the high-temperature transition represents the transition of the partially interdigitated gel phase to the liquid-crystalline phase. You will be left in suspense as to the reason for this behavior, which will be discussed at the end of the section on sphingomyelins (Section III.D).

The normalized transition entropy changes for the low- and high-temperature phase transitions of C(18)C(12)PE are plotted in Figure 8 as points L and H, respectively. The partially interdigitated gel ↔ liquid-crystalline phase transition of C(18)C(12)PE is analogous to the phase transitions of C(18)C(18)PE, C(18)C(16)PE, and C(18)C(14)PE. Accordingly, point H conforms to the same linear trend that was defined by the latter three PEs. The total normalized transition entropy change of the two phase transitions of C(18)C(12)PE would correspond to that of a mixed-interdigitated gel ↔ liquid-crystalline phase transition. Indeed, this value (point S in Figure 8, where S = L + H) is almost identical to the normalized transition entropy change of C(18)C(12)PC.[24]

The behavior of a series of 1-acyl-lysophosphatidylethanolamines (lysoPEs) ranging in chain length from 12 to 18 carbons has been investigated by Slater et al.[62] using calorimetry, Raman spectroscopy, and ^{31}P-NMR spectroscopy. The properties of the lysoPEs were dictated by the tendency of the phosphoethanolamine headgroups to exchange headgroup-water interactions for more energetically favorable headgroup-headgroup hydrogen-bonding interactions.

The behavior of C(16)C(0)PE is representative of this series. Unhydrated samples of this lysoPE undergo a lamellar crystalline → micellar phase transition at 59.2°C (T_H). This transition corresponds to the simultaneous hydration and acyl chain melting of a poorly hydrated crystalline phase. The lysoPE molecules pack in the lamellar crystalline phase in a fully interdigitated orthorhombic lattice, and the phosphoethanolamine headgroups are dehydrated. When micelles are cooled from temperatures above T_H and immediately reheated, a single phase transition is seen at 39.7°C (T_L). This transition is a lamellar gel ↔ micellar phase transition. The lysoPE molecules pack in the hydrated lamellar gel phase in a fully interdigitated hexagonal lattice with rotational disorder of the acyl chains. If C(16)C(0)PE micelles are held at a temperature between T_L and T_H, they become metastable with respect to the crystalline phase, which will form directly from the micellar phase by slow interconversion. If C(16)C(0)PE micelles are supercooled to temperatures below T_L, they rapidly convert into the lamellar gel phase. However, the lamellar gel phase is metastable with respect to the lamellar crystalline phase, and the lamellar gel → lamellar crystalline transformation will occur upon prolonged storage of the sample. Thus, the lamellar crystalline phase may arise directly from the micellar phase or by way of a metastable lamellar gel-phase intermediate.

In summary, the thermotropic behavior of the mixed-chain-length PEs is remarkably similar to that of the corresponding PCs. The effects of chain-length asymmetry on the packing associations of the acyl chains in the bilayer gel phase are, with minor exceptions, identical for both phospholipids. The main difference in the behavior of the two phospholipids arises from the pronounced metastability of the hydrated PE phases and their tendency to form a stable lamellar crystalline phase by dehydration.

C. UNSATURATED MIXED-CHAIN-LENGTH PHOSPHATIDYLCHOLINES

Naturally occurring phospholipids are usually of the sn-1 saturated, sn-2 unsaturated acyl chain variety.[1] Thus, there is considerable interest in the properties of unsaturated mixed-chain phospholipids, particularly those with polyunsaturated acyl chains. Unfortunately, there is a dearth of studies on this class of phospholipids due, in part, to the fact that the phase-transition temperatures of these lipids are well below 0°C. In addition, it would be difficult to determine the effective chain length (CL) of polyunsaturated fatty acids for our analysis. However, the effective chain length of monoenoic fatty acids in the bilayer gel phase can be reliably estimated, which allows the chain inequivalence parameter to be calculated for these simple unsaturated mixed-chain-length phospholipids.

One of the most complete studies of simple unsaturated mixed-chain-length PCs is that of Davis et al.[63] These authors investigated the calorimetric behavior of PCs containing oleic acid (cis-9-octadecenoic acid), specifically C(18:1)C(16)PC, C(16)C(18:1)PC, C(18:1)C(18)PC, and C(18)C(18:1)PC. The gel ↔ liquid-crystalline phase transitions of these phospholipids were found to be broad and poorly cooperative. The associated thermodynamic parameters of the phospholipids are listed in Table 1.

A minimum perturbation of the gel-phase hydrocarbon chain packing can be obtained by the following conformation of the oleic acid chain. If a gauche conformation occurs at the β position on either side of the cis double bond and the adjacent α-carbon rotates about the C=C–C single bond by 30°, then the conformation of the oleic acid chain will be linear except for a small lateral displacement of the double bond.[63,64] This conformation is referred to as the Δtg-coupled isomerization or Δtg kink.[64] This conformation of the oleic acid chain plus an all-trans conformation of the saturated acyl chain would allow the chains to maximize their van der Waals contacts in the bilayer gel phase. The length of an oleic acid chain with the above conformation is about 0.3 carbon-carbon bond lengths shorter than an all-trans stearic acid chain.

With these conformational principles in mind, the chain inequivalence parameters of the oleic acid-containing unsaturated mixed-chain-length PCs can be calculated. The values of $\Delta C/CL$ for the PCs range from 0.013 to 0.192. This places them in Region I as defined in Figure 2. Thus, the PCs are either noninterdigitated or pack with partial (dynamic) interdigitation of their acyl chains across the bilayer center in the gel phase. The shape of the plot of normalized transition entropy change vs. $\Delta C/CL$ for the unsaturated mixed-chain-length PCs shown in Figure 2 is identical to that for the saturated mixed-chain-length PCs, but the curve is displaced lower on the y axis. This most likely arises from the perturbing effect of the double bond on the packing association of the hydrocarbon chains in the bilayer gel phase of the unsaturated PCs. Thus, it may be concluded that for the series of unsaturated mixed-chain-length phospholipids with oleic acid chains the effect of chain-length inequivalence on the packing properties of the bilayer gel phase is essentially the same as that seen in saturated mixed-chain-length phospholipids. Clearly, more studies are required on this interesting class of phospholipids.

D. SPHINGOMYELINS

Sphingomyelins (N-acyl-sphingosine-1-phosphocholines) are another interesting class of naturally occurring phospholipids. The sphingomyelin molecule consists of a phosphocholine headgroup, an N-acyl-linked fatty acid chain, and an 18-carbon amine diol backbone, 1,3-dihydroxy-2-amino-4-octadecene (sphingosine). The sphingosine base has a *trans* double bond between carbons 4 and 5 and a free hydroxyl group at carbon 3. The effective chain length of the sphingosine tail is about 14 carbon units. Sphingomyelins of biological origin are of the D-*erythro* configuration. The hydroxyl group and the amide linkage of sphingomyelin are unique among the phospholipids and afford the sphingomyelins a hydrogen bond donor-acceptor capability that has been invoked to explain many of the physical properties of sphingomyelins.[65] The principal N-acyl groups in natural sphingomyelins are C(16), C(24:1), C(22), and C(24).[65] Thus, a significant proportion of naturally occurring sphingomyelins are characterized by significant chain-length asymmetry.

The thermotropic behavior of several synthetic sphingomyelins with the racemic DL-*erythro* configuration has been examined in some detail.[66-72] These species include N-palmitoylsphingosylphosphocholine [C(16)SPM], C(18)SPM, and C(24)SPM. The calorimetrically determined phase-transition parameters of these sphingomyelins are given in Table 1. C(16)SPM exhibits a reversible gel ↔ liquid-crystalline phase transition at 41.3°C.[66] The behavior of C(18)SPM is more complex. Incubation of a preparation of this sphingomyelin at low temperatures for prolonged periods yields a single endothermic phase transition at 57°C.[67] This phase transition is believed to be a lamellar crystalline → liquid-crystalline phase transition. Cooling the sample results in a supercooled liquid-crystalline phase that undergoes a transition to a metastable gel phase at 44°C. Immediate reheating reveals only a metastable gel ↔ liquid-crystalline phase transition centered at 45°C.[67]

On the basis of the conformation of ceramides and sphingolipids[68,69] determined by X-ray diffraction of single crystals, it can be calculated that an N-acyl chain length of 13.5 carbons would be required to give a value of zero for ΔC. Thus, the chain inequivalence parameters for the sphingomyelins can be calculated. Figure 8 shows that C(16)SPM is about in the middle of Region I, while C(18)SPM is near the end of Region I as defined by the mixed-chain-length PEs. Thus, it can be predicted that both of these sphingomyelins pack with partial (dynamic) interdigitation of their hydrocarbon chains across the bilayer center in the gel phase. In accordance with this prediction, electron density profiles of small unilamellar vesicles of C(16)SPM and C(18)SPM revealed a modest level of interdigitation in these bilayer systems.[71]

The chain inequivalence parameter for C(24)SPM is very close to that for C(18)C(12)PC and C(18)C(12)PE and places this sphingomyelin in the middle of Region II. Thus, one would predict a mixed-interdigitated gel-phase packing for C(24)SPM. DSC reveals two phase transitions for C(24)SPM, one at 40°C and the other at 47.5°C.[70] A Raman spectroscopic study[72] comparing the thermotropic behavior of C(24)SPM to that of C(18)C(12)PC and C(18)C(16)PC indicates that the C(24)SPM transition at 40°C corresponds to a mixed-interdigitated ↔ partially interdigitated gel-phase transition, while the transition at 47.5°C represents the partially interdigitated gel ↔ liquid-crystalline phase transition. The normalized transition entropy change for the low- and high-temperature phase transitions of C(24)SPM are plotted in Figure 8 as points 16L and 16H, respectively. The partially interdigitated gel ↔ liquid-crystalline phase transition of C(24)SPM is analogous to the phase transitions of the mixed-chain-length PEs and sphingomyelins of Region I. Accordingly, point 16H conforms to the same linear trend as these phospholipids. The total normalized transition entropy change for the two transitions of C(24)SPM corresponds to that of a mixed-interdigitated gel ↔ liquid-crystalline phase transition. Indeed, this value (point 16S in Figure 8, where 16S = 16L + 16H) conforms quite well to the profile defined by the mixed-chain-length phospholipids in Region II. Clearly, there is a strong similarity in the thermotropic behaviors of the sphingomyelins and the mixed-chain-length PEs.

It is interesting to compare the thermotropic behaviors of C(18)C(12)PC, C(18)C(12)PE, and C(24)SPM. The adoption of the mixed-interdigitated gel-phase packing comes at the expense of eliminating, or diminishing, energetically favorable intermolecular interactions that originate in the interface or polar headgroup regions of the phospholipids. Also, the expanded bilayer surface of the mixed-interdigitated phase results in exposure of the terminal methyl groups of the longer phospholipid acyl chains to water at the bilayer surface. The energy cost of this exposure has been calculated to be about 1.4 kcal/mol.[33] As the bilayer is heated, the strength of the van der Waals interactions among the hydrocarbon chains of the mixed-interdigitated gel phase will become weakened as the result of thermal expansion. For the C(18)C(12)PE and C(24)SPM mixed-interdigitated bilayers, a critical temperature will be reached where the energy lost by converting from a mixed-interdigitated to a partially interdigitated chain packing conformation is outweighed by the energy gained in eliminating the exposure of the terminal methyl groups to water and forming intermolecular hydrogen bonds in the headgroup [C(18)C(12)PE] or the interface [C(24)SPM] region of the bilayer. In contrast, PCs are devoid of any such intermolecular interactions; thus, the mixed-interdigitated gel phase of C(18)C(12)PC is stable until the bilayer under-goes the gel ↔ liquid-crystalline phase transition.

E. SUMMARY

The phase diagram of Figure 2 was originally derived from the thermotropic behavior of the C(18)C(18)PC through C(18)C(0)PC mixed-chain-length PC series. However, this phase diagram has subsequently been shown to be more general.[25,29] Thus, the thermotropic behavior of any mixed-chain-length PC, irrespective of the distribution of the acyl chains on the glycerol backbone, will conform to this phase diagram. Also, with minor exceptions, sphingomyelins and mixed-chain-length PEs display thermotropic behavior that is consistent with this phase diagram. These conclusions underscore the utility of the chain inequivalence parameter in analyzing the behavior of mixed-chain-length phospholipids. The key concept conveyed by the chain inequivalence parameter is that the thermotropic behavior of the mixed-chain-length phospho-lipids is dictated by the size of the chain mismatch region (ΔC) relative to the length of the hydrocarbon region (CL) of the phospholipid molecule. The distribution of the inequivalent acyl chains on the glycerol backbone, the absolute size of the chain mismatch region, or the absolute lengths of the acyl chains are not of primary importance. For example, mixed-chain-length PCs with a chain inequivalence parameter near 0.5 will adopt a mixed-interdigitated gel-phase packing, irrespective of the molecular weight of the phospholipid molecule.[49]

It is worthwhile to restate the properties of the mixed-chain-length phospholipids in the four regions defined in the phase diagram of Figure 2. Phospholipids in Regions I and II form a lamellar phase at all temperatures. The phospholipids in Region I adopt a noninterdigitated (Figure 7A) or a partially interdigitated (Figure 7B) gel-phase packing, whereas those in Region II form a mixed-interdigitated (Figure 7C) gel-phase packing. The phospholipids in Region III are thought to adopt a cylindrical (rod-like) micellar morphology in the chain-melted state and a partially interdigitated gel-phase packing. The phospholipids in Region IV form discrete, nonspherical micelles in the chain-melted state and adopt a fully interdigitated (Figure 7D) gel-phase packing organization.

If one were to ignore the presence of Region II in the phase diagram of Figure 2, the plot of normalized transition entropy change vs. $\Delta C/CL$ would conform to a well-shaped curve with a minimum near $\Delta C/CL \sim 0.65$ (see the dotted line under Region II). This pattern of behavior can be explained by the following arguments. There exists a "critical value" of $\Delta C/CL$ in the phase diagram near 0.65. Below this value, the extended ends of the longer phospholipid acyl chains that form the partially interdigitated bilayer center (the ΔC region) are short enough that the van der Waals (dispersion) forces between these chain segments are insufficient to stabilize a gel-phase packing against the disruptive effect of thermal energy at temperatures just below T_m. The close proximity of the terminal methyl groups of the chain to the bilayer center will also serve to disrupt the chain packing in the ΔC region.[24] Consequently, the ΔC region will remain disordered while the upper chain segments undergo the liquid-crystalline ↔ gel phase transition. Thus, for phospholipids with chain inequivalence parameters less than 0.65, the normalized transition entropy change will increase in proportion to a decrease in the size of the ΔC region relative to the hydrocarbon thickness (CL) of the bilayer.

Above the critical value of $\Delta C/CL$, the terminal methyl groups of the chain begin to move away from the bilayer center up toward the interface region, where they are less disruptive to chain packing. At the same time, the lengths of the extended chain segments that form the ΔC region are now great enough that the van der Waals forces between these chain segments are sufficiently strong to support a gel-phase packing against the disruptive effect of thermal energy at temperatures just below T_m. Thus, the ΔC region will now participate in the liquid-crystalline ↔ gel phase transition. In addition, the strength of the lateral chain-chain interactions and the intrachain conformational order will increase in the ΔC region in proportion to its relative size. Consequently, for phospholipids with chain inequivalence parameters greater than 0.65, the normalized transition entropy change will increase in proportion to an increase in the size of the ΔC region relative to the hydrocarbon thickness (CL) of the bilayer. Of course the normalized entropy minimum at $\Delta C/CL \sim 0.65$ is never actually realized because the ability of the mixed-chain-length phospholipids to form the mixed-interdigitated gel-phase packing is superimposed upon the effects discussed above.

The different morphologies adopted by the chain-melted phase of the mixed-chain-length phospholipids can be understood by considering the effect of chain-length asymmetry on the molecular geometry of the phospholipids. Israelachvili[73] has derived a "shape factor" that determines the morphology adopted by amphiphiles in aqueous solution. This shape factor is given as v/a_ol_c, where a_o is the optimal headgroup area of the phospholipid, v is the hydrocarbon volume, and l_c is the critical chain length. The shape factor will determine whether the amphiphiles form spherical micelles ($v/a_ol_c < 1/3$), nonspherical micelles ($1/3 < v/a_ol_c < 1/2$), or bilayers ($1/2 < v/a_ol_c < 1$).

For the mixed-chain-length PC series C(18)C(18)PC through C(18)C(0)PC, a reasonable estimate[74] for a_o is 0.6 nm^2, and the value of l_c is held constant at 2.43 nm, which corresponds to the extended length of the stearic acid chain.[75] The value of v is allowed to vary and is estimated from the equation given by Tanford.[74,75]

$$v \sim [0.0274 + 0.0269(n_1 + n_2)] \text{ nm}^3$$

Aqueous dispersions of C(18)C(18)PC through C(18)C(9)PC have been shown to form a lamellar phase at all temperatures,[29] and their shape factors are, indeed, all above 0.5, with the value for C(18)C(9)PC being 0.52. However, as the shape factor decreases in value for these PCs, the average size of the liposomes is observed to decrease.[76] In addition, the distribution of liposomal sizes in the dispersions becomes more bimodal, with the dispersions consisting of both larger multilamellar liposomes and smaller single-walled vesicles.[32,76] The value of the shape factor for C(18)C(8)PC is 0.49; this suggests that this phospholipid will form a micellar chain-melted phase, which is observed to be the case. The shape factors for C(18)C(8)PC through C(18)C(0)PC are all between 0.5 and 0.33, which suggests that these phospholipids will form nonspherical micellar structures in their chain-melted phases. However, the shape factors for C(18)C(2)PC and C(18)C(0)PC are both close to 0.33, which is consistent with their forming discrete micelles that are nonspherical in shape. The shape factors for C(18)C(8)PC through C(18)C(4)PC are in the range of 0.49 to 0.42, which will impart a "truncated cone" shape to these phospholipids. This molecular geometry is consistent with the formation of cylindrical micelles.[73]

IV. APPLICATIONS TO LIPOSOME TECHNOLOGY

The studies described in this review on the physical properties of mixed-chain-length phospholipid assemblies have demonstrated that considerable manipulation of the structure and dynamic properties of the bilayer phase can be achieved through the use of mixed-chain-length phospholipids. These properties include the bilayer thickness, the degree of bilayer interdigitation, the phospholipid headgroup area, the temperature of the main bilayer phase transition, and the ability of the bilayer to undergo transitions between lamellar and nonlamellar phases. The bulk solution properties of the mixed-chain-length phospholipid suspension, such as viscosity and optical transparency, can also be varied. Thus, I would like to conclude this review with a brief discussion of how the properties of mixed-chain-length phospholipids could be exploited in a few selected areas of liposome technology.

It has been demonstrated that bilayers can act as a unique reaction medium capable of enhancing reaction rates by several millionfold over those in aqueous solution.[77] The solubilization of reactants on vesicle surfaces can also lead to altered reaction routes and stereochemistries. These effects are related to the bilayer association/dissociation constants of both reactants and products and their orientation on the bilayer surface.[78] The key properties of the bilayer surface responsible for these reactant/product-bilayer interactions are believed to be amphiphile headgroup packing, surface charge density, and degree of surface hydration.[78,79] In symmetric-chain-length phospholipid bilayers, varying the headgroup packing without varying the bilayer phase or headgroup type is difficult to achieve. Headgroup packing variations have been most commonly produced through radius-of-curvature effects by varying vesicle size.[78] For example, the reaction rate for the catalytic bromination of *trans*-stilbene by C(16)C(16)PC vesicles increased with decreasing vesicle size.[80]

Mixed-chain-length phospholipids and interdigitated bilayers offer a way to vary headgroup spacing and packing properties without altering vesicle size. For example, a headgroup area of 45 Å² is observed for the C(18)C(0)PC lamellar phase (two chains per headgroup) while a headgroup area of 63 Å² is observed for the C(18)C(10)PC mixed-interdigitated gel phase (three chains per headgroup). The fully interdigitated bilayer formed by the ether phospholipid 1,2-di-*O*-hexadecyl-*sn*-glycero-3-phosphocholine has a gel-phase headgroup area of 79 Å² (four chains per headgroup).[81] In contrast, the variation in headgroup area observed in the bilayer gel phase of symmetric-chain-length PCs as a function of chain length is only about 15 Å². Our studies with the phosphatidic acid derivative C(18)C(10)PA indicate that a mixed-interdigitated bilayer gel phase is formed by this phospholipid. Consequently, mixed-chain-length ionic phospholipids could be employed to vary the charge density of the bilayer surface. Also, the elimination of strong headgroup-headgroup interactions in the mixed-interdigitated bilayer could facilitate the interaction of the phospholipid headgroups with reactants or ions in solution. Thus, the use of mixed-chain-length phospholipids to alter the surface properties of liposomes could make a significant contribution to this important area of synthetic chemistry.

Recently, bilayers have been shown to be capable of serving as supporting matrices for semiconductor particles.[5] The asymmetric formation of semiconductor particles, such as In$_2$S$_3$, on one side of a bilayer can lead to the appearance of a photovoltage that results from the vectorial transfer of charges in the direction opposite to that of the bilayer potential. Thus, these colloidal semiconductor particles behave like current-rectifying diodes.[5,82] These model systems have many potential applications, particularly in photochemical solar energy conversion and storage.[5,83] For these model systems to function, the semiconductor particle must penetrate into the bilayer to a sufficient depth that transmembrane electron transfer is possible. The electrical properties of bilayer semiconductors are thus related to the effective resistance and capacitance of the bilayer membrane, both of which are related to bilayer thickness.[5]

The ability to vary the thickness of the bilayer-semiconductor junction would provide a means of regulating the properties of these model systems. The variation in bilayer thickness that can be achieved with symmetric-chain-length phospholipids is limited to varying the phospholipid chain length within the fixed noninterdigitated bilayer structure. In contrast, mixed-chain-length phospholipids, with their ability to form various types of interdigitated bilayers, offer the ability to produce bilayers with a much greater range and variety of hydrocarbon thicknesses. For example, stable bilayers cannot be formed from PCs with chain lengths shorter than 12 carbons [C(12)C(12)PC]. This PC corresponds to a noninterdigitated hydrocarbon thickness of 24 carbons. Stable mixed-interdigitated bilayers have been formed from C(16)C(9)PC, which corresponds to a hydrocarbon thickness of just 16 carbons. In addition, the preparation of a mixed-chain-length phospholipid with the desired gel-phase packing and bilayer thickness is easily achieved by selecting the appropriate values of ΔC and CL. Thus, mixed-chain-length phospholipids could be exploited to great advantage in the related areas of semiconductor film technology, biosensors, and artificial photosynthesis.

The main gel $\leftrightarrow$ liquid-crystalline phase transition of phospholipid bilayers leads to pronounced changes in bilayer properties such as membrane fluidity, permeability, and fusibility.[1] The properties of many membrane proteins are also affected by the main bilayer phase transition. Thus, the phospholipid phase transition can be employed as a thermally induced switch to regulate the behavior of liposomes. For example, the increase in bilayer thickness accompanying the mixed-interdigitated gel $\leftrightarrow$ liquid-crystalline phase transition could serve as a thermal switch to turn off a bilayer semiconductor, as described above, or to activate a membrane protein. Thus, the ability to fabricate liposomes with specific phase-transition temperatures would be extremely useful. Unfortunately, symmetric-chain-length PCs offer a limited choice of transition temperatures within a practical range. However, mixed-chain-length phospholipids offer a much wider choice of transition temperatures due to their structural variability and variety of chain packing conformations.

The relationship of chain-length asymmetry to the temperature of the main gel $\leftrightarrow$ liquid-crystalline phase transition has been studied in detail by Huang and co-workers.[84,85] For example, they have shown[84] that the phase-transition temperatures of the PCs in Region I can be related to ΔC and CL by the following equation:

$$T_m = 153.67 - 1.73\Delta C - 1498.03(1/CL) - 139.66\ (\Delta C/CL)$$

The transform of this equation can be used to determine the phospholipid structure (ΔC and CL) required to produce a bilayer with the desired phase-transition temperature. Equations have also been derived for the PCs in region II.[85]

In summary, the great variability in bilayer structure and properties afforded by mixed-chain-length phospholipids holds promise for their practical use in the various fields of liposome technology.

ACKNOWLEDGMENTS

I would like to thank Timothy J. O'Leary for helpful discussions during the preparation of this manuscript and Ching-hsien Huang for his friendship and advice throughout my career.

REFERENCES

1. **Cullis, P. R. and Hope, M. J.**, Physical properties and functional roles of lipids in membranes, in *Biochemistry of Lipids and Membranes*, Vance, D. E. and Vance, J. E., Eds., Benjamin/Cummings Publishing Company, Menlo Park, CA, 1985, chap. 2.

2. **Thompson, T. E. and Huang, C.**, Dynamics of lipids in biomembranes, in *Physiology of Membrane Disorders*, Andreoli, T. E., Hoffman, J. F., and Fanestil, D. D., Eds., Plenum Publishing, New York, 1987, chap. 2.

3. **Mabrey, S. and Sturtevant, J. M.**, Investigation of phase transitions of lipids and lipid mixtures by high-sensitivity differential scanning calorimetry, *Proc. Natl. Acad. Sci. U.S.A.*, 73, 3862, 1976.

4. **Hiff, T. and Kevan, L.**, Effects of the addition of alcohols, cryoprotective agents, and salts on the photoionization yield of chlorophyll *a* in frozen vesicle solutions with and without electron acceptors, *J. Phys. Chem.*, 93, 3227, 1989.

5. **Zhao, X. K., Herve, P. J., and Fendler, J. H.**, Magnetic particulate thin films on bilayer lipid membranes, *J. Phys. Chem.*, 93, 908, 1989.

6. **Zhao, X. K., Baral, S., and Fendler, J. H.**, Electrochemical characterization of bilayer lipid membrane-semiconductor junctions, *J. Phys. Chem.*, 94, 2043, 1990.

7. **Baumann, G., Easton, G., Quint, S. R., and Johnson, R. N.**, Molecular switching in neuronal membranes, in *Molecular Electronic Devices II*, Carter, F. L., Ed., Marcel Dekker, New York, 1987, chap. 4.

8. **Haga, M., Sugawara, S., and Itagaki, H.**, Drug sensor: liposome immunosensor for theophylline, *Anal. Biochem.*, 118, 286, 1981.

9. **Drexler, K. E.**, *Nanosystems*, John Wiley & Sons, New York, 1992.

10. **Fendler, J. H.**, Reactivity control and synthetic applications, in *Membrane Mimetic Chemistry*, John Wiley & Sons, New York, 1982, chap. 11.

11. **Kirby, C. J.**, Controlled delivery of functional food ingredients: opportunities for liposomes in the food industry, in *Liposome Technology*, Vol. II, 2nd ed., Gregoriadis, G., Ed., CRC Press, Boca Raton, FL, 1993, chap. 13.

12. **Hauser, H., Pascher, I., Pearson, R. H., and Sundell, S.**, Preferred conformation and molecular packing of phosphatidylethanolamine and phosphatidylcholine, *Biochim. Biophys. Acta*, 650, 21, 1981.

13. **Hitchock, P. B., Mason, R., Thomas, K. M., and Shipley, G. G.**, Structural chemistry of 1,2-dilauroyl-DL-phosphatidylethanolamine: molecular conformation and intermolecular packing of phospholipids, *Proc. Natl. Acad. Sci. U.S.A.*, 71, 3036, 1974.

14. **Pearson, R. H. and Pascher, I.**, The molecular structure of lecithin dihydrate, *Nature*, 281, 499, 1979.

15. **Huang, C.**, Roles of carbonyl oxygens at the bilayer interface in phospholipid-sterol interactions, *Nature*, 259, 242, 1976.

16. **Lee, A. G., Birdsall, N. J. M., Metcalfe, J. C., Warren, G. B., and Roberts, G. C. K.**, A determination of the mobility gradient in lipid bilayers by ^{13}C nuclear magnetic resonance, *Proc. R. Soc. London Ser. B*, 193, 253, 1976.

17. **Hauser, H., Guyer, W., Pascher, I., Skrabal, P., and Sundell, S.**, Polar headgroup conformation of phosphatidylcholine. Effect of solvent and aggregation, *Biochemistry*, 19, 366, 1980.

18. **Gally, H.-U., Niederberger, W., and Seelig, J.**, Conformation and motion of the choline headgroup in bilayers of dipalmitoyl-sn-glycero-3-phosphocholine, *Biochemistry*, 14, 3647, 1975.

19. **Wohlgemuth, R., Waespe-Šarčević, N., and Seelig, J.**, Bilayers of phosphatidylglycerol. A deuterium and phosphorous nuclear magnetic resonance study of the headgroup region, *Biochemistry*, 19, 3315, 1980.

20. **Gaber, B. P., Yager, P., and Peticolas, W. L.**, Conformational nonequivalence of chains 1 and 2 of dipalmitoylphosphatidylcholine as observed by Raman spectroscopy, *Biophys. J.*, 24, 677, 1978.

21. **Seelig, J. and Seelig, A.**, Lipid conformation in model membranes and biological membranes, *Q. Rev. Biophys.*, 13, 19, 1980.

22. **Zaccai, G., Büldt, G., Seelig, A., and Seelig, J.**, Neutron diffraction studies on phosphatidylcholine model membranes, *J. Mol. Biol.*, 134, 693, 1979.

23. **Browning, J. L.**, NMR studies of the structural and motional properties of phospholipids in membranes, in *Liposomes: From Physical Structure to Therapeutic Applications*, Knight, C. G., Ed., Elsevier/North-Holland, New York, 1981, 189.

24. **Mason, J. T., Huang, C., and Biltonen, R. L.**, Calorimetric investigations of saturated mixed-chain phosphatidylcholine bilayer dispersions, *Biochemistry*, 20, 6086, 1981.

25. **Mason, J. T. and Stephenson, F. A.**, Thermotropic properties of saturated mixed acyl phosphatidylethanolamines, *Biochemistry*, 29, 590, 1990.

26. **Mason, J. T., Broccoli, A. V., and Huang, C.,** A method for the synthesis of isomerically pure saturated mixed-chain phosphatidylcholines, *Anal. Biochem.*, 113, 96, 1981.

27. **Ali, S. and Bittman, R.,** Mixed-chain phosphatidylcholine analogues modified in the choline moiety: preparation of isomerically pure phospholipids with bulky head groups and one acyl chain twice as long as the other, *Chem. Phys. Lipids*, 50, 11, 1989.

28. **Nicholas, A. W., Khouri, L. G., Ellington, J. C., and Porter, N. A.,** Synthesis of mixed-acid phosphatidylcholines and high pressure liquid chromatographic analysis of isomeric lysophosphatidylcholines, *Lipids*, 18, 434, 1983.

29. **Huang, C. and Mason, J. T.,** Structure and properties of mixed-chain phospholipid assemblies, *Biochim. Biophys. Acta*, 864, 423, 1986.

30. **Phillips, M. C., Williams, R. M., and Chapman, D.,** On the nature of hydrocarbon chain motions in lipid liquid crystals, *Chem. Phys. Lipids*, 3, 234, 1969.

31. **Huang, C., Mason, J. T., and Levin, I. W.,** Raman spectroscopic study of saturated mixed-chain phosphatidylcholine multilamellar dispersions, *Biochemistry*, 22, 2775, 1983.

32. **Hui, S. W., Mason, J. T., and Huang, C.,** Acyl chain interdigitation in saturated mixed-chain phosphatidylcholine dispersions, *Biochemistry*, 23, 5570, 1984.

33. **McIntosh, T. J., Simon, S. A., Ellington, J. C., and Porter, N. A.,** New structural model for mixed-chain phosphatidylcholine bilayers, *Biochemistry*, 23, 4038, 1984.

34. **Mattai, J., Sripada, P. K., and Shipley, G. G.,** Mixed-chain phosphatidylcholine bilayers: structure and properties, *Biochemistry*, 26, 3287, 1987.

35. **Chen, S. C., Sturtevant, J. M., and Gaffney, B. J.,** Scanning calorimetric evidence for a third phase transition in phosphatidylcholine bilayers, *Proc. Natl. Acad. Sci. U.S.A.*, 77, 5060, 1980.

36. **Serrallach, E. N., de Haas, G. H., and Shipley, G. G.,** Structure and thermotropic properties of mixed-chain phosphatidylcholine bilayer membranes, *Biochemistry*, 23, 713, 1984.

37. **Tardieu, A., Luzzati, V., and Reman, F. C.,** Structure and polymorphism of the hydrocarbon chains of lipids: a study of lecithin-water phases, *J. Mol. Biol.*, 75, 711, 1973.

38. **Wang, Z.-q., Lin, H.-n., and Huang, C.,** Differential scanning calorimetric study of a homologous series of fully hydrated saturated mixed-chain $C(X):C(X+6)$ phosphatidylcholines, *Biochemistry*, 29, 7072, 1990.

39. **Lewis, R. N. A. H., Mak, N., and McElhaney, R. N.,** A differential scanning calorimetry study of the thermotropic phase behavior of model membranes composed of phosphatidylcholines containing linear saturated fatty acyl chains, *Biochemistry*, 26, 6118, 1987.

40. **Lin, H.-n., Wang, Z.-q., and Huang, C.,** Differential scanning calorimetry study of mixed-chain phosphatidylcholines with a common molecular weight identical with diheptadecanoylphosphatidylcholine, *Biochemistry*, 29, 7063, 1990.

41. **Huang, C.,** Mixed-chain phospholipids and interdigitated bilayer systems, *Klin. Wochenschr.*, 68, 149, 1990.

42. **Bultmann, T., Lin, H.-n., Wang, Z.-q., and Huang, C.,** Thermotropic and mixing behavior of mixed-chain phosphatidylcholines with molecular weights identical with that of L-α-dipalmitoylphosphatidylcholine, *Biochemistry*, 30, 7194, 1991.

43. **Flory, P. J.,** *Statistical Mechanics of Chain Molecules*, John Wiley & Sons, New York, 1969, chap. 1.

44. **Mason, J. T. and Huang, C.,** Chain length dependent thermodynamics of saturated symmetric-chain phosphatidylcholine bilayers, *Lipids*, 16, 604, 1981.

45. **Yellin, N. and Levin, I. W.,** Hydrocarbon chain trans-gauche isomerization in phospholipid bilayer gel assemblies, *Biochemistry*, 16, 642, 1977.

46. **Zhu, T. and Caffrey, M.,** Thermodynamic, thermomechanical, and structural properties of a hydrated asymmetric phosphatidylcholine, *Biophys. J.*, 65, 939, 1993.

47. **Wong, P. T. T. and Huang, C.,** Structural aspects of pressure effects on infrared spectra of mixed-chain phosphatidylcholine assemblies in D_2O, *Biochemistry*, 28, 1259, 1989.

48. **Xu, H. and Huang, C.,** Scanning calorimetric study of fully hydrated asymmetric phosphatidylcholines with one acyl chain twice as long as the other, *Biochemistry*, 26, 1036, 1987.

49. **Lin, H.-n., Wang, Z.-q., and Huang, C.,** The influence of acyl chain-length asymmetry on the phase transition parameters of phosphatidylcholine dispersions, *Biochim. Biophys. Acta*, 1067, 17, 1991.

50. **Boggs, J. M. and Mason, J. T.,** Calorimetric and fatty acid spin label study of subgel and interdigitated gel phases formed by asymmetric phosphatidylcholines, *Biochim. Biophys. Acta*, 863, 231, 1986.

51. **Lewis, R. N. A. H., McElhaney, R. N., Österberg, F., and Gruner, S. M.,** Enigmatic thermotropic phase behavior of highly asymmetric mixed-chain phosphatidylcholines that form mixed-interdigitated gel phases, *Biophys. J.*, 66, 207, 1994.

52. **Halladay, H. N., Stark, R. E., Ali, S., and Bittman, R.,** Magic-angle spinning NMR studies of molecular organization in multibilayers formed by 1-octadecanoyl-2-decanoyl-*sn*-glycero-3-phosphocholine, *Biophys. J.*, 58, 1449, 1990.

53. **Shah, J., Sripada, P. K., and Shipley, G. G.,** Structure and properties of mixed-chain phosphatidylcholine bilayers, *Biochemistry*, 29, 4254, 1990.

54. **Cullis, P. R. and de Kruijff, B.,** Polymorphic phase behavior of lipid mixtures as detected by ^{31}P NMR, *Biochim. Biophys. Acta*, 507, 207, 1978.

55. **Wu, W.-G., Huang, C., Conley, T. G., Martin, R. B., and Levin, I. W.,** Lamellar-micellar transition of 1-stearoyllysophosphatidylcholine assemblies in excess water, *Biochemistry*, 21, 5957, 1982.

56. **Wu, W.-G. and Huang, C.,** Kinetic studies of the micellar to lamellar phase transition of 1-stearoyllysophosphatidylcholine dispersions, *Biochemistry,* 22, 5068, 1983.

57. **Hui, S. W. and Huang, C.,** X-ray diffraction evidence for fully interdigitated bilayers of 1-stearoyllysophosphatidylcholine, *Biochemistry,* 25, 1330, 1986.

58. **Wu, W., Stephenson, F. A., Mason, J. T., and Huang, C.,** A nuclear magnetic resonance spectroscopic investigation of the headgroup motions of lysophospholipids in bilayers, *Lipids,* 19, 68, 1984.

59. **Huang, C., Mason, J. T., Stephenson, F. A., and Levin, I. W.,** Raman and ^{31}P NMR spectroscopic identification of a highly ordered lamellar phase in aqueous dispersions of 1-stearoyl-2-acetyl-*sn*-glycero-3-phosphorylcholine, *J. Phys. Chem.,* 88, 6454, 1984.

60. **Boggs, J. M.,** Intermolecular hydrogen bonding between lipids: influence on organization and function of lipids in membranes, *Can. J. Biochem.,* 58, 755, 1980.

61. **Xu, H., Stephenson, F. A., and Huang, C.,** Binary mixtures of asymmetric phosphatidylcholines with one acyl chain twice as long as the other, *Biochemistry,* 26, 5448, 1987.

62. **Slater, J. L., Huang, C., Adams, R. G., and Levin, I. W.,** Polymorphic phase behavior of lysophosphatidylethanolamine dispersions, *Biophys. J.,* 56, 243, 1989.

63. **Davis, P. J., Fleming, B. D., Coolbear, K. P., and Keough, K. M. W.,** Gel to liquid-crystalline transition temperatures of water dispersions of two pairs of positional isomers of unsaturated mixed-acid phosphatidylcholines, *Biochemistry,* 20, 3633, 1981.

64. **Huang, C. and Mason, J. T.,** Complementary packing of phosphoglyceride and cholesterol molecules in the bilayer, in *Membranes and Transport,* Vol. 1, Martonosi, A. N., Ed., Plenum Publishing, New York, 1982, 15.

65. **Barenholz, Y. and Thompson, T. E.,** Sphingomyelins in bilayers and biological membranes, *Biochim. Biophys. Acta,* 604, 129, 1980.

66. **Barenholz, Y., Suurkuusk, J., Mountcastle, D., Thompson, T. E., and Biltonen, R. L.,** A calorimetric study of the thermotropic behavior of aqueous dispersions of natural and synthetic sphingomyelins, *Biochemistry,* 15, 2441, 1976.

67. **Estep, T. N., Calhoun, W. I., Barenholz, Y., Biltonen, R. L., Shipley, G. G., and Thompson, T. E.,** Evidence for metastability in stearoylsphingomyelin bilayers, *Biochemistry,* 19, 20, 1980.

68. **Pascher, I.,** Molecular arrangements in sphingolipids. Conformation and hydrogen bonding of ceramide and their implication on membrane stability and permeability, *Biochim. Biophys. Acta,* 455, 433, 1976

69. **Pascher, I. and Sundell, S.,** Molecular arrangements in sphingolipids. The crystal structure of cerebroside, *Chem. Phys. Lipids,* 20, 175, 1977.

70. **Sripada, P. K., Maulik, P. R., Hamilton, J. A., and Shipley, G. G.,** Partial synthesis and properties of a series of N-acyl sphingomyelins, *J. Lipid Res.,* 28, 710, 1987.

71. **Maulik, P. R., Atkinson, D., and Shipley, G. G.,** X-ray scattering of vesicles of *N*-acyl sphingomyelins, *Biophys. J.,* 50, 1071, 1986.

72. **Levin, I. W., Thompson, T. E., Barenholz, Y., and Huang, C.,** Two types of hydrocarbon chain interdigitation in sphingomyelin bilayers, *Biochemistry,* 24, 6282, 1985.

73. **Israelachvili, J.,** *Intermolecular and Surface Forces,* 2nd ed., Academic Press, New York, 1992, chap. 17.

74. **Cevc, G. and Marsh, D.,** *Phospholipid Bilayers,* John Wiley & Sons, New York, 1987, chap. 2.

75. **Tanford, C.,** *The Hydrophobic Effect,* John Wiley & Sons, New York, 1980, chap. 2

76. **Mason, J. T., Huang, C., and Biltonen, R. L.,** Effect of liposomal size on the calorimetric behavior of mixed-chain phosphatidylcholine bilayer dispersions, *Biochemistry,* 22, 2013, 1983.

77. **Grieser, F. and Drummond, C. J.,** The physiochemical properties of self-assembled surfactant aggregates as determined by some molecular spectroscopic probe techniques, *J. Phys. Chem.,* 92, 5580, 1988.

78. **Kawamuro, M. K., Chaimovich, H., Abun, E. B., Lissi, E. A., and Cuccovia, I. M.,** Evidence that the effects of synthetic amphiphile vesicles on reaction rates depend on vesicle size, *J. Phys. Chem.,* 95, 1458, 1991.

79. **Carmona-Ribeiro, A. M. and Hix, S.,** pH effects on properties of dihexadecyl phosphate vesicles, *J. Phys. Chem.,* 95, 1812, 1991.

80. **Mizutani, T. and Whitten, D. G.,** Bromination of surfactant and hydrophobic *trans*-stilbenes in aqueous micelles and vesicles. Evidence for wide variation of solubilization/reaction sites in microheterogeneous media, *J. Am. Chem. Soc.,* 107, 3621, 1985.

81. **Ruocco, M. J., Siminovitch, D. J., and Griffin, R. G.,** Comparative study of the gel phases of ether- and ester-linked phosphatidylcholines, *Biochemistry,* 24, 2406, 1985.

82. **Chang, A.-C., Pfeiffer, W. F., Guillaume, B., Baral, S., and Fendler, J. H.,** Preparation and characterization of selenide semiconductor particles in surfactant vesicles, *J. Phys. Chem.,* 94, 4284, 1990.

83. **Katz, J. J. and Hindman, J. C.,** in *Photochemical Conversion and Storage of Solar Energy,* Connolly, J. S., Ed., Academic Press, New York, 1981, 27.

84. **Huang, C., Li, S., Wang, Z.-Q., and Lin, H.-N.,** Dependence of the bilayer phase transition temperatures on the structural parameters of phosphatidylcholines, *Lipids,* 28, 365, 1993.

85. **Huang, C., Wang, Z.-Q., Lin, H.-N., and Brumbaugh, E. E.,** Calorimetric studies of fully hydrated phosphatidylcholines with highly asymmetric acyl chains, *Biochim. Biophys. Acta,* 1145, 298, 1993.

Chapter 9

Aggregation Properties of Gangliosides: Micelles and Vesicles

Mario Corti and Laura Cantù

CONTENTS

I. INTRODUCTION

In aqueous solvents, the simultaneous tendency to preserve and to avoid contact with water leads amphiphilic molecules to organize themselves as to create distinct hydrophilic and hydrophobic domains, like the surface and the core of a globular micelle or the outer and inner parts of a bilayer. The packing of amphiphiles in an aggregate obeys some simple rules following the consideration that, except for surface roughness, no water can exist inside the hydrophobic domain, so that one of its dimensions cannot exceed twice the monomer hydrophobic length, while contact with water is allowed by the hydrophilic headgroup. This requirement is very stringent and can be handled theoretically in the frame of the "opposing forces" model.[1] According to this model, surface tension and hydrophobic forces are summarized in an attractive interaction at the hydrocarbon-water interface. The repulsive forces among headgroups, electrostatic, hydration and steric, are also added up to an equivalent repulsive interaction acting at the same interface. Therefore, two net opposite forces are applied at this surface, the one tending to decrease, the other to increase the interfacial area per headgroup exposed to the aqueous phase: the two forces balance in correspondence of an "optimal surface area" a_0, which is assumed to be the area at the interface required by the monomer in the aggregated structure. Following these observations, it is quite useful to describe the packing of amphiphilic molecules into an aggregate by a simple geometrical model.[2]

The hydrophobic part of an amphiphile is schematized as occupying inside the aggregate a place with the shape of a truncated cone, identified by three parameters: volume V, length ℓ, and the area a_0 at the interface, Figure 1. They are summarized in a packing parameter or shape factor $P = V/(a_0\ell)$ which assumes the limiting values 1/3 for a true cone and 1 for a cylinder. The corresponding aggregated structures are spheres and bilayers. Vesicles, liposomes and in general membrane-like structures are formed when $1/2 < P < 1$. Micelles are formed when $1/3 < P < 1/2$. In between these values, higher Ps identify larger and more asymmetrical micelles. An example of an amphiphile which forms rather spherical micelles is the the 12-carbon-chain sodium dodecyl sulphate (SDS) surfactant in water. Its packing parameter[2] is quite close to 1/3, that is P = 0.37 with V = 350 Å³, ℓ = 16.7 Å and a_0 = 57 Å². On the other hand, the amphiphile which typically forms vesicles and bilayers is egg lecithin which has a packing parameter P= 0.85, with V = 1063 Å³, ℓ = 17.5 Å and a_0 =70 Å².

These geometrical arguments seem rather simple and convincing, mostly for what concerns the limiting values of 1/3 and 1 for the packing parameter: the spherical micelle case with small aggregation number (<100) and the bilayer case, for most of the well known liposome-forming amphiphiles. It is, however, a purpose of this paper to present these geometrical ideas in more detail. This is accomplished experimentally. by studying amphiphiles which have packing parameters close to 1/2, the value at which the transition between globular and bilayer-type structure occurs. It is clear that, in this region, the final

0-8493-4731-9/96/$0.00+$.50

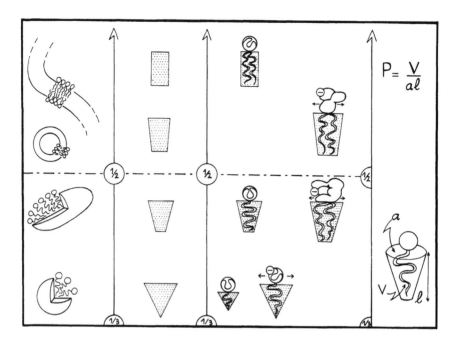

Figure 1 Pictorial sketch of the relationship between the aggregate (first lane) and monomer (third lane) geometries through the packing parameter P. The second lane shows the evolution of the packing shape as P increases. The dashed line at P = 1/2 marks the abrupt change from micellar to vesicular aggregation. The third lane shows how some monomer properties can play a role in determining the packing shape.

shape of the aggregate is very sensitive to small changes in the characteristics of the individual molecules so that even minor geometrical effects can be detected.

Gangliosides have been shown to be a good example of such amphiphiles, since they have large hydrophobic and hydrophilic parts in the same molecule.[3] Gangliosides are natural glycosphingolipids occurring in plasma membranes.[4] They are double-tailed amphiphilic molecules, like phospholipids, in which a ceramide lipid portion, constituted by a sphingosine and a fatty acid with roughly 20 carbons each, carries a rather bulky headgroup made up of several sugar rings, some of which are sialic acid residues (Figure 2). The sugars in the main chain are ordered according to a fixed sequence and the positions available for branching of sialic acid residues are fixed, so that each molecule can be obtained from the other by adding or subtracting a given sugar ring. This means that not only the number of sugar units, but also the intramolecular interactions can be changed when considering different gangliosides.

The availability of a considerable number of high-purity gangliosides, prepared by specific chromatographic procedures in order to obtain molecules containing similar lipid moieties, allows, therefore, a significant test of the influence of the headgroup geometry on the final shape of the aggregate.

II. STRUCTURE OF GANGLIOSIDES

Table 1 shows the structure of the oligosaccharide portion of some gangliosides. The oligosaccharide chains of GM3, GM2, GM1, GD1a, GalNAc-GD1a, GD1b, and GT1b[4] are linked to the ceramide moiety through a glucosidic linkage and the structure of these gangliosides can be schematically obtained by sequential addition of neutral sugars and sialic acid units to the trisaccharide α-Neu5Ac-(2-3)-β-Gal-(1-4)-β-Glc belonging to the GM3 structure. The ganglioside GM4 is like GM3 without the Glc (Glucose) group. GD1b-L is the lactone parent of GD1b with the carboxyl group of the external sialic acid esterifying the C9 primary hydroxyl group of the inner sialic acid unit. GD1b-L occurs in the mammalian nervous system naturally and can be also prepared by a simple synthetic procedure starting from GD1b.[5] Neutral sugars are present in the β configuration and sialic acids in the α one.

"Sialic acid"[6] is a code to indicate all the derivatives of neuraminic acid (5-amino-3,5-dideoxy-D-glycero-D-galacto-nonulosonic acid). The most represented structures are the N-acetyl- and the N-glycolyl-derivatives.

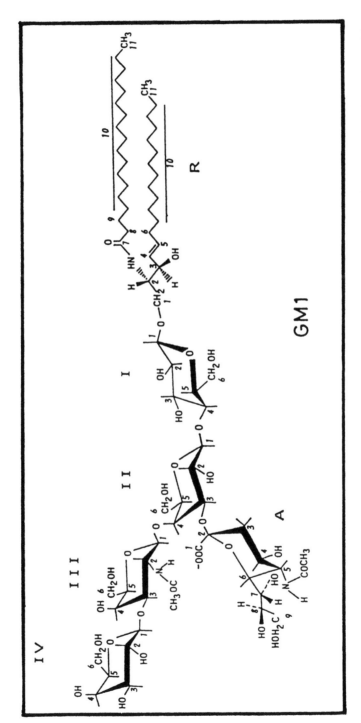

Figure 2 Chemical structure of the GM1 ganglioside. GM2 is like GM1 without the galactose IV. GM3 is like GM1 without the galactose IV and the *N*-acetylgalactosamine III. A is the *N*-acetylneuraminic acid (sialic acid).

Table 1 Structure of Various Gangliosides

GM3
II³Neu5AcLacCer
α-Neu5Ac-(2-3)-β-Gal-(1-4)-β-Glc-(1-1)-Cer

GM2
II³Neu5AcGgOse₃LacCer
β-GalNAc-(1-4)-[α-Neu5Ac-(2-3)]-β-Gal-(1-4)-β-Glc-(1-1)-Cer

GM1, GM1(Neu5Ac)
II³Neu5AcGgOse₄Cer
β-Gal-(1-3)-β-GalNAc-(1-4)-[α-Neu5Ac-(2-3)]-β-Gal-(1-4)-β-Glc-(1-1)-Cer

GM1, GM1(Neu5Gc)
II³Neu5GcGgOse₄Cer
β-Gal-(1-3)-β-GalNAc-(1-4)-[α-Neu5Gc-(2-3)]-β-Gal-(1-4)-β-Glc-(1-1)-Cer

GD1a
IV³Neu5AcII³Neu5AcGgOse₄Cer
α-Neu5Ac-(2-3)-β-Gal-(1-3)-β-GalNAc-(1-4)-[α-Neu5Ac-(2-3)]-β-Gal-(1-4)-β-Glc-(1-1)-Cer

GalNAc-GD1a
IV³GalNAcIV³Neu5AcII³Neu5AcGgOse₄Cer
β-GalNAc-(1-4)-[α-Neu5Ac-(2-3)]-β-Gal-(1-3)-β-GalNAc-(1-4)-[α-Neu5Ac-(2-3)]-β-Gal-(1-4)-β

GD1b
II³(Neu5Ac)₂GgOse₄Cer
β-Gal-(1-3)-β-GalNAc-(1-4)-[α-Neu5Ac-(2-8)-α-Neu5Ac-(2-3)]-β-Gal-(1-4)-β-Glc-(1-1)-Cer

GD1b-L, GD1b-lactone
II³[α-Neu5Ac-(1-9,2-8)-α-Neu5Ac]GgOse₄Cer
β-Gal-(1-3)-β-GalNAc-(1-4)-[α-Neu5Ac-((1-9,2-8)-α-Neu5Ac-(2-3)]-β-Gal-(1-4)-β-Glc-(1-1)-Cer

GT1b
IV³Neu5AcII³(Neu5Ac)₂GgOse₄Cer
α-Neu5Ac-(2-3)-β-Gal-(1-3)-β-GalNAc-(1-4)-[α-Neu5Ac-(2-8)-α-Neu5Ac-(2-3)]-β-G al-(1-4)-β-Glc-(1-1)-Cer

Note: Nomenclatures are in accordance with the recommendations introduced by the IUPAC-IUB Commission on Biochemical Nomenclature in 1977.

The lipid moiety of gangliosides is called ceramide and is constituted by a long chain amino alcohol, generally called sphingosine, connected to a fatty acid by an amide linkage. The long chain bases are 18 or 20 carbon atom structures, mainly containing a trans double bond at position 3-4. Fatty acids vary from 14 to 26 carbon atoms.

Gangliosides can be prepared homogeneously in both the oligosaccharide and ceramide portions by normal-phase-chromatography followed by reversed-phase-chromatography.[4] Semisynthetic procedures have been applied to the preparation of gangliosides containing a desired ceramide structure.[7]

III. CRITICAL MICELLE CONCENTRATION (cmc) OF GANGLIOSIDES

Gangliosides are amphiphiles with a strong hydrophobic character, due to the presence of an extended double tail. Also their hydrophilic character is highly pronounced, progressively increasing as the complexity of the headgroup increases, corresponding to the addition of sugar rings in an extended and ramificated structure.

The hydrophilic-hydrophobic balance is then the competition between two opposite strong requirements, any variation of which sensibly affects the aggregative behavior. Of course, gangliosides, being amphiphilic compounds, are present in dilute solutions as aggregates of high molecular weight above a given concentration, the critical micelle (aggregate) concentration (cmc). Thermodynamics[1] predicts cmc values in terms of the chemical-potential difference for an individual molecule to be free in solution or inside an aggregate. The higher this difference, the lower is the cmc and also the slower is the exchange process of individual molecules from aggregates to solution and vice versa.[8,9] For example, a strong hydrophilic character, with respect to the hydrophobic one, will be reflected into a high cmc value as well as in a high number of free monomers which can quickly exchange in solution. Any change in the

Table 2 Qualitative Connection Among Different Physical Properties of Amphiphiles

Hydrophobic character	Hydrophilic character	cmc	Fraction of free monomers	Exchange rate	Aggregate dimension
High	Low	Low	Low	Slow	Large
Low	High	High	High	Quick	Small

Table 3 Partial Specific Volumes, Refractive Index Increments and cmc Values of Various Gangliosides

		Experimental		Calculated		
	XL:XS	v cm³/g	dn/dc cm³/g	v cm³/g	dn/dc cm³/g	cmc 10⁻⁸ M
GT1b	0.26:0.74	0.7440	0.155	0.7445	0.1553	3.9
GD1a	0.30:0.70	0.7665	0.152	0.7661	0.1518	2.8
GM1	0.36:0.64	0.7976	0.146	0.7985	0.1465	2.0
GM2	0.40:0.60			0.8201	0.1429	1.1
GM3	0.53:0.47		0.138	0.8902	0.132	0.34

hydrophobic or hydrophilic characteristics of the individual ganglioside molecules will affect the equilibrium properties in solution in a constant trend, as summarized in Table 2.

The cmc values of gangliosides have been measured by different techniques,[10-12] but today much confusion still exists in the literature. The main problem is that cmc values are quite low, of the order of 10^{-8} M, due to the presence of two long hydrophobic chains in the ganglioside molecule. This makes eperimental determinations quite difficult since ganglioside adsorption at surfaces and the presence of impurities may affect results enormously. An attempt to overcome the "experimental confusion" is to apply well established thermodynamic concepts to the calculation of cmc values of a series of gangliosides[13] starting from the two properly measured cmc values of GM1[11] and GM1(LCB:1,C2).[14] GM1(LCB:1,C2) is the GM1 synthetic derivative with an acetyl group substituting the fatty acid. Table 3 reports these estimated cmc values for the gangliosides series GM3, GM2, GM1, GD1a, GT1b, and GM1(LCB:1,C2).

IV. DENSITY AND REFRACTIVE INDEX OF GANGLIOSIDE SOLUTIONS

Density measurements on amphiphiles in solution give data on their partial specific volume, which is a parameter used in the interpretation of sedimentation as well as X-ray or neutron scattering measurements. Some picnometric data have been reported.[15] With a completely different technique, measurements have been performed with an 8-digit Anton Paar DMA601 density meter on solutions of GM1, GD1a, and GT1b in the range 0.025–0.2% concentration by weight at T = 25°C.[16] The measured values, plotted vs. concentration, lie on a straight line the slope of which, $d\rho/dc$, is related to the partial specific volume of the ganglioside via the equation:

$$\bar{v} = (1 - d\rho/dc)/\rho_0 \tag{1}$$

where ρ_0 is the density of the solvent. The values of $\bar{v}$ obtained for the three gangliosides are reported in Table 3. Data show a trend which can be theoretically reproduced, following the suggestion of Curatolo et al.,[17] by assuming that the contributions coming from the lipidic and saccharidic parts of the molecule, $\bar{v}_l$ and $\bar{v}_s$, sum up to give the overall v, being weighted by the weight fractions of the two moieties in the single ganglioside molecule, X_l and X_s, that is:

$$\bar{v} = \bar{v}_l X_l + \bar{v}_s X_s \tag{2}$$

With a statistical treatment of the data obtained, the values $\bar{v}_l = 1.144 \pm 0.015$ cm³/g and $\bar{v}_s = 0.604 \pm 0.0065$ cm³/g were calculated. Table 3 reports the calculated values also for GM2 and GM3, which have not been measured.

The refractive index increment with concentration, dn/dc, is an important parameter for molecular weight determination with the light scattering technique.[18] Table 3 reports the dn/dc values measured on the same samples used for density measurements, performed with a Chromatix KMX 16 differential refractometer. The regular trend shown by the data suggests that they could eventually be theoretically reproduced following the same procedure used for the partial specific volume, that is, to attribute to the lipidic and to the saccharidic part of the ganglioside molecule a contribution to the total dn/dc which scales with their weight fractions:

$$dn/dc = (dn/dc)_l X_l + (dn/dc)_s X_s \tag{3}$$

A statistical treatment of the collected data gives the values $(dn/dc)_l = 0.08987 \pm 0.00764$ cm³/g and $(dn/dc)_s = 0.1783 \pm 0.00357$ cm³/g. By using these values the theoretical dn/dc's have been calculated, as reported in Table 3.

The temperature dependence of dn/dc has been measured for GM1. Its behavior can be described by the experimental equation

$$(dn/dc)(T) = 0.1465 - (3.804 \times 10^{-4}) \times (T(°C) - 25) \tag{4}$$

In the absence of more specific data, the same behavior can be assumed to be valid for the other gangliosides.

V. MICELLES OF GANGLIOSIDES

Within the ganglioside series, GM3, GM2, GM1, GD1a, GalNAcGD1a, GD1b, GD1b-L, and GT1b, all having the same hydrophobic portion and displaying a steric packing of extended and ramified different headgroups, the geometrical packing properties can be qualitatively attributed to the hydrophilic moiety. In fact, different gangliosides will require an interfacial area inside an aggregate as large as to provide in the hydrophilic layer a place wide enough to host the sugar rings and their hydration water.

In principle, the extended double tailed hydrophobic part assigns gangliosides to the family of membrane-forming amphiphiles. In fact, GM4 and GM3, those with smaller headgroups in the series, have been seen to form vesicles.[19] Nevertheless, for gangliosides of higher complexity than GM3, the role played by the hydrophilic part of the molecule becomes very important. In fact, from GM2 on, the sugar headgroup is so extended that, despite the presence of a double hydrophobic tail, like phospholipids, micelles are formed instead of vesicles. More than that, since the micellar structure is more strictly defined than the vesicular one, due to the fact that the hydrophobic domain of a micelle has to arrange as a globular liquid drop, small changes in the hydrophilic portion will affect sensibly the aggregation properties. In other words, the headgroup repulsive contribution is so high and the hydrophilic-hydrophobic balance is so delicate that the aggregate geometry, which is the resulting effect of such balance, will be rather sensitive to even small details in the headgroup interactions.

In Table 4 the aggregate physical parameters of different gangliosides are reported. They have been deduced by static and dynamic laser light scattering observations and sometimes confirmed by neutron and X-ray scattering experiments in the millimolar range of concentration.[3,5,12,19-26] A small amount of NaCl, to a final 30 mM ionic strength, has been added to the solutions in order to screen the intermicellar electrostatic repulsion, which does not affect the sterically driven packing behavior of gangliosides.[27,28]

For the monosialogangliosides GM3, GM2 and GM1 (Figure 3) which progressively add an external sugar in the main chain, the primary effect is that of enhancing the hydrophilic character of the amphiphilic molecule. The resulting influence on the aggregative properties can be appreciated by looking at Table 4. A decrease of the aggregation number and even a dramatic change from vesicular to micellar structure is observed in the sequence. These features are expected and can be accounted for by simple considerations. As the number of sugar units increases, a stronger hydrophilic repulsive contribution to the "opposing forces" balance is given and then a larger interfacial area is required by the molecule in the aggregate, which becomes more curved and smaller. Therefore, the aggregation number decreases.

Some more complex considerations can be made by looking at gangliosides belonging to the series GM1, GD1a, GD1b, GD1b-L, and GT1b (Figure 3). These molecules carry the whole 4-sugars neutral main chain and differ in the number and disposition of the branched sialic acid residues, the contribution of which to the packing in the aggregate is merely due to their physical dimension. The simple number-of-sugars rule is obeyed for the the gangliosides GM1, GD1a, and GT1b. In fact, more and more

Table 4 Physico-Chemical Parameters of Different Gangliosides

	MW	S	N	a_o (Å²)	P	R_H (Å)	Q (e.u.)
GM4*	1015	2	18,000	80	>.5	270	—
GM3*	1195	3	14,000	80	>.5	250	—
GM2	1398	4	529	92	0.440	66	100
GM1	1560	5	301	95.4	0.428	58.7	48
GM1$_{acetyl}$	1336	5	76	64.8	0.370	34	—
GM1S	1560	5	339	96.3	0.433	61.2	—
GD1a	1851	6	226	98.1	0.416	58	60
GD1b	1851	6	170	100.8	0.405	52	—
GD1b-L	1851	6	229	97.6	0.418	57	—
GT1b	2142	7	176	100.8	0.405	53.2	—
GalNAc-GD1a	2072	7	246	97	0.421	60	—

Note: Monomer molecular weight MW, number of sugars in the headgroup S, aggregation number N, interface area per monomer a_o, packing parameter P, hydrodynamic radius R_H, charge Q.

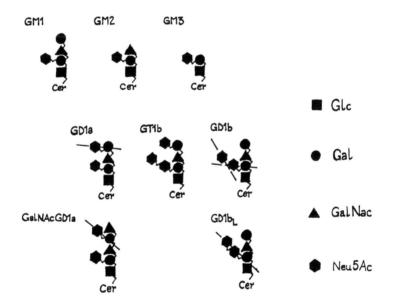

Figure 3 Schematic representation of the saccharidic headgroup of various gangliosides. Glc stands for glucose, Gal stands for galactose, GalNac stands for *N*-acetylgalactosamine, Neu5Ac stands for *N*-acetylneuraminic acid (sialic acid).

hydrophilic repulsive contribution is given by 5, 6, and 7 sugar units, respectively, resulting in progressively smaller aggregates, with lower aggregation number.

Some confusion arises when considering GD1b and GD1b-L, which stand out from the simpler rules described above. GD1b, which has the same number of sugars, both neutral and dissociating, as GD1a, behaves like GT1b, a more hydrophilic ganglioside. GD1b-L, which displays a small chemical modification with respect to its parent GD1b, is similar to GD1a. Moreover, the more complex ganglioside GalNAc-GD1a, which carries seven sugar units as GT1b, packs tighter than GD1a, which carries only six. This means that a level has been reached at which the extended and ramified structure of the ganglioside headgroup can no longer be underestimated as in a rough hydrophilicity evaluation. In the frame of the model described above, this is equivalent to notice that the steric repulsion among headgroups plays some additional role, resulting in the fact that different gangliosides displace the "opposing forces" surface at different distances below the hydrophilic–hydrophobic interface. The whole ganglioside molecule can be seen to occupy, inside an aggregate, a solid angle which, besides being dictated by the ceramide hydrophobic core general rules, is sensibly affected by the headgroup conformation.

This leads to the identification, along the headgroup extension, of "hot" or "cool" environments according to whether significant or negligible changes in the solid angle requirements are brought about by modifications in such positions. With these considerations in mind, some remarks can be made on GD1b. According to the "hydrophilicity" rule, GD1b micelles should have the same aggregation number as GD1a, having the same sugars both neutral and dissociating; but the different disposition of the second sialic acid residue plays an important role. The second sialic ring, which in GD1a is located in an external position, bound to the last neutral sugar of the main chain, in GD1b is branched to the first sialic, closer to the hydrophilic-hydrophobic interface. More room is then needed by the monomer at this level. This means that a wider solid angle per GD1b molecule is required inside the micelle, which is reflected in a smaller aggregation number.

The relevant influence of the structure of the headgroup in the first-sialic environment is confirmed by comparing the micellar properties of GD1b and GT1b. Although carrying one more sugar, a sialic ring placed at the end of the main chain, and then having a higher hydrophilic character, GT1b forms micelles with the same aggregation number as GD1b. This means that the solid angle required by the two-sialic group of sugars in the inner position is wide enough to host an additional sialic in an external "cooler" position.

Some deeper considerations can be made by looking at the aggregation properties of GD1b-L as compared with GD1b. GD1b-L, the monolactone derivative of GD1b, shows a chemical modification in the two-sialic group, with the formation of an inner ester. The aggregation number of GD1b-L micelles is higher than GD1b and close to GD1a: the increase of the solid angle to locate a second sialic acid in an inner position is no more needed. By proton nuclear magnetic resonance (NMR), using nOe effect, the monomer conformation can be directly evaluated with confidence. The comparison of GD1b and GD1b-L 3D-structures[29] reveals that the chemical modification induces a spatial rearrangement of the two-sialic group, forcing it to be more lined up with the neutral chain, reducing the angle between the neutral chain and the axis of the first-sialic acid. The GD1b/GD1b-L case is then a nice example of how ganglioside micellar parameters are sensitive to the secondary structure of the headgroup, and then be used, once a proper model is followed, to support direct observations at the molecular level.

A second interesting example is provided by GalNAc-GD1a, although it requires some more considerations. The trio of sugars, made up by the second and third sugars in the neutral chain and the inner sialic acid, namely the group GalNAc-Gal-Neu5Ac, is thought to form a strongly interacting bulky superunit. NMR observations on GM3,[30] as compared with GM1, suggest that the presence of GalNAc sensibly limits the rotational degrees of freedom of Neu5Ac around its bond axes to Gal, while the lack of GalNAc should destroy the superunit and release Neu5Ac from a tight-to-the-neutral-chain position. As a result, a ganglioside molecule lacking a GalNAc group is likely to require a wider solid angle in an aggregated structure. This behavior has been verified by comparing the aggregative properties of GalNAc-GD1a, which carries a second GalNAc-Gal-Neu5Ac trio on top of the first one, to those of the micelle-forming GD1a, which lacks the external GalNAc. Far from following the number-of-sugars rule, the aggregation number of GalNAc-GD1a is slightly higher than GD1a. This is explained by considering that the less mobile external trio of sugars requires a smaller solid angle and hence allows a larger micelle.

The influence of the hydrophobic portion on the aggregation properties of gangliosides is not evidenced as far as natural molecules are concerned, because of their limited variability in the lipid moiety. Some differences in the micellar parameters have been observed in GM1$_S$ in relation to the release of the double bond in the sphingosine, by catalytic hydrogenation of natural GM1. As shown in Table 4, the molecular weight of GM1$_S$ micelles is 18% higher, with a corresponding larger hydrodynamic radius and asymmetry, with respect to the natural GM1. A quite small increase in the packing parameter is sufficient to justify such a variation in the micellar size, since the packing parameter is close to the transition value of 0.5. The change may come from the maximum chain length ℓ. In GM1 the double bond close to the headgroup favors the parallel orientation of the axes of the two hydrocarbon chains. The saturation of the double bond tends to straighten up the hydrocarbon chain of the sphingosine, the axis of which can now be at an angle with the axis of the fatty acid. This may cause an overall reduction of the maximum chain length. A value of the order of 0.5 Å is reasonable and is just what is needed to explain the slight increase in packing parameter from GM1 to GM1$_S$.

The extent of hydrophobic contribution to the aggregation properties comes out drastically when the semisynthetic GM1$_{acetyl}$ ganglioside is observed.[14] The almost complete removal of the second lipid chain, corresponding to the substitution of an extended fatty acid with an acetyl group reduces the hydrophobic volume of the molecule from 965 to 566 Å^3. The packing parameter becomes smaller, since the headgroup is unchanged and the hydrophobic chain length, mainly determined by the sphingosine, is also the same. Consequently, $GM1_{acetyl}$ forms small and rather spherical micelles. There is a huge variation of the micellar parameters of $GM1_{acetyl}$ with respect to GM1 (Table 4).

VI. CHARGE OF GANGLIOSIDE MICELLES

Gangliosides may hydrolyze in solution, giving rise to charged micelles which interact electrostatically with each other if the ionic strength of the solution is not so high as to screen completely the Coulomb repulsion.

If observed with light or neutron scattering techniques, the charge of the micelles can be inferred by performing experiments in which the ganglioside concentration is kept constant, while the ionic strength of the solution is modified by adding NaCl. In fact, the neutron scattering pattern at low angles is influenced by the presence of a structure factor, which has a characteristic oscillating behavior in the presence of interactions and levels off progressively as salt is added to the solution. If the light scattering technique is used, the same structure factor has the effect of depressing the scattered intensity to an extent which depends on the range of repulsive interactions: as salt is added to the solution the scattered intensity increases until it reaches a saturation value when repulsion is completely screened. Light and neutron scattering results as a function of the ionic strength of the solution can be theoretically reproduced by constructing the radial distribution function g(r) of the particles in the HNC approximation,[31] using for the pair interaction potential a form consisting of a hard core repulsion plus a screened Coulomb potential. The charge Q of the micelles generating the interaction potential is a parameter, the value of which is going to be determined during the fitting procedure.

The charge determination has been performed on GM1, GM2, and GD1a.[32,33] The experimental data are reproduced by using a constant charge in the full range of ionic strengths: this can be assumed to be the charge of the micelle, as it has been observed that the aggregation number does not change on salt addition.[28] Results are reported in Table 4. It can be observed that only a fraction, between 10 and 20%, of the micellized ganglioside is dissociated, giving rise to a micellar charge which is much lower than the aggregation number.

VII. MIXED SYSTEMS

Mixed micelles may be formed with gangliosides. When a synthetic surfactant, the nonionic $C_{12}E_8$, is added to a GM1 solution, mixed micelles are formed with a molecular weight which depends on their molar ratio. Figure 4 shows the dependence of the mixed micelles molecular weight when different amounts of $C_{12}E_8$ are added to a 0.8 mM solution of GM1. The molecular weight monotonically decreases from the value of the pure GM1 micelle to that of the pure $C_{12}E_8$ micelle.[34] The full line is the prediction of a thermodynamic model which involves regular mixing of the two amphiphiles in the micelle.[35] The agreement is quite good. This means that the nonionic surfactant mixes randomly with GM1 in the micelle.

This may not be the case when two gangliosides are mixed. In fact, mixed micelles of GM2 and GT1b do not follow the predictions of ideal mixing. For instance, for a mole fraction of GT1b, X = 0.2, the mixed micelle molecular weight is even larger than the ones of both the individual micelles. Nonrandom mixing means obviously that some clustering occurs in the mixed micelle. A careful comparison of light and neutron scattering results[21] suggests that GT1b preferentially locates in regions of larger curvature in the mixed micelle, namely at the edges, of the globular nonspherical mixed micelle, while GM2 is more able to fit in the flatter aggregate regions.

Some geometrical effects can also be observed when a ganglioside is mixed with phospholipids. Figure 5 shows the molecular weight and hydrodynamic radius of egg-yolk lecithin vesicles with increasing amounts of GM1.[22] The presence of GM1 allows smaller vesicles, since it has a larger headgroup area than lecithin. Vesicles do not form any more when the percentage of GM1 is higher than about 20%.

The study of the time of formation of mixed micelles gives strong support to the theoretical prediction that natural ganglioside cmc values are low (10^{-9}–10^{-8} M) and within roughly one order of magnitude across the whole series of gangliosides. Figure 6 shows the evolution towards equilibrium of mixed micelles of GM2 and GT1b starting from individual micelles.[36] The final configuration is reached in about 10 h. The full line represents the theoretical behavior calculated according to a model which attributes

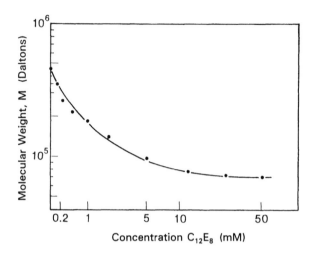

Figure 4 Molecular weight of mixed GM1-$C_{12}E_8$ micelles in 25 mM sodium phosphate/5 mM Na$_2$EDTA (pH 7.0) at 25°C as a function of $C_{12}E_8$ molecular concentration with fixed 0.8 mM GM1 concentration. (From Corti, M. et al., *J. Phys. Chem.*, 86, 2533, 1982. With permission.)

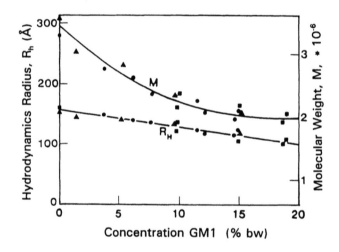

Figure 5 Molecular weight M and hydrodynamic radius R_H as a function of the GM1 molar fraction for EPC/GM1 mixed monolamellar vesicles in 25 mM Tris-HCl, pH 7.0, buffer at 37°C. The various symbols correspond to different series of preparations. (From Masserini, M. et al., *Chem. Phys. Lipids,* 37, 83, 1985. With permission.)

the mixing process to monomer transfer from one micelle to another via free molecules in solution. The equilibrium time is then dictated by the exchange rate between aggregates and solution, and in particular it is expressed as a bilinear function of the exchange rates of the two molecular species, so that the quicker one will dominate.[36] To obtain long equilibration times, it is then necessary that both the rates are slow and quite similar. Because the exchange rate is closely connected to the cmc of the amphiphile, the observed behavior is an indirect proof that both GM2 and GT1b cmc values are low, say, within one order of magnitude. An additional support to this conclusion is that when mixing GM2 with much higher cmc amphiphiles, like Triton X®-100 or sodium cholate, no time dependence is observed.[36]

VIII. VESICLES OF GANGLIOSIDES

So far, gangliosides, with their peculiar property of having a packing parameter close to the transition value of 0.5, due to their bulky hydrophobic and extended and ramified hydrophilic parts, have been useful to study geometric effects in micellar aggregation. It is now interesting to discuss how such features are reflected in the physical properties of the ganglioside vesicles, which are formed above the micelle to vesicle borderline, Figure 1, as compared with those of the common lecithin vesicles.

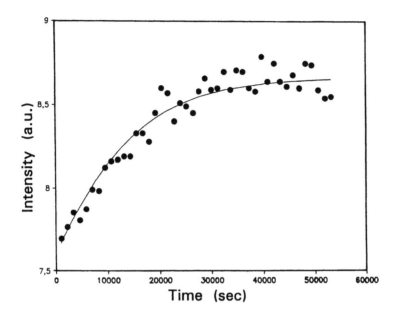

Figure 6 Scattered light intensity vs. time during the mixed-micelle formation of GM2 and GT1b. The full line is the model prediction. (From Cantú, L., Corti, M., and Salina, P., *J. Phys. Chem.*, 95, 5981, 1991. With permission.)

Unlike other structures, vesicles are generally believed to be in a nonequilibrium state of matter, since preparation methods normally involve a supply of external energy, like sonication and pressure filtration, or chemical treatments like detergent depletion or reversed-phase evaporation.[37] Spontaneous formation of small unilamellar vesicles[38] has nevertheless been reported in rather special cases, either by mixing two ionic surfactants with oppositely charged headgroups,[39,40] or for unusual pH conditions,[41,42] or by dilution of sponge phases in multicomponent systems.[43,44] It is, then, a rather peculiar behavior displayed by the ganglioside GM3, which, simply dissolved in water at normal pH, spontaneously forms vesicles made of a single amphiphile, in thermodynamic equilibrium.[23] This last case is quite different from the one reported in References 39 and 40, where the interaction between the two different amphiphiles gives rise to a spontaneous curvature of the bilayer and hence the vesicle formation is due to a clear enthalpic effect. In fact, for the single component system, the spontaneous curvature of the bilayer is strictly zero since the two monolayers, made of the same amphiphile, have spontaneous curvatures of the same magnitude but opposite in sign. Therefore, spontaneous vesicle formation can only be ascribed to entropic effects in conjunction with a quite low bending elasticity of the bilayer.

After the preliminary observation by electron microscopy that GM3 forms vesicles in water solution,[19] careful static and dynamic, both polarized and depolarized, laser light scattering experiments are used to characterize the system and, then, to verify the important finding that in solution vesicles are in equilibrium with a small amount of large aggregates, mostly of lamellar type.[24,45-46] The laser light scattering technique is not intrusive and, therefore, quite suitable to study equilibrium properties.

The ganglioside GM3, prepared as sodium salt, is dissolved in 30 mM NaCl water solution at a concentration of 0.1 mM, which is very dilute (0.119 mg/cm^3). NaCl is added to shield Coulomb interactions among vesicles.[28,29] Vesicles form spontaneously also in pure water or at different salt concentrations. As a test of the fact that GM3 solutions reach thermodynamic equilibrium spontaneously, it has been checked that sonicated and unsonicated solutions gave exactly the same results. Solutions were filtered on 0.4 μm polycarbonate filters prior to measurements. Aggregates in solutions are studied as a function of temperature, between 7 and 50°C, as temperature affects the aggregate distribution. No hysteresis effects have been observed along temperature scans.

The whole set of the light-scattering data are fitted together in a rather unique way by assigning almost all the ganglioside material to a vesicle distribution and a small percentage of it to flat bilayers and multilamellar liposomes. Thermodynamic equilibrium of the different species is quite evident, since large aggregates, after removal by microporous filtration, reform immediately at the expense of the vesicle population present in solution.

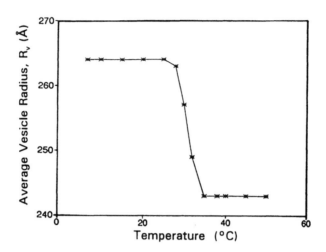

Figure 7 Average radius of the pure GM3 vesicles as a function of temperature.

The vesicle distribution is assumed to be the one predicted theoretically by Helfrich[47] for vesicles in thermodynamic equilibrium:[48]

$$W_v(r) = N(2/<N>)^2 exp(-N/<N>) \tag{5}$$

where $N = M_v/m$ is the number of monomers in a vesicle. The only free parameter is the average vesicle aggregation number $<N>$. Flat bilayers, schematized as very thin oblate ellipsoids with an axial ratio of 100, have an average size of 10,000 Å. The liposomes are treated as spheres of 10,000 Å diameter. The relative concentrations are reported in Table 4 for the two extreme temperatures. Good consistency in the fit of the absolute scattered intensity data is obtained with a value of 19 Å for the hydrophobic layer thickness ℓ.

The light-scattering result that the hydrophobic layer thickness of GM3 vesicles is small is quite interesting, as it can be due to chain interdigitation. In fact, the best-fit value is close to the length of the hydrophobic part of a single GM3 ganglioside molecule and not twice as large, as would have been expected if GM3 vesicles behaved like normal phospholipids ones. Small-angle X-ray measurements[49] performed on the same GM3 vesicle solution seem to confirm such a small bilayer thickness.

The temperature dependence of the average vesicle radius is reported in Figure 7. The corresponding average GM3 vesicle aggregation number ranges from 14,300 to 17,100.

Of course, the extremely simple distribution of large aggregates, made of two δ-functions, may not be too realistic, but it is surely indicative of the presence of two types of large objects in solution, the flat and the bulky ones.

The bulky aggregates should be present, otherwise absolute intensities and dynamic data could not match together. They represent only about 5×10^{-4} of the total GM3 concentration, Table 5, which in terms of volume fraction of GM3 in solution, is an extremely small quantity, of the order of 10^{-8}.

On the other hand, a strong verification that also flat objects are present is given by the existence of a depolarized component of the scattered light, which cannot be due to centrosymmetrical aggregates like vesicles. In fact, depolarized light-scattering theory predicts that for thin spherical vesicles, even if made up of anisotropic molecules, no depolarization can occur, due to internal cancellation effects.[50] On the other hand, if the shape of the aggregate is axisymmetric, the optical anisotropy of the molecule gives rise to a depolarization effect. This is the case with an amphiphilic cylindrically symmetric molecule embedded into a cylindrically symmetric aggregate like a disc. Depolarization measurements are therefore quite useful to indicate fragments of bilayers in the presence of a large population of spherical vesicles because vesicles do not contribute to them.

It may be argued that the existence of a small population of discs in equilibrium with vesicles is due to the presence in solution of some impurities of amphiphilic nature which could shield from water the unfavorable edges of the discs. In order to stabilize the discs, these molecules must have a smaller packing parameter[2] than GM3, that is, a larger ratio of the head-to-tail cross sections. This point has been checked by adding to the GM3 solution a second amphiphile, the ganglioside GM1, which has the same

Table 5 Concentration Values of GM3, in Percent by Weight, for the Different Types of Aggregates at the Two Extreme Temperatures, as Obtained from the Global Fit of the Light-Scattering Data

T(°C)	Vesicles	Discs	Spheres
7	96	3.9	0.04
50	98	1.9	0.06

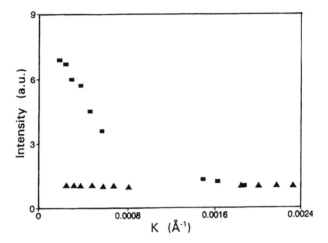

Figure 8 Angular dependence of the scattered intensity of 30 mM NaCl water solutions of pure GM3 (rectangles) and of mixed GM3 and GM1, 65:35 molar ratio (triangles).

hydrophobic part of GM3 but a larger headgroup. In fact, GM1 by itself forms micelles in solution with a small radius of curvature. The effect is that discs are not at all stabilized, but they gradually disappear as the GM1 concentration is increased, until a pure vesicle solution is obtained for a GM1 to GM3 ratio of 35 to 65. This is shown in Figure 8, where the angular distribution of the scattered light intensity of the mixed ganglioside solution is shown together with the pure GM3 solution, for comparison. The large scattered intensity at low angles, which is a clear indication of the existence of large aggregates in solution, disappears when GM1 is added. Besides, the scattered depolarized light intensity is observed to drop down completely, which is a further verification that discs are no longer present in solution. GM3-GM1 solutions have been prepared first in chloroform-methanol, so that gangliosides are dissolved in monomeric form, dried under vacuum, and then dissolved in water.

Absence of impurities in the biologically prepared ganglioside GM3 is also confirmed by the fact that test experiments performed on a small sample of synthetically prepared GM3[51] give exactly the same scattering curves as the ones obtained with the biological sample.

For a discussion of the GM3 vesicle properties, in conjunction with the ganglioside monomer characteristics, the most interesting light-scattering results can be summarized as follows: (1) vesicles form spontaneously in a single amphiphile water solution and are thermodynamically stable; (2) the bilayer thickness is smaller than twice the GM3 molecular length, indicating chain interdigitation; (3) different populations of aggregates are found in solution, that is, a few percent of the amphiphilic molecules do not go into vesicles, but form larger aggregates mainly of disc-like shape; and (4) the addition of a second amphiphile, the ganglioside GM1, with a larger headgroup than GM3 does not stabilize discs, as would be expected if GM1 provided a better coverage of the disc edges. Instead, the large aggregates progressively disappear as the GM1 content is increased, and finally only vesicles are found to be present in the mixed solution.

The behavior of the pure GM3 ganglioside solution can be qualitatively discussed in terms of the low bending elasticity of the ganglioside bilayer. The fact that GM3 vesicles form spontaneously in solution is a clear indication that the energy cost to form a vesicle is rather small, of the order of the thermal energy.

For a symmetric bilayer the bending energy per unit area,[52] to the approximation of no higher terms than second order in the mean curvature $H = c_1 + c_2$ contributing to it, may be written as $g = 1/2\kappa H^2 + \bar{\kappa} K$, where $K = c_1 c_2$ is the Gaussian curvature of the bilayer, c_1 and c_2 being the principal curvatures. There are two elastic moduli, the bending rigidity κ and the modulus of Gaussian curvature, $\bar{\kappa}$. For the vesicles of the ganglioside GM3, the spontaneous curvature term is absent, for symmetry reasons, since the two monolayers are made of the same amphiphile. The total bending energy of a sphere ($c_1 = c_2 = 1/r$) is $4\pi(2\kappa + \bar{\kappa})$. It may be regarded as the minimum energy required to form a vesicle from a planar bilayer and it is independent of the vesicle radius, within the above approximation. The bending rigidity modulus κ is positive, while $\bar{\kappa}$ may be positive or negative. When two monolayers made of the same amphiphilic molecules are stuck together to form the bilayer, each monolayer feels frustrated for any finite value of the spontaneous curvature of the monolayers. As a consequence of the frustration of the monolayers, the modulus of Gaussian curvature $\bar{\kappa}$ of the bilayer is found, in general, to be finite[53] and its sign is just the opposite of that of the molecular asymmetry, defined by the difference of the mean head and chain areas of the amphiphilic molecule, as demonstrated in Reference 54. For the GM3 ganglioside molecule the area occupied by the headgroup, made of three sugar rings, with a linear dimension of approximately 4 Å each, arranged in a ramified structure, is undoubtedly larger than the hydrophobic area determined by the chain spacing, of the order of 4.2 Å.[55] The modulus of Gaussian curvature is therefore expected to be negative for the ganglioside bilayer. There are also good reasons to believe that the bending rigidity itself is small as compared to the one reported normally for phospholipids.[56] First of all, the large headgroup dimension may cause disorder in the hydrophobic chains, then the hydrophobic thickness is small. Both effects are known to reduce the bending rigidity of the bilayer.[57] Finally, bending rigidity can also be reduced by protrusion effects.[58] In fact, the ganglioside GM3 does not have a sharp transition from the hydrophilic to the hydrophobic part along the molecule, due to the presence of the NH, CO, and OH groups in the sphingosine, Figure 2, which may allow more freedom to its motion normal to the bilayer surface. A low value of the bending rigidity modulus, of the order of $k_B T$, has also been found experimentally for large vesicles of glucosidic surfactants.[59] The bending energy of the GM3 vesicles can therefore be quite small, due to the combined effect of a small rigidity modulus and a negative κ. It is clear that the vesicle phase can be thermodynamically stable in this case, since translational entropy can balance the small positive energy required for the vesicle formation. Renormalization of the bending rigidity due to thermal undulations[60] gives a slight dependence of the vesicle energy from its size. The vesicle size distribution, Equation 5, used in the fit of the GM3 experimental data, is calculated with an effective rigidity which takes care of this renormalization effect. The average vesicle radius determined experimentally decreases with temperature, going from 265 Å at 7°C to 245 Å at 50°C. This could be due to an increased fluidity of the hydrophobic chains of GM3 or, more generally, to the reduction of the effective bending rigidity of the bilayer with temperature.

Also, the vesicle dimension should be connected to the rigidity parameters of the bilayer. In fact, the value of the average diameter of the spontaneously forming GM3 vesicles should be of the order of magnitude of the persistence length[61] of the GM3 bilayer, or, similarly, of the length scale L at which the effective rigidity κ' becomes very small:

$$\kappa' = \kappa - [\alpha k_B T/(4\pi)] \ln(L/a) << k_B T \qquad (6)$$

where the factor α is predicted to be either 1, see Reference 60, or 3, see Reference 62, and a is the bilayer thickness. In fact, an undulated piece of membrane will be "floppy" and start to make contacts with itself, and eventually close up in a vesicle, at effective rigidities near $k_B T$, but still positive. By taking L equal to 530 Å, the GM3 vesicle average diameter, and $a = 35$ Å its bilayer thickness, the effecive rigidity κ' becomes zero for $\kappa = 0.22\ k_B T$, with $\alpha = 1$, or $\kappa = 0.65\ k_B T$ when $\alpha = 3$. These values of κ are small, but not unreasonable.

The average vesicle radius has been predicted to be a weak function of the surfactant volume fraction for spontaneous vesicle formation,[43] although this dependence has been measured to be even weaker. The same behavior has been observed also for GM3 vesicles, for which no appreciable radius variation has been detected over two orders of magnitude of concentration values, that is, from 10^{-5} to 10^{-3} in volume fraction of surfactant. The GM3 situation could indeed be similar to the multicomponent system described in Reference 43. If the formalism developed in Reference 43, which calculates the average vesicle radius by taking care of renormalization effects both in κ and in $\bar{\kappa}$, is used, a value of $(2\kappa + \bar{\kappa}) = 2.7 k_B T$ is found for GM3 vesicles, which, again, is a rather reasonable value.

GM3 vesicles form spontaneously in a situation of zero spontaneous curvature of the bilayer. In fact, the two monolayers have exactly the same composition, since the solution contains a single amphiphile. The small amount of NaCl and of free monomers of GM3 (of the order of 10^{-8} M) present in solution cannot alter this picture.

Furthermore, it is found that not all amphiphilic molecules go into vesicles. A percentage of them form larger aggregates (see Table 5), which are in equilibrium with vesicles and slightly change their distribution as temperature is varied. This fact should somehow be connected to the frustration of the bilayer in the vesicles. The thermodynamic equilibrium is out of question, since large aggregates reform very rapidly, after removal by microporous filtration.

It was not possible to quantify the lifetime of this kinetic effect with the light scattering apparatus used in the experiments on GM3 solutions, but it is surely shorter than 20–30 sec, the time required by the scattered intensity to reach its asymptotic value after filtration.

The thermodynamic understanding of the presence of such large aggregates is not at all clear at this moment. Anyway, some energetic considerations can be done. The existence of membrane bunches in solution seems not to be a theoretical problem.[63,64] The bulky spheres could therefore represent onions of GM3 bilayers which can be predicted if a fourth order term in the mean curvature is added to the usual bending energy of a closed bilayer of spherical shape.[64] The number n of skins in the onion is then predicted to be inversely proportional to the usual term $(2\kappa + \bar{\kappa})$, containing the elastic moduli of the GM3 layers. Since it has been shown above that this term has to be small in GM3 solutions, onion-like stacking could be feasible.

It is by far more difficult to accept the idea that lamellar aggregates are thermodynamically stable in solution. The problem comes from the unfavorable edge energy due to water exposure of the hydrophobic chains. In principle, if the elastic energy required to close a vesicle equals the edge energy of the bilayer disc of equivalent surface, the thermodynamic equilibrium between discs and vesicles could be possible.[65] For instance, the edge energy of the GM3 vesicle-equivalent disc of 500 Å radius and 19 Å of hydrophobic layer thickness becomes 10 $k_B T$ for an edge tension of 2.5×10^{-8} erg/cm, which is about a factor of 100 smaller than what is reported for lecithin vesicles.[65] Such a low value makes the single-vesicle single-disc coexistence quite unfeasible. If, on the other hand, one considers the ensemble of discs and vesicles, the overall edge energy required by discs for a fixed amount of amphiphile going into discs (say, the 2–4% of GM3 of the present case), is inversely proportional to the disc radius. Therefore "infinite" lamellae could be in equilibrium with vesicles at the limit. In other words, the total positive elastic energy required by the GM3 vesicles, due to the lack of spontaneous curvature of the bilayer, may be equal to the total edge energy that is needed by the lamellar aggregates in the water solution. As already mentioned before, this unfavorable energy condition could be balanced by translational entropy.

It remains to be discussed however why discs of finite dimensions, average radius of 5000 Å, are stable in solution. Again, entropic effects could play a role: of course, there is an entropic gain if an "infinite" lamella breaks up into smaller discs.

It is interesting however to consider the possibility for the discs of having a really small edge energy due to the peculiar features of the GM3 bilayer. The ganglioside molecule itself has a large saccharidic headgroup which is normally believed to be quite hydrated and mobile. It is therefore plausible that the sugar groups could rearrange themselves in order to better shield the exposure to water of the hydrophobic tails at the disc edges. This is even more plausible considering the fact that the hydrophobic layer of GM3 bilayers is already found to be thinner than for normal phospholipids, due to chain interdigitation.

The problem of the energy balance among vesicles with no spontaneous curvature and finite lamellae with exposed edges is overcome when GM1 is added to GM3. In this case the large aggregates disappear completely from the GM3 solution, leaving vesicles only. This can be easily understood in terms of a theoretical model developed for mixed copolymer bilayers.[66] The model predicts that curvature energy favors an inhomogeneous distribution of the two gangliosides among the layers. Therefore, spontaneous curvature readjustments via demixing can give rise to a finite spontaneous curvature of the bilayer, energetically favoring vesicles towards other structures.

Furthermore, the observation that discs disappear when GM1 is added is a clear indication that the simple criticism of disc stabilization by amphiphilic impurities is not appropriate. In fact, amphiphilic molecules with larger headgroups could, in principle, segregate at the disc edges to shield hydrocarbons more easily. Indeed, GM1 is one of such amphiphilic molecules, but its effect on the GM3 solution is just the opposite.

For the pure GM3 solution, the fast equilibration times of the vesicle distribution as temperature is varied and when aggregates reform after filtration is an interesting indication that monomer exchange is not the effect responsible of the equilibrium among the different aggregates. It is known, in fact, that monomer exchange in ganglioside micellar solutions is an extremely slow process[36] connected to the very low solubility of ganglioside monomers (cmc, in micellar language). The experimental observations indicate then that the floppy pure GM3 vesicle should form and break rather rapidly, thus allowing reequilibration among the different aggregates in solution.

The reason why gangliosides give such a low bending elasticity can again be qualitatively understood by simple geometrical arguments. The oligosaccharide chain headgroup of gangliosides has a lateral extension which is larger than the corresponding one of the phospholipid headgroups, phosphatidylcholine, for example. On the other hand, the hydrophobic dimensions are about the same. It is reasonable, therefore, to think that in the bilayer the sugar headgroups exert a greater lateral force than the corresponding choline groups, which do not exceed the hydrocarbon chain lateral dimension. This can reduce the bending elasticity, since it is known that the compression elasticity of the hydrocarbon chains is quite high. In other words, the restoring force due to the sugar headgroups compression upon bending is smaller than the one coming from the hydrocarbon chains. A confirmation for this is given by the comparison of the interfacial area for GM3, which is 80 Å^2, see Table 3, and the one for egg lecithin, which is only 70 Å^2.

The strong chain interdigitation can also be qualitatively understood by considering the large lateral extension of the sugar headgroups which are capable of shielding more hydrophobic surface from water. In order to obtain similar interdigitation effects in phosphatidylcholine-containing bilayers, it is necessary to add some other molecules which adsorb slightly at the hydrophilic-hydrophobic interface.[67] The added molecules provide the extra surface area needed to shield the hydrocarbons in the interdigitated phase, which cannot be shielded by the small phosphatidylcholine groups.

It is interesting to note that the mismatch in packing areas of the polar head and the hydrocarbon chains may also justify some anharmonic contributions to the energy of the bilayers. In the case of low bending elasticity, and hence large thermal fluctuations, a small anharmonic term may favor finite size lamellar structures.[46] This can help in understanding the existence of a small lamellar population in equilibrium with vesicles in the single amphiphile solution of the ganglioside GM3.

In summary, the large dimensions of the saccharidic headgroups and the reduced thickness of the bilayer are the main factors which justify the low bending elasticity of GM3 vesicles.

As a final remark, it is interesting to note that the packing parameter for GM3 is 0.63, which is larger than 1/2, as it should be for a vesicle, but sensibly smaller than 0.85, the one of egg lecithin. It is therefore reasonable to think that increasing values of packing parameters in the range $1/2 < P < 1$ characterize bilayers with increasing mechanical rigidity.

REFERENCES

1. **Tanford, C.,** *The Hydrophobic Effect*, Wiley, New York, 1980.
2. **Israelachvili, J. N.,** *Intermolecular and Surface Forces*, Academic Press, New York, 1990.
3. **Cantù, L., Corti, M., Sonnino, S. and Tettamanti, G.,** Light scattering measurements on gangliosides: dependence of micellar properties on molecular structure and temperature, *Chem. Phys. Lipids*, 41, 315, 1986.
4. **Tettamanti, G., Sonnino, S., Ghidoni, R., Masserini, M. and Venerando, B.,** Chemical and functional properties of gangliosides, in *Physics of Amphiphiles: Micelles, Vesicles and Microemulsions*, Degiorgio, V. and Corti, M., Eds., North-Holland, Amsterdam, 1985, 607.
5. **Cantù, L., Corti, M., Casellato, R., Acquotti, D. and Sonnino, S.,** Aggregation properties of GD1a, II^3 Neu5Ac$_2$ GgOse$_4$ Cer, and of GD1b-lactone, II^3 [α-Neu5Ac-(2->8,1->9) α-Neu5Ac]GgOse$_4$ Cer, in aqueous solution, *Chem. Phys. Lipids*, 60, 111, 1991.
6. **Schaurer, R.,** *Sialic Acid: Chemistry, Metabolism and Functions*, Springer-Verlag, Wien, 1982.
7. **Sonnino, S., Kirschner, G., Ghidoni, R., Acquotti, D. and Tettamanti G.,** Preparation of GM1 ganglioside molecular species having homogeneous fatty acid and long chain base moieties, *J. Lipid Res.*, 26, 248, 1985.
8. **Israelachvili, J. N., Mitchell, D. J. and Ninham, B. W.,** Theory of self-assembly of hydrocarbon amphiphiles into micelles and bilayers, *J. Chem. Soc., Faraday Trans. II*, 72, 1525, 1976.
9. **Israelachvili, J. N., Marcelja, S. and Horn, R. G.,** Physical principles of membrane organization, *Q. Rev. Biophys.*, 13, 2, 121, 1980.
10. **Ulrich-Bott, B. and Wiegant, H.,** Micellar properties of glycosphingolipids in aqueous media, *J. Lipid Res.*, 25, 1233, 1984.

11. **Corti, M. and Degiorgio, V.,** Micellar properties of gangliosides, in *Solution Behaviour of Surfactants*, Vol. 1, Mittal, K. L. and Fendler, E. J., Eds., Plenum Press, New York, 1982, 573.

12. **Corti, M., Cantù, L., Sonnino, S. and Tettamanti, G.,** Aggregation properties of gangliosides in aqueous solutions, in *New Trends in Ganglioside Research: Neurochemical and Neuroregenerative Aspects*, Ledeen, R. W., Hogan, E. L., Tettamanti, G., Yates, A. J. and Yu, R. K., Eds., Fidia Research Series, Vol.14, Liviana Press, Padova,1988,79.

13. **Sonnino, S., Cantù, L., Corti, M., Acquotti, D. and Venerando, B.,** Aggregative properties of gangliosides in solution, *Chem. Phys. Lipids*, 71, 21, 1994.

14. **Sonnino, S., Cantù, L., Corti, M., Acquotti, D., Kirschner, G. and Tettamanti, G.,** Aggregation properties of semisynthetic GM1 ganglioside containing an acetyl group as acyl moiety, *Chem. Phys. Lipids*, 56, 49, 1990.

15. **Mraz, W., Schwarzmann, G., Sattler, J., Momoi, T., Seeman, B. and Wiegandt, H.,** Aggregate formation of gangliosides at low concentrations in aqueous media, *Hoppe-Seyler's Z. Physiol. Chem.*, 361, 177, 1980.

16. **Corti, M., Cantù, L. and Salina, P.,** Aggregation properties of biological amphiphiles, *Adv. Colloid Interface Sci.*, 36, 153, 1991.

17. **Curatolo, W., Small, D. M. and Shipley, G. G.,** Phase behaviour and structural characteristics of hydrated bovine brain ganglioside, *Biochim. Biophys. Acta*, 468, 11, 1977.

18. **Corti, M.,** The laser light scattering technique and its application to micellar solutions, in *Physics of Amphiphiles: Micelles, Vesicles and Microemulsions*, Degiorgio, V. and Corti, M., Eds., North-Holland, Amsterdam, 1985, 59.

19. **Sonnino, S., Cantù, L., Acquotti, D., Corti, M. and Tettamanti, G.,** Aggregation properties of GM3 ganglioside (II^{3}Neu5AcLacCer) in aqueous solutions, *Chem. Phys. Lipids*, 52, 231, 1990.

20. **Corti, M., Degiorgio, V., Ghidoni, R., Sonnino S. and Tettamanti G.,** Laser-light scattering investigation of the micellar properties of gangliosides, *Chem. Phys. Lipids*, 26, 225, 1980.

21. **Cantù, L., Corti, M. and Degiorgio, V.,** Mixed micelles of gangliosides, *J. Phys. Chem.*, 94, 793, 1990.

22. **Masserini, M., Sonnino, S., Giuliani, A., Tettamanti, G., Corti, M., Minero, C. and Degiorgio, V.,** Laser-light scattering study of size and stability of ganglioside-phospholipid small unilamellar vesicles, *Chem. Phys. Lipids*, 37, 83, 1985.

23. **Cantù, L., Corti, M., Musolino, M. and Salina, P.,** Spontaneous vesicles formed from a single amphiphile, *Europhys. Lett.*, 13(6), 561, 1990.

24. **Cantù, L., Corti, M. and Musolino, M.,** Micelles and vesicles of gangliosides, in *The Structure and Conformation of Amphiphilic Membranes*, Lipowsky, R., Richter, D. and Kremer, K., Eds., Springer Proceedings in Physics, Vol. 66, Springer-Verlag, Berlin, 1992, 185.

25. **Corti, M., Cantù, L., Del Favero, E. and Boretta, M.,** The effect of temperature on interacting micelles of gangliosides in water, *J. Physics: Cond. Matt.*, 6A, 363, 1994.

26. **Cantù, L., Corti, M., Zemb, T. and Williams, C.,** Small angle X-ray and neutron scattering from ganglioside micellar solutions, *J. Phys. (France) IV*, 3, 221, 1993.

27. **Cantù, L., Corti, M. and Degiorgio, V.,** Static and dynamic properties of solutions of strongly interacting ionic micelles, *Europhys. Lett.*, 2, 673, 1986.

28. **Degiorgio, V., Cantù, L., Corti, M., Piazza, R. and Rennie, A.,** Light and neutron scattering studies of strongly interacting ionic micelles, *Colloid and Surface*, 38, 169, 1989.

29. **Acquotti, D., Fronza, G., Ragg, E. and Sonnino, S.,** Three dimensional structure of GD1b and GD1b-monolactone gangliosides in dimethylsulphoxide: a nuclear Overhauser effect investigation supported by molecular dynamics calculations, *Chem. Phys. Lipids*, 59, 107, 1991.

30. **Siebert, H. C., Reuter, G., Schauer, R., von der Lieth, C. W. and Dabrowsky, J.,** Solution conformations of GM3 gangliosides containing different sialic acid residues as revealed by NOE-based distance mapping, molecular mechanics and molecular dynamics calculations, *Biochemistry*, 31, 6962, 1992.

31. **Goldstein, D.,** *States of Matter*, Prentice-Hall, New York, 1975.

32. **Cantù, L., Corti, M. and Degiorgio, V.,** Static and dynamic light scattering study of solutions of strongly interacting ionic micelles, *Faraday Disc. Chem. Soc.*, 83, 287, 1987.

33. **Cantù, L., Corti, M., Degiorgio, V., Piazza, R. and Rennie, A.,** Neutron scattering from ganglioside micelles, *Prog. Colloid Polym. Sci.*, 76, 216, 1988.

34. **Corti, M., Degiorgio, V., Ghidoni, R., and Sonnino, S.,** Mixed micelles of GM1 ganglioside and a nonionic amphiphile, *J. Phys. Chem.*, 86, 2533, 1982.

35. **Clint, J.,** Micellization of mixed nonionic surface active agents, *J. Chem. Soc.*, 71, 1327, 1975.

36. **Cantù, L., Corti, M. and Salina, P.,** Direct measurement of the formation time of mixed micelles, *J. Phys. Chem.*, 95, 5981, 1991.

37. **Lasic D.,** The mechanism of vesicle formation, *Biochem. J.*, 256, 1, 1988.

38. **Lasic D.,** On the thermodynamic stability of liposomes, *J. Colloid Interface Sci.*, 140, 302, 1990.

39. **Kaler, E. W., Murthy, A. K., Rodriguez, B. E. and Zasadzinski, J. A. N.,** Spontaneous vesicle formation in aqueous mixtures of single-tailed surfactants, *Science*, 245, 1371, 1989.

40. **Safran, S. A., Pincus, P. A., Andelman, D. and MacKintosh, F. C.,** Stability and phase behaviour of vesicles in surfactant mixtures, *Phys. Rev. A*, 43, 1071, 1991.

236

41. **Hauser, H., Gains, N. and Muller, M.,** Vesiculation of unsonicated phospholipid dispersions containing phosphatidic acid by pH adjustment: physicochemical properties of the resulting unilamellar vesicles, *Biochemistry*, 22, 4775, 1983.
42. **Talmon, Y., Evans, D. F. and Ninham, B. W.,** Spontaneous vesicles formed from hydroxide surfactants: evidence from electron microscopy, *Science*, 221, 1047, 1983.
43. **Herve, P., Roux, D., Bellocq, A. M., Nallet, F. and Gulik-Krzywicki, T.,** Dilute and concentrated phases of vesicles at thermal equilibrium, *J. Phys. II (France)*, 3, 1225, 1993.
44. **Munkert, U., Hoffmann, H., Thunig, G., Meyer, H. W. and Richter W.,** Perforated vesicles in ternary surfactant systems of alkyl-dimethylaminoxides, cosurfactant and water, *Prog. Colloid Polymer Sci.*, 93, 137, 1993.
45. **Cantù, L., Corti, M., Lago, P. and Musolino, M.,** Characterization of a vesicle distribution in equilibrium with large aggregates by accurate static and dynamic light scattering measurements, in *Photon Correlation Spectroscopy: Multicomponent Systems,* Schmitz, K. S., Ed., *Proc. SPIE 1430,* 144, 1991, ISBN 0-8194-0520-5.
46. **Cantù, L., Corti, M., Del Favero, E. and Raudino, A.,** Vesicles as an equilibrium structure of a simple surfactant-water system, *J. Phys. II (France)*, 4, 1585, 1994.
47. **Helfrich, W.,** Size distribution of vesicles: the role of the effective rigidity, *J. Phys. France,* 4, 321, 1986.
48. Vesicle distributions different from the one of Reference 47 have also been proposed, see for instance Eriksson, J. C., Bergstrom, M. and Ljunggren, S., Size distribution of equilibrated surfactant vesicles, *Prog. Colloid Polymer Sci.,* 93, 225, 1993. The precision of the light-scattering data treatment is not sufficient to distinguish among them.
49. **Cantù, L., Corti, M., Del Favero, E. and Zemb, T.** (to be published).
50. **Aragon, S. R. and Elwenspoek, M.,** Mie scattering from thin spherical bubbles, *J. Chem. Phys.,* 77, 3406, 1982.
51. This sample has been kindly supplied by FIDIA Spa.
52. **Helfrich, W.,** Elastic properties of lipid bilayers: theory and possible experiments, *Z. Naturforsch.,* 28c, 693, 1973.
53. **Porte, G., Appell, J., Bassereau, P. and Marignan, J.,** L_α to L_3: a topology driven transition in phases of infinite fluid membranes, *J. Phys. (France),* 50, 1335, 1989.
54. **Petrov, A. G., Mitov, M. D. and Derhansky, A.,** Saddle splay instability in lipid bilayers, *Phys. Lett.,* 65a, 374, 1978.
55. **Luzzati, V.,** X-ray diffraction studies of lipid-water systems, in *Biological Membranes,* Chapman, D., Ed., Academic Press, New York, 1968, 71.
56. **Meleard, P., Mitov, M. D., Faucon, J. F., and Bothorel, P.,** Dynamics of fluctuating vesicles, *Europhys. Lett.,* 11, 355, 1990.
57. **Safran, S. A.,** Microemulsions: an ensemble of fluctuating interfaces, in *Structure and Dynamics of Strongly Interacting Colloids and Supramolecular Aggregates in Solution,* Chen, S. H. et al., Eds., Kluwer Academic Publishers, Holland, 1992, 237.
58. **Lipowsky, R. J. and Grotehans, S.,** Hydration vs. protrusion forces between lipid bilayers, *Europhys. Lett.,* 3, 599, 1993.
59. **Muts, M. and Helfrich, W.,** Bending rigidities of some biological model membranes as obtained from Fourier analysis of contour sections, *J. Phys. (France),* 51, 991, 1990.
60. **Helfrich, W.,** Effect of thermal undulations on the rigidity of fluid membranes and interfaces, *J. Phys. (France),* 46, 1263, 1985.
61. **Langevin, D.,** Micelles and microemulsions, *Annu. Rev. Phys. Chem.,* 43, 341, 1992.
62. **Peliti, L. and Leibler, S.,** Effects of thermal fluctuations on systems with small surface tension, *Phys. Rev. Lett.,* 54, 1690, 1985.
63. **Lipowsky, R.,** Stacks and bunches of fluid membranes, *J. Phys.: Cond. Matt.,* 6A, 409, 1994.
64. **Helfrich, W.,** Lyotropic lamellar phases, *J. Phys.: Cond. Matt.,* 6A, 79, 1994.
65. **Fromherz, P.,** Lipid-vesicle structure: size controlled by edge-active agents, *Chem. Phys. Lett.,* 94, 259, 1983.
66. **Dan, N. and Safran, S. A.,** Spontaneous curvature of mixed copolymer bilayers, *Europhys. Lett.,* 21, 975, 1993.
67. **McIntosh, T. J., McDaniel, R. V. and Simon, S. A.,** Induction of an interdigitated gel phase in fully hydrated phophatidylcholine bilayers, *Biochem. Biophys. Acta,* 731, 109, 1983.

Chapter 10

Effect of Solutes on the Membrane Lipid Phase Behavior

Boris Tenchov and Rumiana Koynova

CONTENTS

I. INTRODUCTION

Due to their applications as experimental biomembrane models, the lipid dispersions in water are at present one of the most thoroughly studied lyotropic liquid crystalline systems. The investigations on the lipid phase behavior are a long known example of a very productive application of these systems.[1] Numerous studies in this field have revealed that membrane lipids display rich lyotropic and thermotropic polymorphism characterized by a large set of lamellar and non-lamellar phases. Frequently occurring among them are the lamellar subgel L_c and gel L_β phases, which are stable at low temperatures, and the lamellar liquid crystalline L_α phase, which forms at ambient temperatures. The biological significance of the latter phase is well known — it is presently accepted that the cellular membranes are liquid crystalline bilayers with proteins embedded in them.[2] At elevated temperatures, many lipids can also form non-lamellar liquid crystalline phases represented by several phases of cubic symmetry (Q_i) and the inverted hexagonal phase H_{II}.[3-6] The ability of membrane lipids to form non-lamellar phases is in contrast to the traditional view of their role as an inert matrix. A number of new concepts suggest their participation in various cellular processes. These concepts usually take advantage of the fact that formation of non-lamellar loci in a biomembrane destroys its integrity and barrier properties with all consequences thereof. Non-lamellar lipid patterns are believed to play important role in processes of cell fusion, cell division, cell-cell and cell-virus interactions, transport of macromolecules across membranes, anaesthetic action, cell damage caused by extreme physiological conditions such as high and low temperatures, dehydration, etc.[5-14]

In fully hydrated membrane lipids the temperature-induced phase transitions occur in the following order:

$$L_C \leftrightarrow L_\beta \leftrightarrow L_\alpha \leftrightarrow Q_i \leftrightarrow H_{II} \tag{1}$$

where temperature increases from left to right.

Most lipids, however, do not exhibit the general sequence (Equation 1) in temperature scans. Non-bilayer forming lipids such as phosphatidylethanolamines often display a reduced sequence $L_C \rightarrow L_\alpha \rightarrow H_{II}$ and do not form L_β and cubic phases, at least in first heating scans. Also, some glycoglycerolipids were found to display direct $L_\beta \rightarrow H_{II}$ or $L_C \rightarrow H_{II}$ transitions in heating scans, skipping all intermediate phases.[15,16] Due to slow transition kinetics and formation of metastable phases, the cooling phase sequences in many lipid-water systems differ from the heating ones.[17] A comprehensive description of the lipid phase behavior is rather difficult to develop as it must involve quantitative evaluation of several different interactions contributing to the stability of any given phase (hydrophobic effect, steric, electrostatic, van der Waals interactions, inter- and intramolecular hydrogen bonding, etc.). A number of studies have demonstrated the important role of the intermolecular interactions in the lipid aggregates, described in terms of lipid packing and spontaneous surface curvature, in the lamellar to non-lamellar transformations.[6,11,12,18] The data about solute effects summarized in this article show that the composition of the aqueous phase is another important factor strongly modulating the relative stability of the lipid lamellar and non-lamellar phases.

It is long recognized that the interactions of the polar groups of the amphiphiles with the water molecules give a big contribution to the energy balance in a lipid phase. A relatively high hydration, characteristic of membrane lipids, is responsible in particular for their ability to form liquid crystalline bilayers well separated by aqueous spaces. A number of lipid phase transformations (chain melting, lamellar to inverted hexagonal transitions) occur with large changes in lipid surface area and respective changes in the amount of bound water. It is thus clear that the lipid-water interactions are an important determinant also for the thermodynamic stability and existence ranges of the lipid phases. The lipid hydration is determined by the chemical structure of the polar groups, but it is important to note that in fully hydrated systems, where water is in excess, the extent of hydration of the polar groups depends also on the state of bulk water, via the equal chemical potentials of bound and free water. On the other side, low-molecular water-structure makers (kosmotropic solutes) and water-structure breakers (chaotropic solutes) strongly modulate the structure of bulk water. It is thus possible that, even without direct interactions with the lipid polar groups, solutes can influence the properties of the lipid interface and, consequently, the lipid phase behavior. Indirect solute effects of such kind, generally termed the Hofmeister effect, have been found in many studies of different interfaces (see, for example, Reference 19). The Hofmeister effect originates from the competition between interface groups and solutes for water molecules. By stabilizing the structure of bulk water, kosmotropic solutes tend to reduce surface hydration and interface area. By destabilizing the structure of bulk water, the chaotropic solutes have the opposite effect. Here we summarize the published data about solute effects on the lipid phase behavior and point out that a consistent explanation of these effects can be given by considering them as another manifestation of the Hofmeister effect. This summary also outlines means to induce or suppress formation of the lamellar liquid-crystalline phase by regulating the composition of the aqueous environment at constant temperature.

II. INFLUENCE OF SOLUTES ON THE PHASE TRANSITION TEMPERATURE

Upward or downward shifts of the transition temperature are the most pronounced and easily detectable effect of solutes on the lipid phase behavior. It is illustrated by Figure 1 which demonstrates the effects of NaSCN, Gu.HCl and sucrose on the $L_\beta \to L_\alpha$ and $L_\alpha \to H_{II}$ transitions in fully hydrated dihexadecylphosphatidylethanolamine. The other thermodynamic characteristics of the lipid phase transitions — enthalpy, transition width, and maximum specific heat — appear to be weakly influenced by solutes. The data in References 20–22 seem to show that addition of Na_2SO_4, NaF, NaCl or $CaCl_2$ leads to increased enthalpy of the $L_\alpha \to H_{II}$ phase transitions, accompanied by a slight enthalpy decrease and broadening of the $L_\beta \to L_\alpha$ transitions. Our calorimetric data show a similar effect of sucrose on the dihexadecylphosphatidylethanolamine transition parameters, while NaSCN broadens both the $L_\beta \to L_\alpha$ and $L_\alpha \to H_{II}$ transitions and decreases the enthalpy of the latter transition.

It is appropriate to describe the dependence of transition temperature on solute concentration by the ratio dT_{tr}/dc, where dT_{tr} is the transition shift induced by an increase of solute concentration with dc. It is important to note that dT_{tr}/dc is a thermodynamic quantity similar to the quantity dT_{tr}/dp in the Clapeyron-Clausius equation, which describes transition shifts with pressure increase. An equation relating dT_{tr}/dc to the parameters of the lipid-water systems is given elsewhere.[23] It is used further in this chapter for analysis of the experimental data.

Linear shifts of T_{tr} with solute concentration are characterized by a single value of the derivative dT_{tr}/dc in the whole concentration range (Figure 2). When this dependence is not linear at higher solute concentrations (see Figures 1 and 3 for examples) it is appropriate to use the initial value of dT_{tr}/dc at c $\to$ 0. This limit is representative for dilute (ideal) solutions. Experimental data about solute effects are available at present about three classes of lipids — phosphatidylethanolamines,[20-22,24-30] phosphatidylcholines,[26,31-36] and glycolipids (R. Koynova et al., submitted for publication). From these data we have determined the initial transition shifts ΔT_{tr} normalized per 1 M of solute. Tables 1 through 3 summarize the values obtained in this way.

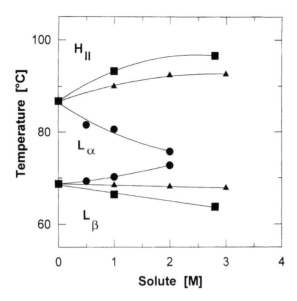

Figure 1 Phase transition temperatures of dihexadecylphosphatidylethanolamine in NaSCN (■), GuHCl (▲), and sucrose (●) solutions.

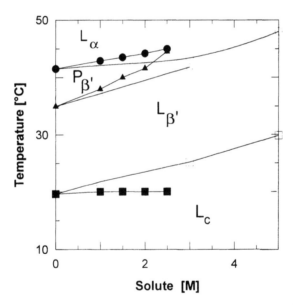

Figure 2 Phase transition temperatures of dipalmitoylphosphatidylcholine in sucrose (full symbols) and proline (open symbols) solutions. (Data from Tsvetkova, N. et al., *Chem. Phys. Lipids*, 60, 51, 1991.)

A. PHOSPHATIDYLETHANOLAMINES

Best studied at present are the effects of chaotropic and kosmotropic solutes on the phase behavior of phosphatidylethanolamines (Table 1). The solutes used in these studies include inorganic salts, mono- and disaccharides, sugar alcohols, some polyols, proline, and urea. It is evident from Table 1 that the solute effects obey the following general rules:

1. The solute-induced temperature shifts of the $L_\beta \rightarrow L_\alpha$ and $L_\alpha \rightarrow H_{II}$ transitions in phosphatidylethano-lamines are of opposite sign, i.e., an upward shift of one of these two transitions is always combined with a downward shift of the other transition.

2. The $L_\beta \rightarrow L_\alpha$ transition is much less sensitive to solutes than the $L_\alpha \rightarrow H_{II}$ transition. On average, the temperature shifts of the former transition are several times smaller compared to those of the latter one.

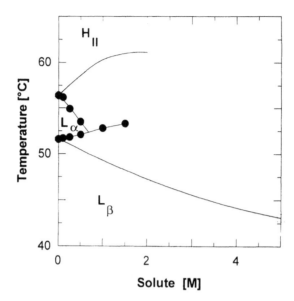

Figure 3 Phase transition temperatures of ditetradecylglucosylglycerol in NaSCN (□), and sucrose (●) solutions.

3. Chaotropic solutes (NaI, Gu.HCl, NaSCN, urea) induce upward shift of the $L_\alpha \rightarrow H_{II}$ transition, while kosmotropic solutes (sugars, sugar alcohols, polyols, proline) induce downward shift of this transition. At the same time, the $L_\beta \rightarrow L_\alpha$ transition moves in the opposite directions. Thus, chaotropic solutes favor formation of L_α at the expense of L_β and H_{II}, while kosmotropic solutes reduce the existence range of the L_α phase.

A remarkable manifestation of the solute effect is that a high enough concentration of sucrose or NaCl fully suppresses the L_α phase (Figure 1). Consequently, direct $L_\beta \rightarrow H_{II}$ transitions take place in dihexadecylphosphatidylethanolamine and distearoylphosphatidylethanolamine under these conditions.[25,30] In dimyristoylphosphatidylethanolamine, high NaCl concentration induces formation of an H_{II} phase, which does not appear in pure water.[21] On the other hand, 1 M of NaSCN eliminates the H_{II} phase in soybean phosphatidylethanolamine.[37]

According to their effect on the $L_\alpha \rightarrow H_{II}$ transition of palmitoyloleoylphosphatidylethanolamine and dielaidoylphosphatidylethanolamine (Table 1), the sodium salts arrange as follows:

$$SO_4^{2-} > OAc^- > Cl^- > Br^- > I^- > SCN^- \qquad (2)$$

This is a characteristic lyotropic (Hofmeister) series for anions. It can be also obtained by many other methods, for example, by the order in which these anions elute from a Sephadex® G-10 column.[19] For phosphatidylethanolamines, the sign inversion point between "kosmotropic" and "chaotropic" sides of the series (Equation 2) is between Br⁻ and I⁻, while this point is found to be at about the position of the chloride anion in other systems (for example, Sephadex® column, protein solutions, etc.).[19]

B. PHOSPHATIDYLCHOLINES

Saturated phosphatidylcholines of equal chain length are bilayer-forming lipids characterized by the phase sequence:

$$L_c \leftrightarrow L_{\beta'} \leftrightarrow P_{\beta'} \leftrightarrow L_\alpha \qquad (3)$$

Solute effects have been studied mainly for the pretransition ($L_{\beta'} \rightarrow P_{\beta'}$) and the main transition ($P_{\beta'} \rightarrow L_\alpha$) in dipalmitoylphosphatidylcholine.[31-35] There are single measurements of the $L_c \rightarrow L_{\beta'}$ transition of dipalmitoylphosphatidylcholine in proline and sucrose solutions (Table 2). It has been reported that 1 M of KSCN causes chain interdigitation in the gel phase of dipalmitoylphosphatidylcholine.[36] However, the initial slopes dT_{tr}/dc cannot be calculated in this case since no measurements at lower KSCN concentrations have been carried out. The transition shifts ΔT_{tr} per 1 M of solute for dipalmitoylphosphatidylcholine

Table 1 Influence of Solutes on the Phase Transition Temperatures of Phosphatidylethanolamines

Lipid	Solute	Transition shift $\Delta T_{tr}/1\ M$ solute		Ref.
		$L_\beta - L_\alpha$	$L_\alpha - H_{II}$	
Inorganic Salts				
DMPE[a]	NaCl	1.5	−8.0	20,21
		(1–4 M)	(3–6 M)	
DPPE	NaCl	2.0	−7.4	20
	CaCl$_2$	3.4		28
POPE	Na$_2$SO$_4$		−16.0	22
	NaCl		−7.3	22
	NaBr		−3.5	22
	NaI		6.0	22
	GuHCl		8.0	22
	NaSCN		27.0	22
DSPE	NaCl	1.3	−4.3	20
DEPE	Na$_2$SO$_4$	1.7	−9.9	29
	NaOAc	1.5	−9.1	29
	NaCl	1.3	−5.6	29
	NaSCN	−3.3	20.0	29
	GuHCl	−4.6	5.0	29
	GuHSCN	−6.5	50.0	29
DTPE	NaCl	1.2	−3.8	21
DHPE	NaCl	1.3	−2.5	21
	NaSCN	−2.3	6.5	b
	GuHCl	−0.3	3.2	b
Monosaccharides				
DEPE	Glucose		−0.2	24
	Galactose		−0.6	24
	Fructose		−2.5	24
Disaccharides				
POPE	Sucrose		−16.5	22
	Raffinose		−30.0	22
DEPE	Sucrose		−12.6	24
	Trehalose		−10.8	24
	Lactose		−10.0	24
	Maltose		−9.0	24
DPPE	Sucrose	2.7	−19.6	b
DSPE	Sucrose	1.4	−8.4	25
	Trehalose	1.1	−8.7	25
DHPE	Sucrose	2.0	−5.6	b
	Trehalose	1.5	−8.6	b
Sugar alcohols				
POPE	Sorbitol		−9.0	22
DEPE	Sorbitol		−5.4	24
	myo-Inositol		−5.5	24
Other polyols				
POPE	Glycerol		−3.3	22
DSPE	Glycerol	0.9	−1.0	27
DEPE	Glycerol		−2.8	24
	PEG		0.2	24
Amino acids				
DHPE	Proline	0.9	−1.8	26
Others				
POPE	Urea		1.3	22

[a] PE, phosphatidylethanolamine; DMPE, dimyristoyl PE; DPPE, dipalmitoyl PE; DSPE, distearoyl PE; DTPE, ditetradecyl PE; DHPE, dihexadecyl PE; DEPE, dielaydoyl PE; POPE, palmitoyl oleolyl PE.

[b] Data from Koynova et al., submitted for publication.

Table 2 Influence of Solutes on the Phase Transition Temperatures of Dipalmitoylphosphatidylcholine

Solute	Transition shift $\Delta T_{tr}/1~M$ solute			Ref.
	$L_c - L_{\beta'}$	$L_{\beta'} - P_{\beta'}$	$P_{\beta'} - L_\alpha$	
Proline	2.0	2.3	0.7	26
4-Hydroxyproline		0.8	0.6	31
Glycine		1.0	0.4	31
Betaine		2.2	0.8	31
Trehalose		1.3	0.7	31
Sucrose	0.4	3.8	1.4	a
LiCl		3.9	0.8	32
NaCl		1.9	0.5	32
KCl		1.3	0.4	32
RbCL		0.7	0.1	32
CsCl		−0.1	0.0	32

[a] Data from Koynova et al., submitted for publication.

Table 3 Influence of Solutes on the Phase Transition Temperatures of Glycoglycerolipids[a]

Lipid	Solute	Transition shift, $\Delta T_{tr}/1~M$ solute				
		$L_C - L_\beta$	$L_C - L_\alpha$	$L_\beta - L_\alpha$	$L_\beta - H_{II}$	$L_\alpha - H_{II}$
14-Glc[b]	Sucrose		1.2	1.0		−5.8
	Trehalose			0.4		−10.0
	NaSCN		−2.0	−2.2		3.9
16-Glc	NaSCN	−0.8		−1.1		1.9
16-Man	NaSCN			−0.7	−0.7	1.5

[a] Data from Koynova et al., submitted for publication.
[b] 14-Glc, ditetradecylglucosylglycerol; 16-Glc, dihexadecylglucosylglycerol; 16-Man, dihexadecylmannosylglycerol.

are summarized in Table 2. It shows that the pretransition and main transition move in the same direction with the solute concentration, and that the former transition is more sensitive to solutes than the latter one. Similarly to the chain-melting transition in phosphatidylethanolamines, the main transition in dipalmitoyl-phosphatidylcholine shifts to higher temperatures in the presence of kosmotropic solutes. A noteworthy consequence of the greater sensitivity of the pretransition to solutes is that it merges with the main transition at high enough solute concentrations. Sucrose and proline in concentrations of 2.4 and 3 M, respectively, fully suppress the rippled $P_{\beta'}$ phase, as demonstrated in Figure 2.

On the basis of their effect on the $L_{\beta'} \rightarrow P_{\beta'}$ and $P_{\beta'} \rightarrow L_\alpha$ transitions of dipalmitoylphosphatidylcholine, the alkali chlorides arrange in the order:

$$Li^+ > Na^+ > K^+ > Rb^+ > Cs^+ \tag{4}$$

Identical Hofmeister series for cations have been obtained by numerous other experiments (chromato-graphic behavior, ionic entropy of dilution, etc.).[19] According to the data in Table 2, the kosmotropic/chaotropic sign inversion point for the cation series (Equation 4) in the case of dipalmitoylphosphatidyl-choline is at the position of Cs[+]. In other systems, this point was found to be near the position of Na[+].[19] A comparison of Tables 1 and 2 shows that the anion series (Equation 2) modify the chain-melting temperature of the lipids in a much broader range than the cation series (Equation 4).

In contrast to alkali chlorides, NO_3^- and ClO_4^- alkali salts do not produce a measurable effect on the phase transitions of dipalmitoylphosphatidylcholine.[32] No systematic effect on the transition temperatures has been detected also in the presence of up to 50 mM $CaCl_2$.[33] According to published data, the enthalpy of the dipalmitoylphosphatidylcholine $L_{\beta'} \rightarrow P_{\beta'}$ and $P_{\beta'} \rightarrow L_\alpha$ transitions is not sensitive to addition of solutes.[32,33]

Ethylene glycol and its dimers, trimers, and polymers very slightly modulate the phase behavior of dimyristoylphosphatidylcholine, dipalmitoylphosphatidylcholine, and distearoylphosphatidylcholine.[34,35] These data suggest that ethylene glycol and diethylene glycol might be classified as weak chaotropes, while triethylene glycol and polyethylene glycol appear to act as weak kosmotropes.

C. GLYCOLIPIDS

Solute effects on the glycolipid phase behavior have been recently measured in our laboratory by differential scanning calorimetry for several synthetic ether-linked glycoglycerolipids with saturated chains of equal length. The transition shifts given in Table 3 show that the solute effects in glycolipids are governed by the same general rules as in the case of phosphatidylethanolamines — chaotropic solutes favor formation of L_α at the expense of L_β and H_{II}, while kosmotropic solutes reduce the temperature range of the L_α phase. For example, 0.3 M of trehalose eliminates the L_α phase and brings about a direct $L_\beta \rightarrow H_{II}$ transformation in ditetradecylglucosylglycerol (Figure 3). On the other hand, high NaSCN concentrations suppress the formation of the H_{II} phase. Another noteworthy effect of the chaotropic solute NaSCN is the induction of intermediate L_α phases in dihexadecylglucosylglycerol and dihexadecylmannosylglycerol, undergoing direct $L_\beta \rightarrow H_{II}$ transitions in pure water (R. Koynova et al., submitted). However, the effect of NaSCN on the $L_\alpha \rightarrow H_{II}$ transitions in glycolipids is several times smaller than the corresponding effect in phosphatidylethanolamines (compare Tables 1 and 3).

III. THERMODYNAMIC THEORY OF THE SOLUTE EFFECT

As noted above, the derivative dT_{tr}/dc is a thermodynamic quantity which is related to the other thermodynamic variables of the lipid-water system. This relation represents an extended version of the Clapeyron-Clausius equation. According to Reference 23, it reads:

$$\frac{dT_{tr}}{dc} = \frac{kT_{tr}^2}{\Delta H}\left(1 - c^L/c\right)\left(x'' - x'\right)$$ (5)

where k is the Boltzmann constant; ΔH is transition enthalpy; x' and x'' are numbers of bound water molecules per lipid molecule in the low- and high-temperature phases, respectively; c^L and c denote the solute concentration in the bound and bulk water, respectively. It is assumed in Equation 5 that the aqueous solutions are diluted and behave as ideal ones, and that the solute concentrations in the bound water of the two coexisting phases are equal ($c^L = c' = c''$).

Several important features of the solute effects can be readily deduced from Equation 5:

1. Solutes can alter the transition temperature only between lipid phases with different hydration ($x' \neq x''$). Their effect is proportional to the change in the lipid hydration during the phase transition.
2. Even solute distribution between bound and free water ($c^L = c$) results in disappearance of the solute effect. One may then define kosmotropes by the inequality $c^L < c$, and chaotropes by the inverse inequality $c^L > c$. For kosmotropes, it follows from Equation 5 that the transition moves upward when the low-temperature phase is the less hydrated one ($x' < x''$) and downward when $x'' < x'$. Vice versa for chaotropes, the slope of T_{tr} is negative when the low-temperature phase is the less hydrated one, and positive in the opposite case.
3. An important consequence of Equation 5 is that the absolute value of dT_{tr}/dc is inversely proportional to the latent heat ΔH of the phase transition. This feature is identical to the standard Clapeyron-Clausius equation where dT_{tr}/dp is also inversely proportional to ΔH.
4. In ideal (dilute) solutions, the ratio c^L/c depends on the pressure and temperature only. Hence, at low solute concentration the transition temperature T_{tr} is a linear function of c ($\Delta T_{tr} \sim \Delta c$). Deviations from linearity at higher solute concentrations might result from non-ideal behavior of the aqueous solution.

IV. CORRELATIONS BETWEEN EXPERIMENT AND THERMODYNAMIC THEORY

The experimental results shown in Tables 1 through 3 and Figures 1 through 3 and the thermodynamic description of the solute effects represented by Equation 5 correlate in several important points:

1. Transitions of smaller latent heat such as $L_\alpha \to H_{II}$ and $L_{\beta'} \to P_{\beta'}$ are more sensitive to solutes than transitions of larger latent heat such as the chain-melting transitions. This effect is a direct consequence of the inverse proportionality between dT_{tr}/dc and ΔH in Equation 5.

2. The experimentally observed linear change of the transition temperature at small solute concentration ($\Delta T_{tr} \sim \Delta c$) follows from Equation 5 under the assumption of dilute (ideal) solutions. The deviations from linearity at higher solute concentrations (Figures 1 and 3) might indicate that the ideal solution assumption is not valid at these concentrations.

3. Kosmotropic and chaotropic solutes have opposite effects on the phase transition temperatures. According to Equation 5, these effects are due to uneven solute distribution between bulk and bound water. Kosmotropes tend to be excluded from the interface ($c^L < c$), while chaotropes tend to accumulate at the interface ($c^L > c$). The tendency of Hofmeister solutes to unevenly distribute between bulk water and interfaces is well known from other studies.[19,29,38] Taking into account solute distribution, Equation 5 correctly predicts the sign of dT_{tr}/dc in the different transitions. For example, chain-melting transitions move upwards in the presence of kosmotropic solutes because both terms in the right side of Equation 5, $(1 - c^L/c)$ and $(x'' - x')$, are positive in this case. A similar argument shows that chaotropes must induce an upward shift of the $L_\alpha \to H_{II}$ transitions, as observed experimentally.

Compared to the L_β and H_{II} phases, the lamellar liquid-crystalline phase L_α has the maximum surface area per lipid molecule. Since kosmotropes favor minimization of the lipid area exposed to water, they tend to destabilize the L_α phase and to reduce its existence range. On the other hand, chaotropic solutes tend to expand the lipid interface area and the stability range of the L_α phase. These conclusions are in complete agreement with the predictions of Equation 5. However, the lack of reliable data about the solute distribution defined by c^L/c in Equation 5, and also about the lipid hydration in the various lipid phases, precludes quantitative comparisons between experimentally determined transition shifts and those calculated from Equation 5.

V. SUMMARY

Data about solute effects on the phase transition temperatures are available at present about three classes of lipids — phosphatidylethanolamines, phosphatidylcholines, and glycoglycerolipids.

It is clear from Tables 1 through 3 that, according to their effect on the lipid phase transitions, the low-molecular, water-soluble additives fall into two categories: (1) solutes favoring formation of lamellar liquid crystalline phase L_α at the expense of the inverted hexagonal phase H_{II} and (2) solutes favoring formation of H_{II} at the expense of the L_α phase. The disparate effects of the different solutes on the lipid mesophase behavior are interpreted as a manifestation of the Hofmeister effect. It relates to their ability to modify water structure and to unevenly distribute between bound and free water. For example, disaccharides, polyols, and sugar alcohols are water-structure stabilizing compounds, also referred to as kosmotropic solutes. Denaturants, such as NaSCN and Gu.HCl, represent the chaotropic or water-structure breaking agents. The modification of water structure by Hofmeister solutes has an impact on the properties of the lipid-water interface. Depending on the nature of the solute, it results in alteration of the relative stability of the lyotropic mesophases. Kosmotropic solutes tend to minimize the area of the lipid-water contact. They favor formation of the H_{II} phase since, in comparison to the L_α phase, it has smaller surface area in contact with water. At high enough concentrations of kosmotropic solutes, the L_α phase may completely disappear from the phase diagram. This is precisely what is seen with sucrose, trehalose, proline, and some salts and is consistent with the opposite effect caused by chaotropic solutes. These effects can be described by an extension of the Clapeyron-Clausius relation derived in this case as a relation between phase transition temperature and solute concentration.[23]

In comparison to other measures of the Hofmeister series with inorganic salts, the lipid phase transition temperatures give identical anion and cation rank ordering. However, the sign inversion point, for both anions and cations, is located in the "chaotropic" side of the series. Another noteworthy feature is that anions have a greater effect on the transition temperature than cations.

There are experimental indications that addition of specific solutes can induce the appearance of missing phases. Our calorimetric studies on glycolipids have shown that the denaturing agent NaSCN converts the direct $L_\beta \to H_{II}$ transitions into $L_\beta \to L_\alpha \to H_{II}$ sequences with a stable L_α intermediate appearing at the expense of the L_β and H_{II} phases.

To our knowledge, no systematic studies on the effects of chaotropic and kosmotropic solutes have been carried out for membrane lipid-water phases forming cubic phases and also for lyotropic systems other than those formed by membrane lipids. Such studies can be expected to result in methods for regulation of the phase behavior of the lyotropic liquid crystals in a broad range of temperatures by controlling the composition of the aqueous phase. Concerning this goal, it is noteworthy that the experiments with membrane lipids have shown that the boundaries between lamellar and non-lamellar phases can be shifted by several tens of degrees in either direction, merely by addition of appropriate chaotropic or kosmotropic solutes to the aqueous phase.

ACKNOWLEDGMENT

The authors gratefully acknowledge support from grants K-2/91 and K-21/91 of the Bulgarian National Science Fund.

REFERENCES

1. **Ladbrooke, B. D. and Chapman, D.,** *Chem. Phys. Lipids,* 3, 304, 1969.
2. **Singer, S. and Nickolson, G.,** *Science,* 175, 720, 1972.
3. **Luzzati, V.,** in *Biological Membranes, Physical Fact and Function,* Vol. 1, D. Chapman, Ed., Academic Press, New York, 1968, 71.
4. **Small, D. M.,** *Handbook of Lipid Research,* Vol. 4, *The Physical Chemistry of Lipids from Alkanes to Phospholipids,* Plenum Press, New York, 1987.
5. **Lindblom, G. and Rilfors, L.,** *Biochim. Biophys. Acta,* 988, 221, 1989.
6. **Seddon, J. M.,** *Biochim. Biophys. Acta,* 1031, 1, 1990.
7. **Shinitzky, M.,** in *Physiology of Membrane Fluidity,* M. Shinitzky, Ed., CRC Press, Boca Raton, FL, 1984, chap. 1.
8. **Verkleij, A. J.,** *Biochim. Biophys. Acta,* 779, 43, 1984.
9. **Culis, P. R., Hope, M. J., DeKruijff, B., Verkleij, A. J., and Tilcock, C. P. S.,** in *Phospholipids and Cellular Regulation,* J. F. Kuo, Ed., CRC Press, Boca Raton, FL, 1985, 1.
10. **Larsson, K.,** *Langmuir,* 4, 215, 1988.
11. **Siegel, D. P.,** *Biophys. J.,* 49, 1155, 1986.
12. **Siegel, D. P.,** *Biophys. J.,* 49, 1171, 1986.
13. **Quinn, P. J.,** *Cryobiology,* 22, 28, 1985.
14. **Ellens, H., Siegel, D. P., Alford, D., Yeagle, P., Boni, L., Lis, L. J., Quinn, P., and Bentz, J.,** *Biochemistry,* 28, 3692, 1989.
15. **Hinz, H.-J., Kuttenreich, H., Meyer, R., Renner, M., Frund, R., Koynova, R., Boyanov, A., and Tenchov, B.,** *Biochemistry,* 30, 5125, 1991.
16. **Mannock, D. and McElhaney, R.,** *Biochem. Cell Biol.,* 69, 863, 1991.
17. **Tenchov, B. G.,** *Chem. Phys. Lipids,* 57, 165, 1991.
18. **Gruner, S.,** *J. Phys. Chem.,* 93, 7562, 1989.
19. **Collins, K. D. and Washabaugh, M. W.,** *Q. Rev. Biophys.,* 18, 323, 1985.
20. **Harlos, K. and Eibl, H.,** *Biochemistry,* 20, 2888, 1981.
21. **Seddon, J. M., Cevc, G., and Marsh, D.,** *Biochemistry,* 22, 1280, 1983.
22. **Sanderson, P., Lis, L. J., Quinn, P. J., and Williams, W. P.,** *Biochim. Biophys. Acta,* 1067, 43, 1991.
23. **Brankov, J. G. and Tenchov, B. G.,** *Prog. Colloid Polymer Sci.,* 93, 153, 1993.
24. **Bryszewska, M. and Epand, R.,** *Biochim. Biophys. Acta,* 943, 485, 1988.
25. **Koynova, R., Tenchov, B., and Quinn, P. J.,** *Biochim. Biophys. Acta,* 980, 377, 1989.
26. **Tsvetkova, N., Koynova, R., Tsonev, L., Quinn, P. J., and Tenchov, B.,** *Chem. Phys. Lipids,* 60, 51, 1991.
27. **Williams, W. P., Quinn, P. J., Tsonev, L., and Koynova, R.,** *Biochim. Biophys. Acta,* 1062, 123, 1991.
28. **Doerfler, H.-D., Miethe, P., and Meyer, H. H.,** *Chem. Phys. Lipids,* 54, 171, 1990.
29. **Epand, R. and Bryszewska, M.,** *Biochemistry,* 27, 8776, 1988.
30. **Marsh, D. and Seddon, J. M.,** *Biochim. Biophys. Acta,* 690, 117, 1982.
31. **Rudolph, A. S. and Goins, B.,** *Biochim. Biophys. Acta,* 1066, 90, 1991.
32. **Tolgyesi, F., Gyorgyi, S., and Sugar, I. P.,** *Mol. Cryst. Liquid Cryst.,* 128, 263, 1985.
33. **Lis, L. J., Tamura-Lis, W., Mastran, T., Patterson, D., Collins, J. M., Quinn, P. J., and Qadri, S.,** *Mol. Cryst. Liquid Cryst.,* 178, 11, 1990.
34. **Yamazaki, M., Ohshika, M., Kashiwagi, N., and Asano, T.,** *Biophys. Chem.,* 43, 29, 1992.
35. **Yamazaki, M., Kashiwagi, N., Miyazu, M., and Asano, T.,** *Biochim. Biophys. Acta,* 1109, 43, 1992.
36. **Cunningham, B. A., Lis, L. J., and Quinn, P. J.,** *Mol. Cryst. Liquid Cryst.,* 141, 361, 1986.
37. **Yeagle, P. and Sen, A.,** *Biochemistry,* 25, 7518, 1986.
38. **Arakawa, T. and Timasheff, S. N.,** *Biochemistry,* 23, 5912, 1984.

Chapter 11

Interaction of Membranes in the Presence of Divalent Cations

S. Marčelja

CONTENTS

I. INTRODUCTION

Extensive experimental evidence accumulated over the past 20 years indicates that the presence of divalent cations in the aqueous phase has a dramatic effect on the interaction of bilayer membranes.[1,2] Here we wish to use current theoretical knowledge of the diffuse double layer interaction in order to examine the origin of this sensitivity to divalent ions. While many details, most notably the effects of the structure of the aqueous solvent still need to be studied, recently improved understanding of the double layer interaction is now capable of explaining the specific role of divalent ions in membrane interaction.

Double layer properties can be studied with models which include the full structure of the aqueous solvent or with the so-called primitive model of electrolytes, where aqueous solvent is treated as a dielectric continuum. The solutions obtainable with full molecular models[3] have not yet reached the accuracy where they could predict forces between surfaces, colloidal particles, or vesicles. We shall therefore use the primitive model for all calculations. Nevertheless, work with molecular models will be useful in understanding the limitations of the primitive model and the general trends of the behavior in aqueous solvents.

In particular, we wish to examine adhesion of vesicles induced by the addition of divalent cations. As our goal is limited to presenting the physical mechanism involved, we can explicitly mention only a small representative sample chosen from the comprehensive experimental work on this topics. More complete listing can be found in comprehensive reviews.[1,2]

II. ATTRACTIVE REGIME OF THE DOUBLE LAYER INTERACTION

The electrical double layer structure is commonly described with the Gouy-Chapman (GC) or Gouy-Chapman-Stern (GCS) picture. Extensive testing has confirmed its suitability for the description of the electrostatic potential and ion density in the vicinity of membranes.[4] The associated theory of colloid stability is named DLVO theory, after Derjaguin, Landau, Verwey and Overbeek. Again, extensive testing has confirmed the validity of the DLVO theory.[5] Nevertheless, notable exceptions were found with minerals like mica or TiO_2 in solutions of divalent electrolytes[6,7] where anomalous behavior was observed.

The GCS theory makes two drastic approximations: it neglects the correlation between positions of individual ions (i.e., ion density depends only on the mean electrostatic potential) and the size of the ions. With respect to the evaluation of the interaction between the surfaces, these defects are not important as long as electrostatic potentials are small (<25 mV) and there are no divalent ions in the system. Outside of this weak coupling regime, the GCS theory still provides good description of the variation of the mean electrostatic potential or the average ion density, but not the interaction or the association of colloidal particles or vesicles. GCS theory has thus been very successful in describing the physical behavior involving charged surfaces in monovalent solutions and most behavior in 2:1 and 2:2 solutions. More detailed expositions of regimes where different approximations are applicable can be found in recent work on membrane undulations.[8]

The physical quantities that cannot always be simplified to the level of the GCS theory are the interaction energy or the force between charged surfaces. The reason for this failure is a strong attractive contribution arising under typical solution conditions from the correlation between divalent ions in the solution. To get an approximate measure, it is convenient to introduce the Bjerrum length, $(ze)^2/\varepsilon k_B T$, a distance at which electrostatic interaction between ions is equal to the thermal energy. For divalent ions in water the Bjerrum length is about 28 Å, and for distances shorter than the Bjerrum length the correlation dominates thermal disorder. When surface separation is small with respect to that scale, cross-surface correlation becomes significant and the attractive term is important.[9]

The force due to ion correlation is a part of the van der Waals interaction, understood here more broadly to include the slow ion movement in addition to the very rapid electromagnetic fields arising mostly from the UV part of the spectrum. While this is not a new insight,[10] its importance for double layer interaction was not fully appreciated until the initial Monte Carlo simulations.[11]

The attractive term in the double layer interaction has by now been studied in detail for the primitive model of electrolytes. The two major methods used in these investigations, Monte Carlo simulation[11-14] and hypernetted chain (HNC) integral equations[14-15] for inhomogenous electrolyte produce results in perfect agreement with each other. We can therefore say that the behavior of the primitive model electrolytes is well understood.

Details of ion layering and counterion adsorption are not available from primitive model studies, as they do depend on discrete aqueous solvent. However, insight obtained from the work with molecular solvent[3,16] shows that the short-range features reflecting the aqueous structure are largely additive to the longer-ranged electrostatic forces. The primitive model is still useful, and improvements can be obtained by including into the calculation specific information obtained from molecular models, for example the hydrated ion radii or more accurate configuration of surface layers.

The results in the following examples were obtained using the HNC integral equations as described in our earlier work.[15,17,18] The inhomogenous electrolyte between planar surfaces is equilibrated with the bulk reservoir of a given concentration, and the force between the surfaces is evaluated by adding the conventional osmotic term, the ion correlation term and the hard-core repulsion. Adsorption of counterions can be included by adding a constant term to the energy of ions in contact with the surface.

The charge residing on a membrane surface is discrete, and there is a discontinuity in the effective dielectric constant at the boundary between the aqueous and the hydrophobic environment. Perhaps surprisingly, both experiments[19] and calculations[12,20] show that the corresponding refinements of the basic model have little effect on the electrostatic potential or the double layer interaction. In the examples shown below, discreteness of surface charge and electrostatic images are neglected, although a fully quantitative comparison with experiments should include the required short-ranged corrections.

In Figure 1 we illustrate the region of the attractive regime of the double layer interaction in salt-free systems, where the only ions present are divalent counterions balancing the surface charge. The situation is very similar in 2:1 electrolytes, because very few monovalent coions will enter the space between closely apposed interacting surfaces. It is also not very different in mixed 1:1 and 2:1 electrolytes, discussed in more detail below, because the divalent counterion will be the preferred species between the surfaces.

III. ASYMPTOTIC BEHAVIOR

Some distance away from a surface, the electrostatic potential is small and counterions are not crowded. Under such conditions mean field approximation applies and the behavior of all physical quantities follows exponential variations of the GC theory.[21] Although a dense layer of ions near a surface is more complicated, from a distance it can always be considered to form a part of some "effective" surface charge. Alternatively, the surface potential can be understood to have some "effective" value, which is lower than the actual surface potential. The same applies with respect to the screening cloud surrounding each individual ion.

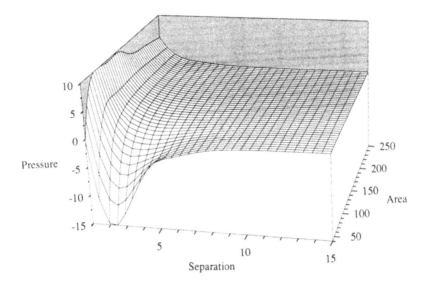

Figure 1 In the presence of divalent counterions, diffuse double layer force can be either attractive or repulsive. Attraction is found for small surface separations and moderately high or high surface charge. Separation between the surfaces is measured in angstroms, area per unit charge in square angstroms and pressure between the surfaces in MN/m^{-2}. While for the calculation of this figure only divalent counterions were assumed to be present in the gap between the surfaces, the situation is similar in 2:1 or 2:2 electrolytes. For this and subsequent figures, calculations were performed with the anisotropic HNC integral equations as described in the text, with all ion radii set to 2.125 Å. Separation is measured from the centers of ions in contact with a surface, i.e., zero separation corresponds to one layer of ions in contact with both surfaces.

If one now looks at a surface from a distance where GC laws apply, the localized ion accumulation screening the surface charge appears as a part of an appropriate "effective" Stern layer. An empirical association constant will then at the same time take into account both the adsorbed counterions and counterions resident in the initial part of the diffuse layer. Within such a picture, GCS formalism will provide an adequate description of the long-range behavior. The approximation is very good because the strong initial screening is restricted to one or two layers away from the surface.

The asymptotic forms for physical quantities can be formally derived from the liquid state theory,[22] and the expressions for the calculation of "effective" quantities are in some cases already available. Naturally, the actual calculation in the dense interacting regime is complicated, and has so far been performed in only a few cases.[23]

IV. INTERACTION OF ADSORBED CHARGE

Adsorption of counterions to the surface decreases diffuse double layer repulsion. Such charge does not contribute to the dominant repulsive term which is proportional to the density of free ions in the middle of the double layer.

If adsorbed charges have some lateral freedom of movement (as is the case in fluid membranes), they will rearrange their positions in order to correlate with charges on the apposing surface. In case of interacting membranes, negatively charged lipid headgroups and associated positively charged adsorbed counterions partially neutralize each other, but remain capable of interaction across the fluid gap separating the surfaces. The result of such correlation interaction is an attractive contribution similar to that described for counterions resident in the diffuse layer.[24]

While this effect is usually neglected, in some cases it may provide an important contribution to the overall interaction.[25] This is particularly true in cases where strong adsorption of counterions compensates or nearly compensates the original surface charge. The simplest model to study is that of adsorbed ions completely free to move on the surface. Detailed calculations of the resulting interaction have been performed and both the analytical asymptotic forms and numerical methods have been derived.[26] We illustrate this interaction later in Figure 4.

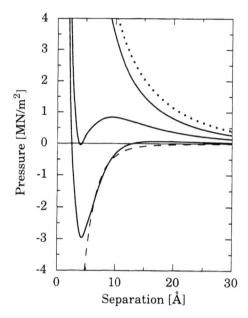

Figure 2 Double layer interaction of charged planar surfaces (90 Å²/unit charge) in 0.1 M of 1:1 salt and varying amounts of 2:1 salt (where the counterion is divalent). HNC calculations are shown as full lines. The top lines correspond to 0.1 M 1:1 electrolyte only. The GC calculation (dotted line) is more repulsive than the corresponding HNC result. Upon addition of 1 mM of 2:1 salt, the repulsion at small separations is almost wholly removed (middle full line). Addition of 10 mM of 2:1 salt (lower full line) changes short distance behavior into attraction. The attraction is similar in magnitude to that calculated for mobile adsorbed divalent ions compensating the surface charge (bottom dashed line). The calculations shown in the figures do not include van der Waals attractive interaction, which should be added before comparison is made with experiments.

V. INTERACTION OF CHARGED MEMBRANES

Membranes or vesicles formed from charged lipids will have sufficiently high surface charges to make electrostatic interaction in the presence of divalent counterions attractive at small surface separations. Corresponding experiments with flat membranes have been performed by both the osmotic pressure[27] and surface force apparatus techniques.[28] As an illustration, let us consider in some detail the measurements of Marra[28] on the interaction free energy between PG membranes adsorbed on mica in the surface force apparatus.

At large separations, force measurements show double layer repulsion and the results can be fitted to the GCS theory. Yet in the presence of CaCl$_2$, inconsistencies appear when the results are compared to measurements obtained using other techniques. A typical example is provided by DMPG in 0.1 M NaCl and 1 mM CaCl$_2$. Using the GCS theory to interpret electrophoretic mobility data, Lau et al.[29] find the intrinsic association constant for calcium of 8.5 M^{-1}, while the force measurement fit using the same GCS theory gives 40 M^{-1}. In terms of surface charge, the higher value corresponds to only about 10% of the lipid being charged. At smaller separations, surfaces in the apparatus jump into adhesive contact sooner than expected on the basis of the GCS theory.

The behavior at both large and small surface separations can be understood on the basis of more accurate calculations of the double layer interaction. A typical example is shown in Figure 2. Considering first the small separations, we see that the electrostatic correlation attraction becomes significant at surface separations of some 15 Å. This effect applies in addition to the conventional van der Waals interaction, resulting in an increase of the overall attraction. Experiments therefore indicate an early jump into primary contact.

The observed exponential repulsion at large separations corresponds to the asymptotic regime of the interaction (Figure 3). In this regime, both GCS and HNC theories can be made to fit the data, but the surface charge or potential needed to fit the data in each theory is very different. The strength of the repulsion in the GC calculation is related to the effective surface charge, which is normally smaller than the real charge. (It can be larger only if counterions are very large.) The effect is small when counterions are monovalent, but it becomes very significant in the case of substantial surface charge and divalent

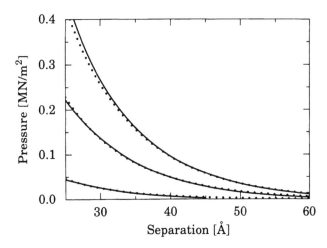

Figure 3 The interaction for the system shown in Figure 2 at larger surface separations. This detail shows the HNC calculation of repulsive tails (full lines), with the GC double layer repulsion of similar magnitude shown as dotted lines. The asymptotic behavior of the accurate HNC calculation is the same as that of a the GC calculation for a smaller "effective" surface charge. In this example, GC calculations shown in the figure were performed for the area per unit surface charge of 195, 340, and 800 Å² for the addition of 0, 1, and 10 mM, of 2:1 salt respectively. As the amount of divalent salt increases, the effective surface charge becomes very much lower than the real charge. Adsorption of divalent counterion to the surface (not included in the calculation) would further enhance this trend.

counterions. In the example of Figure 3, the effective surface charge for 0.1 M 1:1 salt and 1 mM 2:1 salt is about one fourth of the value of the real charge. When viewed from a distance, interpretation of the force, ion density or potential within the GCS theory corresponds to a smaller, *effective* surface charge.

VI. INTERACTION OF PARTIALLY CHARGED MEMBRANES

In case of weakly charged membranes, addition of divalent cation is per se not sufficient to turn the double layer interaction into attraction. Nevertheless, compared to the expectations based on the GCS theory the repulsion is significantly reduced (Figure 4).

To understand the attraction between partially charged vesicles or multilayers observed when a small quantity of CaCl$_2$ is added to the solution it may be important to include the correlation interaction between the charge adsorbed on each surface. The following example, discussed in more detail in Reference 30 is intended as an illustration of the principles involved.

We select a surface charge of 1 unit charge per 250 Å². The force between such membranes in a 0.1 M 1:1 electrolyte is repulsive (Figure 4). However, we know from electrophoretic measurements[31] that calcium strongly adsorbs to phospholipid membranes. With increasing adsorption, repulsion at shorter range is gradually reduced. Long-range repulsive tail remains present until the surface charge is fully compensated. However, for higher levels of adsorption the repulsive tail is very weak. In case of vesicles, the repulsion may become sufficiently weak so that it can be overcome by thermal motion. When surface charge is fully compensated the electrostatic interaction is attractive, as shown in Figure 4.

Full quantitative comparisons depend on details of adsorption, hydrated ion radius, etc., that still need to be determined. We have therefore restricted the discussion to semi-quantitative illustration of possible trends in the behavior. Electrostatic double layer interaction at low surface charge is weaker and the calculation is therefore more sensitive to details assumed in the adopted model. Work currently in progress with molecular model solvents should be able to shed additional light on the problem of adhesion of weakly charged vesicles upon addition of divalent counterions.

VII. CONCLUDING REMARKS

We have discussed here some general features of the electrostatic double layer interaction between membrane surfaces. Although the continuum solvent model used in the calculations is not accurate in details of the short-range structure and forces, the general physical mechanisms and trends associated with

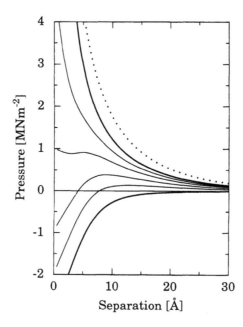

Figure 4 Electrical double layer interaction of weakly charged planar surfaces in the presence of divalent salt. The surfaces were assumed to carry one unit charge per 250 Å² and the electrolyte consisted of 0.1 *M* of 1:1 salt and 1 m*M* of 2:1 salt. In this case, the electrical double layer interaction is repulsive (top, heavy full line) although the repulsion is weaker than the corresponding GC calculation (top, dotted line). With varying degrees of counterion adsorption to the surfaces, double layer repulsion weakens even further, and an attractive regime appears at smaller surface separations (middle full lines). The bottom, heavy full line corresponds to very strong adsorption, where surface charge is fully compensated by adsorbed divalent counterions. The adsorbed counterions are assumed to be mobile within the plane of adsorption, and their correlation with adsorbed charges on the opposing surface leads to the attractive electrostatic interaction.

electrostatic interaction of charged particles in high dielectric constant background are correctly described. Molecular solvent terms consist of dipolar and higher-order multipolar electrostatic interactions and hard-core interactions. All of them have a range shorter than the dominant Coulomb force. They will affect asymptotic behavior through a "renormalization" of charges and the decay length, similar to the renormalization necessitated by a transition from the GCS theory to the more accurate HNC theory. The corrections will not be as large as those described in this work, and it is unlikely that they will qualitatively change the behavior determined by the longer-ranged 1/r Coulomb potential between ions.

The generally accepted picture[32] where adhesion of vesicles made of charged or partially charged lipids is controlled by the double layer electrostatic interaction and van der Waals interaction is thus retained, with corrections due to correlation and size effects. The ion correlation term could be thought of as being a part of either the electrostatic or van der Waals force, as the two are intrinsically linked.[33] Compared to the conventional osmotic repulsion, the attractive term is significant only in the presence of divalent ions. The actual event of fusion cannot be understood without a specific reference to the kind of divalent ion (e.g., Mg and Ca have a very different effect[34,35]) and depends on the details of the interaction mechanism at the membrane surface. Nevertheless, we believe that sufficient evidence now exists to show that the electrostatic forces in the presence of divalent ions provide a means of controlling the macroscopic behavior with only a small absolute change in the overall ionic concentration of a colloidal or vesicle dispersion.

REFERENCES

1. Nir, S., Bentz, J., Wilschut, J., and Düzgünes, N., Aggregation and fusion of phospholipid vesicles. *Prog. Surface Sci.,* 13, 1, 1983.
2. Ohki, S., Doyle, D., Flanagan, T. D., Hui, S. W., and Mayhew, E., Eds., *Molecular Mechanisms of Membrane Fusion,* Plenum Press, New York, 1988.

3. **Patey, G. N. and Torrie, G. M.,** Molecular solvent models of electrical double layers, *Electrochim. Acta,* 36, 1677, 1991.

4. **McLaughlin, S.,** The electrostatic properties of membranes, *Annu. Rev. Biophys. Chem.,* 18, 113, 1989.

5. **e.g., Hunter, R. J.,** *Foundations of Colloid Science,* Clarendon Press, Oxford, 1987.

6. **Kekicheff, P., Marčelja, S., Senden, T. J., and Shubin, V. E.,** Charge reversal seen in electrical double layer interaction of surfaces immersed in 2:1 calcium electrolyte, *J. Chem. Phys.,* 99, 6098, 1993.

7. **Mange, F., Couchot, P., Foissy, A., and Pierre, A.,** Effect of sodium and calcium ions on the aggregation of titanium dioxide, at high pH, in aqueous dispersions, *J. Colloid and Interface Sci.,* 159, 58, 1993.

8. **Pincus, P., Joanny, J.-F., and Andelman, D.,** *Europhys. Lett.* 11, 763, 1990; Higgs, P. G., and Joanny, J.-F., *J. Phys. (France),* 51, 2307, 1990.

9. **Stevens, M. J. and Robbins, M. O.,** Density functional theory of ionic screening: When do like charges attract? *Europhys. Lett.,* 12, 81, 1990.

10. **Mahanty, J. and Ninham, B. W.,** *Dispersion Forces,* Academic Press, London, 1976.

11. **Guldbrand, L., Jönsson, B., Wennerström, H., and Linse, P.,** Electrical double layer forces. A Monte Carlo study, *J. Chem. Phys.,* 80, 2221, 1984.

12. **Bratko, D., Jönsson, B., and Wennerström, H.,** Electrical double layer interactions with image charges, *Chem. Phys. Lett.,* 128, 449, 1986.

13. **Woodward, C. E., Jönsson, B., and Åkesson, T.,** The ionic correlation contribution to the free energy of spherical double layers, *J. Chem. Phys.,* 89, 5145, 1988.

14. **Kjellander, R., Åkesson, T., Jönsson, B., and Marčelja, S.,** Double layer interactions in mono- and divalent electrolytes. A comparison of the anisotropic hypernetted chain theory and Monte Carlo simulations, *J. Chem. Phys.,* 97, 1424, 1992.

15. **Kjellander, R. and Marčelja, S.,** Inhomogenous Coulomb fluids with image interactions between planar surfaces, *J. Chem. Phys.,* 82, 2122, 1985.

16. **Wei, D., Patey, G. N., and Torrie, G. M.,** Double-layer structure at an ion-adsorbing surface, *J. Phys. Chem.,* 94, 4260, 1990.

17. **Kjellander, R. and Marčelja, S.,** Surface interactions in simple electrolytes, *J. Phys. (France),* 49, 1009, 1988.

18. **Marčelja, S.,** Interactions between interfaces in liquids. In *Liquids at Interfaces,* Charvolin, J., Joanny, J. F., and Zinn-Justin, J., Eds. North-Holland, Amsterdam, 1990, pp. 100–134.

19. **Langner, M., Cafiso, D., Marčelja, S., and McLaughlin, S.,** Electrostatics of phosphoinositide bilayer membranes, theoretical and experimental results, *Biophys. J.,* 57, 335, 1986.

20. **Kjellander, R. and Marčelja, S.,** Correlation and image charge effects in electric double layers, *Chem. Phys. Lett.,* 112, 49, 1984.

21. At very low ionic strengths, ion pairing in 2:2 and 2:1 electrolytes becomes significant and ion pairs have to be considered as separate species. See Haymet, A. D. J., Integral equation theory for associating liquids: Dimer and trimer concentrations for model 1:3 electrolytes, *J. Chem. Phys.,* 100, 3767, 1994 and references therein.

22. **Kjellander, R. and Mitchell. D. J.,** An exact but linear and Poisson-Boltzmann-like theory for electrolytes and colloid dispersions in the primitive model, *Chem. Phys. Lett.,* 200, 76, 1992.

23. **Ennis, J., Kjellander, R., and Mitchell, D. J.,** Dressed ion theory for bulk symmetric electrolytes in the restricted primitive model, *J. Chem. Phys.,* 102, 975, 1995.

24. **Kjellander, R. and Marčelja, S.,** Interactions between ionic surface layers, *Chemica Scripta,* 25, 112, 1985.

25. **Attard, P., Mitchell, D. J., and Ninham, B. W.,** The attractive force between polar lipid bilayers, *Biophys. J.,* 53, 457, 1988.

26. **Attard, P., Mitchell, D. J., and Ninham, B. W.,** Beyond Poisson-Boltzmann: images and correlations in the electric double layer. II. Symmetric electrolyte, *J. Chem. Phys.,* 89, 4358, 1988.

27. **Leikin, S., Parsegian, V. A., Rau, D. C., and Rand, R. P.,** Hydration forces, *Annu. Rev. Phys. Chem.,* 44, 319, 1993, and references therein.

28. **Marra, J.,** Direct measurement of the interaction between phosphatidylglycerol bilayers in aqueous electrolyte solutions, *Biophys. J.,* 50, 815, 1986.

29. **Lau, A., McLaughlin, A., and McLaughlin, S.,** The adsorption of divalent cations to phosphatidylglycerol bilayer membranes, *Biochim. Biophys. Acta,* 645, 279, 1981.

30. **Marčelja, S.,** Electrostatics of membrane adhesion, *Biophys. J.,* 61, 1117, 1992.

31. **McLaughlin, S., Mulrine, N., Gresalfi, T., Vaio, G., and McLaughlin, A.,** Adsorption of divalent cations to bilayer membranes containing phosphatidylserine, *J. Gen. Physiol.,* 77, 445–473, 1981.

32. **Evans, E. and Needham, D.,** Physical properties of surfactant bilayer membranes, thermal transitions, elasticity, rigidity, cohesion and colloidal interactions, *J. Phys. Chem.,* 91, 4219, 1987.

33. **Attard, P., Kjellander, R., Mitchell, D. J., and Jönsson, B.,** Electrostatic fluctuation interaction between neutral surfaces with adsorbed, mobile ions or dipoles, *J. Chem. Phys.,* 89, 1664, 1988.

34. **Papahadjopoulos, D., Vail, W. J., Newton, C., Nir, S., Jacobson, K., Poste, G., and Lazo, R.,** Studies on membrane fusion III. The role of calcium-induced phase change, *Biochim. Biophys. Acta,* 465, 579, 1977.

35. **Nir, S.,** A model for cation adsorption in closed systems: application to calcium binding to phospholipid vesicles, *J. Colloid Interface Sci.,* 102, 313, 1984.

Chapter 12

Use of Liposomes to Model Aggregation and Fusion Phenomena

Shlomo Nir

CONTENTS

I. INTRODUCTION

The advantage in using liposomes in studying aggregation and fusion phenomena stems from the possibility of controlling and manipulating their compositions and sizes. In addition, a variety of substances can be encapsulated in vesicles or incorporated in their membranes, which forms the basis for fusion assays based on aqueous contents or membrane mixing. Variations in membrane composition have been utilized in understanding factors affecting vesicle-vesicle fusion as well as virus-vesicle fusion, and consequently have enabled to make projections for the more complex fusing systems, such as virus-cell, or the fusion of intracellular vesicles. Thus, studies on the effect of Ca^{2+} and other ions on fusion of acidic liposomes[1-10] are relevant since Ca^{2+} is involved in the control of exocytosis and myoblast fusion. Certain cytoplasmic proteins, such as synexin and clathrin were shown to modulate the fusion of liposomes.[11-17] Several studies were reported on the fusion of liposomes with intracellular membranes, such as secretory granules,[18,19] Golgi[20,21] membranes and endosomal vesicles.[22] The purpose of this chapter is to give a brief review with an emphasis on recent results.

II. MATHEMATICAL MODELING OF AGGREGATION AND FUSION

Derivations and solutions of the differential equations of the kinetics of fusion of vesicles and kinetics and equilibrium of reversible aggregation of vesicles or general colloidal particles have been presented.[5,6,23-25] The mass action equations describing the aggregation process are essentially based on the ideas of Smoluchowski[26] and Fuchs.[27] The equations for the kinetics of vesicle-vesicle or vesicle-virus fusion reduce to the equations describing aggregation in the limit of no fusion.[5,6,23,25] These equations are based on the approach that the process of membrane fusion consists of three main steps: (1) Close approach or point contact between two membranes. In a system of vesicles, this is the aggregation step. (2) Membrane destabilization. The two apposed membranes which are ordered structures must undergo a transient disturbance in order that a new structure be formed later. (3) Membrane merging. The destabilized membranes fuse to form a single continuous membrane.

The mathematical procedure based on a mass action model, has enabled us to distinguish between two steps in the overall fusion reaction, the close approach or aggregation, which involves a collision between particles and is of second order in particle concentrations, and the subsequent first order fusion reactions, which involves membrane destabilization and merging.

Ordinarily, the minimal set of rate constants required to describe the kinetics of fusion includes $f(s^{-1})$, the rate constant of fusion per se, $C(M^{-1}s^{-1})$, the rate constant of aggregation or adhesion, and $D(s^{-1})$, the rate constant of dissociation. The programs[6,23-25] enable the user to employ different rate constants for higher order aggregation-fusion products, but it is always desirable to employ a minimal number of parameters in describing any physical process. The full set of equations is given in References 23 through 25. Here, we will present the initial stages where only an aggregate of two particles, or dimer $[A = A(1,1)]$

0-8493-4731-9/96/$0.00+$.50
© 1996 by CRC Press, Inc.

is formed and is fused to a doublet [F = F(1,1)]. In this case, the chain of reactions is given schematically by

$$L + V \leftrightarrow A(1,1) \rightarrow F(1,1) \qquad (1)$$

In Equation 1, L and V may denote liposomes and virions, or labeled and unlabeled liposomes. The indices in the matrices A and F denote the numbers of particles of types 1 and 2 in the aggregation or fusion products, respectively.

The kinetic equations are

$$dL/dt = -C \cdot L \cdot V + D \cdot A \qquad (2)$$

$$dV/dt = -C \cdot L \cdot V + D \cdot A \qquad (3)$$

$$dA/dt = C \cdot L \cdot V - f \cdot A - D \cdot A \qquad (4)$$

$$dF/dt = f \cdot A \qquad (5)$$

In Equation 4 the term $C \cdot L \cdot V$ is of second order in particle concentration, and gives the rate of generation of aggregates, A, whereas the first order terms, $f \cdot A$ and $D \cdot A$ give the rate of annihilation of A due to fusion and dissociation.

By focusing on a very dilute system, where the aggregation will be the rate limiting step and fusion will occur relatively fast, the analysis of kinetics of fusion yields the rate constant of aggregation, C. From the results obtained with a more concentrated suspension and the knowledge of C, it is possible to determine the fusion rate constant, f, by relying mostly, but not exclusively, on the initial stages of the fusion process. Initially, the value of D affects the results minimally, since most of the particles are not aggregated.

In certain cases, the rate constant f can be determined directly by pre-aggregating vesicles, e.g., by NaCl[28] or by a protein,[17,29] and then initiating fusion by adding Ca^{2+} or another protein.[30]

From the simulation of the kinetics of fusion as monitored by a fluorescence assay, it is possible to determine the dependence of the rate constants on pH, Ca^{2+} concentration, temperature, peptide/lipid ratio, etc., and test a variety of hypotheses.[9,10,25]

A. AGGREGATION

According to the Smoluchowski-Fuchs[27] theory, the forward rate constants in Equations 2 to 4 can yield information on the potential barriers for close approach, irrespective of their nature, e.g., electrostatic repulsion between charged surfaces, or short-range hydration forces. The experimental problem is to determine the state of aggregation directly by counting particles or indirectly from certain spectroscopic procedures. The initial stages of aggregation of small phosphatidylserine (PS) vesicles were determined from light scattering measurements, and the mass action kinetic model yielded both the coefficients C and D, hence the potential barrier for close approach of vesicles and the depth of the potential well holding the particles together in the presence of 650 mM NaCl.[31] As the aggregate sizes increase, accounting for interference effects in the light scattered from different parts of the aggregated particles becomes complicated. In a recent study on the same system, we circumvented this difficulty by focusing on theoretical and experimental determination of the time of equilibration, $\bar{t}$ of the aggregation process.[32]

Estimates were calculated from Smoluchowski-type kinetic equations, extended to include dissociation reactions. An extension of the analytic solution for the monomer-dimer system yielded a closed form expression for $\bar{t}$ in the general case. Equilibration times could also be estimated from a single equation, which expresses the rate of change of the number of composite particles, using averaged values for the rate constants. Sodium-induced aggregation of PS vesicles was studied by means of turbidity and energy transfer measurements. The reversibility of the aggregation process has been verified with both procedures, confirming also no exchange of the fluorescent probes used between vesicles in an aggregate. The calculations could predict the decrease in $\bar{t}$ values with an increase in vesicle concentration and gave rough estimates for $\bar{t}$ values. The kinetics of turbidity increase was similar to that of fluorescence increase of acceptor molecules. With 200 μM lipid + 650 mM NaCl at 23°C the turbidity and fluorescence intensity reached equilibrium values in approximately 2 min. At lower temperatures both the extent of aggregation and $\bar{t}$ increased in accord with estimates.[32] Good estimates for $\bar{t}$ were also found for the process of vesicle binding to cells.[33]

We have recently studied the aggregation of dipalmitoyl phosphatidylcholine (DPPC)/cholesterol (2/1) liposomes induced by antibodies.[34] The membranes of the vesicles contained 0.25 to 2 mol % of FG_3P, a fluorescein hapten anchored to the bilayer through a tripeptide (gly_3) linkage to dipalmitoyl-phosphatidylethanolamine (PE). The vesicles with or without the hapten molecules did not aggregate. Aggregation was induced by the addition of anti-fluorescein monoclonal antibodies (MABs). Vesicle aggregation occurred by inter-vesicle binding, where the antibody crosslinks hapten molecules on two liposomes. This process mimics the membrane-membrane interactions via specific ligand and receptors as well as the agglutination reaction of antigens by antibodies, which is part of the defense mechanism in the immune system. It also has direct applications to immunoassays based on liposome aggregation.

Vesicle aggregation by the antibody was studied by turbidimetry and freeze-fracture electron micros-copy of samples frozen throughout the course of the aggregation. Rapid freezing was achieved with a double propane-jet apparatus. The aggregate morphologies and the time evolution of the aggregate size distribution were obtained from the two-dimensional fracture views with a stereological correction. The aggregation kinetics were simulated by considering dynamical aggregation according to a mass-action model with two parameters, the rate constants for antibody-mediated vesicle aggregation and disaggre-gation.

In the latter work, we introduced a notation A(I,J) to denote the molar concentration of aggregate of (I + J) vesicles, formed from I vesicles with one or more antibodies available for binding to other vesicles and J vesicles that do not contain antibodies. In our formalism, an aggregate can contain no more than one antibody-free vesicle, i.e., J is either 0 or 1. After some experimentation with rules for dissociation of aggregates, dissociation was allowed for A(I,1) and A(2,0) aggregates, but not for A(I,0), I > 2. In the case of trimers and larger aggregates of the type A(I,0), cyclization is possible, and this multivalent interaction should lend increased stability.

The best fit of the calculated to experimental results gave $C = 9 \times 10^4 M^{-1} \cdot S^{-1}$ and $D = 0.008 \, S^{-1}$ for the case where one of the particles forming the aggregate has a free antibody, assuming that 30% of the MAB bind bivalently intravesicularly and are removed from the aggregation process. This value of C is several orders of magnitude below $3 \times 10^9 M^{-1} \cdot S^{-1}$, which is the value of C in diffusion controlled aggregation at room temperature.[25,26] The rate and extent of aggregation increased dramatically when the number of MAB per vesicle increased from 1.1 to 4.5.[34]

In the last examples, the rate constants C and D were determined by simulations of aggregation experiments. In most cases, these parameters were determined from the analysis of kinetics of fusion.[25] We emphasize that our approach is not based on finding a functional form that is the best in terms of curve fitting, but rather, the starting point is a theoretical equation. Once the rate constants have been deter-mined, a crucial test is the ability to yield predictions for the kinetics of fusion and/or aggregation for other particle concentrations and ratios between labeled and unlabeled particles. In several cases,[4,35] we also demonstrated that the equations could yield predictions for quantities which were not considered in the simulations of the fusion curves. The employment of fast-freezing electron microscopy to visualize the progress of the Ca^{2+}-induced fusion of PS/PE (1:3) liposomes, demonstrated the ability of the calculations to yield predictions for the vesicle sizes and the distribution of aggregation-fusion products.[35] Initially, most of the vesicles were in monomers. One minute after the addition of Ca^{2+}, the fraction of monomers was significantly reduced, and dimers, trimers, tetramaers, etc., were observed. However, after 15 min the monomers became once again the predominant component, whereas the fraction of oligomers dropped significantly, due to fusion of aggregated particles resulting in the formation of larger monomers, in accord with the calculations. A list of selected rate constants is given in Table 1.

The employment of vesicles also enabled a demonstration of reversible aggregation of PS vesicles in a secondary minimum.[36] This demonstration relied on the ability of producing vesicles of desired sizes and calculating cation binding to PS vesicles.[37] The application of binding equations for a closed system[38] showed that the binding coefficients of Ca^{2+} to fusing PS or PS/PE liposomes increased by several orders of magnitude when Ca^{2+}-induced fusion occurred, whereas no changes were observed in aggregating systems.

B. FUSION

We will limit this section to reviewing recent results on virus-liposome fusion. A recent general review[39] also covers other topics. Studies of final extents and kinetics of virus-liposome fusion have elucidated the

Table 1 Rate Constants of Vesicle-Vesicle Aggregation and Fusion[a]

Composition, size and references	Fusion or aggregation inducing agents and conditions	Rate constant of aggregation $C(M^{-1} \cdot s^{-1})$	Rate constant of fusion $f(s^{-1})$	Rate constant of deaggregation[b]	Fusion or aggregation assay and comments
DPPC/Cholesterol[34]	Aggregation was induced by antibody-hapten interaction	$9 \cdot 10^4$	No fusion	0.008	Fast-freezing EM
Chromaffin granule ghosts[57]	Synexin and pH 6, 37°C	$5 \cdot 10^9$	>10	0.5	Membrane mixing
PS/PE(1:3)LUV[35]	$4\ mM\ Ca^{2+}$	$3.2 \cdot 10^9$	1.3	0.5	Contents mixing
		$4 \cdot 10^7$	0.04[c]	0	Contents mixing
					Size distribution determined by fast-freezing EM
POPC SUV[58]	GALA peptide at pH 5 and lipid/peptide ratio 50/1	$2.5 \cdot 10^6$	0.08	≤0.005	Membrane mixing
	lipid/peptide 100/1	$1.3 \cdot 10^6$	0.03		Vesicle size determined by dynamic light scattering
	HIV_{arg} at lipid/peptide ratio 50/1	d	0.005	d	

[a] The uncertainties in the estimates of the rate constants are about 20, 25, and 50% for C, f, and D, respectively. More details can be found in the cited references. Whenever not stated explicitly, the measurements were performed at room temperature, pH 7.4.

[b] The value 0 indicates that D << f.

[c] $f_{ii} = f_{116}$.

[d] Vesicles were preaggregated by $5\ mM\ Ca^{2+}$. See Reference 59 for more cases.

effect of liposome composition in the target membrane on the fusion characteristics of influenza virus,[23,40,41] Sendai virus,[24,42-44] HIV-1,[45] and SIV.[46] For instance, the increase in PC contents in liposomes results in reduced fusion activity of all viruses. On the other hand, complete fusion activity was found for influenza virus fusing with PS liposomes, whereas only partial fusion activity was found for Sendai virus and HIV-1 toward these liposomes. Several studies established the particular role of cholesterol[47-49] and sphingomyelin[50] in the fusion of Semliki Forest virus with target membranes. The fusion of immunodeficiency viruses[45,46] with several types of liposomes is enhanced by the presence of Ca^{2+}, unlike influenza or Sendai viruses.

Glycolipids have been viewed as receptors for virions. For influenza virus fusing with liposomes at 37°C, it was found that the ganglioside G_{D1a} has little effect on the fusion rate constants, or on the final extents of fusion, but enhances threefold the rate constant of aggregation.[41] In a recent study,[51] it was concluded that when the HA protein of influenza virus was bound to ganglioside it became incapable of participating in fusion. In that study, optimal surface concentrations for virus-liposome fusion were observed. The ganglioside also reduced the low pH inactivation of the virus. In another recent study,[52] it was found that G_{D1a} also affects lipid polymorphism as determined by ^{31}P nuclear magnetic resonance. The ganglioside promotes the formation of isotropic structures in monomethyl DOPE. G_{D1a} also raises the bilayer to hexagonal phase transition temperature of this lipid. The effects of G_{D1a} on the kinetics of viral fusion can be understood on the basis of its role in facilitating the binding of Sendai virus to target membranes as well as its effects on membrane physical properties. Fusion of Sendai virus with liposomes composed of egg PE is particularly sensitive to the presence of ganglioside. In the absence of ganglioside, no fusion is observed due to the absence of virus binding to the target membrane. Between 2 and 6 mol % G_{D1a} in egg PE liposomes there is a marked increase in the rate constant of binding of the virus to the liposome but a decrease in the fusion rate constant.

Analysis of the results of final extents of virus-liposome fusion led to the conclusion that the fusion products consist of a single virus and several liposomes. This conclusion was found to hold for influenza virus.[23,41] All three viruses studied, influenza virus, Sendai virus, and HIV-1, exhibit a phenomenon of partial fusion activity toward most liposomes. As discussed,[53] one possible explanation for these findings is that the virus preparation is heterogeneous. Specifically, it is possible that a certain fraction of virions cannot form sufficiently close contacts with given liposomal membranes, and cannot induce target membrane destabilization, although binding can occur. Alternatively, instead of dividing the virus population into active and inactive virions, the membrane of each virion might include a limited number of fusion-capable sites, consisting of well-defined combinations of glycoproteins at various spatial arrangements (e.g., the F and HN glycoproteins of Sendai virus). If virus binding to a liposome does not occur at an "active" (viral) site, fusion will not occur. If binding is essentially irreversible under the given conditions, as the analysis has indicated for the case of Sendai virus[24] and HIV-1,[45] all of the virions may be bound but a certain fraction will remain unfused. Experiments were carried out to resolve this issue of the gross mechanism of partial fusion activity in the cases of Sendai virus[54] and recently influenza virus.[55]

In the first study,[54] the experimental procedure was designed to separate the nonfused Sendai virus from the liposomes and fusion products after a long period of incubation at pH 7.4 and 37°C. At that stage, the final extent of fusion, or R_{18} fluorescence increase, had been attained. Although the binding of the fusion-inactive virus to the liposomes was known to be essentially irreversible, the nonfused virus could be separated from the liposomes by employing sucrose gradients and prolonged centrifugation. The idea was that, if the collected unfused virions were inherently fusion-inactive, then their subsequent incubation with liposomes should not result in their fusion. In order to optimize the chances for viral fusion, liposomes were added to R_{18}-labeled virions in large excess. Liposomes consisting of PS were initially chosen since only 25% of the virions were fusion-active at neutral pH.

We found that the collected "fusion-inactive" virions fused with added PS liposomes, the percent of virions fusing in the second round being close to that in the first round. Results with liposomes composed of PS/DOPE/Cholesterol 40/20/40 (mol %) exhibited the same phenomenon, that the "fusion-inactive" virions are capable of fusing, although the percent of virions that fused in the second round was about two thirds of the value found in the first round. A similar conclusion was recently reached for the case of influenza virus interacting with PC/PE (2/1), PC/PE/Chol (2/1/1), and PC/PE/PS/Chol (1/1/1/1) LUV.[55]

These results demonstrate that, in the case of Sendai virus and influenza virus fusing with liposomes of a variety of compositions, the existence of a majority of unfused virions does not imply that most of the individual virions are incapable of fusing. Apparently, in most of the encounters between the virus and liposomes, attachment of the virus to the liposome occurs at an "inactive" site on the viral membrane, such that the virus does not fuse, although it binds to the liposome. This binding is essentially irreversible under normal conditions. If such an attached virus is released from the liposome (e.g., by means of a sucrose gradient), it has the same chance as any other virus in the population to fuse with the liposomes by forming an attachment via an active site consisting of a proper combination and arrangement of glycoproteins.

We found that the final extents of fusion of Sendai virus[56] with didodecyl phosphate (DDP) vesicles and those of influenza virus interacting with liposomes[55] increase with temperature. The analysis of kinetics of the overall fusion process[56] indicated a dramatic increase in the rate constant of dissociation in this temperature range. It may be that, at higher temperatures, some fraction of the improperly bound virions could be released and thus have an increased chance of fusing in later encounters with the vesicles.

It is of interest to examine to what extent the general concept of irreversible binding of the virus at an inactive site on the virus, or at the target membrane can be operative in partial fusion activity in virus-cell systems.

ACKNOWLEDGMENT

The expert typing of Mrs. Sue Salomon is gratefully acknowledged.

REFERENCES

1. **Papahadjopoulos, D., Vail, W. J., Newton, C., Nir, S., Jacobson, K., Post, G., and Lazo, R.,** Studies on membrane fusion. III. The role of calcium-induced phase changes, *Biochim. Biophys. Acta,* 465, 579, 1977.
2. **Wilschut, J., Duzgunes, N., Fraley, R., and Papahadjopoulos, D.,** Studies on the mechanism of membrane fusion: kinetics of Ca^{2+}-induced fusion of phosphatidylserine vesicles followed by a new assay for mixing of aqueous vesicle contents, *Biochemistry,* 19, 6011, 1980.
3. **Wilschut, J., Duzgunes, N., and Papahadjopoulos, D.,** Calcium/magnesium specificity in membrane fusion: kinetics of aggregation and fusion of phosphatidylserine vesicles and the role of bilayer curvature, *Biochemistry,* 20, 3126, 1981.
4. **Nir, S., Bentz, J., and Wilschut, J.,** Mass action kinetics of phosphatidylserine vesicle fusion as monitored by coalescence of internal vesicle volumes, *Biochemistry,* 19, 6030, 1980.
5. **Bentz, J., Nir, S., and Wilschut, J.,** Mass action kinetics of vesicle aggregation and fusion, *Colloids Surfaces,* 6, 333, 1983.
6. **Nir, S., Bentz, J., Wilschut, J., and Duzgunes, N.,** Aggregation and fusion of phospholipid vesicles, *Prog. Surf. Sci.,* 13, 1, 1983.
7. **Wilschut, J., Nir, S., Scholma, J., and Hoekstra, D.,** Kinetics of Ca^{2+}-induced fusion of cardiolipin-phosphatidylcholine vesicles: correlation between vesicle aggregation, bilayer destabilization, and fusion, *Biochemistry,* 24, 4630, 1985.
8. **Bental, M., Wilschut, J., Scholma, J., and Nir, S.,** Ca^{2+}-induced fusion of large unilamellar phosphatidylserine/cholesterol vesicles, *Biochim. Biophys. Acta,* 898, 239, 1987.
9. **Duzgunes, N., Hong, K., Baldwin, P. A., Bentz, J., Nir, S., and Papahadjopoulos, D.,** Fusion of phospholipid vesicles induced by divalent cations and protons. Modulation by phase transitions, free fatty acids, monovalent cations, and polyamines, in *Cell Fusion,* Sowers, A.E., Ed., Plenum Press, New York, 1987, 241.
10. **Papahadjopoulos, D., Nir, S., and Duzgunes, N.,** Molecular mechanisms of calcium-induced membrane fusion, *J. Bioenerg. Biomembr.,* 22, 157, 1990.
11. **Hong, K., Duzgunes, N., and Papahadjopoulos, D.,** Modulation of membrane fusion by calcium-binding proteins, *Biophys. J.,* 37, 296, 1982.
12. **Hong, K., Duzgunes, N., Ekerdt, R., and Papahadjopoulos, D.,** Synexin facilitates fusion of specific phospholipid vesicles at divalent cation concentrations found intracellularly, *Proc. Natl. Acad. Sci. U.S.A.,* 70, 4942, 1982.
13. **Blumenthal, R., Henkart, M., and Steer, C. J.,** Clathrin-induced pH dependent fusion of phosphatidylcholine vesicles, *J. Biol. Chem.,* 258, 3409, 1983.
14. **Hong, K., Yoshimura, T., and Papahadjopoulos, D.,** Interactions of clathrin with liposomes: pH-dependent fusion of phospholipid membranes induced by clathrin, *FEBS Lett.,* 191, 17, 1985.
15. **Yoshimura, T., Maezawa, S., and Hong, K.,** Exposure of hydrophobic domains of clathrin in its membrane fusion-inducible pH region, *J. Biochem.,* 101, 1265, 1987.

16. **Maezawa, S., Yoshimura, T., Hong, K., Duzgunes, N., and Papahadjopoulos, D.**, Clathrin-induced fusion of phospholipid vesicles associated with its membrane binding and conformational change, *Biochemistry*, 28, 1422, 1989.

17. **Meers, P., Bentz, J., Alford, D., Nir, S., Papahadjopoulos, D., and Hong, K.**, Synexin enhances the aggregation rate but not the fusion rate of liposomes, *Biochemistry*, 27, 4430, 1988.

18. **Bental, M., Lelkes, P. I., Scholma, J., Hoekstra, D., and Wilschut, J.**, Ca^{2+}- independent, protein-mediated fusion of chromaffin granule ghosts with liposomes, *Biochim. Biophys. Acta*, 774, 296, 1984.

19. **Meers, P., Ernst, J. D., Duzgunes, N., Hong, K., Fedor, J., Goldstein, I. M., and Papahadjopoulos, D.**, Synexin-like proteins from human polymorphonuclear leukocytes. Identification and characterization of granule-aggregating and membrane-fusing activities, *J. Biol. Chem.*, 262, 7850, 1987.

20. **Kobayashi, T. and Pagano, R. E.**, ATP-dependent fusion of liposomes with the Golgi apparatus of perforated cells, *Cell*, 55, 797, 1988.

21. **Kagiwada, S., Murata, M., Hishida, R., Tagaya, M., Yamashina, S., and Ohnishi, S.**, In vitro fusion of rabbit liver Golgi membranes with liposomes, *J. Biol. Chem.*, 268, 1430, 1993.

22. **Vidal, M. and Hoekstra, D.**, Characterization of endocytic vesicle-liposome fusion using a lipid mixing assay, presented in Nato Advanced Study Inst. on *Trafficking of Intracellular Membranes: From Molecular Sorting to Membrane Fusion*, Espinho, Portugal, 1994.

23. **Nir, S., Stegmann, T., and Wilschut, J.**, Fusion of influenza virus and cardiolipin liposomes at low pH: mass action analysis of kinetics and extent, *Biochemistry*, 25, 257, 1986.

24. **Nir, S., Klappe, K., and Hoekstra, D.**, Mass action analysis of kinetics and extent of fusion between Sendai virus and phospholipid vesicles, *Biochemistry*, 25, 8261, 1986.

25. **Nir, S.**, Modeling of aggregation and fusion of phospholipid vesicles, in *Cellular Membrane Fusion*, Wilschut, J. and Hoekstra, D., Eds., Marcel Dekker, New York, pp. 127–153, 1990.

26. **Smoluchowski, M. V.**, Investigation into a mathematical theory of the kinetics of coagulation of colloidal solutions, *Z. Physik Chem. (Leipzig)*, 92, 129, 1917.

27. **Fuchs, N. A.**, *The Mechanics of Aerosols*, Pergamon Press, New York, 1964.

28. **Braun, G., Lelkes, P. I., and Nir, S.**, Effect of cholesterol on Ca^{2+}-induced aggregation and fusion of sonicated phosphatidylserine-cholesterol vesicles, *Biochim. Biophys. Acta*, 812, 688, 1985.

29. **Gad, A. E., Bental, M., Elyashiv, G., Weinberg, H., and Nir, S.**, Promotion and inhibition of vesicle fusion by polylysine, *Biochemistry*, 24, 6277, 1985.

30. **Poulain, F.R., Nir, S., and Hawgood, S.**, Kinetics of phosholipid membrane fusion induced by surfactants apoproteins A and B, unpublished data, 1994.

31. **Bentz, J. and Nir, S.**, Aggregation of colloidal particles modeled as a dynamical process, *Proc. Natl. Acad. Sci. U.S.A.*, 78, 1634, 1981.

32. **Peled, R., Braun, G., and Nir, S.**, Time of equilibration in reversible aggregation of particles, *J. Colloid Interface Sci.*, 169, 204, 1995.

33. **Nir, S., Peled, R., and Lee, K. D.**, Analysis of particle uptake by cells: binding to several receptors, equilibration time, endocytosis, *Colloids Surfaces*, 89, 45, 1994.

34. **Lee, K. D., Kantor, A. B., Nir, S., and Owicki, J. C.**, Aggregation of Hapten- bearing liposomes mediated by specific antibodies, *Biophys. J.*, 64, 905, 1993.

35. **Hui, S., Nir, S., Stewart, T. F., Boni, L. T., and Huang, S. K.**, Kinetic measurements of fusion of phosphatidylserine-containing vesicles by electron microscopy and fluorometry, *Biochim. Biophys. Acta*, 941, 130, 1988.

36. **Nir, S., Bentz, J., and Duzgunes, N.**, Two modes of vesicle aggregation: particle size and the DLVO theory, *J. Colloid Interface Sci.*, 84, 266, 1981.

37. **Nir, S., Newton, C., and Papahadjopoulos, D.**, Cation binding to phosphatidylserine vesicles, *Bioelectr. Chem. Bioener.*, 5, 116, 1978.

38. **Nir, S.**, Cation adsorption in closed systems. Application to calcium binding to phospholipd vesicles, *J. Colloid Interface Sci.*, 102, 313, 1984.

39. **Duzgunes, N. and Nir, S.**, Liposomes as tools for elucidating the mechanisms of membrane fusion. In *Liposomes*, Schuber, F. and Filippot, J., Eds., CRC Press, Boca Raton, FL, (in press).

40. **Stegmann, T., Hoekstra, D., Scherphof, G., and Wilschut, J.**, Kinetics of pH-dependent fusion between influenza virus and liposomes, *Biochemistry*, 24, 3107, 1985.

41. **Stegmann, T., Nir, S., and Wilschut, J.**, Membrane fusion activity of influenza virus. Effects of gangliosides and negatively charged phospholipids in target liposomes, *Biochemistry*, 28, 1698, 1989.

42. **Klappe, K., Wilschut, J., Nir, S., and Hoekstra, D.**, Parameters affecting fusion between Sendai virus and liposomes. Role of viral proteins, liposome composition, and pH, *Biochemistry*, 25, 8252, 1986.

43. **Amselem, S., Barenholz, Y., Loyter, A., Nir, S., and Lichtenberg, D.**, Fusion of Sendai virus with negatively charged liposomes as studied by pyrene-labelled phospholipid liposomes, *Biochim. Biophys. Acta*, 860, 301, 1986.

44. **Cheetham, J. J., Epand, R. M., Andrews, M., and Flanagan, T. D.**, Cholesterol sulfate inhibits the fusion of Sendai virus to biological and model membranes, *J. Biol. Chem.*, 265, 12404, 1990.

45. **Larsen, C. E., Nir, S., Alford, D. R., Jennings, M., Lee, K. D., and Duzgunes, D.,** Human immunodeficiency virus type 1 (HIV-1) fusion with model membranes: kinetic analysis and the role of lipid composition, pH and divalent cations, *Biochim. Biophys. Acta,* 1147, 223, 1993.

46. **Larsen, C. E., Alford, D. R., Young, L. J. T., McGraw, T. P., and Duzgunes, N.,** Fusion of simian immunodeficiency virus with liposomes and erythrocyte ghost membranes: effects of lipid compositions, pH and calcium, *J. Gen. Virol.,* 71, 1947, 1990.

47. **White, J. and Helenius, A.,** pH-dependent fusion between the Semliki Forest virus membrane and liposomes, *Proc. Natl. Acad. Sci. U.S.A.,* 77, 3273, 1980.

48. **Kielian, M. and Helenius, A.,** Role of cholesterol in fusion of Semliki Forest virus with liposomes, *J. Virol.,* 52, 281, 1984.

49. **Duzgunes, N.,** Cholesterol and membrane fusion, in *Biology of Cholesterol,* Yeagle, P. L., Ed., CRC Press, Boca Raton, FL, 197, 1988.

50. **Nieva, J. L., Bron, R., Corver, J., and Wilschut, J.,** Membrane fusion of Semliki Forest virus requires sphingolipids in the target membrane, *EMBO J.,* 13, 2797, 1994.

51. **Alford, D., Ellens, H., and Bentz, J.,** Fusion of influenza virus with sialic acid bearing target membranes, *Biochemistry,* 33, 1977, 1994.

52. **Epand, R. M., Nir, S., Parolin, M., and Flanagan, T. D.,** The role of the ganglioside G_{D1a} as a receptor for Sendai virus, *Biochemistry,* 34, 1084, 1995.

53. **Nir, S., Stegmann, T., Hoekstra, D., and Wilschut, J.,** Kinetics and extents of fusion of influenza virus and Sendai virus with liposomes. In *Molecular Mechanisms of Membrane Fusion,* Ohki, S., Doyle, D., Flanagan, T. D., Hui, S. W., and Mayhew, E., Eds., Plenum Press, NY, 451–465, 1988.

54. **Nir, S., Duzgunes, N., Pedroso de Lima, M. C., and Hoekstra, D.,** Fusion of enveloped viruses with cells and liposomes, *Cell Biophys.,* 17, 181, 1990.

55. **Ramalho-Santos, J., Pedroso de Lima, M. C., and Nir, S.,** unpublished data, 1994.

56. **Fonteijn, T. A. A., Engberts, J. B. F. N., Nir, S., and Hoekstra, D.,** Asymmetric fusion between synthetic DI-N-Dodecyl phosphate vesicles and virus membranes, *Biochim. Biophys. Acta,* 1110, 185, 1992.

57. **Nir, S., Stutzin, A., and Pollard, H. B.,** Effect of synexin on aggregation and fusion of chromaffin granule ghosts at pH 6, *Biochim. Biophys. Acta,* 903, 309,1987.

58. **Parente, R. A., Nir, S., and Szoka, F. C., Jr.,** pH-dependent fusion of phosphatidylserine small vesicles, *J. Biol. Chem.,* 263, 4724, 1988.

59. **Nieva, J. L., Nir, S., Muga, A., Goni, F. M., and Wilschut, J.,** Interaction of the HIV-1 fusion peptide with phospholipid vesicles: Different structural requirements for fusion and leakage, *Biochemistry,* 33, 3201, 1994.

Chapter 13

Effect of Polymers Attached to Lipid Headgroups on Properties of Liposomes

Vladimir P. Torchilin

CONTENTS

I. INTRODUCTION

For a long time, the modification of the liposome surface was considered mainly from the point of view of the potential use of liposomes as drug carriers. Since liposomes, almost independently of their composition, size, and charge are almost quantitatively captured by cells of the reticuloendothelial system within the first hour after their intravenous administration,[1] the main goals for liposome surface modification were (1) to increase liposome longevity and stability in the circulation, (2) to change liposome biodistribution, (3) to achieve targeting effect, and (4) to impart to liposomes some "unusual" properties (such as pH- or thermo-sensitivity).

The most evident approaches to modify surface properties of liposomes are (1) to vary liposome phospholipid composition (resulting in variation of the liposome charge and phase state), or (2) to attach some nonphospholipid compounds to the liposome surface. Various nonphospholipid modifiers were suggested for controlling liposome biodistribution and properties *in vivo*. The most important and well studied among them are

1. Antibodies and their fragments
2. Proteins
3. Mono-, oligo-, and polysaccharides
4. Chelating compounds (such as EDTA, DTPA or deferoxamine)
5. Soluble synthetic polymers

Liposome modification with antibodies and other specific ligands (such as sugar moieties) leads to drastic changes in liposome biodistribution without necessary changes in the liposome surface properties: all the targeting phenomena take place because of specific recognition between liposome-immobilized substances and the appropriate target within the body. Similar changes in biodistribution can be found for liposomes of different composition, if they are modified with an identical ligand. Properties of specific ligand-targeted liposomes have already been discussed in detail elsewhere; see, for example, References 2–4, and will not be discussed here.

0-8493-4731-9/96/$0.00+$.50
© 1996 by CRC Press, Inc.

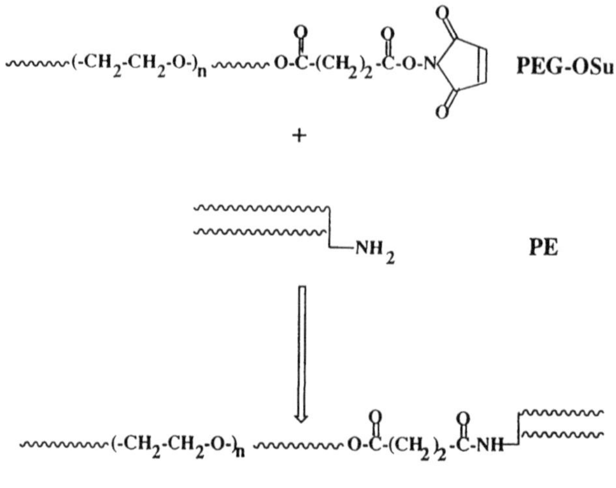

PEG-PE

(for incorporation into the liposomal membrane)

Figure 1 Scheme of PEG derivatization with terminal hydrophobic moiety for incorporation into liposome.

Liposome surface modification with chelating compounds or moieties capable of firm binding heavy metals with liposomes, serves a special application of liposome as diagnostic (imaging) agents and will be discussed here in some detail later. We will start, however, with the discussion of some phenomena arising as a result of the liposome surface modification with certain polymers.

Independently of the liposome designation, liposome modification with synthetic polymers results in the appearance of interesting theoretical and experimental (practical) problems connected with the accessibility of the liposome surface for liposome-interacting substances. Moreover, the presence of some polymers on the liposome surface can influence physicochemical (for example, spectral) properties of the liposome membrane constituents or some other membrane-bound or membrane-associated moieties.

It is also known that liposome membrane permeability changes (reflecting intra-membrane phase separation, variations of membrane components lateral diffusion rates, and some other phenomena) occur when liposome interacts with polyelectrolytes[5-7] (for review, see Reference 8). In this case, however, we deal with preliminary adsorption of polyelectrolyte onto the liposome surface, whereas we rather prefer to consider polymers anchored in the liposome membrane via hydrophobic substitutients.

General regularities underlying major changes of liposome properties upon the liposome surface modification with lipid-attached polymers are very important for the understanding of both physicochemical properties of surface-modified liposomes and their biological behavior.

II. PROPERTIES OF POLYMER-GRAFTED LIPOSOMES — ACCESSIBILITY OF LIPOSOME SURFACE COATED WITH FLEXIBLE POLYMER

A. PEG-COATED LIPOSOMES — GENERAL CONSIDERATIONS

While attempting to prepare biologically stable liposomes, an important breakthrough was achieved when long-circulating liposomes were first described.[9-11] One of the most popular and successful methods of their preparation involves liposome coating with poly(ethylene glycol), or PEG.[12-15] For this purpose, the reactive derivative of hydrophilic PEG is terminus-modified with hydrophobic moiety according to the scheme presented in Figure 1. The amphiphilic PEG obtained can be incorporated into the liposomal membrane in the process of liposome preparation by, for example, detergent dialysis method.[12] The explanation of the mechanism of PEG protective effect on liposomes involves the participation of PEG in the repulsive interactions between PEG-grafted membranes and other particles,[16] the role of surface charge and hydrophilicity of PEG-coated liposomes,[17] and generally speaking, the decreased rate of plasma proteins (opsonins) adsorption on the hydrophilic surface of PEGylated liposomes.[18]

In an attempt to further understand what peculiarities of PEG behavior on the liposome surface underlay its ability to prevent liposome opsonization or, in more general terms, to change surface

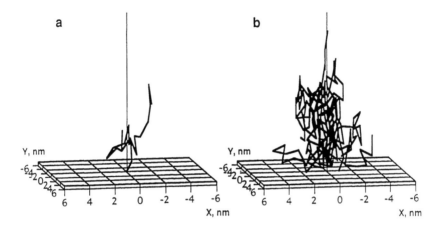

Figure 2 Computer simulation of conformational "cloud" formation by surface-attached polymer. Polymer molecule is conditionally assumed to consist of 20 segments, 1 nm each. Produced by random flight simulation. Unrestricted segment motion assumed. Space restriction: Z > 0. (a) One random conformation. (b) Superposition of eleven random conformations. (From Torchilin, V.P., Omelyaneko, V.G., Bogdanov, A.A., Jr., Papisov, M.I., Trubetskoy, V.S., Herron, J.N., and Gentry, C.A., *Biochim. Biophys. Acta,* 1195(1), 11, 1994. With permission.)

properties of liposomes, we have tried to consider the phenomenon from a physicochemical point of view, and hypothesized that the molecular mechanism of PEG protective action is determined by the properties of a flexible polymer molecule (free rotation of individual units around inter-unit linkages) in solution, and includes the formation of dense polymeric "cloud" over the liposome surface even at relatively low polymer concentrations.[19,20]

The very simple quantitative criteria can be introduced to describe the interaction of the polymer-protected liposome with a macromolecule (protein) from an external solution. Considering the diffusional movement of any macromolecule molecule toward the liposome surface as the initial step of a molecule-to-liposome interaction, we can express the degree of liposome protection as the probability of macromolecule collision with a polymer (P_{pol}) instead of liposome (P_{lip}). Thus, if the macromolecule is completely unable to reach the liposome surface ($P_{lip} = 0$), $P_{pol} = 1 - P_{lip} = 1$. When $P_{pol} = 0$, no protection is achieved.

To forecast possible P values, we can describe the behavior of liposome-grafted polymer molecule in terms of statistical physics, e.g., applying simplified model of a polymer solution.[21] We can consider it, for example, as a three-dimensional network, where each cell may be occupied either with a polymer unit or with a solvent (water) molecule. From this point of view, the more flexible the polymer is, i.e., the more independent is the motion of any polymeric unit relative to the neighboring one, the larger is the total number of its possible conformations and the higher is the transition rate from one conformation to another. It means, that water-soluble flexible polymer statistically exists as a distribution ("cloud") of probable conformations. Figure 2 shows how this cloud is formed (under some simple assumptions) and its density and uniformity increase with the increase in the number of possible conformations. The polymer flexibility correlates with its ability to occupy with high frequency many cells in solution, temporarily squeezing water molecules out of them (i.e., making them impermeable for other solutes). To reach the liposome surface, macromolecules from the solution have to penetrate the whole cloud, formed by the liposome-attached polymer molecule. Thus, relatively small numbers of water-soluble and very flexible polymer molecules can create sufficient numbers of high density conformational "clouds" over the liposome surface protecting the latter from the interaction with "foreign" macromolecules. These grafted polymer molecules form protective "umbrellas" on the liposome surface, and P_{lip} value depends on their amount (N_P), effective square (S_P) and "reliability" (P^*), expressed as:

$$P^* = \frac{1}{S_P} \int_S \frac{dP_{pol}}{dS_P}$$

where P^* is the average P_{pol} within the "umbrella" volume (see Figure 3). It is easy to see that the average P_{pol} value for the entire liposome is

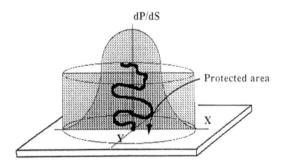

dP/dS

Protected area

X

Y

Figure 3 The distribution of P_{pol} density in the vicinity of polymer molecule attached to the liposome surface (bell-shaped curve) and the average P_{pol} value within the "umbrella" volume. (From Torchilin, V.P., and Papisov, M.I., *J. Liposome Res.*, 4(1), 725, 1994. With permission.)

$$P_{pol} = \frac{N_P S_P}{S_{lip}} P^*, \qquad \text{or} \qquad P_{lip} = 1 - \frac{N_P S_P}{S_{lip}} P^*$$

where S_{lip} is the liposome surface square. The last equation can be transformed into:

$$P_{pol} = \gamma \frac{S_P}{S_l} P^*$$

where γ is the molar ratio polymer/lipid in the outer monolayer, and S_l — the average area occupied by single lipid molecule (for the given liposome size and composition). At high γ values, polymer will be "stretched" out of the liposome forming a dense "brush".[22] Therefore, γ (S_P/S_l) is always < 1. The maximal protection can be achieved when $\gamma S_P \approx S_l$ (the polymer "clouds" are practically fused) and P^* value is close to 1.

For a rigid chain polymer (where unit motion is hindered), the number of possible conformations for such polymers is lower, besides the conformational transitions proceed with a slower rate than those of a flexible polymer. It means that the density of the conformational "cloud" for a rigid polymer will be very uneven during a single collision act, and the number of water molecules disturbed much smaller. In terms of the "cellular" model, it appears that there should exist a sufficient water space through which the normal diffusion of macromolecules from the external medium (such as blood plasma proteins) toward the liposome surface is still possible. Thus, to protect the liposome, one has to bind much larger number of rigid polymer molecules on the liposome surface. This difference between flexible and rigid polymers may only increase with the polymer MW. Thus, good water-solubility and hydrophilicity alone cannot provide sufficient protective effect.[23] Only when the same polymer combines both hydrophilicity and flexibility, can it serve as an effective liposome protector even at relatively low concentration of the surface-immobilized macromolecules (the most obvious example of such polymers is PEG). Other possible candidates are, for example, poly(acrylamide), poly(vinylpyrrolidone), and poly(vinyl alcohol).

The next question to answer is how can we estimate the size of the area on the liposome surface protected with a single polymer molecule of a given molecular weight? Or, how many polymer molecules do we need to protect the liposome of a given size? Such parameters as the average end-to-end distance of a polymer random coil in solution, $R_{0,sol}$, give us certain insight on the "cloud" density (assuming that PEG forms random coil in solution). A polymer molecule is located mainly in the volume "between the ends" (inside the appropriate sphere for the molecule in solution and inside the hemi-sphere for the surface-immobilized polymer). So, we can assume that $R_{0,sol}$ value nearly corresponds to the radius of the "dense cloud" (R), which can hardly be penetrated with a macromolecule from the external medium. The end-to-end distance for the surface-attached molecule (Figure 4) might be about two times longer than $R_{0,sol}$ (for a flexible polymer, the fixation of one end does not influence molecular mobility significantly). Thus, we can assume the radius of the protected area being from $R_{0,sol}$ to $2R_{0,sol}$, and within this area $P^* \approx 1$. Using this simple approach and $R_{0,sol}$ values for PEG of different molecular weight published in Reference 24, we can estimate the area of the liposome surface that can be protected by single PEG

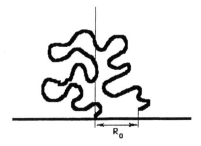

Figure 4 The average distance between the fixed and free ends of a polymer molecule ($R_0 = 2R_{0,sol}$) as a criterion of the size of protected area. Single polymer molecule supposed to form the "umbrella" with the diameter of $2R_0$ (or $4R_{0,sol}$); and the "reliability" parameter, P^*, is close to 1 within the umbrella, and close to 0 outside. (From Torchilin, V.P., and Papisov, M.I., *J. Liposome Res.*, 4(1), 725, 1994. With permission.)

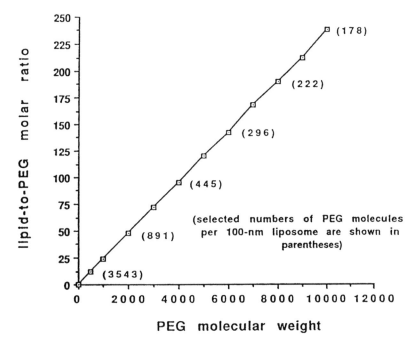

Figure 5 The dependence of the number of PEG molecules required for complete protection of 100-nm liposome on PEG molecular weight.

molecule. Assuming about $4.25 \cdot 10^4$ lipid molecules in the outer monolayer of single-bilayer 100 nm liposome,[25,26] we can calculate the molar ratio PEG-to-lipid required for 100% protection of liposome surface (surface area, S, $3.14 \cdot 10^4$ nm²), see Figure 5. One should mention that the figures calculated match well with published experimental data,[15,27] thus the end-to-end distance may be used as an estimate for the protected square radius. As it clearly follows from the Figure 5, the size of protected area linearly depends on the polymer molecular weight.

B. COMPUTER SIMULATION OF PEG BEHAVIOR ON LIPOSOME SURFACE

1. Computer Model

The further development of our model includes the computer simulation described below.[28] Spatial distribution of short polymers grafted onto the liposome surface was simulated using a three-dimensional random flight model. The polymer was assumed to consist of absolutely rigid segments with free segment rotation around intersegmental conjunction. The program developed using MSExcel™ macrolanguage for the computer simulation of random polymer conformation included following procedures.

The grafting point was assumed to be located at $x = y = z = 0$. At step one, the random direction of the first segment was assumed by generation of two random angles: between the segment axis and axis z, and between the segment axis projection onto xy plain and axis x. At step two, the coordinates of the remote end of the first segment were calculated, conditionally assuming the segment length $l = 1$ nm for a flexible polymer and $l = 5$ nm for a rigid one. If the resulting position of the segment terminus was invalid (segment location below the liposome surface, i.e., $z < 0$), the result was omitted. The resulting coordinates were used as a starting point for the calculation of the next segment position using the same procedure. In this model, the segment mass is assumed to concentrate at the remote segment terminus. This assumption distorts the profile of mass distribution for the very first segment. The distribution of segments, starting from the third one, is practically insensitive to this assumption.

For better visual representation of the polymer density distribution, the data arrays describing conformation of the rigid polymer were produced assuming that its segment is formed of five "frozen" segments of the flexible polymer; and a slice of the polymer cloud ($y = 0 \pm 0.25$), rather than the entire data array, can be shown (compare Figures 2 and 5).

Corresponding to our model, the degree of the liposome surface protection by polymer molecules should correlate with the effective polymer density per surface square unit. To gain insight into the possible pattern of surface density (i.e., the shape of the polymer "umbrella"), the relative polymer densities for the flexible and rigid polymers were calculated. First, the distance (r) between the segment end and axis z was calculated for each segment. Then, the number of segment termini located within certain distance interval ($r, r + \Delta r$) was counted for a large number of conformations. The resulting distribution was normalized per square, i.e., divided by the area of the circular annulus of the width Δr and the inner radius equal to r. Normally, 440 simulated random conformations of each polymer were used (8800 segment terminus spatial coordinates).

2. Significance of Polymer Chain Flexibility

To perform our analysis of this approach, we are assuming that (1) protecting polymer contacts with macromolecules from the external medium does not result in the opsonization, (2) protecting polymer is randomly distributed over the liposome surface and does not form any aggregates or clusters.

As we have already mentioned, the water-soluble polymer statistically exists as a distribution ("cloud") of probable conformations (Figure 2). From the computer analysis it follows (Figure 6) that the flexible polymer forms the conformational cloud with very high density in its central part. A rigid polymer of the same length (its segment was conditionally assumed to be five times longer than for the flexible polymer) forms a broad, but loose and thus permeable cloud. Additionally, this cloud is less uniform due to slower translocation of individual segments.

3. Two Types of Active Sites on PEG-Coated Surface

If there exist any reactive centers on the liposome surface (e.g., binding sites for certain macromolecules or other ligands), the presence of PEG in concentrations insufficient for complete coating of the surface should lead to the appearance of two populations of those centers. The first population will consist of reactive moieties excluded from the volume occupied by PEG. Such reactive centers should posses all the properties (including the reactivity) of similar centers on "plain" liposomes. Other centers remaining within the polymer cloud form the second population with sharply decreased ability to participate in "normal" interactions. Figure 7 illustrates this phenomenon, and permits us to suppose that kinetic parameters of the reactive site will depend on its location on the liposome surface partially coated with PEG. In this way, this model permits us to foresee the possibility of simultaneous immobilization of different functional modalities on the liposome surface.[15]

4. Optimal Size of Protective Polymer Molecule

The approach developed opens the opportunity to predict the optimal size (molecular weight) of protective polymer molecules (Figure 8). The assumption that a certain level of polymer cloud density (L) provides sufficient protection to the liposome surface, allows us to suggest the existence of polymer length optimum (Figure 8B). Below the optimum, the polymer cloud will not have enough density even near the attachment point (Figure 8A), whereas very long polymers provide the density much above the necessary one (Figure 8C).

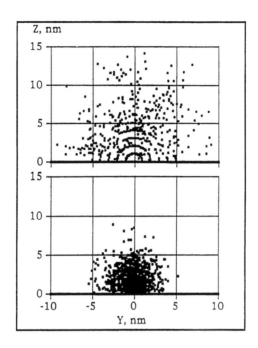

Figure 6 Distribution of polymer conformations in space; slice X = 0 ± .25 nm. Produced by random flight simulation (Z > 0, polymer length 20 nm, 440 conformations). Upper panel: segment length is 5 nm (rigid polymer); lower panel: segment length is 1 nm (flexible polymer). Calculations for the rigid polymer have been done under the assumption that only every fifth 1-nm segment may change the direction. The "rings" on the upper panel are the result of the assumption that the net mass is concentrated at the remote termini of "frozen" shorter segments. (From Torchilin, V.P., Omelyaneko, V.G., Bogdanov, A.A., Jr., Papisov, M.I., Trubetskoy, V.S., Herron, J.N., and Gentry, C.A., *Biochim. Biophys. Acta*, 1195(1), 11, 1994. With permission.)

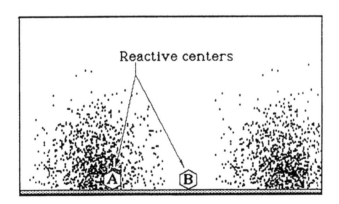

Figure 7 The scheme of reactive site location on the liposome surface at low PEG concentration. The site can be either sterically hindered by polymer "cloud" (A), or exposed and readily available for a variety of interactions (B). Kinetic parameters of chemically identical but differently located sites may differ. (From Torchilin, V.P., Omelyaneko, V.G., Bogdanov, A.A., Jr., Papisov, M.I., Trubetskoy, V.S., Herron, J.N., and Gentry, C.A., *Biochim. Biophys. Acta*, 1195(1), 11, 1994. With permission.)

C. INTERACTION OF FLUORESCENT MARKERS WITH POLYMER-MODIFIED LIPOSOME SURFACE

To confirm the model described experimentally, we have investigated the efficacy of the fluorescence quenching of the liposome-incorporated fluorescent markers (NBD-PE and lipid-derivatized fluorescein) with soluble macromolecules (Rh-OVA and anti-fluorescein antibody, respectively) from the external medium, depending on the type of the liposome-attached polymer.[28]

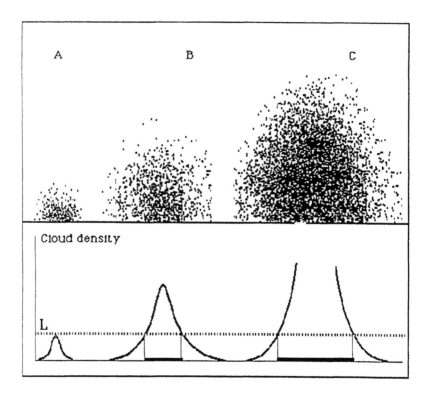

Figure 8 Polymers of different length grafted onto the liposome surface and corresponding densities of polymer "clouds" (non-quantitative computer simulation). L — minimal level of polymer "cloud" density providing sufficient protection for the surface. A — short-chain polymer cannot create sufficiently dense "cloud", even if it is highly flexible. B — polymer with optimal chain length provides sufficient protection and still does not exclude the possibility of surface-located reactive sites to enter normal interactions. C — polymer with excessive molecular weight provides unnecessary high "cloud" density and sterically hinders any reactive site on the liposome surface. (From Torchilin, V.P., Omelyaneko, V.G., Bogdanov, A.A., Jr., Papispov, M.I., Trubetskoy, V.S., Herron, J.N., and Gentry, C.A., *Biochim. Biophys. Acta*, 1195(1), 11, 1994. With permission.)

Commercial fluorescent phospholipid, NBD-PE, was used for incorporation into the liposome membrane. Conjugate for liposomal NBD-PE quenching was prepared from mannose-free ovalbumin and Lissamine™ rhodamine B sulfonyl chloride (Molecular Probes, Inc.) according to manufacturer's instructions. The resultant purified Rh-OVA contained 10–15 sulforhodamine B residues per mol of protein.

Membranotropic derivative of fluorescein, GFl-PE, was synthesized from (5-aminoacetamido) fluorescein and N-glutaryl-dioleoyl-PE (NGPE) in the presence of N-hydroxysulfosuccinimide together with 1-ethyl-3-[3′-(dimethylamino)propyl]-carbodiimide.

Antifluorescein monoclonal antibody 4-4-20 able to quench fluorescein fluorescence was generated through chemically mediated fusion of BALB/C murine splenic lymphocytes with the Sp 2/0-Ag 14 myeloma cell line as described in Reference 29. Hybridoma cell line was obtained from Professor Edward W. Voss, Jr., University of Illinois at Urbana-Champaign. Antibody preparation was characterized by gel electrophoresis and fluorescein quenching assays.[30]

Liposomes were prepared from the mixture of phosphatidyl choline and cholesterol in 7:3 molar ratio by detergent (octyl glucoside) dialysis method.[31] When necessary, 1% mol of the fluorescent label, NBD-PE or GFl-PE, or/and required quantities of PEG-PE (prepared as in Reference 12, PEG MW 5,000), or dextran-SA (prepared by reductive amination as in Reference 32, see Figure 9; dextran MW 6,000) were added to the starting lipid composition. Liposomal suspensions formed were sized by consecutive filtration through 0.6, 0.4, and 0.2 µm polycarbonate filters. According to the data of liposome size measurements using Coulter N4 MD Submicron Particle Size Analyzer, all the liposome obtained were within 160 ± 25 nm range (unilamellar vesicles). For fluorescence quenching experiments liposome preparations (5 mg total lipid/ml) were diluted until it reached the required fluorescence intensity.

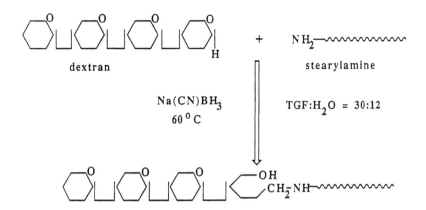

Figure 9 Synthesis of hydrophobic dextran derivative for incorporation into the liposomal membrane.

Fluorescence intensity measurements[28] were made using a photon-counting spectrofluorimeter PC-1. In the case of liposomal NBD quenching with Rh-OVA, initial fluorescence of NBD-containing liposomes was recorded at 25°C. Various quantities of 1 μM Rh-OVA solution were then added, and the residual fluorescence was registered after 10 min incubation. When more than 5% vol of Rh-OVA solution was added to liposome suspension, dilution factor was taken into account. Both NBD and Rh were excited at 485 nm. The emission was monitored using a monochromator with a 2 mm slit (16 nm FWHM band-pass).

Association kinetics of antifluorescein antibody with liposomal fluorescein was measured by recording the decrease in the fluorescence intensity of the fluoresceyl-ligand, as a function of time. Samples were excited at 485 nm using a monochromator with a 2 mm slit width, and emission was monitored through the interference filter 514.5 (Oriel Corporation). To ensure the pseudo-first order kinetics, antibodies were used in large molar excess towards liposomal fluorescein (92 and 10 nM, respectively).

Values for the fraction of free ligand (L/L_0) were calculated using the following formula:[30]

$$L/L_0 = \frac{F_{(t)} F_{(\infty)}}{F_{(0)} - F_{(\infty)}}$$

where $F_{(t)}$ is the fluorescence intensity at a given time (t), and $F_{(0)}$ and $F_{(\infty)}$ are the initial and limiting fluorescence intensity values, respectively. The values for L/L_0 were plotted vs. time and analyzed by linear regression, which yielded the effective reaction rate constant (k_{eff}) from the least squares slope. The bimolecular association constant (k_2) was determined from the effective rate constant using the equation:

$$k_2 = \frac{f_{b(\infty)}}{P} k_{eff}$$

where $f_{b(\infty)}$ is the equilibrium fraction of the liposome-bound fluoresceyl-ligand, and P is the active site concentration (twice antibody concentration since antibody molecule carries two binding sites). The equilibrium fraction of the bound ligand was determined using the equation:

$$f_{b(\infty)} = Q_{(\infty)} / Q_{max}$$

where $Q_{(\infty)}$ is the fluorescence quenching value for the association reaction at equilibrium, and Q_{max} is the maximum quenching value and under the conditions used equals 0.98.[30,33] In a general case, quenching values can be determined from the fluorescence measurements using the equation:

$$Q = (F_0 - F_t) / (F_0 - F_{bkg})$$

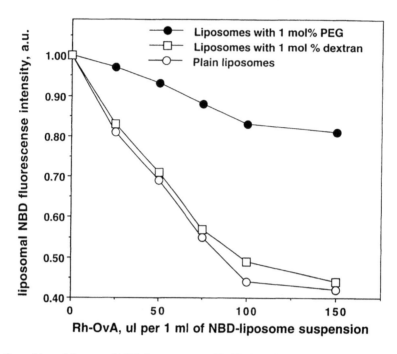

Figure 10 Quenching of liposomal NBD fluorescence with Rh-OVA from the solution at different Rh-OVA concentrations. Empty circles — plain liposomes; filled circles — liposomes with 1% mol PEG-PE; empty squares — liposomes with 1% mol dextran-stearylamine.

where $F_{bkg} = F_0/2$ (assuming fluorescein symmetrical distribution between outer and inner monolayers and taking into account intraliposomal fluorescein).

To study the interaction of polymer-modified NBD-liposomes with Rh-OVA, plain NBD-liposomes and NBD-liposomes containing 1% mol of lipid-conjugated PEG or dextran were treated with increasing quantities of Rh-OVA. NBD fluorescence diminished identically for Rh-OVA-treated "plain" NBD-liposomes and NBD-liposomes with 1% mol of dextran (Figure 10). About 50% of initial fluorescence can be quenched, which evidences even distribution of liposomal NBD between outer and inner monolayers of the membrane (only "outer" NBD is susceptible for quenching). However, NBD fluorescence quenching is drastically hindered in NBD-liposomes containing 1% mol of PEG (theoretical quantity of PEG 5,000 required for complete liposome protection, see Figure 5). Even at maximal Rh-OVA concentration, we still observed about 80% of the initial fluorescence. Since the whole quenching process is limited only by Rh-OAB diffusion from the solution to the liposome surface, it is evident that the presence of PEG on the surface creates diffusional hindrances for this process. The data agree well with our hypothesis on different densities of "protective clouds" for flexible (PEG) and rigid (dextran) polymers.

Another experimental proof for our hypothesis was obtained using the system involving liposome surface-incorporated fluorescein derivative, GFl-PE, and anti-fluorescein antibody (Figures 11 and 12, Table 1). In this case, Q values were equal for "plain" GFl-liposomes and for GFl-liposomes with 1% mol of dextran, k_{eff} and k_2 values were also identical (Figure 12A and 12C, Table 1). This indicates that the presence of dextran in the quantities used did not create any diffusional limitations for antibody-to-fluorescein interactions. The presence of 1% mol PEG in liposomes noticeably decreased both the rate of fluorescein quenching and the quantity of fluorescein residues accessible for the interaction with antibody (Figure 11). Q value was also substantially decreased (from 0.76 to 0.58), as well as both rate constants (by about five-fold). This effect may be explained by strong diffusional limitations for antibody penetration towards the liposome surface imposed by PEG. The density of the protective cloud of PEG is sufficiently high in this case.

Incorporation of smaller quantity of PEG (0.2% mol) into liposomes revealed a somewhat decreased total quenching (Q value number is intermediate between two marginal cases), and the existence of two different fluorescein pools on the liposome surface (Figure 12B). One of them was quenched with the

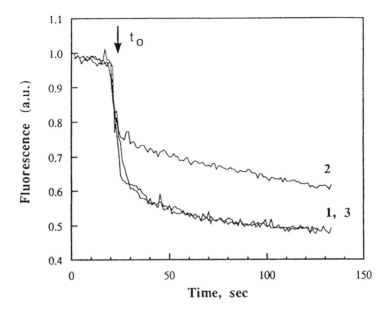

Figure 11 Actual typical time-courses of liposomal fluorescein quenching with anti-fluorescein antibody. 1 — "Plain" fluorescein-liposomes; 2 — Fluorescein-liposomes with 1% mol of PEG; 3 — Fluorescein-liposomes with 1% mol of dextran. t_o — the point of antibody addition to liposomes; this parameter can vary slightly in different runs. (From Torchilin, V.P., Omelyaneko, V.G., Bogdanov, A.A., Jr., Papispov, M.I., Trubetskoy, V.S., Herron, J.N., and Gentry, C.A., *Biochim. Biophys. Acta*, 1195(1), 11, 1994. With permission.)

same time rate as fluorescein on PEG-free liposomes (similar values for both kinetic constants; Table 1), whereas the quenching kinetics for another was close to that for fluorescein on PEG-liposomes with high PEG content. This can reflect different location of fluorescein molecules on the liposome surface — between and inside PEG "clouds", as we predicted on the basis of computer simulation data (Figure 7). Thus, the kinetic parameters of reactive site really depend on its location on the liposome surface when the quantity of PEG on the liposome is not sufficient to form an even protective cloud over the liposome surface. Thus, the role of statistical properties of macromolecules in the solution might be decisive in the behavior of PEG-coated liposomes.

D. ALTERNATIVE POLYMERS FOR STERIC PROTECTION OF LIPOSOMES

The model of flexible polymer behavior on the liposome surface suggested by us and described above, formulates some general requirements towards polymers that can be used for liposome protection. These polymers should be soluble and hydrophilic, and have highly flexible main chain. It was already mentioned that polyacrylamide (PAA) and polyvinylpyrrolidone (PVP) might be considered as the most appropriate candidates among alternative liposome protectors.

Recently, we reported the first experimental data on the possibility of using these polymers for preparation of long-circulating liposomes.[34] Amphiphilic PAA and PVP have been used in our studies, prepared by free-radical polymerization of monomers in organic solvent in the presence of different quantities of long-chain fatty acid chloroanhydride (used as a chain terminator introducing terminal long-chain acyl group into the polymer molecule). Both polymers were prepared with MW values 6,000–8,000 and modified with terminal palmityl group (PAA-P and PVP-P) or dodecyl group (PAA-D and PVP-D).

Egg phosphatidyl choline:cholesterol (7:3) liposomes were prepared by the detergent (octyl glycoside) dialysis with the addition of 2.5 or 6.5% mol of the corresponding amphiphilic polymer. For biodistribution studies, liposomes were labeled with ^{111}In by transchelation method via membrane-incorporated diethylene triamine pentaacetic acid-stearylamine prepared as in Reference 35. Liposome size was between 165 and 190 nm. Biodistribution experiments were performed in BALB/C mice injected with different liposome preparations via the tail vein.

The data presented in Figure 13 clearly demonstrate that amphiphilic derivatives of PAA and PVP can provide effective protection to liposomes *in vivo*. PAA-P, PVP-P, and PEG-PE of similar molecular

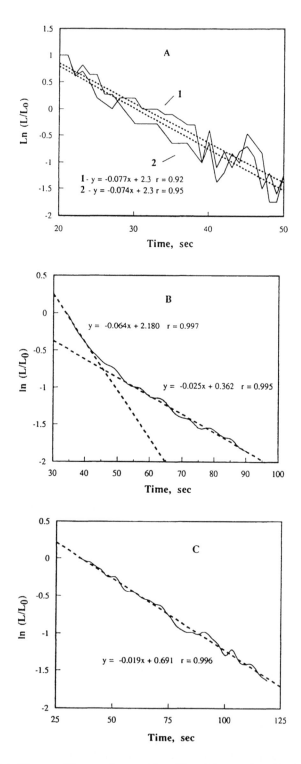

Figure 12 Linearization of liposomal fluorescein quenching with anti-fluorescein antibody for typical experiments (dotted lines). L_0 and L — initial and current fluorescence, respectively. A — "plain" fluorescein-liposomes (1) and fluorescein-liposomes with 1% mol of dextran (2). B — fluorescein-liposomes with 0.2% mol PEG. Two phases on the kinetic curve can be seen — initial fast phase, reflecting the quenching of exposed antigen, and subsequent slow phase, reflecting the quenching of PEG-protected antigen. C — fluorescein-liposomes with 1% of PEG. (From Torchilin, V.P., Omelyaneko, V.G., Bogdanov, A.A., Jr., Papisov, M.I., Trubetskoy, V.S., Herron, J.N., and Gentry, C.A., *Biochim. Biophys. Acta*, 1195(1), 11, 1994. With permission.)

Table 1 Kinetic Parameters of Liposomal Fluorescein Quenching with Antifluorescein Antibody for Different Liposome Preparations

Liposome	Q	k_{off} (sec^{-1})	k_2 (M^{-1}sec^{-1})
"Plain"	0.76 ± 0.05	0.07 ± 0.01	$2.9 \cdot 10^5$
0.2% mol PEG	0.70 ± 0.05	0.06 ± 0.01[a]	$2.3 \cdot 10^{5a}$
		0.02 ± 0.01[b]	$0.9 \cdot 10^{5b}$
1 % mol PEG	0.58 ± 0.05	0.02 ± 0.005	$0.6 \cdot 10^5$
1 % mol dextran	0.76 ± 0.05	0.07 ± 0.01	$2.9 \cdot 10^5$

[a] Kinetic constant values for fast-quenching pool of fluorescein molecules on the liposome surface.
[b] Kinetic constants for slow-quenching fluorescein pool. The data are given as M ± sem for 3 to 6 independent measurements for each sample.

weight (ca. 6,000–8,000) being used in similar concentration, all provide efficient steric protection for liposomes and noticeably increase the residence time of liposomes in the blood (Figure 13A). Half-clearance times for PVP-P-, PAA-P-, and PEG-liposomes with 2.5% mol content of protective polymer are ca. 45, 80, and 80 min, and for PVP-P-, PAA-P-, and PEG-liposomes with 6.5% mol content of protective polymer ca. 120, 140, and 170 min, respectively. Half-clearance time for "plain" liposomes of the same size is only about 10–15 min.

The protective activity of PAA-D and both PVP-D is, however, much lower. Despite the definite increase in the circulation time and some decrease in the liver capture of liposomes, these polymers are still much less effective steric protectors than polymers with longer acyl anchors. This can be easily understood taking into account the supposed energy of interaction between the fatty acyl anchor, which keeps polymer on the liposome surface, and the hydrophobic part of the liposomal membrane. From the thermodynamic point of view, the relatively short dodecyl group is unable to keep a 6–8 kDa polymer molecule on the liposome surface: the energy of the polymeric chain motion is, probably, comparable (or even higher) with the energy of dodecyl group interaction with phospholipid surroundings within the liposomal membrane. As a result, PAA-D and PVP-D might be relatively easily removed from the liposomal membrane, and demonstrate only a slight and transient protective effect. The longer palmityl anchor provides much more firm polymer binding with liposomes (higher energy of interaction with hydrophobic membrane core due to the larger number of membrane-embedded CH$_2$-groups), and thus much better liposome steric protection.

Similar regularities have been found following liposome accumulation in the liver (Figure 13B). Plain liposomes are captured by the liver very fast (more than 50% in 45 min and ca. 70% in 240 min). The longest circulating liposomes containing 6.5% mol of PAA-P, PVP-P, or PEG demonstrate much slower liver uptake: less than 20% of these liposomes are captured in 45 min, and less than 40% in 240 min. Liposomes with 2.5% mol of protective polymer demonstrate intermediate liver uptake.

Thus, the decrease in protecting polymer content and shortening of the hydrophobic anchor facilitate liposome clearance, which can be easily understood in terms of the density of protective polymer grafting on the liposome surface, and interaction energy between grafted polymer and liposomal membrane. Still, the blood clearance and liver accumulation of all polymer-modified liposomes in the liver proceeds slower than for "plain" liposomes.

Very similar data have been obtained by us recently (Torchilin, Veronese, Trubetskoy, submitted) with other new hydrophilic and flexible polymers, such as branched PEG (2PEG) and polyacryloylmorpholine (PAM). Amphiphilic derivatives of these polymers were prepared by chemical modification of activated carboxy-terminal PAM and 2PEG with phosphatidyl ethanolamine and incorporated into liposome surface by detergent dialysis method. Sharply increased circulation times of PAM- and 2PEG-coated liposomes in mice confirm used selection criteria and suggested protection mechanism.

Thus, hydrophilic and flexible synthetic polymers, such as linear and branched PEG, PAA, PVP, and PAM, when made amphiphilic by modification at one terminus with long-chain fatty acyl or phospholipid residue, can incorporate into the liposome surface and make liposome long-circulating. The protection effects observed are determined by the conformational behavior of polymer molecules in the solution; the scale of these effects might be interpreted in terms of the balance between the energy of hydrophobic anchor interaction with the membrane core and the energy of polymer chain motion in the water.

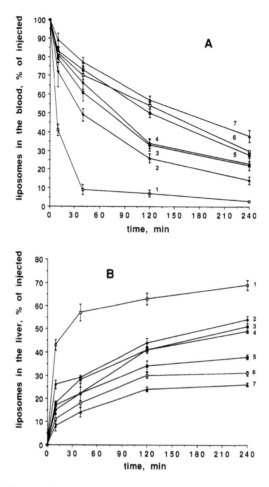

Figure 13 A — liposome clearance from the blood of experimental mice; B — liposome accumulation in the liver. 1 — "plain" liposomes; 2 — PVP-P-liposomes (2.5% mol PVP); 3 — PAA-P-liposomes (2.5% mol PAA); 4 — PEG-liposomes (2.5% mol PEG); 5 — PVP-P-liposomes (6.5% mol PVP); 6 — PAA-P-liposomes (6.5% mol PAA); 7 — PEG-liposomes (6.5% mol PEG). (From Torchilin, V.P., Shtilman, M.I., Trubetskoy, V.S., Whiteman, K., and Milstein, A.M., *Biochim. Biophys. Acta*, 1195(1), 181, 1994. With permission.)

III. MAGNETIC PROPERTIES OF POLYMER-MODIFIED LIPOSOMES

A. PEG ON THE SURFACE OF Gd-CONTAINING LIPOSOMES

The presence of grafted polymer (PEG) on the liposome surface can affect not only liposome ability to interact with external macromolecules (and, as a consequence, *in vivo* longevity), but also some other important characteristics of liposomal preparations. Thus, for example, taking into account the ability of PEG molecule to tightly bind water molecules, we can expect that liposome-bound PEG might affect relaxivity properties of liposome membrane-incorporated paramagnetic labels. If so, it might be quite important for potential use of liposomes as diagnostic agents, for example, in magnetic resonance (MR) imaging. Liposome coating with polymers can help to find optimal performance composition in the delivery of paramagnetic label-loaded liposomes to their targets, such as, for example, lymph nodes. MR visualization of lymph nodes with paramagnetic label-loaded subcutaneously administered liposomes is quite important from the practical point of view, because the imaging of lymph nodes plays a major role during the early detection of neoplastic involvement in cancer patients.[36] Liposomes, as any other particulate, has been shown to accumulate in lymph nodes with the lymph flow upon intraperitoneal or subcutaneous injection.[37] Tagging of the vesicles with different reporting groups might detect the abnormalities within the node architecture and, hence, be helpful in diagnosis of cancer. As far as we are discussing in this volume "nonmedical" application of liposomes, I shall not concentrate on pure

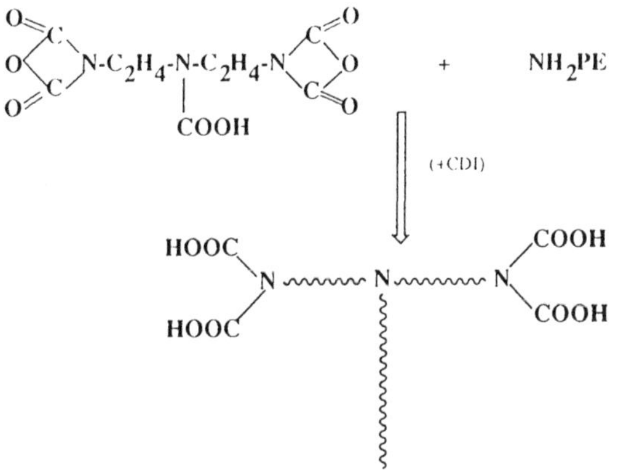

Figure 14 Scheme of DTPA derivatization with hydrophobic "tail" for incorporation into liposomes.

diagnostic aspects of the problem, but rather on the changes in liposome paramagnetic properties caused by grafting liposomes with PEG.

"Plain" liposomes have been already used as carriers for imaging agents for MR-tomography. For this purpose, liposomes were loaded with corresponding imaging agents, such as Gd complex with strong chelating agent diethylenetriamine pentaacetic acid (DTPA).[38,39] The loading can be performed by the entrapment of the corresponding agent into water interior of the liposome or by the attachment of the agent (usually heavy metal ion) to the liposome surface via different chelating groups. For better incorporation into the liposomal membrane, a chelator can be preliminary modified with fatty acid or phospholipid residue.[40] In this particular case, the additional modification of the liposome surface with polymers aims to improve liposome accumulation in lymph nodes or/and to enhance the signal of liposome-associated label.

To follow possible polymer-provoked changes in paramagnetic liposome properties, we have used NMR-spectroscopy with Gd-containing liposomes. As previously mentioned, the coating of MR-active Gd-liposomes with PEG can permit to change Gd water surroundings due to the presence of water molecules tightly associated with PEG molecule, and to increase thus the possible signal. The amphiphilic chelator DTPA-PE was synthesized using egg PE and DTPA anhydride according to Reference 40; see the reaction scheme on Figure 14. The membranotropic chelator was saturated with Gd ions in pyridine,[41] dialyzed against the deionized water, and freeze-dried. In this study we have used plain (nonmodified) Gd-containing liposomes, as well as liposomes modified with dextran and PEG.[42] Prior to the incorporation into the liposomal membrane, dextran (MW 6,000) and PEG (PEG sucinimidyl succinate, activated PEG, MW 5,000) were "hydrophobized" with stearylamine[32] and PE,[12] respectively; see also Figures 1 and 9. The liposomes have been prepared from egg PC, cholesterol, Gd-DTPA-PE and the corresponding amphiphilic modifier (60:25:10:5 molar ratio). The organic solvent was evaporated from the solution of the liposome components in chloroform, and the lipid film formed was additionally dried in the high vacuum for 2 h. After this, HBS was added to a lipid film and vortexed. The lipid suspension formed was extruded consequently through the set of polycarbonate filters with the pore diameter of 0.6, 0.4, and 0.2 μm. 200 nm vesicles were obtained as was determined by quasi elastic laser light scattering. The relaxation parameters of all preparations were measured in HBS using a 5 MHz RADX NMR proton spin analyzer. Gd determinations were performed on a commercial basis by Galbraith Laboratories, Inc. (Knoxville, TN).

PEG-coated liposomes in this particular case have to combine two important features provided by the grafted polymer: macrophage-evading properties and enhanced MR-signal. The relaxivity ($1/T_1$) measurements of liposome preparations used (Figure 15) demonstrate that $1/T_1$ values of PEG-Gd-liposomes are about two times higher than the corresponding parameters for plain Gd-liposomes and dextran-Gd-liposomes. This fact might be explained by the presence of increased amount of PEG-associated water protons in the close vicinity of chelated Gd ions located on the liposomal membrane. This might decrease the residence time of inner sphere water molecules, which results in relaxivity increase. From the physiological point of view, upon lymph node uptake Gd-liposomes are compartmentalized within

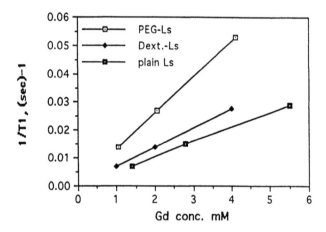

Figure 15 Molar relaxivities of different Gd-liposome preparations: (□) PEG-Gd-liposomes; (♦), dextran-Gd-liposomes; (■) plain Gd-liposomes.

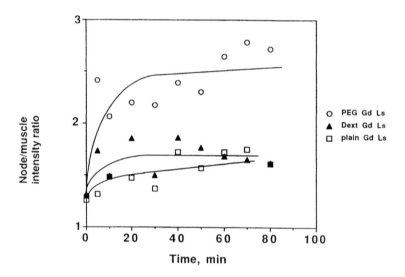

Figure 16 The kinetics of the relative axillar lymph node-to-muscle MR signal intensity ratio after the subcutaneous administration of different Gd-containing liposomes in rabbits: (○), PEG-Gd-liposomes; (▲) dextran-Gd-liposomes; (□) plain Gd-liposomes; (20 mg of total lipid, 0.5 ml HBS).

phagocytic cell lysosomes. As mentioned previously,[43] this phenomenon theoretically might lead to the decrease of liposome relaxivity in the intranodal tissue because not all tissue water is available for the interaction with paramagnetic ions. If this is true, this effect should be less expressed in case of PEG-Gd-liposomes which "bring" their own PEG-associated water to cell organelles, reducing thus possible tissue relaxivity decline.

The *in vivo* imaging of the axillary lymph node area was performed using 1.5 Tesla GE Signa MRI scanner (T_1 weighted pulse sequence, fat suppression mode) during 2 h after the subcutaneous administration of the liposomal preparation (0.5 ml, 20 mg of total lipid) into the paw of an anesthetized rabbit. Animal transverse slice scans from the MRI instrument were analyzed by the image processing software ("Khoros", University of New Mexico) in order to determine the relative target-to-nontarget (lymph node/muscle) pixel intensity. For the actual liposome delivery measurements, the liposomes of the same composition were labeled with 100–200 μCi of ^{111}In-DTPA-SA[35] and subcutaneously injected into the rabbit paw, all conditions being as for MRI experiments. Animals were sacrificed by pentobarbital overdose 2 h after injection. Axillar lymph nodes were removed, weighed and gamma-counted for ^{111}In radioactivity.

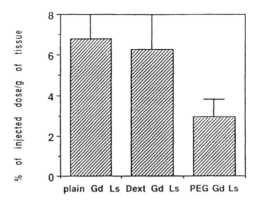

Figure 17 The accumulation of ^{111}In-labeled Gd-liposomes in the axillar lymph node 2 h after the subcutaneous administration (20 mg of total lipid, 0.5 ml HBS, 100–200 µCi/animal). (From Trubetskoy, V.S. and Torchilin, V.P., *J. Liposome Res.*, 4(2), 961, 1994. With permission.)

The kinetics of the relative MR signal intensity of axillar lymph node after the subcutaneous administration of different Gd-containing liposomal preparations is presented on Figure 16. It is evident that the plain (non-modified) liposomes only slightly enhance the signal from the lymph node with time, node-to-muscle intensity ratio being around 1.5 even after 80 min of observation. Dextran-coated liposomes do not differ much from the plain ones. At the same time PEG-coated Gd-containing vesicles provide fast and effective development of the signal from the axillar lymph node: node-to-muscle signal ratio reaches about 2.5 already in 5 to 10 min.

The measurements of the actual intranodal delivery of liposomes using the surface-bound ^{111}In radiolabel demonstrated the decreased accumulation of PEG-Gd-liposomes in the lymph node under the study (Figure 17). This finding seems quite expected, taking into consideration the macrophage-evading properties of PEG-modified liposomes, and the fact that the lymphoid tissue contains considerable amounts of macrophages which serves as scavengers for filtered particulates from the local lymph.[44] Thus, the lesser amount of the contrast material provides the greater signal enhancement because of higher relaxivity of the preparation caused by the presence of grafted PEG on the liposome surface.

B. POLYCHELATING POLYMERS ON LIPOSOME SURFACE

Another possible utilization of liposome modification with polymers for improvement of paramagnetic liposome relaxivity is connected with polymer use for enhanced loading of liposomes with paramagnetic metal ions. As it was already mentioned, liposomes loaded with chelated paramagnetic ions have been demonstrated to be useful as MRI contrast agents for visualization of macrophage-rich tissues like liver and spleen.[45] Among different paramagnetic ion-containing compounds, Gd-DTPA complexes were the first to be incorporated inside the liposomal aqueous compartment.[38,39] However, a low-molecular weight water-soluble paramagnetic probe might leak out from liposomes upon the contact with body fluids, which destabilize most liposomal membranes.[1] Moreover, it has been shown that when high concentrations of Gd-DTPA are encapsulated inside liposomes for better potency, the relaxivity is no longer increasing with the increase in Gd concentration. One of the possible explanations of this phenomenon is decreased residence time of water molecules in liposome interior.[45] The next step in the development of usable Gd-based liposomal contrast media was the creation of membranotropic chelating agents such as DTPA-SA,[35] DTPA-PE,[40] and amphiphilic acylated paramagnetic complexes of Mn and Gd.[46] The increased relaxivity of these macromolecular or particle-bound chelated Gd as compared with low-molecular weight Gd chelates is believed to be connected with the increased rotational correlation time of paramagnetic moieties.[45] All paramagnetic probes described to date have contained one chelated metal atom per molecule. The main idea behind the further improvement of liposome paramagnetic properties is to increase the number of chelated paramagnetic atoms attached to a single lipid anchor. Upon incorporation into the liposome membrane, such probes can sharply increase the load of the membrane-bound Gd ions per vesicle. For this purpose, a novel Gd-containing amphiphilic polychelator, N,α-polylysyl-DTPA-*N*-glutaryl-phosphatidyl ethanolamine (Gd-poly-NGPE), was synthesized as a further development of membrane-associated contrast agents for MRI.[47]

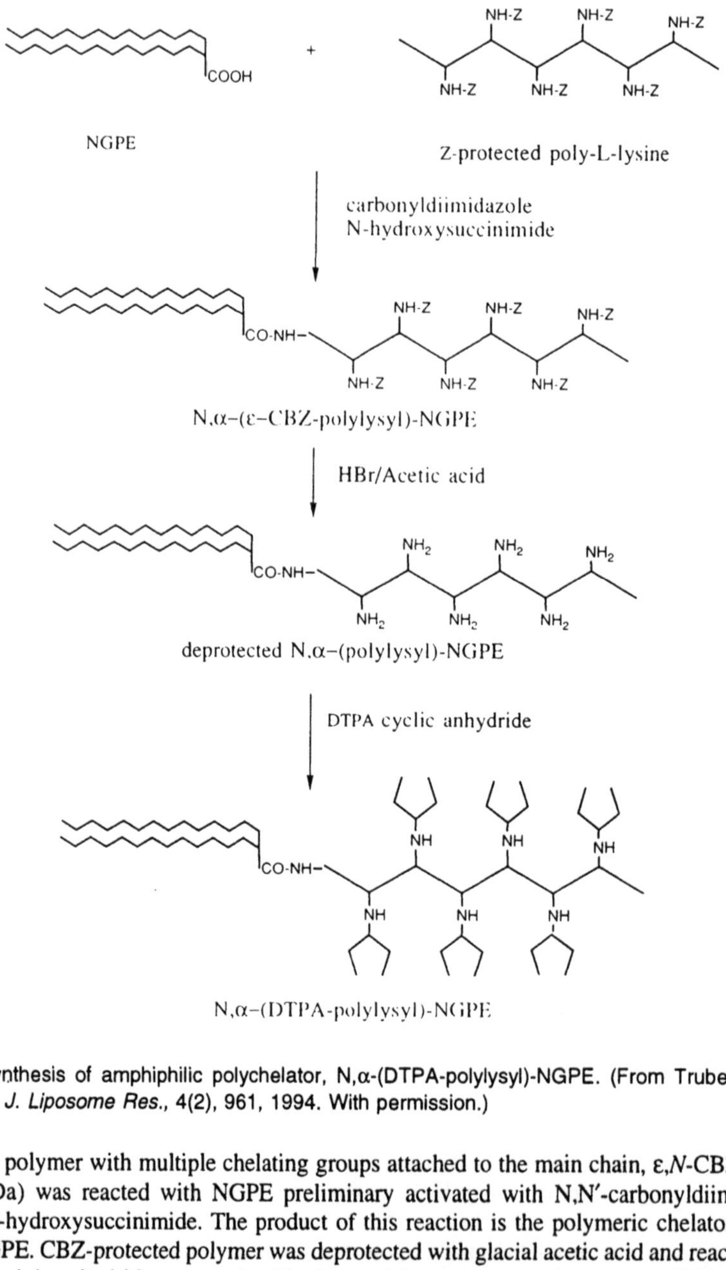

Figure 18 Synthesis of amphiphilic polychelator, N,α-(DTPA-polylysyl)-NGPE. (From Trubetskoy, V.S. and Torchilin, V.P., *J. Liposome Res.*, 4(2), 961, 1994. With permission.)

To prepare polymer with multiple chelating groups attached to the main chain, ε,*N*-CBZ-poly-L-lysine (MW 3,000 Da) was reacted with NGPE preliminary activated with N,N′-carbonyldiimidazole in the presence of *N*-hydroxysuccinimide. The product of this reaction is the polymeric chelator, *N*,α-(ε-CBZ-polylysyl)-NGPE. CBZ-protected polymer was deprotected with glacial acetic acid and reacted with DTPA anhydride. Succinic anhydride was used to block remaining polymer amino groups. Gd-loaded polymeric chelate was prepared by interaction of GdCl$_3$ · 6H$_2$O and DTPA-polylysyl-NGPE in dry pyridine.[41]

Gd-poly-NGPE-containing liposomes were prepared by freeze-drying method and extruded consequently through the polycarbonate filters with 0.6, 0.4, and 0.2 μm pore size. The mean diameter of liposomes obtained was 215 nm (quasi elastic light scattering). DTPA-SA and DTPA-PE were synthesized as in References 35 and 40, respectively, and loaded with Gd as described for Gd-poly-NGPE. Gd-DTPA-SA and Gd-DTPA-PE liposomes were prepared using the same freeze-dry method as for Gd-poly-NGPE liposomes. *In vitro* relaxation parameters of all preparations were measured using 5 MHz RADX NMR Proton Spin Analyzer in HBS. Gd determinations were performed on a commercial basis by Galbraith Laboratories, Inc. (Knoxville, TN).

Although there still is a great deal of discussion on what mode of paramagnetic ion association with the liposome is preferred (membrane-bound vs. entrapped),[45] the recent efforts in this area indicate that

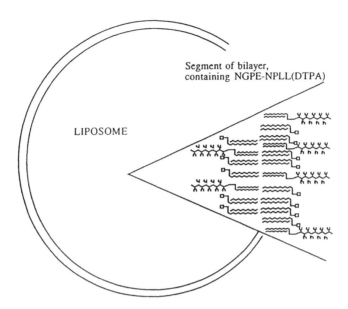

Figure 19 Schematic structure of liposome with membrane incorporated N,α-(DTPA-polylysyl)-NGPE. (From Trubetskoy, V.S. and Torchilin, V.P., *J. Liposome Res.*, 4(2), 961, 1994. With permission.)

the membranotropic chelator-paramagnetic metal complexes are superior to the entrapped ones due to enhanced relaxivity.[46,48] So, the general idea to increase the number of chelated Gd atoms attached to a single lipid anchor, looks quite natural as a remedy to reduce the amount of expensive lipids required for preparation of the diagnostic liposomes.

The use of polylysine *N*-terminus modification chemistry (Figure 18) originates from our previous works where a similar technique was employed for the design of the chelating polymer-antibody conjugates.[49,50] Amphiphilic polymeric modifiers of the type shown, where a hydrophilic polymer is tail-to-head bound to a lipid anchor are known and widely used in liposome research, for example, PEG-PE[12] or neoglycolipids.[51] Upon the incorporation into the bilayer, the NGPE anchor capped with a chelating polymer should form a "coat" of chelated metal atoms on both sides of the liposomal membrane (Figure 19). These metal atoms are directly exposed to the both interior and exterior water environment, which might enhance the relaxivity of the paramagnetic ions.

Gd content determination in Gd-poly-NGPE demonstrated that the latter contains approximately 40% (w/w) Gd, which corresponds to 8–10 metal atoms per single lipid-modified polymer molecule assuming its molecular weight to be 3,500–4,000 Da. Since we have used poly-ε-CBZ-L-lysine with polymerization degree 11 for the synthesis of polychelator, one can theoretically introduce up to 11 metal atoms into it, if all the polylysine ε-amino groups are modified with DTPA residues. This is evidently superior to one metal atom per one lipid "tail" for previously used Gd-DTPA-PE and Gd-DTPA-SA probes. The higher Gd content leads to the better relaxivity parameters. To prove this we have performed proton relaxivity measurements for the different liposomal preparations each containing 3% mol of the individual amphiphilic Gd-containing probe (Figure 20). The results have demonstrated that polychelator-containing liposomes have higher relaxation influence on water protons than conventional liposomal preparations at the same phospholipid content.

To investigate the dependence of probe membrane density on the preparation relaxivity, the inverse T_1 response on amphiphilic chelator content has been studied. Gd-DTPA-PE was found to have an optimum relaxivity at approximately 15 mol % for egg lecithin/cholesterol liposomes. This finding is consistent with the results of Grant et al.[40] who have found that liposomes with 12.5% mol of Gd-DTPA-PE demonstrate the maximal relaxivity. These authors have explained the phenomenon observed by the closeness of Gd atoms to one another at elevated Gd-DTPA-PE concentrations. It is interesting to note that Gd-poly-NGPE liposomes do not possess a relaxivity maximum at least within the concentration range studied (0–20 mol %), suggesting an increase in inter-metal atom distances on the liposome membrane.

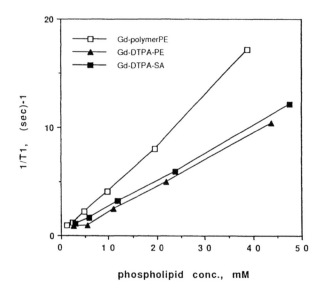

Figure 20 Molecular relaxivities (T₁) of Gd-poly-NGPE-, Gd-DTPA-PE- and Gd-DTPA-SA-containing liposomes (RADX Proton Spin Analyzer, 5 MHz, room temperature, 3 mol % of Gd amphiphilic chelator). (From Trubetskoy, V.S. and Torchilin, V.P., *J. Liposome Res.*, 4(2), 961, 1994. With permission.)

To investigate the efficiency of Gd-poly-NGPE liposomes as MR contrast agents we have performed lymph node visualization in rabbit using subcutaneous administration of 20 mg of egg lecithin/cholesterol (75:25) liposomes containing 5% mol of Gd-poly-NGPE in 0.5 ml of HBS. MR images were acquired for 2 h with 10 min intervals using 1.5 Tesla GE Signa MRI instrument. The first image was taken 5 min after administration. Along with the preinjection image, transverse slice MR scans were acquired using fat suppression mode and T₁ weighted pulse sequence. The rabbit transverse scan demonstrates that axillary and subscapular lymph nodes can be seen on the scan taken only 5 min post-injection; such fast response would be valuable for MR lymph node visualization in clinical circumstances. The time between contrast agent administration and image acquisition for any other lymphotropic imaging agent is usually in a range of several hours.

The results presented in this section clearly show that surface grafting with polymers can strongly influence properties of paramagnetic liposomes. The enhanced MR signal can be obtained by two possible approaches: (1) the increase in the quantity of liposome-bound label via liposome grafting with polymeric chelate, and (2) the enhancement of the label signal via changes in label microsurroundings as a result of liposome grafting with hydrophilic polymer. The combination of both approaches is also possible.

IV. CONCLUSION

Liposome modification with various polymers is a powerful tool for changing liposome properties in the desirable direction. Polymers can limit the accessibility of the liposome surface for other macromolecules (such as opsonins) and change physicochemical parameters of other liposome-bound substances, and thus liposomes themselves. Taking into account the well developed chemistry of phospholipid headgroup modification with functional moieties,[52] one can expect the appearance of new generation of liposomes with specific properties and biological behavior that will be determined by surface-grafted polymers.

REFERENCES

1. **Senior, J.H.**, Fate and behavior of liposomes in vivo: a review of controlling factors, *CRC Crit. Rev. Ther. Drug Carrier Syst.*, 3, 123, 1987.
2. **Gregoriadis, G.**, Targeting of drugs, *Nature (London)*, 265, 407, 1977.
3. **Torchilin, V.P.**, Immobilization of specific proteins on liposome surface: systems for drug targeting, in *Liposome Technology*, Gregoriadis, G., Ed., 1st ed, CRC Press, Boca Raton, FL, 1984, v. 3, 75.

4. **Torchilin, V.P.,** Liposomes as targetable drug carriers, *CRC Crit. Rev. Ther. Drug Carrier Syst.*, 2, 65, 1985.

5. **Seki, K. and Tirrell, D.A.,** pH-Dependent complexation of poly(acrylic acid) derivatives with phospholipid vesicle membranes, *Macromolecules*, 17, 1692, 1984.

6. **Tirrel, D.A., Takigawa, D.Y., and Seki, K.,** pH-Sensitization of phospholipid vesicles via complexation with synthetic poly(carboxylic acids), *Ann. N.Y. Acad. Sci.*, 446, 237, 1985.

7. **Petrukhina, O.O., Ivanov, N.N., Feldstein, M.M., Vasil'ev, A.E., Plate, N.A., and Torchilin, V.P.,** The regulation of liposome permeability by polyelectrolyte, *J. Contr. Rel.*, 3, 137, 1986.

8. **Tirrel, D.,** Macromolecular switches for bilayer membranes, *J. Contr. Rel.*, 6, 15, 1987.

9. **Allen, T.M. and Chonn, A.,** Large unilamellar liposomes with low uptake into the reticuloendothelial system, *FEBS Lett.*, 223, 42, 1987.

10. **Blume, G. and Cevc, G.,** Liposomes for the sustained drug release in vivo, *Biochim. Biophys. Acta*, 1029, 91, 1990.

11. **Papahadjopoulos, D., Allen, T.M., Gabizon, A., Mayhew, E., Huang, S.K., Lee, K.D., Woodle, M.C., Lasic, D.D., Redemann, C., and Martin, F.J.,** Sterically stabilized liposomes — improvements in pharmacokinetics and antitumor therapeutic efficacy, *Proc. Natl. Acad. Sci. U.S.A.*, 88, 11460, 1991.

12. **Klibanov, A.L., Maruyama, K., Torchilin, V.P., and Huang, L.,** Amphipathic polyethyleneglycols effectively prolong the circulation time of liposomes, *FEBS Lett.*, 268, 235, 1990.

13. **Mori, A., Klibanov, A.L., Torchilin, V.P., and Huang, L.,** Influence of steric barrier activity of amphipathic poly(ethyleneglycol) and ganglioside GM_1 on the circulation time of liposomes and on the target binding of immunoliposomes in vivo, *FEBS Lett.*, 284, 263, 1991.

14. **Woodle, M.C. and Lasic, D.D.,** Sterically stabilized liposomes, *Biochim. Biophys. Acta*, 1113, 171, 1992.

15. **Torchilin, V.P., Klibanov, A.L., Huang, L., O'Donnell, S., Nossiff, N.D., and Khaw, B.A.,** Targeted accumulation of polyethylene glycol-coated immunoliposomes in infarcted rabbit myocardium, *FASEB J.*, 6, 2716, 1992.

16. **Needham, D., McIntosh, T.J., and Lasic, D.D.,** Repulsive interactions and mechanical stability of polymer-grafted lipid membranes, *Biochim. Biophys. Acta*, 1108, 40, 1992.

17. **Gabizon, A. and Papahadjopoulos, D.,** The role of surface charge and hydrophilic groups on liposome clearance in vivo, *Biochim. Biophys. Acta*, 1103, 94, 1992.

18. **Lasic, D.D., Martin, F.G., Gabizon, A., Huang, S.K., and Papahadjopoulos, D.,** Sterically stabilized liposomes — a hypothesis on the molecular origin of the extended circulation times, *Biochim. Biophys. Acta*, 1070, 187, 1991.

19. **Torchilin, V.P., Trubetskoy, V.S., Papisov, M.I., Bogdanov, A.A., Omelyanenko, V.G., Narula, J., and Khaw, B.A.,** Polymer-coated immunoliposomes for delivery of pharmaceuticals: targeting and biological stability, in *Proc. 20th Int. Symp. Controlled Release of Bioactive Materials*, The Controlled Release Society, Washington, D.C., 1993, 194.

20. **Torchilin, V.P. and Papisov, M.I.,** Why do polyethylene glycol-coated liposomes circulate so long?, *J. Liposome Res.*, 4, 725, 1994.

21. **des Cloizeaux, J. and Jannink, G.,** *Polymers in Solution. Their Modelling and Structure*, Clarendon Press, Oxford, 1990, pp. 63, 280, 539.

22. **Milner, S.T.,** Polymer brushes, *Science*, 251, 905, 1991.

23. **Blume, G. and Cevc, G.,** Bilayer modification and longevity of liposomes, in *Abstr. Conf. 2nd Liposome Research Days*, Leiden University, The Netherlands, 1992, SC19.

24. **Kurata, M. and Tsunashima, Y.,** Viscosity — molecular weight relationships and unperturbed dimensions of linear chain molecules, in *Polymer Handbook*, Brandup, J. and Himmelgut, E.H., Eds., Wiley, New York, 1989, VII/1.

25. **Huang, C. and Mason, J.T.,** Geometric packing constraints in egg phosphatidylcholine vesicles, *Proc. Natl. Acad. Sci. U.S.A.*, 75, 308, 1978.

26. **Enoch, H.G. and Strittmatter, P.,** Formation and properties of 1000-A-diameter, single-bilayer phospholipid vesicles, *Proc. Natl. Acad. Sci. U.S.A.*, 76, 145, 1979.

27. **Allen, T.M. and Hansen, C.,** Pharmacokinetics of stealth versus conventional liposomes: effect of dose, *Biochim. Biophys. Acta*, 1068, 133, 1991.

28. **Torchilin, V.P., Omelyanenko, V.G., Bogdanov, A.A., Jr., Papisov, M.I., Trubetskoy, V.S., Herron, J.N., and Gentry, C.A.,** On the molecular mechanism of liposc ne surface protection by polyethylene glycol, *Biochim. Biophys. Acta*, 1195, 11, 1994.

29. **Kranz, D. and Voss, E.W.,** Partial elucidation of an anti-hapten repertoire in BALB/c mice: comparative characterization of several monoclonal anti-fluorescyl antibodies, *Mol. Immunol.*, 18, 889, 1981.

30. **Herron, J.N.,** in *Fluorescein Hapten: An Immunological Probe*, Voss, E.W., Jr., Ed., CRC Press, Boca Raton, FL, 1984, 49.

31. **Weder, H.G. and Zumbuehl, O.,** The preparation of variably sized homogeneous liposomes for laboratory, clinical and industrial use by controlled detergent dialysis, in *Liposome Technology*, 1st ed., Vol. I, Gregoriadis, G., Ed., CRC Press, Boca Raton, FL, 1984, 79.

32. **Wood, C. and Kabat, E.,** Immunochemical studies of conjugates of isomaltosyl oligosaccharides to lipid, *J. Exp. Med.*, 154, 432, 1981.

33. **Omelyanenko, V.G., Jiscot, W., and Herron, J.N.,** Role of electrostatic interactions in the binding of fluorescein by anti-fluorescein antibody 4-4-20, *Biochemistry*, 32, 10423, 1993.

34. **Torchilin, V.P., Shtilman, M.I., Trubetskoy, V.S., Whiteman, K., and Milstein, A.M.,** Amphiphilic vinyl polymers effectively prolongs liposome circulation time in vivo, *Biochim. Biophys. Acta,* 1195, 181, 1994.

35. **Kabalka, G.W., Davis, M.A., Moss, T.H., Buonocore, E., Hubner, K., Holmberg, E., Maruyama, K., and Huang, L.,** Gadolinium-labeled liposomes containing various amphiphilic Gd-DTPA derivatives: targeted MRI contrast enhancement agents for the liver, *Magn. Res. Med.,* 19, 406, 1991.

36. **Croll, M.N., Brady, L.W., and Dadparvar, S.,** Implications of lymphoscintigraphy in oncologic practice: principles and differences vis-a-vis other imaging modalities, *Sem. Nucl. Med.,* 13, 4, 1983.

37. **Hirano, K. and Hunt, C.A.,** Lymphatic transport of liposome-encapsulated agents: effects of liposome size following intraperitoneal administration, *J. Pharm. Sci.,* 74, 915, 1985.

38. **Kabalka, G.W., Buonocore, E., Hubner, K., Moss, T., Norley, N., and Huang, L.,** Gadolinium-labeled liposomes: targeted MR contrast agents for the liver and spleen containing paramagnetic amphipatic agents, *Radiology,* 163, 255, 1987.

39. **Tilcock, C., Unger, E., Cullis, P., and MacDougall, P.,** Liposomal Gd-DTPA: preparation and characterization of relaxivity, *Radiology,* 177, 77, 1989.

40. **Grant, C.W.M., Karlik, S., and Florio, E.,** A liposomal MRI contrast agent: phosphatidylethanolamine-DTPA, *Magn. Res. Med.,* 11, 236, 1989.

41. **Kabalka, G.W., Buonocore, E., Hubner, K., Davis, M., and Huang, L.,** Gadolinium-labeled liposomes containing amphipatic agents: targeted MRI contrast agents for the liver, *Magn. Res. Med.,* 8, 89, 1988.

42. **Trubetskoy, V.S., Cannillo, J.A., Milstein, A.M., Wolf, G.L., and Torchilin, V.P.,** Controlled delivery of Gd-containing liposomes to lymph nodes: surface modification may enhance MRI contrast properties, *Magn. Res. Imaging,* 13, 31, 1995.

43. **Lauffer, R.B.,** Magnetic resonance contrast media: principles and progress, *Magn. Res. Q.,* 6, 65, 1990.

44. **Takakura, Y., Hashida, M., and Sezaki, H.,** Lymphatic transport after parenteral drug administration, in *Lymphatic Transport of Drugs,* Charman, W.N., and Stella, V.J., Eds., CRC Press, Boca Raton, FL, 1992, 256.

45. **Tilcock, C.,** Liposomal paramagnetic magnetic resonance contrast agents, in *Liposome Technology,* 2nd ed., Vol. II, Gregoriadis, G., Ed., CRC Press, Boca Raton, FL, 1993, 65.

46. **Unger, E., Shen, D.K., and Fritz, T.,** Liposomes as MR contrast agents: pros and cons, *Magn. Res. Med.,* 22, 304, 1991.

47. **Trubetskoy, V.S. and Torchilin, V.P.,** New approaches in the chemical design of Gd-containing liposomes for use in magnetic resonance imaging of lymph nodes, *J. Liposome Res.,* 4, 961, 1994.

48. **Tilcock, C., Ahkong, Q.F., Koenig, S.H., Brown, R.D., III., Davis, M., and Kabalka, G.,** The design of liposomal paramagnetic MR agents: effect of vesicle size upon the relaxivity of surface-incorporated lipophilic chelates, *Magn. Res. Med.,* 27, 44, 1992.

49. **Slinkin, M.A., Klibanov, A.L., and Torchilin, V.P.,** Terminal modified polylysine-based chelating polymers: highly efficient coupling to antibody with minimal loss of immunoreactivity, *Bioconjugate Chem.,* 2, 342, 1991.

50. **Trubetskoy, V.S., Torchilin, V.P., Kennel, S.J., and Huang, L.,** Use of N-terminal modified poly(L-lysine) — antibody conjugate as a carrier for targeted gene delivery in mouse lung endothelial cells, *Bioconjugate Chem.,* 3, 323, 1992.

51. **Stoll, M.S., Mizuochi, T., Childs, R.A., and Feizi, T.,** Improved procedure for the construction of neoglycolipids having antigenic and lectin-binding activities, from reducing oligosaccharides, *Biochem. J.,* 256, 661, 1988.

52. **Torchilin, V.P. and Klibanov, A.L.,** Coupling and labeling of phospholipids, in *Phospholipids Handbook,* Cevc, G., Ed., Marcel Dekker, New York, 1993, 293.

Chapter 14

Liposomes in Electric Fields

Mathias Winterhalter

CONTENTS

I. INTRODUCTION

Natural cells are exceedingly complex systems. In order to gain an understanding of their basic mechanism, it is necessary to simplify them to a model that can be characterized by a minimal number of variables. For example, studies on pure lipid vesicles demonstrate the influence of the lipid matrix alone, while reconstitution of proteins into lipid membranes simulates other perturbations and simplifies conclusions on their function.

Electric fields play a major role in nature.[1] Throughout this chapter we distinguish two classes of electric fields: intrinsic or externally applied. A primary function of intrinsic electric fields in natural cells is that of forming a barrier and so controlling passive transport processes.[2-4] An example of a secondary effect of intrinsic fields is their influence on shape transitions. A striking feature of, e.g., red blood cells, is their softness. A change in the intrinsic electric fields by, e.g., a pH variation, is able to deliver enough energy to trigger shape transitions.[5-10] In artificial model membranes, intrinsic electric fields influence the stability and the size distribution of liposomes.[11-15] Externally applied electric fields constitute a unique tool for characterization and manipulation of liposomes or cells. Important biotechnological applications are, e.g., electrofusion of cells and electroinjection of macromolecules into living cells.[16-17] More recently, pulsed electric fields have been used to treat skin cancer.[18] Electric fields enhance the drug delivery pathway through the skin, a method known as iontophoresis.[18-19] Although the above mentioned applications are in widespread use, little is known about the underlying processes.

This chapter gives a brief introduction into electric field effects in and on lipid membranes. In the first part of the following section we describe intrinsic electric fields and their influence on the physical properties of liposomes. In the second part, we outline the basic features of external electric fields in lipid systems. The third section is devoted to application of external electric fields to liposomes.

II. LIPOSOMES AND ELECTRIC FIELDS

In this section we introduce basic principles of liposomes in electric fields. We distinguish two categories of electric fields: the first are the intrinsic ones. They originate from the self organization of lipid molecules in aqueous solution. In subsection B, we account for effects of externally applied electric fields on membranes.

A. INTRINSIC ELECTRIC FIELDS IN LIPID MEMBRANES

Intrinsic electric fields in membranes have several different sources. The best understood of these are the electric fields, due to charges located in the head group region.[20-22] Their origin is due to dissociated acidic or basic groups in the lipid head groups, or else to condensation of ions to the lipid/water surface. The Gouy-Chapman theory provides a powerful model to describe the distribution of the ions perpendicular to a charged surface. Within this model, ions are considered as point charges that interact only with an averaged potential stemming from the presence of all other ions. The distribution of the ions in the solution is determined by the balance of two forces. Without thermal energy, the electrostatic attraction of opposite charges would cause the counterions to be present in an infinitely thin layer next to the charged interface. However, counterions free to move in solution each possess a thermal energy of 3/2 kT. This would cause a high osmotic pressure of counterions in this infinite thin layer, and so force them to be spread out into the aqueous volume. The Poisson equation relates the electrostatic potential $\varphi(r)$ with the charge distribution $\rho(r)$ of the ions.[20-24]

$$\Delta\varphi(r) = -\frac{\rho(r)}{\varepsilon_w\varepsilon_0}, \qquad \rho(r) = \sum_{i=+,-} ez^i n_0^i \exp\left(\frac{z^i e\varphi(r)}{kT}\right) \tag{1}$$

where $\varepsilon_w = 80$ is the relative permittivity of the aqueous phase, and $\varepsilon_0 = 8.8\ 10^{-12}$ As/Vm is the absolute permittivity of vaccuum. On the RHS z^i is the number of charges per ith ion in units of $e = 1.602\ 10^{-19}$ As and n_0^i the ith ion concentration in the bulk far from the interface. In the denominator of the exponential, $k = 1.381\ 10^{-23}$ J/K is the Boltzmann constant and T the temperature (degrees Kelvin). The effective charge distribution $\rho(r)$ on the right hand side sums the contributions of the individual counterions. The probability of finding an ion of type i at a distance r is assumed to be Boltzmann-distributed with the electrostatic energy $z^i\varphi(r)$ of this ion at r as the Boltzmann factor. In the following we reduce ourselves to the simplified case of a 1:1 electrolyte. A more detailed look into the underlying theory allows estimation of the typical distance from the charged interface required to observe charge distribution similar to that in the bulk phase. This so-called Debye length $\lambda_D = \sqrt{\varepsilon_w\varepsilon_0 kT / 2n_0 e^2}$ becomes smaller as the the bulk ion concentration n_0 increases. Insertion of the ion distribution ρ from Equation 1 into the Poisson equation yields the Poisson-Boltzmann equation, describing in a self-consistent way the spatial distribution of the electrostatic potential due to the presence of a charged interface in an electrolyte. In general, there is no solution in closed form and one has to solve numerically. In the specific case of a symmetrical electrolyte ($z^+ = z^-$), a solution for planar geometry can be deduced.[20-22] If in addition the surface potential at the charged interface is known and the radius of curvature R is larger than the Debye-Hückel-screening length, the spatial variation of the electrostatic potential for cylindrical and spherical surfaces can be given in powers of λ_D/R.[25-26]

Considerable simplification is possible when the potential drop is small, which has the consequence that the electrostatic energy per ion is less than kT (about 25 mV at room temperature).[20-23] In this case, the ion distribution can be developed as a power series which allows linearization of the Poisson-Boltzmann equation. This limit is better known as the Debye-Hückel approximation. In case of a 1:1 electrolyte Equation 1 can be transformed into:

$$\Delta\varphi = -\frac{e}{\varepsilon_w\varepsilon_0}\left[n_+ \exp\left(\frac{e\varphi}{kT}\right) - n_- \exp\left(\frac{-e\varphi}{kT}\right)\right] = \frac{e}{\varepsilon_w\varepsilon_0} n_0\left(1 + \frac{e\varphi}{kT} + \dots - 1 + \frac{e\varphi}{kT}\dots\right) = \kappa^2\varphi \tag{2}$$

Here, $\kappa = 1/\lambda_D$ is the inverse of the Debye-Hückel screening length. The linearized Poisson-Boltzmann equation can be solved explicitly in a few geometries: for planar geometries the solution for the electrostatic potential is[20,23]

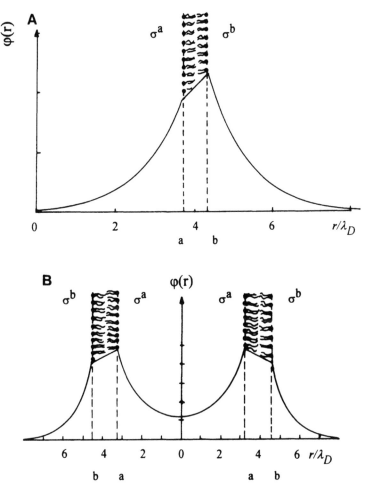

Figure 1 (A) Spatial variation of the electrostatic potential near a uniformly charged planar bilayer. (B) Spatial variation of the electrostatic potential near a uniformly charged vesicle. Note that the electrostatic potential does not decay to zero at $r = 0$. For details see text.

$$\varphi^{(in)}(r) = \frac{\sigma^a}{\varepsilon_w \varepsilon_0} \frac{\exp(-\kappa a)}{\kappa} \exp(\kappa r), \quad \varphi^{(out)}(r) = \frac{\sigma^b}{\varepsilon_w \varepsilon_0} \frac{\exp(\kappa b)}{\kappa} \exp(-\kappa r) \tag{3}$$

for the inside of a planar interface at $r \le a$ and for the outside at $r \ge b$. Figure 1A shows the electrostatic potential $\varphi(r)$ as a function of distance from the interface. If the inside potential $\varphi^{(in)}(a)$ differs from the outer one $\varphi^{(out)}(b)$, the inside of the lipid membrane is not field free. This field is determined by the difference in surface charge densities, thickness of the membrane d and couples both charged interfaces. For spherical geometries (see Figure 1B), one obtains for the solution inside

$$\varphi^{(in)}(r) = \frac{\sigma^a}{\varepsilon_w \varepsilon_0} \frac{a^2}{(\kappa a - 1)ch(\kappa a)} \frac{sh(\kappa r)}{r} \quad \text{for } r < a \tag{4}$$

and outside

$$\varphi^{(out)}(r) = \frac{\sigma^b}{\varepsilon_w \varepsilon_0} \frac{b^2 \exp(\kappa b)}{(1 + \kappa b)} \frac{\exp(-\kappa r)}{r} \quad \text{for } r > b \tag{5}$$

Equation 4 makes the prediction that the electrostatic potential in the middle of a spherical liposome is not zero but has the finite value (develop $sh(\kappa r) \sim \kappa r + ...$ for small κr):

$$\varphi^{(in)}(0) = \frac{\sigma^a}{\varepsilon_w \varepsilon_0} \frac{a^2 \kappa}{(\kappa a - 1) ch(\kappa a)} \qquad (6)$$

According to the RHS of Equation 1, the effective charge distribution is given by a Boltzmann distribution with the electrostatic potential in the exponent. The overlaping electrostatic potential causes an excess of counterions and thereby an additional osmotic stress. In the particular case of a vesicle this increase in osmotic pressure due to the excess counterions at the origin (assuming the Debye-Hückel approximation to hold) is

$$\Pi_{osm}^{(in)}(0) = kT\left(n_+(0) + n_-(0)\right) = kT2n_0\left[1 + \frac{1}{2}\left(\frac{e\varphi^{(in)^2}(r)}{kT}\right)^2\right] = 2n_0 kT + \frac{1}{2}\kappa^2 \varepsilon_w \varepsilon_0 \varphi^{(in)^2}(0) \qquad (7)$$

In the presence of an electric field, the total pressure has to include a contribution stemming from it. Electric fields give rise to the so-called Maxwell stresses which are quadratic in the field strength.[20,21,24] Calculation of the electric field at the origin using Equation 4 gives $E(0) = -\nabla\varphi(0) = 0$ and yields a vanishing contribution to Equation 7. This osmotic pressure from the excess of counterions inside the vesicle tries to increase the volume by causing a dilatation of the liposome area. The relation between the excess osmotic pressure and the lateral tension T in the vesicle membrane is given by the Laplace equation

$$\Delta\Pi_{osm} = \frac{1}{2}\kappa^2 \varepsilon_w \varepsilon_0 \varphi^{(in)^2}(0) = 2\frac{T}{R} \qquad (8)$$

It has been shown that application of tension exceeding a certain critical value causes lysis.[27] Inserting Equations 6 and 7 into 8 shows that this additional osmotic stress has an exponential onset below a critical radius of curvature. Such small radii yields in large mechanical lateral tensions and to lysis. It may be possible that this mechanism prevents the existence of smaller sizes of liposomes — unfortunately, no detailed investigation has yet been made. However, recent data on the phase diagrams of vesicles using charged surfactants suggests such an interpretation.[11-13]

In this context, we should note that Equation 7 demonstrates that the inside of this specific liposome contains excess free charges. However these charges are balanced by the opposite fixed surface charges. The entire inner volume is strictly charge-neutral, as can be readily calculated by integration over the total charge distribution:[20-23]

$$\int_0^a \rho(r)rdr = -\sigma^a \qquad (9)$$

We now discuss limitations of the above models. Surprisingly, the Debye-Hückel-approximation is valid even when the charge density becomes so low that the mean average distance between two surface charges becomes larger than one Debye length. In a theoretical model, single ions were placed on a square lattice,[28] and the counterion distribution was calculated according to the Poisson-Boltzmann theory. It was shown that the difference between the "smeared-out" and the discrete charge models vanishes at a distance beyond one Debye length. If the discrete charges are located at the interface this leads to higher surface potentials than in the Poisson-Boltzmann model. This increase is less pronounced if the surface charges are buried a few angstroms inside the lipid part.

Experimental verification of the counterion distribution opposed to a charged membrane showed more discrepancy to the discrete charge model and a much better agreement with the Poisson-Boltzmann model based on smeared surface charges.[20-23] A possible reason for this astonishing feature could be the rapid local movement of the individual lipid molecules, which gives rise to a smeared-out effective charge density on the static time scale. Poor agreement with the Poisson-Boltzmann theory, however, was found if the charges are placed out of the plane and the electrolyte was able to screen the charges from the backside.[3,29]

Large surface charges, that give surface potentials higher than 25 mV, exceed the opposite limit of the Debye-Hückel approximation, so that the nonlinear Poisson-Boltzmann equation has to be solved. In this case, no general, closed-form solution is known and numerical solutions are required (for a more detailed discussion, see, e.g., Reference 21). The potential drop is much steeper than the exponential one observed in the Debye-Hückel limit, and in this region the agreement with experimental observation becomes less.[3,21] At higher potentials the theoretical ion distribution exceeds the exponential decay causing very high concentrations of counter ions at short distances from the charged surface making the assumptions of the Gouy-Chapman theory critical. In a more realistic model, additional contributions have to be added. Such contributions are, e.g., a hard sphere interaction, which accounts for the deviation from point-like behavior. Furthermore, the approximation that individual ions interact only with the mean field becomes questionable, and direct ion-ion interactions have to be included in the model. A different effect is caused by a decrease in the dielectric constant of the water on approaching the lipid membrane. Fortunately, most of these corrections tend to cancel each other out, and the use of the Poisson-Boltzmann equation usually gives a reasonable estimate. Much progress has been made in the development of more elaborate theories. However, the calculations are rather complex and we refer the reader to the original literature.[30,31]

On the other hand, even within the validity of the Debye-Hückel approximation, many additional complications occur.[3,20] Depending on the boundary conditions during the formation of liposomes the surface charge density might become a function of the surface potential. Several models dealing with a potential-dependent dissociation were presented a few years ago.[32-33]

A different effect occurs at the interface. A deviation from the Poisson-Boltzmann ion distribution eventually occurs due to adsorption effects.[21] If the counterions are hindered in their motional freedom this leads to a reduction in osmotic pressure. This layer where the counterions deviate from the Boltzmann distribution is called the Stern layer. Usually, for simplification, a homogeneous charge distribution is assumed. Recently, several models based on the Donnan equilibrium were proposed.[34-35] The exact potential distribution on a molecular level is not only crucial for transport processes of ions across a membrane binding of macromolecules, but also for the function of receptors. One approach is to idealize the membrane by more and more complex models. Much research is done for "comparatively" well defined solid/electrolyte interfaces. It was found that on a molecular level the electron density in both phases is crucial. In contrast to such solid interfaces, lipid membranes show a rapid local movement, making the interface on a molecular level almost three-dimensional. Having the complexity of the solid interface in mind it is difficult to believe that local theories can be applied successfully to membranes.

Many experiments were done to verify the presented models. In bilayers, the surface potentials at the membrane/water interface are only indirectly accessible. Most of the methods used are based on the spatial variation of the ion concentration.[21,22,36-41] The latter is measured by addition of pH-sensitive labels. Typically, the difference in the partition coefficient of this label to that of the bulk can be recorded either by fluorescence, ESR, or NMR signals and is afterwards related to the electrostatic potential.[21,22,36-41]

Zeta potential measurements constitute a further method to investigate surface potentials.[21,42] The surface charges of a liposomes are hidden behind a cloud of counterions. Including this cloud the liposome is electrostatically neutral: however, an external electric field causes a deformation of the counter-ion cloud. The liposome will tend to migrate in one direction and the counterions in the opposite one. The disadvantage of this method is that the measured surface potential is not that at the lipid/water interface. Due to hydrodynamic friction the plane of reference is some (unknown) distance ζ away from that of the lipid/water interface and the measured potential is that of the phenomenologically determined plane shifted by ζ from the lipid/water interface. Due to the rapid decrease of the potential with distance (see Equation 5 and Figure 1A) the values observed by this method are much smaller than that at the lipid/water surface. An estimate for ζ is obtained using Equation 5 and inserting the measured zeta potential at an unknown distance and the measures surface potential at the lipid/water interface obtained by another method.

The previous section described mainly the electric fields directed from the lipid/water interface towards the aqueous phase. Inside the membrane a different type of electrostatic potential is present, the so-called dipole potentials.[43-53] The exact origins of these potentials are not yet clear. Earlier interpretation associated the dipole potential of a lipid membrane with the macroscopic orientation of various dipoles present in an individual lipid molecule. The measured potential V_{dipole} is related to the total perpendicular dipole moment $\mu_{\perp}$ per lipid molecule:

$$\Delta V_{dipole} = \frac{\mu_\perp}{\varepsilon_w \varepsilon_0 A_{lipid}} \qquad (10)$$

where A_{lipid} as the area per lipid molecule. The Δ in Equation 10 denotes that the measured potentials are relative values referenced to the air/water interface. Experiments on lipid monolayers yield between 300–600 mV with respect to the aqueous phase.[44,47-51] On the other hand, dipole potential measurements in bilayers showed a significantly lower value between 100–300 mV.[46-48,52] To date, there is no model explaining this large discrepancy. One reason might be that in contrast to the monolayer the dipole potential in a bilayer can also decay into the aqueous part of the second lipid/water interface that inhibit the building up of such high surface potentials as found in monolayers. At first, one associates the dipole potential with that present in the zwitterionic head groups. On the contrary, for all common lipids the dipole of the head group is oriented in the opposite way. The interpretation assumes that the total dipole is a sum of different individual dipoles $\mu_\perp = \Sigma \mu_1 + \mu_2 \ldots$ For example, μ_1 represents the head group contribution (opposite to the total potential), μ_2 the contribution from the carbonyl groups, and μ_3 the contribution from the terminal CH_3. The actual understanding attribute the measured dipole potential to all the molecules characterizing the membrane including the interface, reaching from the hydrocarbon interior to the first few water layers.[45]

Several experimental methods are available to test the theoretical predictions.[44-52] A direct measurement of the surface potentials is possible across a lipid monolayer at the air/water interface. In this case two methods are available. The first one uses an air-electrode with an α-emitter to ionize the air above the water sufficiently. The potential is than measured directly with a high impedance voltmeter. A different method is based on the capacity variation of a vibrating plate electrode. Both methods are direct potential measurements and yield roughly the same results:[44-45] both methods also have the disadvantage that the additional potential decay at the electrodes is unknown. This problem is avoided by measuring only relative values, usually the potential with respect to the air/water interface.

It is also possible to measure the inner electrostatic potential indirectly.[47,52] Here, most methods are based on the sensitivity of ion transport across the bilayer to these potentials. A special case of this is the use of lipophilic ions as probes to investigate the electrostatic energy barrier across the membrane. However, this method requires modeling of the barrier which depends crucially on static and dynamic transport characteristic values. The use of foreign molecules as probes, can also be criticized on the basis that the probe molecule itself contributes to the measured barrier.

The actual role of this potential is still unknown. It has been suggested that the dipole potential is related to hydration repulsion between adjacent membranes.[52-54] Although most of the experimental data support such a conclusion, more recently comparative measurements using lipids either with or without the carbonyl dipole contribution show no significant influence on the hydration force. In an indirect way the dipole potentials could also contribute to the bending elasticity.[55]

B. LIPID MEMBRANES IN EXTERNAL ELECTRIC FIELDS

In the previous subsection we briefly discussed the influence of intrinsic electric fields. In the following we will focus on external electric fields which are easily applied using either commercial or homemade electrodes. In this context we recall that application of an electric field across an aqueous solution causes an electrical current and therefore leads to the production of heat. A correct treatment would require use of the formalism of irreversible thermodynamics. Theoretical models based on the minimalization of energy neglect the dissipative flow due to production heat and their predictions have to be considered carefully. A better approach is based on the balance of forces.[56]

External electric fields interact with charged membranes causing either attraction or repulsion and leading eventually to electrophoretic movement. This force is linearly dependent upon the applied field strength, and is widely described in the literature.[57,58] In the following we will restrict ourselves to *a priori* uncharged membranes. In this case, the external field causes polarisation of the material. It is then the induced charges that couple to the external electric field. Obviously, these so-called Maxwell forces are quadratic in the applied electric field strength.[24,56,59,60]

Depending on the geometry of the sample, the external electric field has a specific spatial distribution. In Figure 2 we show three typical samples. The first one contains two chambers separated by a Teflon® wall with a small hole in it. Across this tiny hole of about 1 mm² area, a unilamellar lipid membrane, also

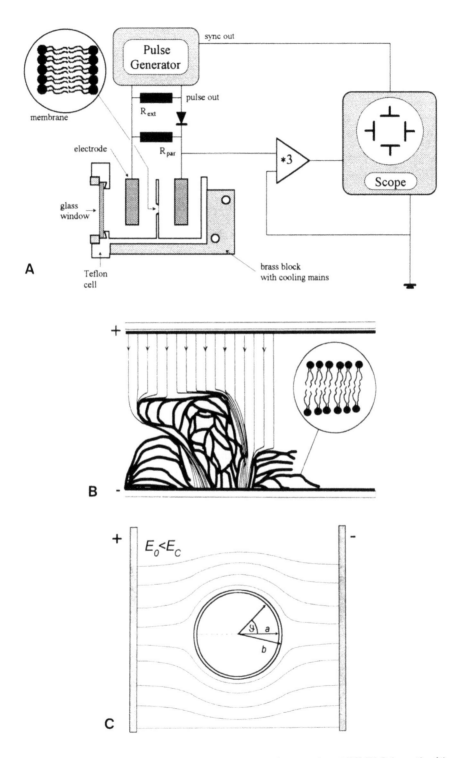

Figure 2 (A) Schematic of a typical bilayer set-up for a charge pulse experiment.[63-65] (B) Schematic of the electric field line distribution for a completely swollen lipid sample as encountered in typical electroswelling samples. Note that the field lines in the low frequency field do not penetrate closed lipid compartments but follow tiny water channels. The smallest channel determines the resistivity of the sample. (C) Schematic of the electric field line distribution around an isolated vesicle in an otherwise homogeneous field.

termed a bilayer, is spanned using the appropriate technique for the desired study.[61,62] The relatively good conductivity of the bulk electrolyte allows the application of a very homogeneous field across the membrane. This set-up provides an appropriate tool to study the electrical properties of large planar unilamellar membranes.[63-65] The conductivity of planar lipid bilayers has been studied for many years. Two different methods or combinations of both are most commonly used: voltage clamp or charge-pulse relaxation. Each has its own advantages and disadvantages. In the first one, externally a constant voltage is applied. This results in large currents under breakdown conditions causing a poorly defined voltage drop at the pore edge. The second method is based on the application of charge-pulses and recording the successive voltage relaxation. This has the advantage of avoiding the large current state. It is much faster and allows investigation within the nano- to millisecond ranges.[63-65] In recent years, bilayer studies became a powerful method to investigate reconstituted membranes.[66,67] A variation of this method is the patch-clamp technique.[68]

Figure 2B shows a bulk sample between, e.g., two metalized glass slides. In such a sample the electric field distribution is rather inhomogeneous and depends also very crucial on the concentration of the lipid material. In a similar device the influence of externally applied field pulses on the orientation of lipid membranes were shown.[69,70] Figure 2C shows a similar set up to the previous one, but interaction with the other liposomes can be neglected. In the following we focus on the latter case of a single free liposome in an otherwise homogeneous electric field as an example of external field effects on membranes.

1. A Single Vesicle in a Homogeneous External Field

The first step in understanding the behavior of a free liposome in an external electric field is to calculate how it perturbs the field. Once the spatial dependency of the field around an uncharged liposome is known the forces stemming from the field are readily calculated.[24,59] For simplicity, we outline the calculation under the assumption that the thickness of a liposome is much larger than the Debye length on the solution. In this case, the counterion broadening due to thermal motion is negligible and the potential drop in the aqueous phase occurs in a negligible small volume and the classical text book electrodynamics can be applied.[24] We further assume that the solution, as well as the lipid membranes, are isotropic and can each be described by a scalar conductivity and static dielectric permittivity. For each homogeneous phase we can say that, due to the absence of free charges on the surface of any small volume element, there are as many field lines directed to the inside as to the outside. In other words, there is no net electric force on this volume element.[24,59] In the absence of free charges, real electric field forces only occur at phase boundaries (e.g., the lipid/water boundary). The detailed calculation of the field lines is lengthy but straightforward.[59] Maxwell's equations have to be fulfilled within the entire chamber: accordingly the electric field strength has to obey two conditions at each phase boundary (see Figure 2C for notation).[57,59,71] The first boundary condition at $r = b$ follows from $curl\ \mathbf{E} = \mathbf{0}$: this causes the electric field lines parallel to the water/lipid interface to be continuous across the interface

$$E_\vartheta^w(b,\vartheta) = E_\vartheta^l(b,\vartheta) \tag{11}$$

where $E_\vartheta^w(b,\vartheta)$ and $E_\vartheta^l(b,\vartheta)$ are the electric field components parallel to the interface in the water (w) and the lipid (l) phases. A similar boundary condition holds for the inner boundary at $r = a$. Due to the possible presence of charges at the interface, and the difference in the relative dielectric permittivities of the lipid $(\varepsilon_l = 2)$ and the aqueous phase $(\varepsilon_w = 80)$, the perpendicular field lines will be discontinuous at the outer water/lipid interface:

$$\varepsilon_w\varepsilon_0 E_r^w(b,\vartheta) - \varepsilon_l\varepsilon_0 E_r^l(b,\vartheta) = \sigma(b,\vartheta) \tag{12}$$

Here, $\sigma(b,\vartheta)$ stands for the surface charge density, $E_r^w(b,\vartheta)$ and $E_r^l(b,\vartheta)$ are the radial field strengths in the aqueous and the lipid media. A similar equation holds at the inner boundary. Because of the conductivity of the aqueous phase, ions start to flow along the potential gradients. In addition to the Maxwell equation, the continuity equation for charge flow:

$$\kappa_w E_r^w(b,\vartheta) = \kappa_l E_r^l(b,\vartheta) \tag{13}$$

must be satisfied at the outer boundary as well as at the inner one. Obviously, the extreme difference, about 10 orders of magnitudes, in membrane conductivity κ_l and water conductivity κ_w does not allow Equations

12 and 13 to be satisfied at the same time for an *a priori* uncharged liposome ($\sigma = 0$). Application of an external field causes accumulation of charges at the membrane/water interface until the boundary conditions are satisfied. For liposomes in aqueous suspension, the time constant of the exponential rise in surface charge density σ (the equilibrium value of which satisfies both equations at the same time) is typically less than 1 μs.[57] As we are interested in macroscopically observable forces, this short time is not observable and we omit further detailed discussion. However, it should be kept in mind that application of AC fields of sufficiently high frequencies reduces the effective potential at the boundary. For simplicity, we assume a spherical geometry of the liposome (see Figure 2C). The solution for the electrostatic potential is obtained via a multipole development in spherical harmonics.[24,59] In the special case of an uncharged vesicle in the absence of an external field, and assuming the entire sample does not contain excess charges, only the first order terms are retained for cylindrical symmetry (represented by a sphere in a plate condenser). The complex ansatz for the electrostatic potential is written:

$$\varphi^{(i)}(r,\vartheta) = \frac{1}{2}\left[\left(a^{(i)}r + \frac{b^{(i)}}{r^2}\right)\cos\vartheta\, e^{-iwt} + CC\right] \tag{14}$$

where CC is the complex conjugate of the left-hand part inside the bracket. The electrostatic potential for each of the three layers (i = inside, lipid or the outer aqueous part) has two complex coefficients $a^{(i)}$ and $b^{(i)}$, which are determined by the four boundary conditions. Two other conditions, namely that the potential has to be finite everywhere, and that it must be consistent with the value imposed by the external electrodes, define the two additional equations required to determine the solution. The calculation of these six coefficients is tedious and we refer to the original literature.[59] Although the individual coefficients are complex quantities, the measurable ones such as the electric potential are again real numbers. Solving the above equations yields for the induced charge density, σ, an angular dependency of the direction of the field, i.e., $\sigma \sim \cos(\vartheta)$, where ϑ is the angle between the electric field and the location on the surface of the vesicle (see Figure 2C). These induced charge densities σ of opposite sign at each water/lipid interface cause the external applied potential to drop across the lipid bilayer.

At this point, we should add a few comments about possible effects due to a finite Debye-length.[72-76] In Section A we demonstrated the potential distribution of a homogeneously charged interface. On the contrary, in the above calculation no surface charges were present *a priori*. The external field induces charges and the thermal motion of these induced charges cause them to be distributed.

2. Electric Field Forces Acting on a Vesicle

Once the electric field distribution around a vesicle is known the forces on it can readily be calculated. The surface force density acting on the interface is obtained by summation of the Maxwell forces over a small volume which includes the interface. Careful inspection of the detailed model yields three different regimes for the electric field-induced forces on vesicles:[59]

- A low frequency regime in which only the ratio of the conductivities of the lipid to the aqueous phases need be taken into account.
- An intermediate regime in which the conductivity of only the aqueous phase, but the dielectric constants of both phases have significant contributions.
- A high frequency regime in which the deformational force of the electric field is determined by the ratio of the dielectric constants of lipid and water. Here, the conductivity of both media can be neglected.

The model presented above considers a lipid vesicle as a single lipid shell within an aqueous medium. In other approaches, the lipid bilayer has been modelled using many shells with different dielectric and conductive properties.[77] However, due to the small volume of each shell, no significant contribution from these forces would be expected. Other models oversimplified the situation of lipid vesicles in an external electric field by assuming an infinitely thin shell.[78]

As schematically indicated in Figure 2C, in the low frequency regime the vesicle interior is shielded by induced charges and is almost field-free. Two major contributions occur. The strongest forces occur due to the fields inside the membrane. The induced charge densities on either side are responsible for a very high screening field inside the membrane of about $E_t^l(b,\vartheta) = (3/2)(b/d)E_0\cos(\vartheta)$ where d is the membrane thickness. At the outer water/membrane boundary this field causes a large Maxwell pressure directed towards the interior of the lipid membrane of approximately:[59]

$$\Pi'_{el}(b,\vartheta) = -\varepsilon_l \cdot \varepsilon_0 \cdot \frac{1}{2}\left[E'_r(b,\vartheta)\right]^2 = -\varepsilon_l \cdot \varepsilon_0 \cdot \frac{9}{8}\left(\frac{b}{d}\right)^2 E_0^2 \cos^2 \vartheta \tag{15}$$

and a similar pressure of vectorially opposite direction at the inner boundary (with the outer vesicle radius b replaced by the inner radius a). The two forces combine to give a pressure perpendicular to the membrane plane, which tends to thin the membrane. A minor force stems from the field lines in the aqueous phase. As the field lines are squeezed out from the vesicle they show a higher density in the external phase at the equator (see Figure 2C) — this gives rise to a pressure on the outside of the vesicle:

$$\Pi^w_{el}(b,\vartheta) = -\left(\frac{9}{8}\right)\cdot \varepsilon_w \cdot \varepsilon_0 \cdot E_0^2 \cdot \sin^2 \vartheta \tag{16}$$

As the aqueous part inside of the vesicle is almost field-free, the contribution from the field on the inside is negligible. The electrostatically induced pressure in Equation 16 has to be balanced by a mechanical pressure. As the inner aqueous volume is regarded as incompressible, a pressure on the equator from the outside cause an elongation of the spherical shape of the vesicle.[59-60,79,80] With increasing field frequency, the thin membrane offers less of a barrier to the applied field, so that application of an external field ceases to cause deformation.[59,60]

3. Stability of a Membrane Under External Electric Fields

Rupture of a membrane can be induced by many techniques. Commonly used ones are osmotic, mechanical (use of shear stress, stretching with micro pipettes), or application of electric field pulses.[27,63-65,81-95] We expect that a similar mechanism underlies all these processes. In the following we discuss the breakdown of lipid membranes induced by electric field in more detail. First, we revisit current static models which permit the stability to be deduced; thereafter we suggest possible extensions with regard to the dynamics of pore formation.

Nothing is known about the early stages of membrane defect formation. Until now, this process could not be resolved by any experimental approach. Earlier investigators suggested an electromechanical compression as the most likely description of the electrical breakdown.[96] In this model it is assumed that the membrane can be regarded as a capacitor filled with a homogeneous elastic dielectric material. The thickness of the dielectric depends on the electric compressive forces caused by the membrane potential, and is counterbalanced by the mechanical elastic restoring forces of the lipids. It was suggested that high electrical fields cause a critical thickness to be reached beyond which electrical breakdown occurs. The critical voltage is dependent upon the elastic compressive modulus perpendicular to the membrane plane, the relative dielectric constant of the membrane interior, and the membrane thickness in the absence of an electric field. However, this model requires that the the membrane be thinned by 30% of the original thickness. Taking into account the incompressibility of the lipid itself, this model would imply a very large area of expansion prior to lysis, which is in contrast to the experimental findings.[27,94] Based on this model, many refinements have been suggested.[97,98] Due to the lack of an observed large area increase at lysis, these models are restricted to the very small range beyond the elastic regime, which has not yet been investigated. In this context we should recall that, when the dielectric surrounding an ion is replaced by a dielectric of lower permittivity, energy must be supplied (the Born energy).[99] This energy hinders the translocation of charged molecules across a lipid membrane. Thinning of the membrane reduces this barrier and enhances the translocation of ions. This effect may favor the formation of the initial defects.

A general model that can describe the mechanical stability of a membrane assumes that a defect has been formed initially.[100] This could be due to thermal fluctuation or to some other perturbation: within such a model the exact process of the initial step does not matter. Once formed, the defect expands reversibly or irreversibly. For large defects (or pores) subject to external electrical fields, this energy contains two opposite contributions

$$E_{pore} = 2\pi \cdot a \cdot \Gamma - \pi \cdot a^2 T \tag{17}$$

The first term represents the energy needed to build up the edge of a pore of radius a where Γ is the line tension. The second term is the elastic energy stored in the defect area: this is due to the surface tension T. Similar models based on statistical mechanics have predicted the energy barrier for pore formation.[101-106] The mechanical lifetime of a membrane was proposed to be a Boltzmann distribution with the energy barrier E_{pore} (given, e.g., by Equation 17) in the exponent.[81,101-111]

Several authors[95,103] have tried to account for electric field effects by adding the condenser energy of a lipid bilayer to Equation 17: they assumed the inside of a lipid pore to be filled with nonconducting water. Under the additional assumption that the electric potential is not altered after a pore is created, the electric field acts simply as a source of additional surface tension.[107-111] However, even for hydrophobic pores, the electric field distribution will be perturbed.[109-111] Moreover, if the extreme difference in conductivity between the membrane and the ionic solution is taken into account, the electric field distribution will be substantially changed.[112,113] Many attempts have been made to overcome this deficiency by modeling detailed structures at the molecular level. A so-called hydrophobic pore was postulated as an initial stage.[85,108-111] A simplified model shows that such effects cause an electrical contribution to both the line tension and the surface tension.[113] If we let φ_0 be the electric potential across the membrane, the electrical line tension $\Gamma_{el} = -\varepsilon_w \cdot \varepsilon_0 \cdot \varphi_0^2/2\pi$ tries to open the pore, and the contribution to the surface tension $T_{el} = -\varepsilon_1 \cdot \varepsilon_0 \cdot \varphi_0^2/2d$ emanating from the field along the membrane surface tends to close it again. Incorporating both contributions into Equation 17 by replacing the purely mechanical parameters with their effective ones we have:

$$E_{pore} = 2\pi \cdot a \cdot \Gamma_{eff} - \pi \cdot a^2 T_{eff} = 2\pi \cdot a \cdot \left(\Gamma - \varepsilon_w \cdot \varepsilon_0 \frac{\varphi_0^2}{2\pi} \right) - \pi \cdot a^2 \cdot \left(T - \varepsilon_1 \cdot \varepsilon_0 \frac{\varphi_0^2}{2d} \right) \qquad (18)$$

Here, d denotes the membrane thickness and φ_0 the applied voltage. Inspection of this shows that two mechanisms of pore formation are possible. Using the applied potential φ_0 as a parameter and increasing it continuously, then at some voltage either $\Gamma_{eff}= 0$ or $T_{eff}= 0$. In the first case, an enhancement of the field reduces the critical radius and promotes rupture. In the second case, an increase of the applied external voltage φ_0 stabilizes the membrane and causes stable pore formation at higher voltage. The measured mechanical properties of membranes suggest the first process, which is described in more detail below. Up to now the second process has not yet been experimentally verified, but might occur in the case of electropermeabilization of flaccid vesicles or cells.

The above model is based on a static view. Once a pore of a given radius a has formed, this model can only predict whether the pore will close again or else grow to infinite size causing membrane rupture. Nothing can be concluded about the dynamics of pore widening or closing from these models. Due to the lack of experimental data on the very early stages of pore formation, none of the models could yet be disproved.

All of these models suggest an energy barrier E_{pore}. Within this framework the lifetime of a membrane is assumed to be distributed according to a Boltzmann function

$$t_{life} = t_0 \cdot e^{-\frac{E_{pore}}{k \cdot T}} \qquad (19)$$

Measurement of the lifetime allows an estimation of the pore energy,[85,91,114] and the surface tension of a bilayer can be obtained by independent measurement. The observed variation in lifetimes with increasing amplitudes of the external field was used to obtain the edge energy as an independent parameter. The observed Boltzmann distributions were very broad and their characterization required many measurements. However, the observed values were in reasonable agreement with those from other methods.[83,95]

4. Kinetics of the Pore Widening

The irreversible breakdown of thin films is a general phenomenon. Bursting of soap films is a typical example, wetting of a liquid on a solid or liquid support another.[115-117] Most investigations in this area have dealt with static properties; relatively few authors have considered the dynamics. The velocity of the phase boundary, or for the purposes of our discussion the velocity of the pore rim, can range from several meters per second down to micrometers per second.[63-65,116,117] The pore opening velocity is a function of the balance between forces driving the opening (or closing) of the pore and those damping the opening. The latter depend upon the velocity and are typically dominated by viscosity, or in fast processes, by inertia.[63-65] The bursting of soap films was carefully investigated some years ago.[117] Large soap films were spanned across a rim and defects were induced by electrical discharges. The kinetics were recorded using a high speed camera. Recently, defects have been induced in planar lipid membranes with electric field pulses. The successive widening of the defects to give large aqueous pores was recorded electrically.[63-65] In this section we will suggest a theoretical model explaining the experimental observations.

An attempt to describe the dynamics of pore widening has been made based upon statistical mechanics.[118] Another procedure started with the static defect model.[119,120] The kinetics are mainly due to diffusion of the lipids and are quantitatively described by the self diffusion coefficient.[119] Unfortunately, such models seem to be in conflict with the experimental findings.[63-65]

Deriving the energy barrier of a defect E_{pore} with respect to the pore radius yields a force which is either directed outward, causing further widening, or inward, causing resealing.[63,113] The resulting excess force will cause a movement of the lipid which is damped by friction or inertia. In general, the movement of a thin film is described by the Navier-Stokes equation, which is quadratic with respect to flow velocity. The prediction of the pertinent dynamics requires knowledge of the experimental boundaries and the exact material flow profile, and a general solution based on this equation is not possible. However, if the range of experimental results are taken into account, the calculations can be simplified to give a solution for the pore velocity over the limited time range of practical interest.

If thermal fluctuations are neglected, then Equation 18 predicts an irreversible breakdown only if the radius a becomes larger than the critical one $a^* = \Gamma_{eff}/T_{eff}$. Inserting typical values of the mechanical edge energies for pure lipids of about $\Gamma = 1 \cdot 10^{-11}$ N requires a voltage of about 400 mV for Γ_{eff} to vanish. Inspection of Equation 18 shows that an applied voltage $\varphi_0 = 400$ mV acting across a bilayer of thickness $d = 5$ nm, can only be responsible for a contribution to the surface tension $T_{el} = 0.3$ mN/m, which is relatively small in comparison to the mechanical tension present in a bilayer of roughly $T = 2$–3 mN/m. We may therefore neglect the electrical contribution to the surface tension and take T_{eff} to be independent of the applied voltage φ_0 with reasonable accuracy. In the following discussion we propose a possible mechanism: an external electric field pulse will decrease the energy barrier for pore formation via a reduction in edge energy. During application of the field pulse the edge energy is low. We assume that defects or pores are formed by some mechanism (exactly what is not critical for these purposes). For a given pore of radius a, the resulting forces will close the pore if $a < a^*$. Larger pores will widen further: the larger the pore becomes, the less important is the influence of the edge energy. If after the application of the field pulse the actual size has not already grown beyond the critical one, resealing will occur. The other case will lead to irreversible rupture. After the initial pore formation, the opening process is driven by the mechanical surface tension alone. On the other hand, reasealing is driven by the mechanical edge energy.

In order to understand the experimental data, we have to model the flow of the lipid material during the irreversible mechanical breakdown.[63-65] Due to its finite surface tension, the lipid is homogeneously stretched like a soap film. Furthermore, we assume that the pore is of cylindrical shape and large enough to neglect effects due to finite thickness.

During the opening process of the pore, lipid flows radial away from the center to the torus of the membrane. As mentioned above, an exact solution of this membrane material flow requires knowledge of the local stress distribution after breakdown. Whatever the local stress distribution, surface tension remains the driving force for the opening process. As a first approximation one would expect that the membrane viscosity would determine the time course of the opening. As shown recently,[63] a damping force acting at the boundary and working against the movement of the pore rim must be added to the mechanical forces derived from Equation 17. If the edge energy of larger pores is neglected, an exponential increase of the pore radius with time is predicted.

$$a(t) = a_0 e^{\left(\frac{T}{4\eta}\right)t} \tag{20}$$

The decay time $t_{vis} = 4\eta/T$ is about $2 \cdot 10^{-6}$ s for reasonable values[121,122] of $\eta \approx 10^{-9}$ Ns/m and of $T = 2 \cdot 10^{-3}$ N/m. Recently, the pore conductivity was recorded in the breakdown experiments over much longer periods than t_{vis}:[63] these experiments could not have failed to detect an exponential increase in the pore conductivity, were this present. Its absence in the case of a planar lipid bilayer indicates that the opening kinetics must be controlled by a different process with a larger degree of damping.[63-65]

A completely different process was proposed some years ago to explain the bursting of soap films.[117] After initiating the irreversible breakdown in a soap film with a spark, the kinetics of the opening were recorded with a high-speed camera. It was assumed that the film is approximately immobile, and that only that lipid that previously covered the hole moves (with constant velocity). The dynamics of the pore widening were obtained by bringing the kinetic energy term into Equation 17:

$$E_c = \pi \cdot a^2 d\rho \cdot (\dot{a})^2 \qquad (21)$$

where ρ is the lipid density. Again, neglecting the electrostatic contribution to the surface tension, as well as the edge effect, we obtain the time dependency of the pore radius[63]

$$a(t) = a_0 + \sqrt{\frac{\Phi \cdot T}{d\rho}} \cdot t = a_0 + u \cdot t \qquad (22)$$

where Φ is a parameter depending on the unknown material flow, as well as on dissipation effects.

The kinetics of pore formation followed by mechanical rupture of lipid bilayers were investigated in detail using the charge pulse technique.[63-65] Several experimental observations suggest that irreversible breakdown is a result of the widening of a single pore. These observations include:

- The statistically broad distribution of lifetimes. A large number of pores would yield a rather narrow distribution of lifetimes.
- After carefully raising the applied voltage to a critical level, a series of narrow distributions of opening velocities was observed.
- In a few experiments, the initial slope suddenly doubled. This was interpreted as the opening of a second pore.

These findings contradict theoretical predictions based upon the coalescence of many pores.[119] However, it should be noted that despite the observations above, large supercritical field pulses may create many pores.

A series of investigations devoted to the study of parameters affecting the kinetics of the irreversible breakdown have been performed. Unfortunately, variation of the lipid composition varies several parameters simultaneously, including membrane thickness, surface tension, edge energy, viscosity, etc. A large increase in effective thickness, i.e., an increase of the mass of the membrane, is expected when polymers adsorb to the membranes. Addition of high molecular weight polymers (Pluronic F-68) resulted in a visible decrease of the pore opening velocity.[64] It is not possible to quantitate this measurement because the mass of polymer adsorbed per unit area of membrane is unknown. A control experiment using polyacrylamide (PAA), a polymer presumed not to interact with the membrane, showed no change in the opening velocity.

Quantitative measurements of the thickness effect were made in a series of experiments which used a defined mixture of diphytanoyl phosphatidylcholine (DPhPC) with distearoyl phosphatidylethanolamine-polyethylene glycol (DSPE-PEG Stealth™-lipid).[65] DSPE-PEG lipids consist of a fully saturated DSPE lipid to which a PEG molecule of known molecular weight is covalently bonded. These lipids can be mixed to give a precisely controlled lipid composition, and hence the creation of lipid films of a defined mass per unit area. The experimentally measured pore-opening velocity of the non-PEG lipid alone was used as the starting point for the theoretical prediction. Unknown dissipative effects should not be changed by the presence of the small amount of PEG-polymer: their effects will be included in the opening velocity of the lipid alone. The additional increase in effective mass density due to the presence of the PEG-polymer moiety was then calculated using scaling theory. The latter yields the Flory radius representing the size of the polymer coil.[123] It was assumed that interstitial water has to follow the movement of the polymer coil and therefore accounts for the effective thickness. In a series of experiments the molecular weight of the attached PEG was increased (350, 750, 2000, and 5000). Increasing the polymer length from PEG-350 to PEG-5000 causes the mass per unit area to increase, and consequently the velocity to decrease. The same was found when the molar fraction of the added polymer was increased. In both series of experiments the agreement with the the prediction of Equation 22 was remarkably good.[65]

In a separate series of experiments, CPCl (cetyl pyridinium chloride) was added as surfactant.[64] Capacity measurements showed no reduction in membrane thickness. It is expected that CPCl should reduce the contact angle and therefore reduces the surface tension of the bilayer spanned over the Teflon®

rim, and so reduce the driving force for pore widening. This is in accordance with the observation of a clear reduction of the opening velocity of the pore. Again, a quantitative study of this observation is lacking.

5. Kinetics of the Pore Resealing

The results from electrical breakdown measurements indicate that voltages around 500 mV induce mechanical rupture of lipid bilayer membranes. However, in some studies it has been demonstrated that bilayer membranes made from oxidized cholesterol can also exhibit reversible electrical breakdown, i.e., increase of the membrane conductance, without mechanical rupture.[93,124,125] More recently, similar phenomena have also been observed on azolectin membranes treated with UO_2^{2+}.[85]

To date, no theoretical model has been able to explain the experimental features of reversible electrical breakdown. In the following we will outline general arguments for a possible underlying mechanism for this phenomenon.

According to the viscoelastic model described above, the opening of pores in membranes is driven by the mechanical surface tension. For vesicles, the latter depends upon the volume-to-area ratio. A loss of the internal volume will clearly decrease the mechanical tension, which will relax almost to zero. According to Equation 22 this reduces the opening velocity which may eventually stop (or even reverse). The mechanical tension of the vesicle depends directly on the flow rate as well as on the instantaneous pore radius, which controls the further efflux. This is in contrast to planar bilayer experiments, where the tension is determined by the contact angle at the rim. In all geometries, these forces work against the edge energy, which promotes resealing. As already pointed out, experiments suggest that electrical fields lower the edge energy. Once the expanding pore is sufficiently large, the contribution stemming from the edge becomes negligible with respect to the energy gain due to surface tension. As shown for irreversible breakdown, the resulting force vector between these two forces is balanced by a dynamic force, either viscosity or inertia. When the applied electric field pulse lasts long enough to allow the kinetics to overcome the critical mechanical pore radius $a*$, irreversible breakdown will occur. When this process is too slow to overcome the critical pore size during pulse application, the pore will reseal. Inspection of the dynamics underlying the resealing process shows that edge energy is the driving force. Due to its linear dependency on the pore radius a (Equation 17), this is a slow process in which viscosity should be taken into account. Arguments similar to those in the previous section can be made regarding the time dependency of the pore radius. The time dependency for the resealing of a pore with radius a is readily obtained

$$a(t) = \frac{\Gamma}{T} - \left(a* - a_0\right) e^{(T/4\eta)t} \qquad (23)$$

This equation is consistent with observations of the resealing process of long-lived pores, which take many seconds. In contrast to the lifetime of the pore in a liposome in the open state, the final resealing process is over within a few milliseconds.[95]

III. LIPOSOME MANIPULATION WITH ELECTRIC FIELDS

External electric fields offer a unique tool for cell characterization or manipulation. They have the great advantage that they are readily controlled from outside the area where they are applied. In contrast to mechanical or chemical methods of perturbation, electric fields may be applied very specifically and can be reversibly switched on and off with high frequency. Prominent methods in biotechnology are, e.g., electrophoresis for cell separation, electric field-induced fusion of cells, and electroinjection of DNA necessary in the field of genetic engineering.[16,17,58,126,127] In the following a short introduction to a few prominent applications to liposomes are given.

A. ELECTROSWELLING

Liposomes are a convenient model system for the study of the physical properties of natural cells. In many cases, the size distribution and the lamellarity of the liposomes is important. Although the production of small unilamellar vesicles is fairly predictable, recipes for large unilamellar vesicles (size >5 μm) are poorly reproducible.[15] In most vesicle formation procedures, the first stage is the formation of large films using organic solvents. After solvent removal, the film is hydrated: this cause a slow swelling of the film which then spontaneously forms onion-like structures which sometimes break into large multilamellar structures. In order to destabilize these intermediate structures, liposome preparations inject some energy into the lipid system, e.g., by sonication. It has been suggested that the breakage of the lipid film is a function of the available energy: larger pieces should receive more energy and therefore break earlier. The individual membrane leaflets round up and form vesicles of relatively homogeneous size.[15]

A further method for the preparation of large unilamellar vesicles is the application of low frequency electric fields. Several procedures to obtain an optimal distribution of large unilamellar vesicles have been proposed.[128-130] Typically, lipid dissolved in chloroform/methanol (9:1 v/v) is spread on a metalized glass slide and dried. After hydration an AC field (e.g., 10 Hz, 1 V across the 300 μm thick sample) was applied for a duration of a few minutes up to 2 h. The yield of large unilamellar vesicles was found to be a function of the thickness of the lipid film. The thinner the film (in these experiments about five layers on average), the better the yield of large unilamellar vesicles. The underlying process is still unclear,[128-130] and various effects may play a role. External fields can couple to intrinsic ones and influence the interaction with adjacent membrane layers. On the other hand, external fields could cause motion of the electrolyte as well as of the lipid. The electric field-induced movement may have a similar mechanism to that of sonication. A different explanation might be that the electric field causes rupture of the film, causing defects that facilitate the rounding up. More recently this method has been optimized and successfully applied for the production of large unilamellar vesicles.[130]

B. LASER TRAPPING

The previous sections of this chapter dealt with DC and AC fields of relatively low frequencies. Optical tweezers are an example of the application of very-high frequency electric fields and have been used as a tool for micromanipulation.[131] In the early part of this century it was known that light beams carry a momentum which can be transferred to particles.[132] In the case of liposomes or cells the incident light is scattered at the water/membrane interface. Part of the electric and magnetic field components do not penetrate the lipid. The contribution from the scattered part exerts a radiation pressure onto the surface and pushes the particle along the direction of the incident beam. This forces decrease with increasing transparency. For the trapping of the particle itself, a second effect is used which takes advantage of the Gaussian-profile of intensity of most laser beams. A small particle having higher optical density than the surrounding medium is pulled into the center of the beam where the field intensity is highest (dielectrophoresis). Provided that this gradient force is larger than the scattering force, a stable trap is possible by focusing the laser beam in the direction of the light. However, if the optical density of the particle is less than that of the surroundings, no stable position exists.[131-133] On the other hand, the latter can be trapped in a laser beam which is profiled to have minimun electric field density in the middle. In this reverse case, the dielectrophoretic force pushes the particle into the region of low field present inside.

In recent years these methods were modified and applied successively to the manipulation of latex beads, liposomes, and even of cells.[131,133-146]

C. ELECTROROTATION

A less commonly known method for the electric-field manipulation of liposomes (or cells) is the application, using three or more electrodes, of an external rotating field.[143-146] A particle between the electrodes is polarized according to the instantaneous field distribution at a given time, which can be described by equations similar to those outlined in Section II.B. However, the field direction changes with time according to the applied frequency and the electrode geometry. This time-dependent field causes a torque on the previously-induced dipole at the particle surface: the decay-time of the dipole depends on the electric properties of the cell. Experimentally, one first records the frequency of the rotating cell vs. the frequency of the applied field. This spectrum is compared to a model spectrum of a rotating cell built up of various shells each described by a conductivity and dielectric permittivity. In many cases (single-shelled objects, or good approximations to these), it then suffices to measure the field frequency that gives rise to fastest rotation, for which a special technique was developed.[145,146] Electrorotation allows non-destructive measurements of the membrane and surface properties and, at least theoretically, of the conductivity inside a cell.

Liposomes were subjected to rotating fields soon after the development of electrorotation:[147,148] liposome rotation was reported to be opposite to the field rotation at kHz frequencies (where the membrane-charging effect dominates the overall polarization), but in the same direction as the field above 0.5 MHz (where the conductivity excess of the liposome probably dominates the polarization).

Later measurements on liposomes detected rotation in response to frequencies below 1 kHz:[148] this was at first associated with the so-called "alpha"-dispersion (possibly an ionic double-layer relaxation), but it was later suggested that the phenomenon could be an electrophoretic effect.[148-151] The power of the method for the characterisation of the membrane properties of unilamellar vesicles was demonstrated by quantitative work[152] on ionophore-treated and on hydrophobic ion-treated thylakoid "blebs" (hypo-osmotically-swollen chloroplasts with diameters of 10–18 μm).

D. DEFORMATION

According to Section II.B, the presence of a liposome in a homogeneous field causes a distortion of the field lines. In the absence of free charges, Maxwell stresses can act only at the membrane/water boundaries and must be balanced by mechanical stresses or movement.[60,83,153-165] At low field strengths, electric field effects are submerged in the Brownian motion and can not be easily seen, if at all. Increasing the amplitude of the external field causes orientation of non-spherical vesicles.[60,83,153-155] In the case of an inhomogeneous externally applied field, dielectrophoresis is observed.[57,161] Particles of higher dielectric permittivity (or higher conductivity in low frequency fields) than the external solution move along the gradient of the Maxwell force into the region of higher field strength, whereas particles of lower permittivity move away. If several liposomes are present in a homogeneous field, alignment to give pearl-chains is observed.[159] In the low frequency regime, flaccid vesicles undergo an elongation parallel to the field lines.[56,137] In this case the Maxwell pressure given by Equation 7 acts against the bending moment. Minimization yields a deformation amplitude:

$$s_2 = \frac{1}{32} \frac{\varepsilon_w \varepsilon_0}{k_c} b^4 E_0^2 \qquad (24)$$

where s_2 as the increase of the radius b of the vesicle and k_c is the bending rigidity.[59] Inserting reasonable values into Equation 24 ($k_c = 10^{-19}$ J and $E_0 = 3 \cdot 10^4$ V/m) shows that significant deformation is only expected for vesicles with radii larger than $b = 1$ μm. The above equation is only valid for small deviations from the spherical shape. Larger deformations influence the field distribution around the vesicles in such a way that the perturbations of the field lines due to their presence decrease,[162,163] so reducing the deformational force of the field. Direct measurement of the deformation amplitude is not precise enough to allow estimation of the bending rigidity. A much better method is to use the electric field-induced stress to smooth out thermal undulations of the vesicle membrane.[79]

Attempts have been made to include the effects of an external electric field[78,162-165] in the shape calculation. It is assumed that the Maxwell force, which tries to deform the vesicle, is balanced against the bending elasticity. For larger deformations, a correction of the field distribution due to the vesicle deformation must be taken into account. Such corrections are rather complex and require numerical solutions to the problem.

Electric fields also influence the shape of liposomes in an indirect way. A few years ago it was suggested that surface charges increase the bending elasticity modulus (see, e.g., Reference 14). It could be shown experimentally that the electrostatic contribution to the effective bending rigidity was negligible.[166] However, this model also predicts vesiculation as a function of increasing surface charge density:[14,167] beyond a critical surface charge density, fission into several smaller liposomes is energetically favorable. This model also predicts fusion of a highly charged vesicle with low or uncharged ones under specific conditions.[168,169]

E. ELECTROPORATION AND ELECTROFUSION

Application of the theory of irreversible breakdown outlined in Section II.B to vesicles shows there to be a considerable potential for reversibility. Although we again expect that surface tension can be the driving force for pore opening, the creation of pores allows the surface tension to relax through efflux of internal volume. Several years ago, large pores were induced in giant vesicles and observed under phase contrast microscopes.[83] The pores were stable as long as the AC-field was applied, and resealed after switch-off. Larger field amplitudes caused a tube-like structure to form. The stability of this structure has been explained by a balance of bending energy and edge energy.

A few years later, it was shown experimentally that vesicles can be permeabilized by short electric field pulses.[84] The relation of applied field strength to the subsequent efflux of tracer molecules has been studied in detail. High electric fields can also be used to prepare large unilamellar vesicles.[168,169] Vesicles of small size were aligned with low AC fields and fused with a series of pulses of higher amplitudes.

Recently, the stability of vesicles during electric field pulses has been investigated.[94] For these experiments, single vesicles were held under a fixed lateral mechanical tension T using micro pipettes. The influence of additional field pulses is seen as a decrease of the critical tension for lysis. A preliminary analysis of the experimental results suggests that the electric field acts as an additional tension and leads to an effective total tension:

$$T_{\text{eff}} = T + \frac{3}{8} \cdot \varepsilon_l \cdot \varepsilon_0 \cdot \left(\frac{b}{d}\right)^2 \cdot E_0^2 \cdot d \qquad (25)$$

When T_{eff} exceeded the critical tension T_{lys}, lysis was observed. However, according to Equation 7, one would expect an opposite sign for the electric field-induced contribution to the tension. In other words, this pressure acts against the mechanical stress. Under other experimental conditions dealing with free vesicles, an increase in area attributed to electric field pulses was observed.[79] In fact, such an area increase has not been observed if the the vesicles are clamped with a fixed suction pressure. We would like to suggest the following explanation of this apparent discrepancy. Under the experimental conditions used, the electric field pulses cause an increase of the area, which is reduced by the aspiration pressure in the micro pipette.[94] After removal of the field pulses, the ensuing increase of tension causes lysis.

Following our discussion we suggest instead of Equation 26 the effective tension

$$T_{\text{eff}} = T + T_{el} + T_{osm} \qquad (26)$$

where T is an externally applied tension (e.g., by micro pipettes), $T_{el} = -\frac{3}{8} \cdot \varepsilon_l \cdot \varepsilon_0 \cdot (b/d)^2 \cdot E_0^2 \cdot d$ is the electrical tension and $T_{osm} = \Delta\Pi_{osm} \cdot b/2$ represents an additional tension due to the presence of an osmotic gradient across the membrane. Again, if the effective tension exceeds the critical point, vesicle lysis would be expected.

The stabilization of large pores by balancing the mechanical forces against electric field pulses has recently been demonstrated using the micro pipette technique.[95] Single vesicles were held under fixed tension applied by micro pipettes. In this series of experiments sub-critical tensions were used. Careful permeabilization yields sub-critical pores which are stabilized by the outflow of the internal volume. If the tension was reduced below a critical threshold the pore began to close. Although the pores were stable for several seconds, the resealing process itself was finished within a few milliseconds. These measurements allowed an independent determination of the edge energy. Investigation on the dynamics of the resealing process in cells showed much slower closing times as the viscosity is by several orders larger.[170,171]

IV. CONCLUSIONS

This chapter has outlined several different sources for electric fields in membranes. Intrinsic fields do not only appear at the lipid/water interface due to dissociation of charges, but also due to macroscopic orientation of the individual lipid molecule into a mono- or a bilayer. These intrinsic electric fields may be responsible for the shape of the liposome or for triggering the shape transition. External electric fields offer a unique tool for liposomal or cell manipulation and for measurements. Application of low frequency, low amplitude fields enhances the formation of large unilamellar liposomes. Once liposomes have been formed, electric fields allow their manipulation or characterization. Beyond a critical field strength, the electric fields induce defects in the membrane. The membrane becomes permeable and fusion between adjacent liposomes may occur. The use of micromanipulators combined with laser technology presents a new range of possibilities for single-vesicle manipulation.

Electric fields have the astonishing feature that they can be considerably amplified when applied across a cell. Although a potential of 1 V is usually considered harmless, a field strength of 250 kV/m (attained in lightning) results when this voltage is applied across a bilayer. This amplification is due to the extreme difference in conductivities between water and lipid, or (in the high frequency regime) to the significant difference in the respective dielectric permittivities. However, the use of externally applied electric fields for *in vivo* work is limited because they do not penetrate deeply into the human body.[1] By contrast, magnetic fields are able to permeate tissue without significant loss. The design of vesicles which might be manipulated by magnetic fields, in ways parallel to those possible in electric fields, has been attempted for some years. In general, it can be said that the magnetic permeability of the liposome must be increased in order to decrease the magnetic transparency. If a significant difference in the magnetic permeability can be attained, the magnetic part of the Maxwell forces may act in a similar way to the electric part. There has been promising recent progress in this direction.[171,172]

ACKNOWLEDGMENTS

The author would like to thank W. M. Arnold, R. Benz, K.-H. Klotz, and V. A. Parsegian for helpful discussions. This work was supported by grants of the Deutsche Forschungsgemeinschaft (Graduiertenkolleg "Magnetische Kernresonanz" and project B7 of the Sonderforschungsbereich 176).

REFERENCES

1. **Polk, C. and Postow, E.** *Handbook of Biological Effects of Electromagnetic Fields.* CRC Press, Boca Raton, FL, 1986.
2. **Sakmann, B. and Neher, E.** *Single-Channel Recording.* Plenum Press, New York, 1983.
3. **McLaughlin, S.** The electrostatic properties of membranes. *Annu. Rev. Biophys. Biophys. Chem.*, 18, 113, 1989.
4. **Läuger, P., Benz, R., Stark, G., Bamberg, E., Jordan, P.C., Fahr, A., and Brock, W.** Relaxation studies on ion transport systems in lipid bilayer membranes. *Q. Rev. Biophys.*, 14, 513, 1981.
5. **Bessis, M., Weed, R.I., and Lebland, P. (Eds.).** *Red Cell Shape.* Springer-Verlag, New York, 1973.
6. **Deuling, H.J. and Helfrich, W.** Red blood cell shapes as explained on the basis of curvature elasticity. *Biophys. J.*, 16, 861, 1976.
7. **Glaser, R.** Echinocyte formation by potential changes of human red blood cells. *J. Memb. Biol.*, 66, 79, 1982.
8. **Sackmann, E., Duwe, H.P., and Engelhardt, H.** Membrane bending elasticity and its role for shape fluctuations and shape transformations of cells and vesicles. *Faraday Dis. Chem. Soc.*, 81, 281, 1986.
9. **Grebe, R., Peterhänsel, G., and Schmid-Schönbein, H.** Change of local charge density by change of local mean curvature in biological bilayer membranes. *Mol. Cryst. Liq. Cryst.*, 152, 205, 1987.
10. **Farge, E. and Devaux, P.F.** Shape change of giant liposomes induced by an asymmetric transmembrane distribution of phospholipids. *Biophys. J.*, 61, 347, 1992.
11. **Talmon, Y., Evans, D.F., and Ninham, B.W.** Spontaneous vesicles formed from hydroxide surfactants. *Science*, 221, 1047, 1983.
12. **Hauser, H., Gains, N., and Müller, M.** Vesiculation of unsonicated phospholipid dispersions containing phosphatidic acid by pH adjustment. *Biochemistry*, 22, 4775, 1983.
13. **Kaler, E.W., Murthy, A.K., Rodriguez, B.E., and Zasadzinski, J.A.N.** Spontaneous vesicle formation in aqueous mixture of single railed surfactants. *Science*, 245, 1371, 1989.
14. **Winterhalter, M. and Helfrich, W.** Bending elasticity of electrically charged bilayers. *J. Phys. Chem.*, 96, 327, 1992 and references herein.
15. **Lasic, D.D.** *Liposomes: From Physics to Applications.* Elsevier, Amsterdam, 1993.
16. **Chang, D.C., Chassy, B.M., Saunders, J.A., and Sowers, A.E. (Eds.).** *Guide to Electroporation and Electrofusion.* Academic Press, New York, 1992.
17. **Borrebueck, C. and Hagen, I. (Eds.).** *Electromanipulation in Hybridoma Technology.* Stockton Press, New York, 1989.
18. **Orlowski, S. and Mir, L.M.** Cell electropermeabilisation. *Biochim. Biophys. Acta*, 1154, 51, 1993.
19. **Wearley, L. and Chien, Y.W.** Enhancement of the in vitro skin permeability of azidothymidine (AZT) via iontophoresis and chemical enhancer. *Int. J. Pharm.*, 55, 1, 1990.
20. **Verwey, E.J.W. and Overbeek, J.T.G.** *Theory of the Stability of Lyophobic Colloids.* Elsevier, Amsterdam, 1948.
21. **Cevc, G. and Marsh, D.** *The Phospholipid Bilayer.* Wiley, New York, 1985.
22. **Israelachvili, J.N.** *The Intermolecular and Surface Forces.* Academic Press, New York, 1985.
23. **Hunter, R.J.** *Foundation of Colloid Science.* Vol. I, Clarendon Press, Oxford, U.K., 1987.
24. **Landau, L.D. and Lifshitz, E.M.** *Electrodynamics of Continuous Media.* Pergamon Press, Oxford, 1963.
25. **Natarajan, R. and Schechter, R.S.** The solution of the nonlinear Poisson-Boltzmann equation for thin, spherical double layers. *J. Coll. Interface Sci.*, 99, 50, 1984 and references herein.
26. **Mitchell, D.J. and Ninham, B.W.** Electrostatic curvature contributions to the interfacial tension of micellar and microemulsion phases. *J. Phys. Chem.*, 87, 2996, 1983.
27. **Needham, D. and Nunn, R.S.** Elastic deformation and failure of lipid bilayer membranes, *Biophys. J.*, 58, 997, 1990.
28. **Nelson, A.P. and McQuarrie, D.A.** The effect of discrete charges on the electrical properties of a membrane. *J. Theor. Biol.*, 55, 13, 1975.
29. **Winiski, A.P., McLaughlin, A.C., McDaniel, R.V., Eisenberg, M., and McLaughlin, S.** An experimental test of the discreteness-of-charge effect in positive and negative lipid bilayers. *Biochemistry*, 25 8206, 1986.

30. E.g., in: **Schmickler, W. and Henderson, D.** New models for the structure of the electrochemical interface. *Prog. Surf. Sci.*, 22, 323, 1988.
31. **Marčelja, S.** Electrostatics of membrane adhesion. *Biophys. J.*, 61, 1117, 1992.
32. **Mille, E. and Vanderkooi, G.** Electrochemical properties of spherical polyelectrolytes. *J. Colloid Interface Sci.*, 61, 455, 1977.
33. **Ninham, B.W. and Parsegian, V.A.** Electrostatic potential between surfaces bearing ionizable groups in ionic equilibrium with physiologic saline solution. *J. Theor. Biol.*, 31, 405, 1971.
34. **Ohshima, H. and Ohki, S.** Donnan potential and surface potential of a charged membrane. *Biophys. J.*, 47, 673, 1985.
35. **Ohshima, H., Makino, K., and Kondo, T.** Potential distribution across a membrane with a surface potential charge layer. *J. Colloid Interface. Sci.*, 113, 369, 1986.
36. **Cevc, G.** Electrostatic characterisation of liposomes. *Chem. Phys. Lipids*, 64, 163, 1993.
37. **Gross, D., Loew, L.M., and Webb, W.W.** Optical imaging of cell membrane potential changes induced by applied electric fields. *Biophys. J.*, 50, 339, 1986.
38. **Ehrenberg, B., Farkas, D.L., Fluhler, E.N., Lojewska, Z., and Loew, L.M.** Membrane potential induced by external electric field pulses can be followed with a potentiometric dye. *Biophys. J.*, 51, 833, 1987.
39. **Kinoshita, K., Ashikawa, I., Saita, N., Yoshimura, H., Itoh, H., Nagayama, K., and Ikegami, A.** Electroporation of cell membrane visualized under a pulsed laser fluorescence microscope. *Biophys. J.*, 53, 1015, 1988.
40. **Hibino, M., Itoh, H., and Kinoshita, K.** Time course of cell electroporation as revealed by submicrosecond imaging of transmembrane potential. *Biophys. J.*, 64, 1789, 1993.
41. **Bouevitch, O., Lewis, A., Pinevsky, I., Wuskell, J.P., and Loew, L.M.** Probing membrane potential with nonlinear optics. *Biophys. J.*, 65, 672, 1993.
42. **McDaniel, R.V., Sharp, K., Brooks, D., McLaughlin, A.C., Winiski, A.P., Cafiso, D., and McLaughlin, S.** Electrokinetic and electrostatic properties of bilayers containing gangliosides G_{m1}, G_{D1a}, G_{T1}. *Biophys. J.*, 49, 741, 1986.
43. **Hunter, R.J.** *The Zeta Potential in Colloidal Science.* Academic Press, New York, 1981.
44. **Gains, G.L.** *Insoluble Monolayers at the Liquid-Gas Interfaces.* Wiley, New York, 1966.
45. **Brockman, H.** Dipole potential of lipid membranes. *Chem. Phys. Lipids*, 73, 57, 1994.
46. **Flewelling, R.F. and Hubbell, W.L.** The membrane dipole potential in a total membrane potential model. *Biophys. J.*, 49, 541, 1986.
47. **Franklin, J.C. and Cafiso, D.S.** Internal electrostatic potentials in bilayers. *Biophys. J.*, 65, 289, 1993.
48. **Pickar, A.D. and Benz, R.** Transport of oppositely charged lipiphilic probe ions in lipid bilayer membranes having various structures. *J. Membr. Biol.*, 44, 353, 1978.
49. **Vogel, V. and Möbius, D.** Local surface potential and electric dipole moments of lipid monolayers. *J. Colloid Interface Sci.*, 126, 408, 1988.
50. **Beitinger, H., Vogel, V., Möbius, D., and Rahmann, H.** Surface potentials and electric dipole moments of ganglioside and phospholipid monolayers. *Biochim. Biophys. Acta*, 984, 293, 1989.
51. **Smarby, J.M. and Brockmann, H.L.** Surface dipole moments of lipids at the argon-water interface. *Biophys. J.*, 58, 195, 1990.
52. **Gawrisch, K., Ruston, D., Zimmerberg, J., Parsegian, V.A., Rand, P.R., and Fuller, N.** Membrane potentials, hydration forces, and the ordering of water at membrane surfaces. *Biophys. J.*, 61, 1213, 1992.
53. **Simon, S.A. and McIntosh, T.J.** Magnitude of the solvation pressure depends on dipole potential. *Proc. Natl. Acad. Sci. U.S.A.*, 86, 9263, 1989.
54. **Cevc, G. and Marsh, D.** Hydration of noncharged lipid bilayer membranes. *Biophys. J.*, 47, 21, 1985.
55. **Winterhalter, M.** On the origin of the bending elasticity. *Prog. Colloid Polymer Sci.*, 98, 1995.
56. **Sauer, F.A.,** in *Interactions between Electromagnetic Fields and Cells.* Chiabreara, A., Nicolini, C., and Schwan, H.P., (Eds.). Plenum Press, New York, 1986, 181.
57. **Pohl, H.A.,** *Dielectrophoresis.* Cambridge University Press, Cambridge, 1978.
58. **Zimm, B.H. and Levence, S.D.** Problems and aspects in the theory of gel electrophoresis of DNA. *Q. Rev. Biophys.*, 25, 171, 1992.
59. **Winterhalter, M. and Helfrich, W.** Deformation of spherical vesicles by external fields. *J. Colloid Interface Sci.*, 122, 583, 1988.
60. **Mitov, M.D., Méléard, P., Winterhalter, M., Angelova, M.I., and Bothorel, P.** Electric-field dependent thermal fluctuation of giant vesicles. *Phys. Rev.*, E 48, 628, 1993.
61. **Coronado, R.** Recent advances in planar lipid bilayer techniques for monitoring ion channels. *Annu. Rev. Biophys. Biophys. Chem.*, 15, 259, 1986.
62. **Montal, M., Darzan, A., and Schindler, H.** Functional reassembly of membrane proteins in planar lipid bilayers. *Q. Rev. Biophys.*, 14, 1, 1991.
63. **Wilhelm, C., Winterhalter, M., Zimmermann, U., and Benz, R.** Kinetics of pore size during irreversible electrical breakdown of lipid bilayer membranes. *Biophys. J.*, 64, 121, 1993.
64. **Klotz, K.-H., Winterhalter, M., and Benz, R.** Use of irreversible electrical breakdown for the study of interaction with surface active molecules. *Biochim. Biophys. Acta*, 1147, 161, 1993.
65. **Winterhalter, M., Klotz, K.H., Lasic, D.D., and Benz, R.** Electric field induced breakdown in lipid membranes. *Prog. Colloid Polymer Sci.*, 98, 1995.

66. **Benz, R.** Permeation of hydrophilic solutes through mitochondrial outer membranes. *Biochim. Biophys. Acta,* 1197, 167, 1994.

67. **Vodyanoy, I. and Bezrukov, S.M.** Probing alamethicin channels with water soluble polymers. *Biophys. J.,* 65, 2097, 1993.

68. **Opsahl, L.R. and Webb, W.W.** Transduction of membrane tension by the ion channel alamethin. *Biophys. J.,* 66, 71, 1994.

69. **Stulen, G.** Electric field effects on lipid membrane structure. *Biochim. Biophys. Acta,* 640, 621, 1981.

70. **Osman, P. and Cornell, B.** The effect of pulsed electric fields on the phosphorus P^{31} spectra of lipid bilayers. *Biochim. Biophys. Acta,* 1195, 197, 1994.

71. **Torza, S., Cox, R.G., and Mason, S.G.** Electrohydrodynamic deformation and burst of liquid drops. *Philos. Trans. R. Soc. (London),* A269, 295, 1971.

72. **Garcia, A., Barchini, R., and Grosse, C.** The influence of diffusion on the permittivity of a suspension of spherical particles with insulating shells in an electrolyte. *J. Phys.,* D18, 1891, 1985.

73. **Garcia, A., Grosse, C., and Brito, P.** On the effects of the volume charge distribution on the Maxwell-Wagner relaxation. *J. Phys.,* D18, 739, 1985.

74. **Grosse, C. and Barchini, R.** On the permittivity of a suspension of charged particles in an electrolyte, *J. Phys.,* D 19, 1113, 1986.

75. **Grosse, C. and Schwan, H.P.** Cellular membrane potentials induced by alternating fields. *Biophys. J.,* 63, 1632, 1992.

76. **Grosse, C. and Barchini, R.** The influence of diffusion on the dielectric properties of suspension of conducting spherical particles in an electrolyte. *J. Phys.,* D25, 508, 1992.

77. **Asami, K. and Irimajiri, A.** Dielectric analysis of mitochondria isolated from rat liver. *Biochim. Biophys. Acta,* 778, 570, 1984.

78. **Hyuga, H., Kinoshita, K., and Wakabayashi, N.** Steady state deformation of a vesicle in alternating fields, *Bioelectrochem. Bioenergetics,* 32, 15, 1993.

79. **Kummrow, M. and Helfrich, W.** Deformation of giant lipid vesicles by electric fields. *Phys. Rev. A,* 44, 8356, 1991.

80. **Bryant, G. and Wolfe, J.** Electromechanical stresses produced in the plasma membranes of suspended cells by applied electric fields. *J. Membr. Biol.,* 96, 129, 1987.

81. **Taupin, C., Dvolaitzky, M., and Sauterey, C.** Osmotic pressure induced pores in phospholipid vesicles. *Biochemistry,* 14, 4771, 1981.

82. **Abidor, I.G., Arakelyan, V.B., Chernomordik, L.V., Chizmadzhev, Y.A., Pastushenko, V.F., and Tarasevich, M.R.** Electrical breakdown of bilayer lipid membranes. *Bioelectrochem. Bioenergetics,* 6, 37, 1979.

83. **Harbich, W. and Helfrich, W.** Alignment and opening of giant vesicles by electric fields. *Z. Naturf.,* 34 A, 1063, 1979.

84. **Teissie, J. and Tsong, T.Y.** Electric field induced transient pores in phospholipid bilayer vesicles. *Biochemistry,* 20, 1548, 1981.

85. **Abidor, I.G., Chernomordik, L.V., Chizmadhev, Y.A., and Sukharev, S.I.** Reversible electrical breakdown of lipid bilayer membranes modified by uranyl ions. *Sov. Electrochem.,* 17, 1543, 1981 and *Bioelectrochemistry,* 9, 141, 1982.

86. **Sukharev, S.I., Arakelyan, V.V., Abidor, I.G., Chernomordik, L.V., and Pastushenko, V.F.** BLM destruction as a result of electrical breakdown. *Biofizika,* 28, 756, 1982.

87. **Chernomordik, L.V., Abidor, I.G., Chizmadhev, Y.A., and Sukharev, S.I.** Mechanism of reversible electrical breakdown of lipid bilayer membranes in presence of uranyl ions. *Sov. Electrochem.,* 18, 88, 1982.

88. **Chernomordik, L.V., Abidor, I.G., Kushnev, V.V., and Sukharev, S.I.** Pore development during the reversible electrical breakdown of lipid bilayer membranes. *Sov. Electrochem.,* 19, 1097, 1983.

89. **Chernomordik, L.V., Abidor, I.G., Chizmadhev, Y.A., and Sukharev, S.I.** Reversible breakdown of lipid bilayer membranes in the electric field. *Biol. Membr.,* 1, 229, 1984.

90. **El-Mashak, E.M. and Tsong, T.Y.** Ion selectivity of temperature induced and electric field induced pores in dipalmitoylphosphatidylcholine vesicles. *Biochemistry,* 24, 2884, 1985.

91. **Chernomordik, L.V., Kozlov, M.M., Melikyan, G.B., Abidor, I.G., Markin, V.S., and Chizmadhev, Y.A.** The shape of lipid molecules and monolayer membrane fusion. *Biochim. Biophys. Acta,* 812, 643, 1985.

92. **Leikin, S.L., Glaser, R.W., and Chernomordik, L.V.** Mechanism of pore formation under electrical breakdown of membranes. *Biol. Membr.,* 6, 944, 1986.

93. **Glaser, R.W., Leikin, S.L., Sokirko, A.I., Chernomordik, L.V., and Pastushenko, V.F.** Reversible electrical breakdown of lipid bilayer membranes. *Biochim. Biophys. Acta,* 940, 275, 1988.

94. **Needham, D. and Hochmuth, R.M.** Electro-mechanical permeabilisation of lipid vesicles. *Biophys. J.,* 55, 1001, 1989.

95. **Zhelev, D. and Needham, D.** Tension stabilized pores in giant vesicles. *Biochim. Biophys. Acta,* 1147, 89, 1993.

96. **Crowley, J.M.** Electrical breakdown of bimolecular lipid membranes as an electromechanical instability. *Biophys. J.,* 13, 711, 1973.

97. **Dimitrov, S.D.** Electric field induced breakdown of lipid bilayer and cell membranes. *J. Membr. Biol.,* 78, 53, 1984.

98. **Dimitrov, S.D. and Jain, R.K.** Membrane stability. *Biochim. Biophys. Acta,* 779, 437, 1984.

99. **Parsegian, V.A.** Energy of an ion crossing a low dielectric membrane. Solution to four relevant electrostatic problems. *Nature (London)*, 221, 844, 1969.

100. **Deryagin, B.V. and Gutop, Y.V.** Theory of the breakdown of free films. *Colloid J. (USSR)*, 24, 370, 1962.

101. **Bivas, I.** Molecular theory of the lifetime of black lipid membranes. *J. Colloid Interface Sci.*, 144, 63, 1991.

102. **Sugar, I.P. and Neumann, E.** Stochastic model for the electric field-induced membrane pores. *Biophys. Chem.*, 19, 211, 1984.

103. **Sugar, I.P., Förster, W., and Neumann, E.** Model of cell electrofusion. *Biophys. Chem.*, 26, 321, 1987.

104. **Derjaguin, B.V. and Prokhorov, A.V.** On the theory of the rupture of black films. *J. Colloid Interface Sci.*, 81, 108, 1981.

105. **Prokhorov, A.V. and Derjaguin, B.V.** On the generalized theory of the bilayer film rupture. *J. Colloid Interface Sci.*, 125, 111, 1988.

106. **Kashchiev, D. and Exerova, D.** Bilayer lipid membrane permeation and rupture due to hole formation. *Biochim. Biophys. Acta*, 732, 133, 1983.

107. **Weaver, J.C. and Mintzer, R.A.** Decreased bilayer stability due to transmembrane potentials. *Phys. Lett.*, 86 A, 57, 1981.

108. **Petrov, A.G., Mitov, M.D., and Derzhanski, A.** Edge energy and pore stability in bilayer membranes. In *Advances in Liquid Crystal Research and Applications*. Bata, L. (Ed.). Pergamon, Oxford, 1980, 2, 695.

109. **Pastushenko, V.F. and Petrov, A.G.** Electro-mechanical mechanism of pore formation in bilayer lipid membranes. *Proc. 7th School on Biophysics of Membrane Transport*, Poland, 1984.

110. **Pastushenko, V.F. and Chizmadhev, Y.A.** Stabilisation of conducting pores in blm by electric current. *Gen. Physiol. Biophys.*, 1, 43, 1985.

111. **Chizmadhev, Y.A. and Pastushenko, V.F.** Electrical stability of biological and model membranes. *Biol. Membr.*, 6, 1013, 1989.

112. **Barnett, A.** The current voltage relation of an aqueous pore in a lipid bilayer membrane. *Biochim. Biophys. Acta*, 1025, 10, 1990.

113. **Winterhalter, M. and Helfrich, W.** Effect of voltage on pores in membranes. *Phys. Rev.*, 36A, 5874, 1987.

114. **Genco, I., Gliozzi, A., Relini, A., Robello, M., and Scalas, E.** Electroporation in symmetric and asymmetric membranes. *Biochim. Biophys. Acta*, 1149, 10, 1993.

115. **DeGennes, P.G.** Wetting. *Rev. Mod. Phys.*, 57, 827, 1985.

116. **Redon, C., Brochard, F., and Rondelez, F.** Dynamics of dewetting. *Phys. Rev. Lett.*, 66, 715, 1991.

117. **Frankel, S. and Mysels, K.J.** The bursting of soap films. *J. Phys. Chem.*, 73, 3028, 1969.

118. **Sugar, I.P.** Stochastic model of electric field induced membrane pores. In *Electroporation and Electrofusion in Cell Biology*. Neumann, E., Sowers, A., and Jordan, C.A. (Eds.). Plenum Press, New York, 1989, chap. 6, p. 97.

119. **Barnett, A. and Weaver, J.C.** Electroporation. *Bioelectrochem. Bioenergetics*, 25, 163, 1991.

120. **Freeman, S.A., Wang, M.A., and Weaver, J.C.** Theory of electroporation of planar lipid bilayer membranes. *Biophys. J.*, 67, 42, 1994.

121. **Hochmuth, R.M. and Evans, E.A.** Extensional flow of erythrocyte membrane from cell body to elastic tether. *Biophys. J.*, 39, 71, 1982.

122. **Waugh, R.E.** Surface viscosity measurements from large bilayer vesicle tether formation. *Biophys. J.*, 38, 19, 1982.

123. **DeGennes, P.G.** *Scaling Concepts in Polymer Physics*. Cornell University Press, Ithaca, NY, 1985.

124. **Benz, R., Beckers, F., and Zimmermann, U.** Reversible electrical breakdown of lipid bilayer membranes. *J. Membr. Biol.*, 48, 181, 1979.

125. **Benz, R. and Zimmermann, U.** The resealing process of lipid bilayer membranes after reversible electrical breakdown. *Biochim. Biophys. Acta*, 640, 169, 1981.

126. **Neumann, E., Schaefer-Ridder, M., Wang, Y., and Hofschneider, P.H.** Gene transfer into mouse lyoma cells by electroporation in high electric fields. *EMBO J.*, 1, 841, 1982.

127. **Hofmann, G.A. and Evans, G.A.** Electronic genetic-physical and biological aspects of cellular electromanipulation. *IEEE Engin. Med. Biol.*, 5, 6, 1989.

128. **Angelova, I.A. and Dimitrov, D.S.** Liposome electroformation. *Faraday Disc. Chem. Soc.*, 81, 303, 1986.

129. **Angelova, I.A. and Dimitrov, D.S.** Swelling of charged lipids and formation of liposomes on electrode surfaces. *Mol. Cryst. Liq. Cryst.*, 152, 89, 1987.

130. **Wick, R., Angelova, M.I., Walde, P., and Luisi, P.L.** Microinjection into giant vesicles and light microscopy investigation of enzyme mediated vesicle transformation. *Proc. Nat. Acad. Sci. U.S.A.*, 92, 1995.

131. **Block, S.M.** Optical tweezers: a new tool for biophysics. In *Noninvasive Techniques in Cell Biology*. Wiley & Sons, New York, 375, 1990.

132. **Svoboda, K. and Block, S.M.** Biological applications of optical forces. *Ann. Rev. Biophys. Biol. Struct.*, 23, 247, 1994.

133. **Debye, P.** Der Lichtdruck auf Kugeln von beliebigem Material. *Annalen der Physik*, IV, 57, 1909.

134. **Askin, A. and Gordon, J.P.** Stability of radiation pressure particle traps. *Optics Lett.*, 8, 511, 1983.

135. **Visscher, K. and Brakenhoff, G.J.** Single beam optical trapping integrated in a confocal microscope for biological application. *Cytometry*, 12, 486, 1991.

136. **Roosen, G.** A theoretical and experimental study of the stable equilibrium positions of spheres levitated by two horizontal laser beams. *Optic Commun.*, 21, 189, 1977.

137. **Ashkin, A.** Application of laser radiation pressure. *Science*, 210, 1081, 1980.

138. **Buican, T.N., Smyth, M.J., Crissman, H.A., Salzman, G.C., Stewart, C.C., and Martin, J.C.** Automated single cell manipulation and sorting by laser light trapping. *Appl. Optics*, 26, 5311, 1987.

139. **Kuo, S.C. and Sheetz, M.P.** Force of single kinesin molecules measured with optical tweezers. *Science*, 260, 232, 1993.

140. **Streubing, R.W., Cheng, S., Wright, W.H., Numajiri, Y., and Berns, M.W.** Laser induced cell fusion in combination with optical tweezers. *Cytometry*, 12, 505, 1991.

141. **Pouligny, B., Martinot-Lagarde, G., and Angelova, M.I.** Encapsulation of solid microspheres by bilayers. *Prog. Colloid Polym. Sci.*, 95, 1995.

142. **Eigen, M. and Rigler, R.** Sorting single molecules. *Proc. Nat. Acad. Sci. U.S.A.*, 91, 5740, 1994.

143. **Arnold,W.M. and Zimmermann, U.** Rotating-field-induced rotation and measurement of the membrane capacitance of single mesophyll cells of *Avena sativa. Zeitschr. Naturf.*, 37c, 908, 1982.

144. **Fuhr, G. and Kuzmin, P.I.** Behaviour of cells in a rotating electric fields with account to surface charges and cell structures. *Biophys. J.*, 50, 789, 1986.

145. **Arnold, W.M. and Zimmermann, U.** Electro-rotation: development of a technique for dielectric measurements on individual cells and particles. *J. Electrostatics*, 21, 151, 1988.

146. **Arnold, W.M.** Analysis of optimum electro-rotation technique. *Ferroelectrics*, 86, 225, 1988.

147. **Zimmermann, U. and Arnold, W.M.** The interpretation and use of the rotation of biological cells. In *Coherent Excitations in Biological Systems*. Fröhlich, H. and Kremer, F. (Eds.), Springer-Verlag, Berlin, 1983, 211.

148. **Arnold, W.M. and Zimmermann, U.** Electric field-induced fusion and rotation of cells. In *Biological Membranes*. Vol. 5 (Chapman, D., Ed.), 389–454. Academic Press, London, 1984.

149. **Wicher, D., Gündel, J., and Matthies, H.** Measuring chamber with extended applications of the electro-rotation, alpha and beta dispersion of liposomes, *Stud. Biophys.*, 115, 51, 1986.

150. **Wicher, D., and Gündel, J.** Electrorotation of multilamellar and oligolamellar liposomes. *Bioelectrochem. Bioenergetics*, 21, 279, 1989.

151. **Wicher, D. and Gündel, J.** Existence of a low-frequency limit for electrorotation experiments. *Studia Biophysica*, 134, 223, 1989.

152. **Arnold, W.M., Wendt, B., Zimmermann, U., and Korenstein, R.** Rotation of a single thylakoid vesicle in a rotating electric field. Electrical properties of the photosynthetic membrane and their modification by iono-phores, lipophilic ions and pH. *Biochim. Biophys. Acta*, 813, 117, 1985.

153. **Schwarz, G.** General equation for the mean electrical energy of a dielectric body in an alternating electrical field. *J. Chem. Phys.*, 39, 2387, 1963.

154. **Schwarz, G., Saito, M., and Schwan, H.P.** On the orientation of nonspherical particles in an alternating field. *J. Chem. Phys.*, 43, 3562, 1965.

155. **Svesek, F., Sukharev, S., Svetina, S., and Žekš, B.** The shape of phospholipid vesicles in electric fields as determined by video microscopy. *Studia Biophysica*, 138, 143, 1990.

156. **Jones, T.B. and Kraybill, J.P.** Active feedback controlled dielectrophoretic levitation. *J. Appl. Phys.*, 60, 1247, 1086.

157. **Marszalek, P., Zielinsky, J.J., Fikus, M., and Tsong, T.Y.** Determination of electric parameters of cell membranes by a dielectric method. *Biophys. J.*, 59, 982, 1991.

158. **Kaler, K.V.I.S. and Jones, T.B.** Dielectrophoretic spectra of single cells determined by feedback controlled levitation. *Biophys. J.*, 57, 173, 1990.

159. **Kaler, K.V.I.S., Xie, J.P., Jones, T.B., and Paul, R.** Dual-frequency dielectrophoretic levitation of *Canola* protoplast. *Biophys. J.*, 63, 58, 1992.

160. **Stenger, D.A., Kaler, K.V.I.S., and Hui, S.W.** Dipole interactions in electrofusion. *Biophys. J.*, 59, 1074, 1991.

161. **Takashima, S. and Schwan, H.P.** Alignment of macroscopic particles in electric fields and its biological applications. *Biophys. J.*, 47, 513, 1985.

162. **Žekš, B., Svetina, S., and Pastushenko, V.F.** The shape of phospholipid vesicles in an external electric field. *Studia Biophysica*, 138, 137, 1990.

163. **Žekš, B., Svetina, S., and Pastushenko, V.F.** Theoretical analyse of lipid shape in constant electric field. *Biol. Membr.*, 8, 430, 1991.

164. **Hyuga, H., Kinishita, K., and Wakayashi, N.** Deformation of vesicles under the influence of strong fields. *Jpn. J. Appl. Phys.*, 30, 1141, 1991.

165. **Sokirko, A., Pastushenko, Y., Svetina, S., and Žekš, B.** Deformation of a lipid vesicle in an electric field. *Bioelectrochem. Bioenergetics*, 34, 101, 1994.

166. **Song, J. and Waugh, R.E.** Bilayer membrane bending stiffness by the ther formation from mixed PC-PS lipid vesicles. *J. Biomech. Eng.*, 112, 235, 1990.

167. **Winterhalter, M. and Lasic, D.D.** Liposome stability and formation. *Chem. Phys. Lipids*, 64, 35, 1993.

168. **Büschl, R., Ringsdorf, H., and Zimmermann, U.** Electric field induced fusion of large liposomes from natural and polymerizable lipids. *FEBS Lett.,* 150, 38, 1982.

169. **Stoicheva, N.G. and Hui, S.W.** Electrofusion of cell-sized liposomes. *Biochim. Biophys. Acta,* 1195, 31, 1994.

170. **Wu, Y., Sjodin, R.A., and Sowers, A.E.** Distinct mechanical relaxation components in pairs of erythrocyte ghost undergoing fusion. *Biophys. J.,* 66, 114, 1994.

171. **Abidor, I.G. and Sowers, A.E.** Kinetics and mechanism of cell membrane electrofusion. *Biophys. J.,* 61, 1557, 1992.

172. **Menager, C. and Cabuil, V.** Synthesis of magnetic liposomes. *J. Colloid Interface Sci.,* 169, 251, 1995.

Chapter 15

Cryo-Electron Microscopy of Liposomes

P.M. Frederik, M.C.A. Stuart, P.H.H. Bomans, D.D. Lasic

CONTENTS

I. INTRODUCTION

The high-resolution observation of specimens in their natural (hydrated) state has long been a goal of microscopy. In electron microscopy, the dogma has been that extensive stabilization (chemical modification) of the specimen is required to withstand the specimen-beam interactions and high vacuum conditions inside the microscope column. Standard techniques have evolved for specimen preparation which include fixation (cross-linking), dehydration, embedding, sectioning and "staining" with salts of heavy metals as the harshest of "standard" techniques. Also more gentle variants are employed in specimen preparation, such as negative staining (again involving salts of heavy metals) and surface replication (mostly in combination with heavy metal evaporation to augment contrast). In most of these "standard" techniques, the native specimen is seriously modified before it is subjected to imaging with the electron microscope. There has been a quest for more gentle preparative procedures, narrowing the gap between the native specimen and its image.

Cryo-electron microscopy, the observation of specimens at low temperatures, is a relatively young branch of microscopy. Taylor and Glaeser[1] demonstrated that the structure of catalase crystals is excellently preserved in frozen hydrated specimens, as was demonstrated by electron diffraction and later extended by imaging in transmission electron microscopy.[2,3] Although the protein crystals were well preserved, as indicated by the vast number of high-order diffraction spots (indicating a much better preservation of the fine structure than in the standard "negatively" stained specimens), the surrounding aqueous matrix of the crystals was crystalline (hexagonal ice), indicating that water was not vitrified during the cooling of these samples. When it was demonstrated that vitrified water can be formed and observed by cryo-electron microscopy[4] and procedures were developed for the transfer and the observation of vitrified suspensions,[5] a stride forward had been made (reviewed by Dubochet et al.[6,7]).

The relation between a hydrated specimen and its electron microscopic image has since become closer. In the early 1980s, many problems associated with cryoprocedures were addressed, such as the design of a practical cryo-holder,[8] the "clean" vacuum conditions needed during cryo-observation,[9,10] and the possibilities of preparing and observing fully hydrated sectioned materials.[11,12,13] At present, methods seem to be well established for the preparation and cryo- observation of thin aqueous films prepared from suspensions. Although cryo-electron microscopy has been frequently used to solve biomedical questions and problems, an increased interest has been noted in recent years in industrial applications. On the basis of our experience with phospholipid suspensions (to be reported in the next paragraphs), we were requested to investigate a variety of industrial products (e.g., aqueous dispersions of detergents, aqueous acryl suspensions, gels, etc.) mostly involving liposome-like structures. Because cryo-electron microscopy provides an unimpaired (no chemical fixation, no staining/fixation with heavy metal salts) vision

0-8493-4731-9/96/$0.00+$.50
© 1996 by CRC Press, Inc.

of the individual structural elements in a hydrated specimen, it may provide unique information. The fact that images of individual particles are imaged at a high (spatial) resolution is both a strength and a weakness of the cryo-approach. Typically, a few hundred to a few thousand suspended particles are studied in a sample studied by cryo-electron microscopy, which is more than several orders of magnitude less than the number of particles involved in most other physicochemical methods. This implies that one has to be extremely cautious for sampling errors and preparation artifacts before firm conclusions can be drawn on the basis of cryo-electron microscopic observations with regard to the structure of a suspension as a whole. When these precautions are taken into account, cryo-electron microscopy can be rewarding — it provides an "unimpaired" vision of individual structural elements.

II. PRINCIPLES OF CRYO-ELECTRON MICROSCOPY OF SUSPENSIONS

The preparation of a sample for cryo-electron microscopy involves only a few steps as outlined below. Compared to "standard" procedures for electron microscopy that mostly involve chemical fixation (modification) and dehydration (extraction) with organic solvents and often more or other chemical treatments, cryo-electron microscopy offers a tremendous reduction in the number of preparatory steps. Cryo-electron microscopy is based on physical fixation; by fast cooling, structures are eternized in about 10^{-5} s. Physical fixation therefore offers a high temporal resolution and if combined with careful (electron) microscopy, this can be combined with a high spatial resolution. Altogether cryo-electron microscopy is a fast procedure — within 30–60 min a sample (suspension) can be carried through the preparative procedures and studied in the microscope. The following preparatory steps are involved and these steps will be discussed in more detail.

1. Preparation of a thin (usually between 50-nm and 200-nm thick) film from an aqueous suspension.
2. Vitrification of the thin film by rapid cooling through plunging into liquid coolant.
3. Transfer of the vitrified film to a cryo-holder and introducing the holder in the high-vacuum system of the column of the electron microscope.
4. Taking micrographs under low dose conditions and under selected (de)focus conditions, thereby tuning the microscope on the spatial frequencies of interest.

A. PREPARATION OF A THIN FILM

A thin film may form spontaneously between the bars of a specimen grid for electron microscopy when such a grid is dipped and withdrawn from a suspension. The stability of thin aqueous films and the further thinning of the films depends to a large extent on the presence of surface active molecules and the viscosity of the solution/suspension. A potential (accidental) source of surface active material is tenside residues from glassware. The usual procedure for the preparation of a thin film for electron microscopy is to blot excess liquid from the grid with filter paper, thereby creating fresh air-water interfaces. Surface active molecules will adsorb to the surface and orient themselves at the interface. The surface tension may thus drop after blotting and the difference between the dynamic and the static surface tension is known as the Maragoni effect. When the solution/suspension has a high concentration of surface active molecules (and the surfactant has a high critical micelle concentration) the drop in the surface tension may take a few hundreds of a second to reach 90% of its end value (e.g., in the case of sodium dodecyl sulphate). The drop in surface pressure is governed by the kinetics of adsorption and this is an exponential function in which the rate constant is diffusion related. With dilute surfactants of high molecular weight (such as proteins) it may therefore take much longer before a stabilizing interfacial layer has been formed. Thus, by dipping a specimen grid in a suspension and blotting away the excess liquid, the spaces between the grid bars (typically having a diameter of 30 µm when using a 700 mesh grid with a hexagonal pattern) become covered by minute films which can best be compared with tiny soap films. Normally, such a film can be vitrified within a few seconds after its formation and by the subsequent vitrification virtually all molecular movements are instantaneously arrested. From the considerations outlined above it is clear that even in the short lifetime of the film surface effects come into play and these should be considered in the interpretation of the final images. Making a thin film will thus deplete the solution from surface active molecules. It should be further noted that the molecular ratio of components in the surface layer can be different from the ratio in the solution/suspension. A classical example in this respect is toilet soap, where the foam is highly enriched in free fatty acids compared to the liquid. Once a surface active layer is settled at an interface it tends to resist thickness fluctuations. This effect is known as the Gibbs effect and is

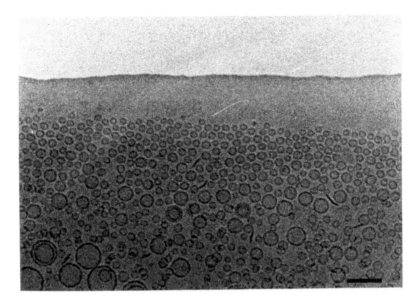

Figure 1 Cryo-electron microscopy of a thin film prepared from sonicated vesicles of DMPC (16 m*M*). Films are spanning between the grid bars of a specimen grid (700 mesh, hexagonal, without supporting film). The film is wedge shaped in its thickness, thin near a hole (top) and without vesicles, gradually increasing in thickness towards the grid bars giving space to more and larger vesicles. Note that during thin film formation (and prior to vitrification) vesicles are sorted by size and concentrated into an array near the thin rim of the film. In the thicker part, vesicles are observed in superposition. Bar represents 100 nm.

explained by the fact that any local depletion of surfactant will increase the surface tension on the same location.

Vertical films as formed between the bars of a specimen grid have the tendency to drain and thus become thinner. Initially the draining follows Poiseuille's law for viscous flow. The two surfaces of the film approach each other, and the film runs through the series of thin film interference colors until it becomes too thin to show any interference at all; the ultimate "black" film. Thinning is thus initially (films thicker than 100 nm) governed by the capillary suction pressure and inversely proportional to the viscosity of the film fluid, but as thinning proceeds (at a thickness of around 100–150 nm) attractive forces between the two surfactant layers at the air-water interfaces come into play. Thinning may proceed until collapse of the film or a stable thin structure may be formed. Such a stable structure can develop if the attractive forces (London-van der Waals interactions) between the interfacial layers are counterbalanced by repulsive forces such as the hydration repulsive force[14] and electrical charges of amphiphilic surfactants.

The concise description of the formation of thin aqueous films (more elaborate descriptions can be found in almost any textbook with a chapter on the physical chemistry of interface formation) is relevant to cryo-electron microscopy since a number of the phenomena can actually be seen in the vitrified thin films observed by cryo-electron microscopy.[15,16] For instance, if a thin film is prepared from sonicated vesicles of DMPC (dimyristoylphosphatidylcholine) dispersed in water, a thin rim of the film is found without any vesicles (see Figure 1). During thinning vesicles are expelled from the central area of the film. Near this edge devoid of vesicles, small vesicles are found aligned in a pseudo-array. During thinning the smallest sonicated vesicles (diameter 15–20 nm) are apparently caught between the interfacial layers and squeezed together. Measurements of the film thickness (densitometry of the negatives) show that the film is wedge shaped and in the thicker parts one can observe vesicles of different sizes sometimes in superposition (also see Figure 1). These observations indicate that the thinning process of a thin film is visualized in the vitrified films. When the same suspension of sonicated DMPC vesicles is used to prepare another film by simply blotting away the excess liquid and allowing the film to stay at room conditions for a few minutes (instead of a split second) before vitrification a completely different image is observed (see Figure 2). The water content of the film is low and the phospholipid is rearranged, instead of the

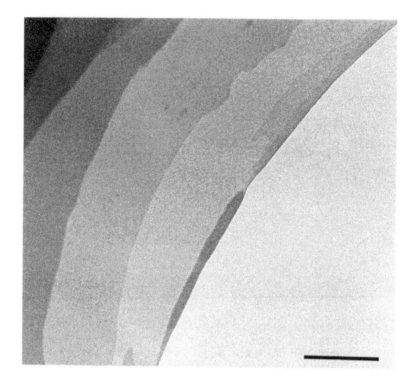

Figure 2 Thin film of DMPC air dried and observed at ambient conditions in the electron microscope. A stepwise increase in thickness is observed and each step represents a thickness increase of the specimen with 4.3 nm corresponding to one bilayer.[15] Note: this is the only image in this chapter where vitrification and low-temperature observation are not involved. Bar represents 500 nm.

vesicles from the original suspension, overlapping sheets of lipid are now observed reminiscent of a paddy field.

Densitometry on the negatives indicates that every "height" step in the image represents a phospholipid bilayer.[15] This observation indicates that lyotropic phases can develop during thinning of the film, apparently the film has lost so much water that the vesicles have collapsed with rearrangement of the lipid. In the lyotropic phase diagram this can be considered as the state where "free" water is absent and the remaining water is associated with the polar headgroups of the phospholipid.

B. VITRIFICATION

Vitrification is the glasslike solidification without the formation of crystals. The vitrified state is thus a metastable state where the viscosity is high (and the thermal motion low) allowing the molecules to remain in a liquid structure instead of in the crystalline structure which would be the preferential state according to the equilibrium phase diagram. For (ultra)structural studies, the vitrified state of water has great advantages because the segregation phenomena associated with crystallization of water are avoided. Most solutes have a low solubility limit in ice (typical in the order of 10^{-4} mol/L with a few exceptions such as NH_4F, HF, HCl) and the ice lattice can therefore accommodate only a limited amount of solutes. The segregation between a crystalline (water rich) phase and a solute rich phase is the cause of freezing artifacts. The characteristics of ice crystal nucleation and growth are such that the ice crystals interfere with the image and cause microscopic damage to the (ultra)structure. Apart from this mechanical damage, also osmotic damage associated with the high solute concentrations that "freeze out" should be considered. Not only solutes may freeze out, but also suspended material is swept together in the interstices left by the ice crystals.[17] Thus, the presence of water in a crystalline state is seriously interfering when the ultrastructure of suspended material in an aqueous environment has to be studied. The preservation of the structure in a vitreous state would therefore be of great help. How can a vitrified state of water (and aqueous solutions) be obtained? In principle, several approaches are possible of which two are commonly applied: at first the addition of "anti-freeze" (cryoprotectant) to obtain a vitrified specimen at moderate cooling velocities, and second, the ultra-rapid cooling of small samples. The last approach is by far preferred since no chemical modifications are required and the material is "physically fixed" in a state as close as possible to the "natural" state.

Cryo-electron microscopy mostly involves ultra-rapid cooling of the specimen although this is not always explicitly mentioned. In the thickness range of specimens studied by cryo-electron microscopy (thinner than 500 nm, preferably thinner than 100 nm for high resolution work), the heat exchange between specimen and coolant is critical whereas with thicker specimens the heat transport through the specimen is (cooling) rate limiting. The specimens that are still transparent to the electron beam as viewed by cryo-electron microscopy are usually in the size range that can be vitrified by plunging into liquid coolant. The heat exchange that is needed for vitrification has been estimated to be in the order of 10^5–10^6°C/s for pure water.[18] A way to obtain pure vitrified water (as judged by X-ray diffraction[19]) is the vacuum condensation of water onto a cold surface. Mayer and Peltzer[20] provided evidence that the solid material thus formed is initially microporous (and thus highly adsorptive), but it will anneal at temperatures still well below the devitrification temperature. Mayer and his group[21,22] have described other methods to obtain vitrified water: spraying through a nozzle on a cold plate placed in vacuum, and jet cooling (through a small orifice) in liquid coolant. The methods to obtain vitrified water rely on small sample dimensions; molecular size in vacuum condensation, μm range in spray, and jet cooling sub-micrometer range in thin film cooling, combined with an optimal heat exchange (high entry velocity in liquid coolant, spreading of drops on a cold surface, etc.) during the vitrification process. Water can thus be vitrified, although it can be argued that some surface active components must be present in the case of thin-film vitrification to ensure the formation and the lifetime (stability) of the film.

For the vitrification of thin films we use a simple apparatus (see Figure 4, from Frederik et al.[23]) as has been described by Dubochet et al.[5] (for an illustration see Reference 7) working according to the guided drop (guillotine) principle. A specimen-grid (∅ 3 mm) held by self-clamping forceps is guided in ethane and cooled to its melting point by liquid nitrogen. At the end of the trajectory braking and bouncing of the rod is damped by a rubber O-ring. In the literature, a number of variants on this simple apparatus can be found: rubber bung/pneumatic driven movement of the specimen into the cryogen, cable/magnetic release mechanisms, etc. Furthermore, the apparatus can be placed in a temperature and humidity controlled environment.[23-25] One of the objectives for such modifications is to obtain time resolution.

C. TIME RESOLUTION

By starting a (chemical) reaction just prior to the vitrification, stages in a dynamic process can be studied. In biology, classical examples of such an approach involve electrical stimulation (nerve ending[26]; heart muscle cells[27]) with a controlled interval between electrical stimulus and vitrification. In these examples the vitrified material had to be processed (cryosectioning, cryofracturing) to disclose details in relatively large specimens. For the study of liposomes, an electrical stimulus is of limited importance (electroporation is a possible exception) for nonmedical applications.

More interesting are the attempts to process material when it is on the grid and already in the form of a thin film. A light driven reaction (flash photolysis of caged compounds[28]), or a temperature driven reaction,[29-31] has a time scale that is better matching the time scale of vitrification than mixing of reactant solutions (e.g., Siegel,[32] Talmon[33]). The basis for studies in dynamic systems must be a careful examination of equilibrium states, a criterion that is not always met. The thermotropic phases of phospholipids are a good test sample for vitrification under equilibrium conditions[15,16] and by placing the vitrification equipment in an incubator at 30 and 24°C respectively, the L_α and $P_{\beta'}$ of DMPC could be visualized.[15] Even handling of thin films at more elevated temperatures has been possible as illustrated by the thermotropic phases of DPPC visualized by starting vitrification at 50°C (L_α-phase, see Figure 3a) or at 39°C ($P_{\beta'}$ phase, see Figure 3b). These images demonstrate that it is not only possible to vitrify water, but also possible to vitrify a thermotropic lipid state. To observe the ripple structure of the $P_{\beta'}$ phases of DMPC and DPPC, we had to use extruded vesicles[34] instead of sonicated vesicles. Sonicated vesicles have a different melting temperature[35] and the ripple period (about 12 nm for DMPC and 15 nm for DPPC) cannot be accommodated in small vesicles. Vesicles vitrified in the L_β phase showed[15,16] no ripples and had a facetted appearance as also observed in the $P_{\beta'}$ vesicles. The facetted appearance of small vesicles in the gel-phase has also been described by Lepault.[36] The examples presented demonstrate that

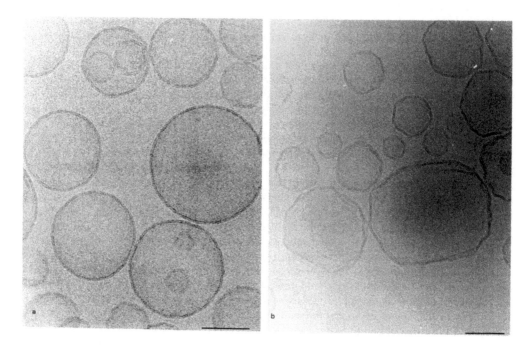

Figure 3 Cryo-electron microscopy of a thin film prepared from (extruded) vesicles of DPPC. The same suspension was vitrified from 50°C (L_α phase, Figure 3A) and from 39°C ($P_{\beta'}$ phase, Figure 3B). The lipid bilayers are clearly resolved and a ripple period (about 15 nm) can be observed in the $P_{\beta'}$ phase. Bar represents 100 nm.

both the high temperature state of water and the state of hydrated lipid at this temperature can be vitrified and observed at low temperature by cryo-electron microscopy. The observation of high temperature states of pure PC prompted us to investigate vesicles from a complex mixture of lipids (DOPE/DOPC/cholesterol[37,38]) and of monomethylated-DOPE. These lipids may adopt an inverted hexagonal (H_{II}) configuration upon warming. When incubated at a pretransition temperature, fusion and rearrangement of bilayers occur. This was investigated by cryo-electron microscopy. A vesicles suspension incubated at 60°C complex structures are found in vitrified films prepared from these suspensions. Multiple bilayer fusions have created a lacework of membranes and when a suspension from monomethylated-DOPE is kept at 60°C for a few hours also this lacework of membranes can be observed in vitrified thin films and, in addition, areas with net-like and honeycomb-like regular structures have been found (see Figure 4). These regular areas suggest that cryo-electron microscopy has imaged a cubic phase in an intermediate stage of its development.

Vitrification as applied in the examples presented is a fast process, fast enough to prevent the relaxation of a high temperature configuration of phospholipids into a low temperature configuration. Also, vitrification is fast enough to prevent ice crystallization in aqueous solutions/suspensions. It even appears that nonaqueous specimens can be vitrified, if these specimens form a thin film (e.g., diesel fuel, images not shown because such a sample is simply amorphous). From our experience we consider vitrification as an ideal and versatile starting point for ultrastructural investigations.

D. CRYO-TRANSFER OF A VITRIFIED SPECIMEN

After vitrification the specimen has to be mounted in a cryo-holder and this assembly will be transferred to the high vacuum inside the column of the electron microscope. During these manipulations, care should be taken that the specimen remains in its original (vitrified) state and remains free from contamination prior to and during exposure to the electron beam. This is a serious problem well known to the pioneers in cryo-electron microscopy. In early days, specimens were often found to be buried in massive contamination layers. If contamination and artifacts in the image are to be minimized, care should be taken to prevent:

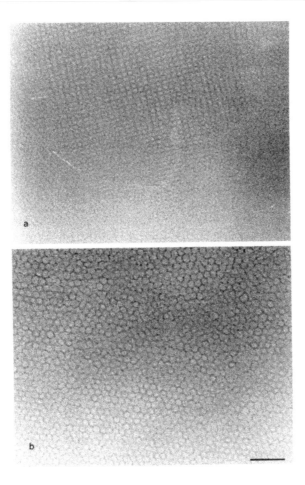

Figure 4 Cryo-electron microscopy of a thin film prepared from monomethylated- DOPE. Suspension was kept at 60°C and at the same temperature a thin film was prepared and vitrified. Complex bilayer arrangements are observed with patches of "ordered" lipid structures such as: (a) net-like and (b) honeycomb-like. Bar represents 100 nm.

- Contamination by frost — hoarfrost deposited on the cold specimen by the condensation of water vapor from the environment. This may occur during mounting of the specimen in the cryo-holder (inside a cold box) and during transfer of the cryoholder with the mounted specimen to the air-lock of the microscope. At this stage, the "first attack" of water vapor takes place usually in the form of sub-micron sized hexagonal ice-crystals.
- Contamination in vacuum — vitreous ice may condense on the cold specimen when the partial vapor pressure of water rises to a critical level (above the equilibrium vapor pressure of water over ice at an ice temperature of the specimen). Upon introduction of a cryo-holder in the vacuum system of the microscope, a transient rise in the vapor pressure of water is to be expected.[10,25] During this "second" attack by water vapor, the specimen should be protected by an efficient shielding system.
- Devitrification by a rise in temperature — the specimen temperature should be kept below –140°C to prevent the devitrification of water resulting in the formation of cubic ice.

On basis of the kinetic theory of gasses, estimates can be made of the rate of condensation/evaporation of water vapor and coolant. With a mass-spectrometer directed toward the specimen site in the microscope column, a measurement of partial vapor pressures involved in a cryotransfer procedure have been obtained[10,25] and these data, when combined with the kinetic gas theory, give an idea of the critical stages in the cryo-transfer procedure. A short description will be presented here.

The starting point for this excursion is a grid carrying a vitrified specimen with solid ethane on both sides. This ethane originates from the liquid in the cooling bath used in vitrification, and since the grid is immediately brought to liquid nitrogen temperature (and kept around this temperature until cryo-observation) the liquid ethane adhering to the grid will solidify (a thin layer is in the order of 50 μm, most layers are 2 to 20 times thicker[10]). This layer may collect frost, but the ethane evaporates (leaving no visible residues or contamination behind) in the air-lock of the microscope. Upon entering the specimen in the column, some ethane was also detected in the vacuum of the microscope column by mass-spectrometry,[10] but the partial vapor pressure of ethane leveled off rapidly. How much water may a cold surface collect when passing through the air? This was estimated to be in the order of 2–200 μm/sec (e.g., during passing of a cryo-holder through the air on its way from the loading station to the air-lock of the microscope, the "first attack" on the specimen by water vapor) and a layer thus collected on cold parts of the equipment is a potential hazard to the specimen. With the cryo-holder in its viewing position inside the microscope, frosted parts may warm up (e.g., by heat influx from warm parts of the microscope) and water may evaporate from these parts. By mass spectrometry, a partial rise in the vapor pressure was observed during the first 10 min after transfer of the cryo-holder into its viewing position. During stages with a high partial vapor pressure of water (4×10^{-3} Pa) water may condense on cold parts with a rate in the order of 10 nm/s. It is clear that the specimen should be protected by shields during this "second attack" by water vapor. When after some time a thermal equilibrium has been established, one should again consider the partial vapor pressure of water around the specimen. This is in the order of 1.2×10^{-6} Pa (ca. 10^{-8} mm Hg, cryo-sorption pump operating in the specimen area) and this may result in a vapor deposition on the cold (100 K) specimen at a rate of 0.16 nm/min (9.6 nm/h). These deposition data indicate that the utmost care should be taken to have a "clean" vacuum system around the specimen with sufficient pumping capacity for water vapor to handle the transient rise in vapor pressure as well as to provide a clean vacuum during the "equilibrium" conditions during EM observation. These conditions are not always fulfilled by standard equipment. The use of additional vacuum pumps or extra cryo-shields (n.b., the partial vapour pressure of water is 10^{-24} mm Hg at the boiling temperature of liquid nitrogen) is therefore often advised and without a thorough analysis of the source of contamination, such advice may result in confusion without solving problems.

E. CRYO-OBSERVATION OF VITRIFIED SPECIMENS

The observation of hydrated specimens is not the sole privilege of cryo-electron microscopy; it can also be accomplished by studying a "wet" specimen in a thin hydration chamber built in the microscope column. These hydration chambers were made with thin windows, or open systems with differential pumping through apertures. Both approaches have the intention to create a small space for the specimen with a controlled environment (temperature, humidity, pressure) at about atmospheric conditions. The demonstration that diffraction spots are better preserved in hydrated catalase crystals than in air dried or negatively stained catalase was first described with the use of an environmental chamber (Matricardi and co-workers[39]). Beam damage and thermal motions within the specimen have been serious drawbacks of this approach and especially in these respects, cryo-observation of vitrified (hydrated) specimens offers a tremendous advantage. Thermal motions are arrested to a large extent, and beam-induced damage is reduced at lower observation temperatures. This can be expressed as the G-value which comes from radiation chemistry and is the number of specific reactions, such as bond rupture, cross linking, or disappearance of original molecules, that are produced by an energy loss of 100 eV. For instance, the photolysis of H_2O expressed as the $G(-H_2O)$ value is 4.5 for the liquid state. The G value decreases to 3.4 in ice at 263 K (–10°C), to 1.0 at 195 K (–78°C) and further to 0.5 at 73 K (–200°C). This protective effect of low temperatures on the water composition has been attributed to an increase in recombination of the water molecule fragments.

Another way to express beam damage is by estimating the disappearance of diffraction spots obtained from protein crystals. A classical example[2] is the fading of diffraction spots from frozen-hydrated catalase; the dose required for complete fading of the diffraction spots at low temperature was 300 e/nm^2, which is 10 times greater than the value for wet catalase at room temperature (300 K). In the current literature, fading is usually expressed as the dose at which diffraction spots have lost 1/e of their initial intensity, but beam damage has also been expressed as a relation between electron dose and the highest resolution at which diffraction spots could still be observed.[3] From the results thus far obtained (a report from a multi-center trial gives a consensus opinion[40]), the low temperature of cryo-em (around 110 K) allows a 10 times higher dose to obtain the same fading of diffraction spots when compared to room temperature observation. A further lowering of the temperature (liquid helium cooled specimen stages) gives a slight further improvement (but with some specimens no improvement at all) in beam protection. The importance of low-temperature protection against beam-induced damage is great for some specimens, it can make the difference between extracting high resolution information from the specimen at low temperature instead of being unable to extract this information because at higher temperatures the specimen is seriously damaged when exposed to the minimum dose needed for recording the (high resolution) information.

The disappearance of diffraction spots is a form of beam damage that can only be observed in periodic specimens (such as two-dimensional crystals), and the first effects are not readily visible in the image: optical diffraction or Fourier transform of digitized images is needed to prove the damaging effect of the electron beam. In nonperiodic specimens diffraction spots are absent and the interaction between the electron beam becomes apparent in another form, amorphous material is leaving the specimen without disturbing the rest of the specimen, or the electron beam creates voids or bubbles in the irradiated area. These voids appear in periodic specimens late; most of the diffraction spots have disappeared before bubbling becomes apparent. Bubbling in a given specimen was found to be dose dependent, e.g., at the lowest irradiation current that allows observation of bubbling phenomena, bubbling starts at the same cumulative dose as with a high current irradiation.[41] In vitrified thin films it was found that the dose at which the first bubbles appeared varied, with high cryoprotectant (glycerol, sucrose, propylene glycol) concentrations bubbling started at a much lower dose. Furthermore, it was found that bubbling depends on the observation temperature and the cooling velocity during the vitrification process. The observation that a specimen is vitrified (or embedded in amorphous ice) is not predictive for its behavior in the electron beam: the thermal history and the (local) structure of water are important parameters in this respect. Considering the destructive effects of beam-specimen interactions, it is clear that the lowest possible irradiation dose should be used in conjunction with efficient recording systems to benefit from the advantages of a vitrified (hydrated) specimen.

The next question is what information the electron-optical system of the microscope will transmit from the specimen. In this respect, a vitrified object is considered as a weak phase object for the electron beam and the contrast transfer function (CTF) of the microscope for these types of objects should be taken into account. The electron-optics transmit only certain spatial frequencies and the characteristics of the transfer function depend on "fixed" instrument parameters (focal length of the objective, chromatic and spheric aberration of the objective lens, coherence of the electron source) and a number of parameters that can be influenced by the operator; (de)focus setting of the objective lens and aperture of the illuminating beam. For a detailed account on the electron optics and mathematics involved, a textbook by Reimer[42] is recommended, and for immediate answers, public domain software is available (Philips Electron Optics, Applications Laboratory, Eindhoven, The Netherlands) presenting the transfer function upon the input of microscope parameters. Figure 5 illustrates the effect of defocus on the contrast transfer function and micrographs represent images taken at the focus settings represented in the graph. It is clear that at certain defocus settings the microscope is blind for some frequencies ("color-blind"). These spatial frequencies in the specimen are transmitted with zero contrast to the image. Fine structural details are represented in the images with roughly the same contrast at optimum defocus ("Scherzer" defocus) but some low spatial frequencies (represented in the overall structure or in the "coarse" structure) in the specimen are transmitted with hardly any contrast or with reversed contrast. Realizing the effect of defocus on the transfer function of the microscope implies that with proper defocus we can tune in on spatial frequencies in the specimen, thereby neglecting other frequencies. When no *a priori* knowledge from the specimen is available, one would like to have a high-contrast image of all the frequencies in the specimen. In theory, this is possible by integrating images taken at different defocus settings, a theory that has been tested recently by using images collected at different defocus settings with a slow-scan CCD camera.[42] Alignment of the images is necessary (to compensate for drift of the specimen holder) and correction for image rotation and magnification change during defocusing were also found to be important corrections for a good quality "final" (integrated) image. The elaborate image processing involved in this image integration makes this approach a complicated task, which is as yet far from routine, but is expected to become routine within a few years when the development of image handling hard- and software keeps its pace.

III. CRYO-ELECTRON MICROSCOPY OF PHOSPHOLIPID AND SURFACTANT SYSTEMS

In the description of the methods involved in cryo-electron microscopy of vitrified thin films we have used illustrations of various lipid/surfactant-based systems. These images are supplemented by a few

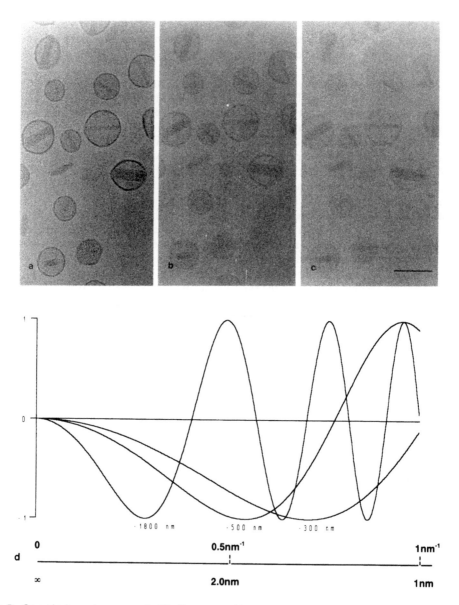

Figure 5 Cryo-electron microscopy of a thin film prepared from doxorubicin loaded liposomes.[47] Cross striations are visible in the core material with a periodicity of 2.6 nm (originating from a doxorubicin gel formed in the presence of sulfate ions). The vesicles (soya bean PC/cholesterol/PEG- DSPE) are shaped by the presence of the core. Through focus series (a) defocus 1800 nm, (b) defocus 500 nm, (c) defocus 300 nm. Note the effect of focus on contrast and resolution. In (d) the contrast transfer function of our microscope is represented for the defocus values used in imaging. The relative contrast (in an arbitrary unit) is given as function of the spatial frequencies (reciprocal scale). Bar represents 100 nm.

additional examples, selected to give a concise overview of phospholipid morphologies. As a starting point a micellar arrangement can be considered (see Figure 6), an example of a lipid based on PE, but with a chemically modified polar head group (extended with polyoxyethylene, MW around 2000). This lipid forms micelles when dissolved in water, and these micelles arrange in (two-dimensional) arrays when a thin film is formed. Sonicated vesicles can be considered as the smallest form of lipids arranged in a self-closed bilayer. With sonicated vesicles we have described the behavior of phospholipids during thin film formation (sorting by size and array formation) as well as the lyotropic phase behavior of phospholipids (Figures 1 and 2). Larger vesicles were required to obtain images of the thermotropic phase

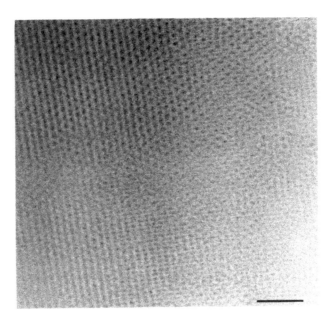

Figure 6 Cryo-electron microscopy of a thin film prepared from a suspension of DSPE-PEG micelles. During thin film formation micelles arrange into (two-dimensional) arrays. Bar represents 100 nm.

behavior of phospholipids as illustrated by Figure 3. These images supplement data obtained by X-ray diffraction[44] as well as obtained by other microscopic techniques (freeze-fracturing[45,46]). Starting from unilamellar vesicles one can investigate more complex structures such as the lace work of membranes that is generated upon heating of certain PE based systems (Figure 4) as well as smaller structures. Detergent/ lipid mixtures give an impression of disintegrating bilayers (Figure 7), with fragments (including "disk-like" micelles) of various sizes. The examples thus far presented give an impression of intrinsic properties (temperature, water content and size are the main variables) of lipids to adopt various shapes. Liposomes can also adjust their shape under the influence of extrinsic factors, such as the presence of gelled material in the interior of the liposome[47] (see Figure 5) or alternatively by the binding of protein (e.g., Annexin V, binding in the presence of Ca^{2+} to negatively charged phospholipids,[47] see also Figure 8) to the exterior of the liposome. The portrait gallery of liposomes thus contains a lot more than perfect spherical vesicles. Also, in nonmedical applications of liposome technologies, cryo-electron microscopy may provide images of the individual structural elements and this should be considered when averaged data (as obtained by most other physicochemical methods) fail short in describing the structure or explaining the characteristics of a suspension.

IV. CONCLUSIONS

Cryo-electron microscopy allows the direct observation of liposomes in their native (hydrated) state without the interference of chemicals. Physical fixation by rapid freezing is fast enough to preserve thermotropic lipid phases as was demonstrated by vitrification of a suspension of DPPC vesicles from the L_α phase (e.g., by initiating vitrification at 50°C) and by vitrification of the same suspension in the $P_{\beta'}$ phase (e.g., vitrification from 39°C). Some examples will be given of interwoven bilayer structures observed upon heating of vesicles in DOPE-based lipid suspensions. Recently, the physical state of drugs loaded into liposomes was investigated by cryo-electron microscopy. The cryo-approach in the study of aqueous suspensions is a versatile method to study individual particles in their "native" environment and may thus provide unique information complementary to information obtained by other physicochemical methods. A short survey will be presented about the methods involved; thin film formation, vitrification, and cryo-electron microscopy, because this may help in the interpretation of the images. The images illustrating the procedures were also chosen to give an overview of lipid morphologies, ranging from micelles to various bilayer forms and including shaping of liposomes from within (a gel in the interior) and from without (protein adsorbed at the outside).

Figure 7 Cryo-electron microscopy of a detergent/lipid mixture (specimen courtesy of J. Bouwstra and B. van de Berg, Center for Bio-Pharmaceutical Sciences, Leiden, The Netherlands). Disk-like bilayer fragments and rod-like micelles can be observed. Bar represents 100 nm.

Figure 8 Cryo-electron microscopy of Annexin V interacting with a vesicle (lipid DOPS/DOPE, molar ratio 20/80 ; Annexin V, 200 μM) in the presence of Ca^{2+} (3 mM). Vesicle is shaped by the formation of two dimensional arrays of Annexin V at the outer surface of the vesicle.[48] Bar represents 100 nm.

REFERENCES

1. **Taylor, K. A. and Glaeser, R. M.,** Electron diffraction of frozen, hydrated protein crystals, *Science,* 186, 1036, 1974.
2. **Taylor, K. A. and Glaeser, R. M.,** Electron microscopy of frozen hydrated biological specimens, *J. Ultrastruct. Res.,* 55, 448, 1976.
3. **Glaeser, R. M. and Taylor, K. A.,** Radiation damage relative to transmission electron microscopy of biological specimens at low temperature: a review, *J. Microsc. (Oxford),* 112, 127, 1978.
4. **Dubochet, J. and McDowall, A. W.,** Vitrification of pure water for electron microscopy, *J. Microsc. (Oxford),* 124, RP3, 1981.
5. **Dubochet, J., Lepault, J., Freeman, R., Berriman, J. A. and Homo, J.-Cl.,** Electron microscopy of frozen water and aqueous solutions, *J. Microsc. (Oxford),* 128, 219, 1982.
6. **Dubochet, J., Adrian, M., Chang, J.-J., Lepault, J. and McDowall A. W.,** Cryoelectron microscopy of vitrified specimens, *Cryotechniques in Biological Electron Microscopy,* Steinbrecht, R. A. and Zierold, K., Eds., Springer-Verlag, Berlin, 1987, 114.

7. **Dubochet, J., Adrian, M., Chang, J.-J., Homo, J.-Cl., Lepault, J., McDowall, A. W. and Schulz, P.,** Cryo-electron microscopy of vitrified specimens, *Q. Rev. Biophys.,* 21, 129, 1988.

8. **Lichtenegger, S. and Hax, W. M. A.,** A cryo-transfer holder for a TEM/STEM system, in *Proc. 7th Eur. Congr. Electron Microscopy,* Brederoo, P. and de Priester, W., Eds., Seventh European Congress on Electron Microscopy Foundation, Leiden, 1980, 652.

9. **Homo, J.-Cl., Booy, F., Labouesse, P., Lepault, J. and Dubochet, J.,** Improved anticontaminator for cryo-electron microscopy with Philips EM 400, *J. Microsc. (Oxford),* 128, 219, 1982.

10. **Frederik, P. M. and Busing, W. M.,** Cryo-transfer revised, *J. Microsc. (Oxford),* 144, 215, 1986.

11. **McDowall, A. W., Chang, J.-J., Freeman, R., Lepault, J., Walter, C. A. and Dubochet, J.,** Electron microscopy of frozen hydrated sections of vitreous ice and vitrified biological samples, *J. Microsc. (Oxford),* 131, 1, 1983.

12. **Frederik, P. M., Busing, W. M. and Persson, A.,** Concerning the nature of the cryosectioning process, *J. Microsc. (Oxford),* 125, 167, 1982.

13. **Frederik, P. M., Busing, W. M. and Hax, W. M. A.,** Frozen-hydrated and drying thin cryo-sections as observed in STEM, *J. Microsc. (Oxford),* 126, RP1, 1982.

14. **Cowley, M., Fuller, N. L., Rand, R. P. and Parsegian, V. A.,** Measurement of repulsive forces between charged phospholipid bilayers, *Biochemistry,* 17, 3163, 1978.

15. **Frederik, P. M., Stuart, M. C. A., Bomans, P. H. H. and Busing, W. M.,** Phospholipid, nature's own slide and cover slip for cryo-electron microscopy, *J. Microsc. (Oxford),* 153, 81, 1989.

16. **Frederik, P. M., Stuart, M. C. A., Schrijvers, A. H. G. J. and Bomans, P. H. H.,** Thin film formation and the imaging of phospholipid by cryo-electron microscopy, *Scand. Microsc.,* Suppl. 3, 277, 1989.

17. **Bachmann, L. and Schmitt, W. W.,** Improved cryofixation applicable to freeze-etching, *Proc. Natl. Acad. Sci. U.S.A.,* 68, 2149, 1971.

18. **Bachmann, L. and Mayer, E.,** Physics of water and ice: implications for cryofixation, in *Cryotechniques in Biological Electron Microscopy,* Steinbrecht, R. A. and Zierold, K., Eds., Springer-Verlag, Berlin, 1987, 3.

19. **Sceats, M. G. and Rice, S. A.,** Amorphous solid water and its relationship to liquid water: A random network model for water, in *Water: A Comprehensive Treatise,* 7, Franks, F., Ed., Plenum Press, New York, 1982, 83.

20. **Mayer, E. and Pletzer, R.,** Astrophysical implications of amorphous ice — a microporous solid, *Nature,* 319, 298, 1986.

21. **Mayer, E.,** Vitrification of pure liquid water, *J. Microsc. (Oxford),* 140, 3, 1985.

22. **Mayer, E. and Brüggeller, P.,** Vitrification of pure liquid water by high pressure jet freezing, *Nature,* 298, 715, 1982.

23. **Frederik, P. M., Stuart, M. C. A., Bomans, P. H. H., Busing, W. M. and Verkleij, A. J.,** Perspective and limitations of cryo-electron microscopy. From model systems to biological specimens, *J. Microsc. (Oxford),* 161, 253, 1991.

24. **Bellare, J. R., Davis, H. T., Scriven, L. E. and Talmon, Y.,** Controlled environment vitrification system; An improved sample preparation technique, *J. Electron Microsc. Technol.,* 10, 87, 1988.

25. **Trinnick, J., Cooper, J., Seymour, J. and Egelmann, E. H.,** Cryo-electron microscopy and three-dimensional reconstruction of actin filaments, *J. Microsc. (Oxford),* 141, 349, 1986.

26. **Heuser, J. E., Reese, T. S., Dennis, M. J., Jan, Y., Jan, L. and Evans, L.,** Synaptic vesicle exocytosis captured by quick freezing and correlated with quantal transmitter release, *J. Cell Biol.,* 81, 275, 1979.

27. **Wendt-Gallitelli, M. F. and Isenberg, G.,** Total and free myoplasmic calcium during a contraction cycle: X-ray microanalysis in guinea-pig ventricular myocytes, *J. Physiol. (London),* 435, 349, 1991.

28. **Ménétret, J.-F., Hofmann, W., Schröder, R. R., Rapp, G. and Goody, R. S.,** Time resolved cryo-electron microscopic study of the dissociation of actinomyosin induced by photolysis of photolabile nucleotides, *J. Molec. Biol.* 219, 139, 1991.

29. **Mandelkow, E.-M., Mandelkow, E. and Milligan, R. A.,** Microtubule dynamics and microtubule caps: a time-resolved cryo-electron microscopy study, *J. Cell Biol.,* 114, 977, 1991.

30. **Chestnut, M. H., Siegel, D. P., Burns, J. L. and Talmon, Y.,** A temperature-jump device for time resolved cryo-transmission electron microscopy, *Microsc. Res. Technol.,* 20, 95, 1992.

31. **Siegel, D. P., Green, W. J. and Talmon, Y.,** The mechanism of lamellar-to-inverted hexagonal phase transitions: a study using temperature jump cryo-electron microscopy, *Biophys. J.,* 66, 402, 1994.

32. **Siegel, D. P., Burns, J. L., Chestnut, M. H. and Talmon, Y.,** Intermediates in membrane fusion and bilayer/ nonbilayer transitions imaged by time-resolved cryo-transmission electron microscopy, *Biophys. J.,* 56, 161, 1989.

33. **Talmon, Y., Burns, J. L., Chestnut, M. H. and Siegel, D. P.,** Time-resolved cryotransmission electron microscopy, *J. Electron Microsc. Technol.,* 14, 6, 1990.

34. **Hope, M. J., Bally, M. B., Wiebb, G. and Cullis, P. R.,** Production of large unilamellar vesicles by a rapid extrusion procedure. Characterization of size distribution, trapped volume and ability to maintain a membrane potential, *Biochim. Biophys. Acta,* 812, 55, 1985.

35. **Lichtenberg, S., Freire, E., Schmidt, C. F., Barenholz, Y., Felgner, P. L. and Thompson, T. E.,** Effect of surface curvature on stability, thermodynamic behaviour and osmotic activity of dipalmitoylphosphatidylcholine single lamellar vesicles, *Biochemistry,* 20, 3462, 1981.

36. **Lepault, J., Pattus, F. and Martin, N.,** Cryo-electron microscopy of artificial biological membranes, *Biochim. Biophys. Acta,* 820, 315, 1985.

37. **Frederik, P. M., Stuart, M. C. A. and Verkleij, A. J.,** Intermediary structures during membrane fusion as observed by cryo-electron microscopy, *Biochim. Biophys. Acta,* 979, 275, 1989.

38. **Frederik, P. M., Burger, K. N. J., Stuart, M. C. A. and Verkleij, A. J.,** Lipid polymorphism as observed by cryo-electron microscopy, *Biochim. Biophys. Acta,* 1062, 133, 1991.

39. **Matricardi, V. R., Moretz, R. C. and Parsons, D. F.,** Electron diffraction of wet proteins: catalase, *Science,* 177, 268, 1972.

40. **Chiu, W., Downing, K. H., Dubochet, J., Glaeser, R. M., Heide, H. G., Knapek, E., Kopf, D. A., Lamvik, M. K., Lepault, J., Robertson, J. D., Zeitler, E. and Zemlin, F.,** (International experimental study group), Cryoprotection in electron microscopy, *J. Microsc. (Oxford),* 141, 385, 1986.

41. **Frederik, P. M., Bomans, P. H. H. and Stuart, M. C. A.,** Matrix effects and the induction of mass loss or bubbling by the electron beam in vitrified hydrated specimens, *Ultramicroscopy,* 48, 107, 1993.

42. **Reimer, L.,** *Transmission Electron Microscopy — Physics of Image Formation and Microanalysis,* 2nd ed., Springer-Verlag, Berlin, 1989, chap. 6.

43. **Typke, D., Hegerl, R. and Kleinz, J.,** Image restoration for biological objects using external TEM control and electronic image recording, *Ultramicroscopy,* 46, 157, 1992.

44. **Janiak, M. J., Small, D. M. and Shipley, G. G.,** Nature of the thermal pretransition of synthetic phospholipids: dimiristoyl- and dipalmitoyllecithin, *Biochemistry,* 15, 4575, 1976.

45. **Verkleij, A. J. and Ververgaert, P. H. J. Th.,** The architecture of biological and artificial membranes as visualized by freeze etching, *Annu. Rev. Phys. Chem.,* 101, 1975.

46. **Larabee, A. L.,** Time dependent changes in the size distribution of distearoylphosphatidylcholine vesicles, *Biochemistry,* 18, 3321, 1979.

47. **Lasic, D. D., Frederik, P. M., Stuart, M. C. A., Barenholz, Y. and McIntosh, T. J.,** Gelation of liposome interior-A novel method for drug encapsulation, *FEBS Lett.,* 312, 255, 1992.

48. **Andree, H. A. M., Stuart, M. C. A., Hermens, W. Th., Reutelingsperger, C. P. M., Hemker, H. C., Frederik, P. M. and Willems, G. M.,** Clustering of lipid-bound Annexin V may explain its anticoagulant effect, *J. Biol. Chem.,* 267, 17907, 1992.

INDEX

A

K

L

M